STUDENT'S SOLUTIONS MANUAL

CHRISTINE VERITY

ELEMENTARY ALGEBRA: CONCEPTS AND APPLICATIONS

EIGHTH EDITION

Marvin Bittinger

Indiana University – Purdue University Indianapolis

David Ellenbogen

Community College of Vermont

Addison-Wesley
is an imprint of

The author and publisher of this book have used their best efforts in preparing this book. These efforts include the development, research, and testing of the theories and programs to determine their effectiveness. The author and publisher make no warranty of any kind, expressed or implied, with regard to these programs or the documentation contained in this book. The author and publisher shall not be liable in any event for incidental or consequential damages in connection with, or arising out of, the furnishing, performance, or use of these programs.

Reproduced by Pearson Addison-Wesley from electronic files supplied by the author.

ISBN-13: 978-0-321-56733-8
ISBN-10: 0-321-56733-1

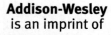

Addison-Wesley
is an imprint of

www.pearsonhighered.com

Contents

Chapter 1

Introduction to Algebraic Expressions

Exercise Set 1.1

1. $10n - 1$ does not contain an equals sign, so it is an expression.

3. $2x - 5 = 9$ contains an equals sign, so it is an equation.

5. $38 = 2t$ contains an equals sign, so it is an equation.

7. $4a - 5b$ does not contain an equals sign, so it is an expression.

9. $2x - 3y = 8$ contains an equals sign, so it is an equation.

11. $r(t + 7) + 5$ does not contain an equals sign, so it is an expression.

13. Substitute 9 for a and multiply.
$$5a = 5 \cdot 9 = 45$$

15. Substitute 4 for r and subtract.
$$12 - 4 = 8$$

17. $\dfrac{a}{b} = \dfrac{45}{9} = 5$

19. $\dfrac{x + y}{4} = \dfrac{2 + 14}{4} = \dfrac{16}{4} = 4$

21. $\dfrac{p - q}{7} = \dfrac{55 - 20}{7} = \dfrac{35}{7} = 5$

23. $\dfrac{5z}{y} = \dfrac{5 \cdot 9}{15} = \dfrac{45}{15} = 3$

25. $bh = (6 \text{ ft})(4 \text{ ft})$
$ = (6)(4)(\text{ft})(\text{ft})$
$ = 24 \text{ ft}^2$, or 24 square feet

27. $A = \dfrac{1}{2}bh$
$ = \dfrac{1}{2}(5 \text{ cm})(6 \text{ cm})$
$ = \dfrac{1}{2}(5)(6)(\text{cm})(\text{cm})$
$ = \dfrac{5}{2} \cdot 6 \text{ cm}^2$
$ = 15 \text{ cm}^2$, or 15 square centimeters

29. $\dfrac{h}{a} = \dfrac{10}{29} \approx 0.345$

31. Let r represent Ron's age. Then we have $r + 5$, or $5 + r$.

33. $6b$, or $b \cdot 6$

35. $c - 9$

37. $6 + q$, or $q + 6$

39. Let m represent Mai's speed. Then we have $8m$, or $m \cdot 8$.

41. $y - x$

43. $x \div w$, or $\dfrac{x}{w}$

45. Let l and h represent the box's length and height, respectively. Then we have $l + h$, or $h + l$.

47. $9 \cdot 2m$, or $2m \cdot 9$

49. Let y represent "some number." Then we have $\dfrac{1}{4}y - 13$, or $\dfrac{y}{4} - 13$.

51. Let a and b represent the two numbers. Then we have $5(a - b)$.

53. Let w represent the number of women attending. Then we have 64% of w, or $0.64w$.

55.
$$\begin{array}{c|c} x + 17 = 42 & \text{Writing the equation} \\ \hline 25 + 17 \;\big|\; 42 & \text{Substituting 25 for } x \\ ? & \\ 42 = 42 & 42 = 42 \text{ is TRUE.} \end{array}$$
Since the left-hand and right-hand sides are the same, 25 is a solution.

57.
$$\begin{array}{c|c} a - 28 = 75 & \text{Writing the equation} \\ \hline 93 - 28 \;\big|\; 75 & \text{Substituting 93 for } a \\ ? & \\ 65 = 75 & 65 = 75 \text{ is FALSE.} \end{array}$$
Since the left-hand and right-hand sides are not the same, 93 is not a solution.

59.
$$\begin{array}{c|c} \dfrac{t}{7} = 9 & \\ \hline \dfrac{63}{7} \;\big|\; 9 & \\ \phantom{\dfrac{63}{7}}? & \\ 9 = 9 & 9 = 9 \text{ is TRUE.} \end{array}$$
Since the left-hand and right-hand sides are the same, 63 is a solution.

61.
$$\begin{array}{c|c} \dfrac{108}{x} = 36 & \\ \hline \dfrac{108}{3} \;\big|\; 36 & \\ \phantom{\dfrac{108}{3}}? & \\ 36 = 36 & 36 = 36 \text{ is TRUE.} \end{array}$$
Since the left-hand and right-hand sides are the same, 3 is a solution.

63. Let x represent the number.

$\underbrace{\text{What number}}_{\;} \; \underbrace{\text{added to}}_{\;} \; 73 \text{ is } 201?$

Translating: $\quad x \qquad + \qquad 73 = 201$

$$x + 73 = 201$$

65. Let x represent the number.

Rewording: 42 times $\underbrace{\text{what number}}$ is 2352?

                  $\downarrow$ $\downarrow$ $\downarrow$ $\downarrow$ $\downarrow$

Translating: 42 $\cdot$ x $= 2352$

$42x = 2352$

67. Let s represent the number of unoccupied squares.

Rewording: $\underbrace{\text{The number of unoccupied squares}}$ $\underbrace{\text{added to}}$ 19 is 64.

                 $\downarrow$ $\downarrow$ $\downarrow$ $\downarrow$ $\downarrow$

Translating: s $+$ $19 = 64$

$s + 19 = 64$

69. Let x represent the total amount of waste generated, in millions of tons.

Rewording: $\underbrace{32\%}$ of $\underbrace{\text{the total amount of waste}}$ is $\underbrace{\text{79 million tons.}}$

             $\downarrow$ $\downarrow$ $\downarrow$ $\downarrow$ $\downarrow$

Translating: 32% $\cdot$ x $=$ 79

$32\% \cdot x = 79$, or $0.32x = 79$

71. The sum of two numbers m and n is $m + n$, and twice the sum is $2(m + n)$. Choice (f) is the correct answer.

73. Twelve more than a number t is $t + 12$. If this expression is equal to 5, we have the equation $t + 12 = 5$. Choice (d) is the correct answer.

75. The sum of a number t and 5 is $t + 5$, and 3 times the sum is $3(t + 5)$. Choice (g) is the correct answer.

77. The product of two numbers a and b is ab, and 1 less than this product is $ab - 1$. If this expression is equal to 48, we have the equation $ab - 1 = 48$. Choice (e) is the correct answer.

79. *Writing Exercise.* A *variable* is a letter that is used to stand for any number chosen from a set of numbers. An *algebraic expression* is an expression that consists of variables, constants, operation signs, and/or grouping symbols. A *variable expression* is an algebraic expression that contains a variable. An *equation* is a number sentence with the verb $=$. The symbol $=$ is used to indicate that the algebraic expressions on either side of the symbol represent the same number.

81. *Writing Exercise.* No; for a square with side s, the area is given by $A = s \cdot s$. The area of a square with side $2s$ is given by $(2s)(2s) = 4 \cdot s \cdot s = 4A \neq 2A$.

83. Area of sign: $A = \dfrac{1}{2}(3 \text{ ft})(2.5 \text{ ft}) = 3.75 \text{ ft}^2$

Cost of sign: $\$120(3.75) = \$450$

85. When x is twice y, then y is one-half x, so

$$y = \frac{12}{2} = 6.$$

$$\frac{x - y}{3} = \frac{12 - 6}{3} = \frac{6}{3} = 2$$

87. When a is twice b, then b is one-half a, so $b = \dfrac{16}{2} = 8$.

$$\frac{a + b}{4} = \frac{16 + 8}{4} = \frac{24}{4} = 6$$

89. The next whole number is one more than $w + 3$:

$$w + 3 + 1 = w + 4$$

91. $l + w + l + w$, or $2l + 2w$

93. If t is Molly's race time, then Joe's race time is $t + 3$ and Ellie's race time is

$$t + 3 + 5 = t + 8.$$

95. *Writing Exercise.* Yes; the area of a triangle with base b and height h is given by $A = \dfrac{1}{2}bh$. The area of a triangle with base b and height $2h$ is given by $\dfrac{1}{2}b(2h) = 2\left(\dfrac{1}{2}bh\right) = 2A$.

Exercise Set 1.2

1. Commutative

3. Associative

5. Distributive

7. Associative

9. Commutative

11. $t + 11$ Changing the order

13. $8x + 4$

15. $3y + 9x$

17. $5(1 + a)$

19. $x \cdot 7$ Changing the order

21. ts

23. $5 + ba$

25. $(a + 1)5$

27. $x + (8 + y)$

29. $(u + v) + 7$

31. $ab + (c + d)$

33. $8(xy)$

35. $(2a)b$

37. $(3 \cdot 2)(a + b)$

39. $s + (t + 6) = (s + t) + 6$
$= (t + 6) + 5$

41. $(17a)b = b(17a)$ Using the commutative law
$= b(a17)$ Using the commutative law again

$(17a)b = (a17)b$ Using the commutative law
$= a(17b)$ Using the associative law

Answers may vary.

43. $(1 + x) + 2 = (x + 1) + 2$ Commutative law
$= x + (1 + 2)$ Associative law
$= x + 3$ Simplifying

45. $(m \cdot 3)7 = m(3 \cdot 7)$ Associative law
$= m \cdot 21$ Simplifying
$= 21m$ Commutative law

47. $2(x + 15) = 2 \cdot x + 2 \cdot 15 = 2x + 30$

49. $4(1 + a) = 4 \cdot 1 + 4 \cdot a = 4 + 4a$

51. $8(3 + y) = 8 \cdot 3 + 8 \cdot y = 24 + 8y$

53. $10(9x + 6) = 10 \cdot 9x + 10 \cdot 6 = 90x + 60$

55. $5(r + 2 + 3t) = 5 \cdot r + 5 \cdot 2 + 5 \cdot 3t = 5r + 10 + 15t$

57. $(a + b)2 = a(2) + b(2) = 2a + 2b$

59. $(x + y + 2)5 = x(5) + y(5) + 2(5) = 5x + 5y + 10$

61. $x + xyz + 1$

The terms are separated by plus signs. They are x, xyz, and 1.

63. $2a + \dfrac{a}{3b} + 5b$

The terms are separated by plus signs. They are $2a$, $\dfrac{a}{3b}$, and $5b$.

65. x, y

67. $4x, 4y$

69. $2 \cdot a + 2 \cdot b$ Using the distributive law
$= 2(a + b)$ The common factor is 2.
Check: $2(a + b) = 2 \cdot a + 2 \cdot b = 2a + 2b$

71. $7 + 7y = 7 \cdot 1 + 7 \cdot y$ The common factor is 7.
$= 7(1 + y)$ Using the distributive law
Check: $7(1 + y) = 7 \cdot 1 + 7 \cdot y = 7 + 7y$

73. $32x + 4 = 4 \cdot 8x + 4 \cdot 1 = 4(8x + 1)$
Check: $4(8x + 1) = 4 \cdot 8x + 4 \cdot 1 = 32x + 4$

75. $5x + 10 + 15y = 5 \cdot x + 5 \cdot 2 + 5 \cdot 3y = 5(x + 2 + 3y)$
Check: $5(x + 2 + 3y) = 5 \cdot x + 5 \cdot 2 + 5 \cdot 3y = 5x + 10 + 15y$

77. $7a + 35b = 7 \cdot a + 7 \cdot 5b = 7(a + 5b)$
Check: $7(a + 5b) = 7 \cdot a + 7 \cdot 5b = 7a + 35b$

79. $44x + 11y + 22z = 11 \cdot 4x + 11 \cdot y + 11 \cdot 2z$
$= 11(4x + y + 2z)$
Check: $11(4x + y + 2z) = 11 \cdot 4x + 11 \cdot y + 11 \cdot 2z$
$= 44x + 11y + 22z$

81. $5n = 5 \cdot n$
The factors are 5 and n.

83. $3(x + y) = 3 \cdot (x + y)$
The factors are 3 and $(x + y)$.

85. The factors are $7, a$ and b.

87. $(a - b)(x - y) = (a - b) \cdot (x - y)$
The factors are $(a - b)$ and $(x - y)$.

89. *Writing Exercise.* No; in general, when subtracting, the result depends on the order in which the operation is performed.

91. Let k represent Kara's salary. Then we have $\frac{1}{2}k$ or $\frac{k}{2}$.

93. *Writing Exercise.* Answers will vary.

95. The expressions are equivalent by the distributive law.
$8 + 4(a + b) = 8 + 4a + 4b = 4(2 + a + b)$

97. The expressions are not equivalent.
Let $m = 1$. Then we have:
$7 \div 3 \cdot 1 = \dfrac{7}{3} \cdot 1 = \dfrac{7}{3}$, but
$1 \cdot 3 \div 7 = 3 \div 7 = \dfrac{3}{7}$.

99. The expressions are not equivalent.
Let $x = 1$ and $y = 0$. Then we have:
$30 \cdot 0 + 1 \cdot 15 = 0 + 15 = 15$, but
$5[2(1 + 3 \cdot 0)] = 5[2(1)] = 5 \cdot 2 = 10$.

101. *Writing Exercise.* $3(2 + x) = 3(2 + 0) = 3 \cdot 2 = 6$;
$6 + x = 6 + 0 = 6$
The result indicates that $3(2 + x)$ and $6 + x$ are equivalent when $x = 0$. (By the distributive law, we know they are not equivalent for all values of x.)

Exercise Set 1.3

1. Since $35 = 5 \cdot 7$, choice (b) is correct.

3. Since 65 is an odd number and has more than two different factors, choice (d) is correct.

5. 9 is composite because it has more than two different factors. They are 1, 3, and 9.

7. 41 is prime because it has only two different factors, 41 and 1.

9. 77 is composite because it has more than two different factors. They are 1, 7, 11, and 77.

11. 2 is prime because it has only two different factors, 2 and 1.

13. The terms "prime" and "composite" apply only to natural numbers. Since 0 is not a natural number, it is neither prime nor composite.

15. Factorizations:

$$1 \cdot 50, \ 2 \cdot 25, \ 5 \cdot 10$$

List all of the factors of 50:

1, 2, 5, 10, 25, 50

17. Factorizations:

$$1 \cdot 42, \ 2 \cdot 21, \ 3 \cdot 14, \ 6 \cdot 7$$

List all of the factors of 42:

1, 2, 3, 6, 7, 14, 21, 42

19. $39 = 3 \cdot 13$

21. We begin factoring 30 in any way that we can and continue factoring until each factor is prime.

$$30 = 2 \cdot 15 = 2 \cdot 3 \cdot 5$$

23. We begin by factoring 27 in any way that we can and continue factoring until each factor is prime.

$$27 = 3 \cdot 9 = 3 \cdot 3 \cdot 3$$

25. We begin by factoring 150 in any way that we can and continue factoring until each factor is prime.

$$150 = 2 \cdot 75 = 2 \cdot 3 \cdot 25 = 2 \cdot 3 \cdot 5 \cdot 5$$

27. We begin by factoring 40 in any way that we can and continue factoring until each factor is prime.

$$40 = 4 \cdot 10 = 2 \cdot 2 \cdot 2 \cdot 5$$

29. 31 has exactly two different factors, 31 and 1. Thus, 31 is prime.

31. $210 = 2 \cdot 105 = 2 \cdot 3 \cdot 35 = 2 \cdot 3 \cdot 5 \cdot 7$

33. $115 = 5 \cdot 23$

35.

$$\frac{21}{35} = \frac{7 \cdot 3}{7 \cdot 5} \qquad \text{Factoring numerator and denominator}$$

$$= \frac{7}{7} \cdot \frac{3}{5} \qquad \text{Rewriting as a product of two fractions}$$

$$= 1 \cdot \frac{3}{5} \qquad \frac{7}{7} = 1$$

$$= \frac{3}{5} \qquad \text{Using the identity property of 1}$$

37. $\dfrac{16}{56} = \dfrac{2 \cdot 8}{7 \cdot 8} = \dfrac{2}{7} \cdot \dfrac{8}{8} = \dfrac{2}{7} \cdot 1 = \dfrac{2}{7}$

39.

$$\frac{12}{48} = \frac{1 \cdot 12}{4 \cdot 12} \qquad \text{Factoring and using the identity property of 1 to write 12 as } 1 \cdot 12$$

$$= \frac{1}{4} \cdot \frac{12}{12}$$

$$= \frac{1}{4} \cdot 1 = \frac{1}{4}$$

41. $\dfrac{52}{13} = \dfrac{13 \cdot 4}{13 \cdot 1} = 1 \cdot \dfrac{4}{1} = 4$

43.

$$\frac{19}{76} = \frac{1 \cdot 19}{4 \cdot 19} \qquad \text{Factoring and using the identity property of 1 to write 19 as } 1 \cdot 19$$

$$= \frac{1 \cdot \cancel{19}}{4 \cdot \cancel{19}} \qquad \text{Removing a factor equal to 1: } \frac{19}{19} = 1$$

$$= \frac{1}{4}$$

45.

$$\frac{150}{25} = \frac{6 \cdot 25}{1 \cdot 25} \qquad \text{Factoring and using the identity property of 1 to write 25 as } 1 \cdot 25$$

$$= \frac{6 \cdot \cancel{25}}{1 \cdot \cancel{25}} \qquad \text{Removing a factor equal to 1: } \frac{25}{25} = 1$$

$$= \frac{6}{1}$$

$$= 6 \qquad \text{Simplifying}$$

47.

$$\frac{42}{50} = \frac{2 \cdot 21}{2 \cdot 25} \qquad \text{Factoring the numerator and the denominator}$$

$$= \frac{\cancel{2} \cdot 21}{\cancel{2} \cdot 25} \qquad \text{Removing a factor equal to 1: } \frac{2}{2} = 1$$

$$= \frac{21}{25}$$

49.

$$\frac{120}{82} = \frac{2 \cdot 60}{2 \cdot 41} \qquad \text{Factoring}$$

$$= \frac{\cancel{2} \cdot 60}{\cancel{2} \cdot 41} \qquad \text{Removing a factor equal to 1: } \frac{2}{2} = 1$$

$$= \frac{60}{41}$$

51. $\dfrac{210}{98} = \dfrac{2 \cdot 7 \cdot 15}{2 \cdot 7 \cdot 7}$ Factoring

$= \dfrac{\cancel{2} \cdot \cancel{7} \cdot 15}{\cancel{2} \cdot \cancel{7} \cdot 7}$ Removing a factor equal to

$1: \dfrac{2 \cdot 7}{2 \cdot 7} = 1$

$= \dfrac{15}{7}$

53. $\dfrac{1}{2} \cdot \dfrac{3}{5} = \dfrac{1 \cdot 3}{2 \cdot 5}$ Multiplying numerators and denominators

$= \dfrac{3}{10}$

55. $\dfrac{9}{2} \cdot \dfrac{4}{3} = \dfrac{9 \cdot 4}{2 \cdot 3} = \dfrac{3 \cdot \cancel{3} \cdot \cancel{2} \cdot 2}{\cancel{2} \cdot \cancel{3}} = 6$

57. $\dfrac{1}{8} + \dfrac{3}{8} = \dfrac{1+3}{8}$ Adding numerators; keeping the common denominator

$= \dfrac{4}{8}$

$= \dfrac{1 \cdot \cancel{4}}{2 \cdot \cancel{4}} = \dfrac{1}{2}$ Simplifying

59. $\dfrac{4}{9} + \dfrac{13}{18} = \dfrac{4}{9} \cdot \dfrac{2}{2} + \dfrac{13}{18}$ Using 18 as the common denominator

$= \dfrac{8}{18} + \dfrac{13}{18}$

$= \dfrac{21}{18}$

$= \dfrac{7 \cdot \cancel{3}}{6 \cdot \cancel{3}} = \dfrac{7}{6}$ Simplifying

61. $\dfrac{3}{a} \cdot \dfrac{b}{7} = \dfrac{3b}{7a}$ Multiplying numerators and denominators

63. $\dfrac{4}{n} + \dfrac{6}{n} = \dfrac{10}{n}$ Adding numerators; keeping the common denominator

65. $\dfrac{3}{10} + \dfrac{8}{15} = \dfrac{3}{10} \cdot \dfrac{3}{3} + \dfrac{8}{15} \cdot \dfrac{2}{2}$ Using 30 as the common denominator

$= \dfrac{9}{30} + \dfrac{16}{30}$

$= \dfrac{25}{30}$

$= \dfrac{5 \cdot \cancel{5}}{6 \cdot \cancel{5}} = \dfrac{5}{6}$ Simplifying

67. $\dfrac{11}{7} - \dfrac{4}{7} = \dfrac{7}{7} = 1$

69. $\dfrac{13}{18} - \dfrac{4}{9} = \dfrac{13}{18} - \dfrac{4}{9} \cdot \dfrac{2}{2}$ Using 18 as the common denominator

$= \dfrac{13}{18} - \dfrac{8}{18}$

$= \dfrac{5}{18}$

71. Note that $\dfrac{20}{30} = \dfrac{2}{3}$. Thus, $\dfrac{20}{30} - \dfrac{2}{3} = 0$.

We can also do this exercise by finding a common denominator:

$\dfrac{20}{30} - \dfrac{2}{3} = \dfrac{20}{30} - \dfrac{20}{30} = 0$

73. $\dfrac{7}{6} \div \dfrac{3}{5} = \dfrac{7}{6} \cdot \dfrac{5}{3}$ Multiplying by the reciprocal of the divisor

$= \dfrac{35}{18}$

75. $\dfrac{8}{9} \div \dfrac{4}{15} = \dfrac{8}{9} \cdot \dfrac{15}{4} = \dfrac{2 \cdot \cancel{4} \cdot \cancel{3} \cdot 5}{\cancel{3} \cdot 3 \cdot \cancel{4}} = \dfrac{10}{3}$

77. $12 \div \dfrac{4}{9} = \dfrac{12}{1} \cdot \dfrac{9}{4} = \dfrac{\cancel{4} \cdot 3 \cdot 9}{1 \cdot \cancel{4}} = 27$

79. Note that we have a number divided by itself. Thus, the result is 1. We can also do this exercise as follows:

$\dfrac{7}{13} \div \dfrac{7}{13} = \dfrac{7}{13} \cdot \dfrac{13}{7} = \dfrac{7 \cdot 13}{7 \cdot 13} = 1$

81. $\dfrac{\frac{2}{7}}{\frac{5}{3}} = \dfrac{2}{7} \div \dfrac{5}{3} = \dfrac{2}{7} \cdot \dfrac{3}{5} = \dfrac{2 \cdot 3}{7 \cdot 5} = \dfrac{6}{35}$

83. $\dfrac{\frac{9}{1}}{\frac{2}{2}} = 9 \div \dfrac{1}{2} = \dfrac{9}{1} \cdot \dfrac{2}{1} = \dfrac{9 \cdot 2}{1 \cdot 1} = 18$

85. *Writing Exercise.* If the fractions have the same denominator and the numerators and/or denominators are very large numbers, it would probably be easier to compute the sum of the fractions than their product.

87. $5(x + 3) = 5(3 + x)$ Commutative law of addition

Answers may vary.

89. *Writing Exercise.* Bryce is canceling incorrectly. The number 2 is not a common factor of both terms in the numerator, so it cannot be canceled. For example, let $x = 1$. Then $(2 + 1)/8 = 3/8$ but $(1 + 1)/4 = 2/4 = 1/2$. The expressions are not equivalent.

91.

Product	56	63	36	72	140	96	168
Factor	7	7	2	36	14	8	8
Factor	8	9	18	2	10	12	21
Sum	15	16	20	38	24	20	29

93. $\dfrac{16 \cdot 9 \cdot 4}{15 \cdot 8 \cdot 12} = \dfrac{\cancel{4} \cdot \cancel{4} \cdot \cancel{3} \cdot \cancel{3} \cdot \cancel{2} \cdot 2}{\cancel{3} \cdot 5 \cdot \cancel{2} \cdot \cancel{4} \cdot \cancel{3} \cdot \cancel{4}} = \dfrac{2}{5}$

95. $\dfrac{45pqrs}{9prst} = \dfrac{5 \cdot \cancel{9} \cdot \cancel{p} \cdot q \cdot \cancel{r} \cdot \cancel{s}}{\cancel{9} \cdot \cancel{p} \cdot \cancel{r} \cdot \cancel{s} \cdot t} = \dfrac{5q}{t}$

97. $\dfrac{15 \cdot 4xy \cdot 9}{6 \cdot 25ab \cdot 15y} = \dfrac{\cancel{15} \cdot \cancel{2} \cdot 2 \cdot \cancel{x} \cdot \cancel{y} \cdot \cancel{3} \cdot 3}{\cancel{2} \cdot \cancel{3} \cdot 25 \cdot \cancel{x} \cdot \cancel{15} \cdot \cancel{y}} = \dfrac{6}{25}$

99. $\dfrac{\frac{15mn}{18bc}}{\frac{25np}{}} = \dfrac{27ab}{15mn} \div \dfrac{18bc}{25np} = \dfrac{27ab}{15mn} \cdot \dfrac{25np}{18bc}$

$= \dfrac{27ab \cdot 25np}{15mn \cdot 18bc} = \dfrac{\cancel{3} \cdot \cancel{9} \cdot a \cdot \cancel{b} \cdot \cancel{5} \cdot 5 \cdot \cancel{n} \cdot p}{\cancel{3} \cdot \cancel{5} \cdot m \cdot \cancel{n} \cdot 2 \cdot \cancel{9} \cdot \cancel{b} \cdot c}$

$= \dfrac{5ap}{2mc}$

101. $\dfrac{5\frac{3}{4}rs}{4\frac{1}{2}st} = \dfrac{\frac{23}{4}rs}{\frac{9}{2}st} = \dfrac{\frac{23rs}{4}}{\frac{9st}{2}} = \dfrac{23rs}{4} \div \dfrac{9st}{2}$

$= \dfrac{23rs}{4} \cdot \dfrac{2}{9st} = \dfrac{23rs \cdot 2}{4 \cdot 9st} = \dfrac{23 \cdot r \cdot \cancel{s} \cdot \cancel{2}}{\cancel{2} \cdot 2 \cdot 9 \cdot \cancel{s} \cdot t} = \dfrac{23r}{18t}$

103. $A = lw = \left(\frac{4}{5}\text{ m}\right)\left(\frac{7}{9}\text{ m}\right)$

$\qquad = \left(\frac{4}{5}\right)\left(\frac{7}{9}\right)(\text{m})(\text{m})$

$\qquad = \frac{28}{45}\text{ m}^2$, or $\frac{28}{45}$ square meters

105. $P = 4s = 4\left(3\frac{5}{9}\text{ m}\right) = 4 \cdot \frac{32}{9}\text{ m} = \frac{128}{9}\text{ m}$, or

$14\frac{2}{9}$ m

107. There are 12 edges, each with length $2\frac{3}{10}$ cm. We multiply to find the total length of the edges.

$12 \cdot 2\frac{3}{10}$ cm $= 12 \cdot \frac{23}{10}$ cm

$\qquad = \frac{12 \cdot 23}{10}$ cm

$\qquad = \frac{\cancel{2} \cdot 6 \cdot 23}{\cancel{2} \cdot 5}$ cm

$\qquad = \frac{138}{5}$ cm, or $27\frac{3}{5}$ cm

Exercise Set 1.4

1. Since $\frac{4}{7} = 0.\overline{571428}$, the correct choice is "repeating."

3. The set of integers consists of all whole numbers along with their opposites, so the correct choice is "integer."

5. A "rational number" has the form described.

7. A "natural number" can be thought of as a counting number.

9. The real number $-10,500$ corresponds to borrowing $10,500. The real number $27,482$ corresponds to the award of $27,482.

11. The real number 136 corresponds to $136°$F. The real number -4 corresponds to $4°$F below zero.

13. The real number -554 corresponds to a 554-point fall. The real number 499.19 corresponds to a 499.19-point gain.

15. The real number 650 corresponds to a $650 deposit, and the real number -180 corresponds to a $180 withdrawal.

17. The real number 8 corresponds to an 8-yd gain, and the real number -5 corresponds to a 5-yd loss.

19. Since $\frac{10}{3} = 3\frac{1}{3}$, its graph is $\frac{1}{3}$ of a unit to the right of 3.

21. The graph of -4.3 is $\frac{3}{10}$ of a unit to the left of -4.

23.

25. $\frac{7}{8}$ means $7 \div 8$, so we divide.

```
     0.8 7 5
8 ⟌ 7.0 0 0
     6 4
     ───
       6 0
       5 6
       ───
         4 0
         4 0
         ───
           0
```

We have $\frac{7}{8} = 0.875$.

27. We first find decimal notation for $\frac{3}{4}$. Since $\frac{3}{4}$ means $3 \div 4$, we divide.

```
     0.7 5
4 ⟌ 3.0 0
     2 8
     ───
       2 0
       2 0
       ───
         0
```

Thus, $\frac{3}{4} = 0.75$, so $-\frac{3}{4} = -0.75$.

29. $\frac{7}{6}$ means $7 \div 6$, so we divide.

```
     1.1 6 6
6 ⟌ 7.0 0 0
     6
     ───
     1 0
       6
       ───
       4 0
       3 6
       ───
         4 0
         3 6
         ───
           4
```

Thus $\frac{7}{6} = 1.1\overline{6}$, so $-\frac{7}{6} = -1.1\overline{6}$.

31. $\frac{2}{3}$ means $2 \div 3$, so we divide.

```
     0.6 6 6 ...
3 ⟌ 2.0 0 0
     1 8
     ───
       2 0
       1 8
       ───
         2 0
         1 8
         ───
           2
```

We have $\frac{2}{3} = 0.\overline{6}$.

33. We first find decimal notation for $\frac{1}{2}$. Since $\frac{1}{2}$ means $1 \div 2$, we divide.

```
     0.5
2 ⟌ 1.0
     1 0
     ───
       0
```

Thus, $\frac{1}{2} = 0.5$, so $-\frac{1}{2} = -0.5$.

35. Since the denominator is 100, we know that $\frac{13}{100} = 0.13$. We could also divide 13 by 100 to find this result.

37.

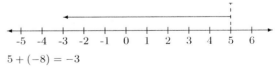

$$\sqrt{5}$$

39.

$$-\sqrt{22}$$

41. Since 5 is to the right of 0, we have $5 > 0$.

43. Since -9 is to the left of 9, we have $-9 < 0$.

45. Since -8 is to the left of -5, we have $-8 < -5$.

47. Since -5 is to the right of -11, we have $-5 > -11$.

49. Since -12.5 is to the left of -10.2, we have $-12.5 < -10.2$.

51. We convert to decimal notation.
$\frac{5}{12} = 0.41\overline{6}$ and $\frac{11}{25} = 0.44$. Thus, $\frac{5}{12} < \frac{11}{25}$.

53. $-2 > x$ has the same meaning as $x < -2$.

55. $10 \le y$ has the same meaning as $y \ge 10$.

57. $-3 \ge -11$ is true, since $-3 > -11$ is true.

59. $0 \ge 8$ is false, since neither $0 > 8$ nor $0 = 8$ is true.

61. $-8 \le -8$ is true because $-8 = -8$ is true.

63. $|-58| = 58$ since -58 is 58 units from 0.

65. $|-12.2| = 12.2$ since -12.2 is 12.2 units from 0.

67. $|\sqrt{2}| = \sqrt{2}$ since $\sqrt{2}$ is $\sqrt{2}$ units from 0.

69. $\left|-\frac{9}{7}\right| = \frac{9}{7}$ since $-\frac{9}{7}$ is $\frac{9}{7}$ units from 0.

71. $|0| = 0$ since 0 is 0 units from itself.

73. $|x| = |-8| = 8$

75. $-83, -4.7, 0, \frac{5}{9}, 2.1\overline{6}, 62$

77. $-83, 0, 62$

79. All are real numbers.

81. *Writing Exercise.* Yes; every integer can be written as $n/1$, a quotient of the form a/b where $b \ne 0$.

83. $3xy = 3 \cdot 2 \cdot 7 = 42$

85. *Writing Exercise.* No; $|0| = 0$ which is neither positive nor negative.

87. *Writing Exercise.* No; every positive number is nonnegative, but zero is nonnegative and zero is not positive.

89. List the numbers as they occur on the number line, from left to right: $-23, -17, 0, 4$

91. Converting to decimal notation, we can write
$$\frac{4}{5}, \frac{4}{3}, \frac{4}{8}, \frac{4}{6}, \frac{4}{9}, \frac{4}{2}, -\frac{4}{3} \text{ as}$$
$0.8, 1.3\overline{3}, 0.5, 0.6\overline{6}, 0.4\overline{4}, 2, -1.3\overline{3}$, respectively. List the numbers (in fractional form) as they occur on the number line, from left to right:
$$-\frac{4}{3}, \frac{4}{9}, \frac{4}{8}, \frac{4}{6}, \frac{4}{5}, \frac{4}{3}, \frac{4}{2}$$

93. $|4| = 4$ and $|-7| = 7$, so $|4| < |-7|$.

95. $|23| = 23$ and $|-23| = 23$, so $|23| = |-23|$.

97. $|x| = 19$

x represents a number whose distance from 0 is 19. Thus, $x = 19$ or $x = -19$.

99. $2 < |x| < 5$

x represents an integer whose distance from 0 is greater than 2 and also less than 5. Thus, $x = -4, -3, 3, 4$.

101. $0.9\overline{9} = 3(0.3\overline{3}) = 3 \cdot \frac{1}{3} = \frac{3}{3}$

103. $7.7\overline{7} = 70(0.1\overline{1}) = 70 \cdot \frac{1}{9} = \frac{70}{9}$

(See Exercise 100.)

105. Nonpositive numbers include zero. Thus, $x \le 0$.

107. Distance from zero can go in the positive or negative direction. Thus, $|t| \ge 20$.

109. *Writing Exercise.* The statement $\sqrt{a^2} = |a|$ for any real number a is true. If a is nonnegative, then $\sqrt{a^2} = a$. If a is negative, $\sqrt{a^2} = -a$ since $\sqrt{a^2}$ must be nonnegative. Thus, for a nonnegative number, the result is the number and, for a negative number, the result is the opposite of the number. This describes the absolute value of a number.

Exercise Set 1.5

1. Choice (f), $-3n$, has the same variable factor as $8n$.

3. Choice (e), 9, is a constant as is 43.

5. Choice (b), $5x$, has the same variable factor as $-2x$.

7. Start at 5. Move 8 units to the left.

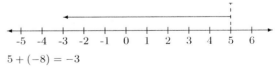

$5 + (-8) = -3$

9. Start at -6. Move 10 units to the right.

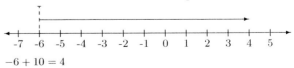

$-6 + 10 = 4$

11. Start at -7. Move 0 units.

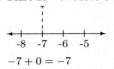

$-7 + 0 = -7$

13. Start at -3. Move 5 units to the left.

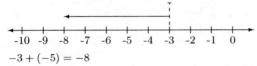

$-3 + (-5) = -8$

15. $-35 + 0$ One number is 0. The answer is the other number. $-35 + 0 = -35$

17. $0 + (-8)$ One number is 0. The answer is the other number. $0 + (-8) = -8$

19. $12 + (-12)$ The numbers have the same absolute value. The sum is 0. $12 + (-12) = 0$

21. $-24 + (-17)$ Two negatives. Add the absolute values, getting 41. Make the answer negative. $-24 + (-17) = -41$

23. $-13 + 13$ The numbers have the same absolute value. The sum is 0. $-13 + 13 = 0$

25. $20 + (-11)$ The absolute values are 20 and 11. The difference is $20 - 11$, or 9. The positive number has the larger absolute value, so the answer is positive. $20 + (-11) = 9$

27. $10 + (-12)$ The absolute values are 10 and 12. The difference is $12 - 10$, or 2. The negative number has the larger absolute value, so the answer is negative. $10 + (-12) = -2$

29. $-3 + 14$ The absolute values are 3 and 14. The difference is $14 - 3$, or 11. The positive number has the larger absolute value, so the answer is positive. $-3 + 14 = 11$

31. $-24 + (-19)$ Two negatives. Add the absolute values, getting 43. Make the answer negative. $-24 + (-19) = -43$

33. $19 + (-19)$ The numbers have the same absolute value. The sum is 0. $19 + (-19) = 0$

35. $23 + (-5)$ The absolute values are 23 and 5. The difference is $23 - 5$ or 18. The positive number has the larger absolute value, so the answer is positive. $23 + (-5) = 18$

37. $-31 + (-14)$ Two negatives. Add the absolute values, getting 45. Make the answer negative. $-31 + (-14) = -45$

39. $40 + (-40)$ The numbers have the same absolute value. The sum is 0. $40 + (-40) = 0$

41. $85 + (-69)$ The absolute values are 85 and 69. The difference is $85 - 69$, or 16. The positive number has the larger absolute value, so the answer is positive. $85 + (-69) = 16$

43. $-3.6 + 2.8$ The absolute values are 3.6 and 2.8. The difference is $3.6 - 2.8$, or 0.8. The negative number has the larger absolute value, so the answer is negative. $-3.6 + 2.8 = -0.8$

45. $-5.4 + (-3.7)$ Two negatives. Add the absolute values, getting 9.1. Make the answer negative. $-5.4 + (-3.7) = -9.1$

47. $\frac{4}{5} + \left(-\frac{1}{5}\right)$ The absolute values are $\frac{4}{5}$ and $\frac{1}{5}$. The positive number has the larger absolute value, so the answer is positive. $\frac{4}{5} + \left(-\frac{1}{5}\right) = \frac{3}{5}$

49. $\frac{-4}{7} + \frac{-2}{7}$ Two negatives. Add the absolute values, getting $\frac{6}{7}$. Make the answer negative. $\frac{-4}{7} + \frac{-2}{7} = \frac{-6}{7}$

51. $-\frac{2}{5} + \frac{1}{3}$ The absolute values are $\frac{2}{5}$ and $\frac{1}{3}$. The difference is $\frac{6}{15} - \frac{5}{15}$, or $\frac{1}{15}$. The negative number has the larger absolute value, so the answer is negative. $-\frac{2}{5} + \frac{1}{3} = -\frac{1}{15}$

53. $\frac{-4}{9} + \frac{2}{3}$ The absolute values are $\frac{4}{9}$ and $\frac{2}{3}$. The difference is $\frac{6}{9} - \frac{4}{9}$, or $\frac{2}{9}$. The positive number has the larger absolute value, so the answer is positive. $\frac{-4}{9} + \frac{2}{3} = \frac{2}{9}$

55. $\quad 35 + (-14) + (-19) + (-5)$
$= 35 + [(-14) + (-19) + (-5)] \quad$ Using the associative law of addition
$= 35 + (-38) \quad$ Adding the negatives
$= -3 \quad$ Adding a positive and a negative

57. $-4.9 + 8.5 + 4.9 + (-8.5)$

Note that we have two pairs of numbers with different signs and the same absolute value: -4.9 and 4.9, 8.5 and -8.5. The sum of each pair is 0, so the result is $0 + 0$, or 0.

59. Rewording: $\underbrace{\text{First increase}}$ plus $\underbrace{\text{decrease}}$

$\qquad\qquad\qquad\qquad\quad \downarrow \qquad\quad \downarrow \qquad \downarrow$

Translating: $15\cancel{c}$ $+$ $(-3\cancel{c})$

plus $\underbrace{\text{second increase}}$ is $\underbrace{\text{change in price.}}$

$\quad\downarrow\qquad\qquad \downarrow \qquad\quad \downarrow \qquad\quad \downarrow$

$\quad +\qquad\qquad 17\cancel{c} \qquad = \quad$ change in price

Since $15 + (-3) + 17$

$\quad = 12 + 17$

$\quad = 29,$

the price rose $29\cancel{c}$ during the given period.

61. Rewording: $\underbrace{\substack{\text{July} \\ \text{bill}}}$ plus $\underbrace{\text{payment}}$ plus

$\qquad\qquad\qquad \downarrow \quad \downarrow \qquad \downarrow \qquad\quad \downarrow$

Translating: -82 $+$ 50 $+$

$$\underbrace{\begin{array}{c}\text{August} \\ \text{charges}\end{array}}_{} \quad \overset{\text{is}}{} \quad \underbrace{\begin{array}{c}\text{new} \\ \text{balance.}\end{array}}_{}$$

$$\downarrow \qquad \downarrow \qquad \downarrow$$

$$(-63) \quad = \quad \begin{array}{c}\text{new} \\ \text{balance}\end{array}$$

Since $-82 + 50 + (-63) = -32 + (-63)$
$$= -95,$$

Chloe's new balance was $95.

63. Rewording: $\underbrace{\begin{array}{c}\text{First try} \\ \text{yardage}\end{array}}_{} \quad \overset{\text{plus}}{} \quad \underbrace{\begin{array}{c}\text{second try} \\ \text{yardage}\end{array}}_{}$

$$\downarrow \qquad \qquad \downarrow \qquad \qquad \downarrow$$

Translating: $(-13) \qquad + \qquad 12$

$$\overset{\text{plus}}{} \quad \underbrace{\begin{array}{c}\text{third try} \\ \text{yardage}\end{array}}_{} \quad \overset{\text{is}}{} \quad \underbrace{\begin{array}{c}\text{total gain} \\ \text{or loss.}\end{array}}_{}$$

$$\downarrow \qquad \downarrow \qquad \downarrow \qquad \downarrow$$

$$+ \qquad 21 \qquad = \qquad \begin{array}{c}\text{total gain} \\ \text{or loss}\end{array}$$

Since $(-13) + 12 + 21$
$$= -1 + 21$$
$$= 20,$$

the total gain was 20 yd.

65. Rewording: $\underbrace{\text{first drop}}_{} \text{ plus } \underbrace{\text{rise}}_{} \text{ plus}$

$$\downarrow \qquad \downarrow \qquad \downarrow \qquad \downarrow$$

Translating: $\left(-\dfrac{2}{5}\right) \quad + \quad 1\dfrac{1}{5} \quad +$

$$\underbrace{\text{drop}}_{} \quad \overset{\text{is}}{} \quad \underbrace{\begin{array}{c}\text{total level} \\ \text{change.}\end{array}}_{}$$

$$\downarrow \qquad \downarrow \qquad \downarrow$$

$$\left(-\dfrac{1}{2}\right) \quad = \quad \begin{array}{c}\text{total level} \\ \text{change}\end{array}$$

Since $-\dfrac{2}{5} + 1\dfrac{1}{5} + \left(-\dfrac{1}{2}\right) = -\dfrac{4}{10} + \dfrac{12}{10} + \left(-\dfrac{5}{10}\right)$
$$= \dfrac{8}{10} - \dfrac{5}{10},$$
$$= \dfrac{3}{10},$$

The lake rose $\dfrac{3}{10}$ ft.

67. Rewording: $\underbrace{\begin{array}{c}\text{Original} \\ \text{balance}\end{array}}_{} \quad \overset{\text{plus}}{} \quad \underbrace{\begin{array}{c}\text{first} \\ \text{payment}\end{array}}_{} \quad \overset{\text{plus}}{}$

$$\downarrow \qquad \downarrow \qquad \downarrow \qquad \downarrow$$

Translating: $-470 \qquad + \qquad 45 \qquad +$

$$\underbrace{\begin{array}{c}\text{new} \\ \text{charges}\end{array}}_{} \quad \overset{\text{plus}}{} \quad \underbrace{\begin{array}{c}\text{second} \\ \text{payment}\end{array}}_{} \quad \overset{\text{is}}{} \quad \underbrace{\begin{array}{c}\text{new} \\ \text{balance.}\end{array}}_{}$$

$$\downarrow \qquad \downarrow \qquad \downarrow \qquad \downarrow \qquad \downarrow$$

$$-160 \qquad + \qquad 500 \qquad = \qquad \begin{array}{c}\text{new} \\ \text{balance}\end{array}$$

Since $-470 + 45 + (-160) + 500$
$$= [-470 + (-160)] + (45 + 500)$$
$$= -630 + 545$$
$$= -85,$$

Logan owes $85 on his credit card.

69. $7a + 10a = (7 + 10)a$ Using the distributive law
$$= 17a$$

71. $-3x + 12x = (-3 + 12)x$ Using the distributive law
$$= 9x$$

73. $4t + 21t = (4 + 21)t = 25t$

75. $7m + (-9m) = [7 + (-9)]m = -2m$

77. $-8y + (-2y) = [-8 + (-2)]y = -10y$

79. $\quad -3 + 8x + 4 + (-10x)$
$$= -3 + 4 + 8x + (-10x) \qquad \begin{array}{l}\text{Using the commuta-} \\ \text{tive law of addition}\end{array}$$
$$= (-3 + 4) + [8 + (-10)]x \qquad \begin{array}{l}\text{Using the distributive} \\ \text{law}\end{array}$$
$$= 1 - 2x \qquad\qquad\qquad\qquad \text{Adding}$$

81. Perimeter $= 8 + 5x + 9 + 7x$
$$= 8 + 9 + 5x + 7x$$
$$= (8 + 9) + (5 + 7)x$$
$$= 17 + 12x$$

83. Perimeter $= 3t + 3r + 7 + 5t + 9 + 4r$
$$= 3t + 5t + 3r + 4r + 7 + 9$$
$$= (3 + 5)t + (3 + 4)r + (7 + 9)$$
$$= 8t + 7r + 16$$

85. Perimeter $= 9 + 6n + 7 + 8n + 4n$
$$= 9 + 7 + 6n + 8n + 4n$$
$$= (9 + 7) + (6 + 8 + 4)n$$
$$= 16 + 18n$$

87. *Writing Exercise.* Answers may vary. One possible explanation follows.

Consider performing the addition on a number line. We start to the left of 0 and then move farther left, so the result must be a negative number.

89. $7(3z + y + 1) = 7 \cdot 3z + 7 \cdot y + 7 \cdot 1 = 21z + 7y + 7$

91. *Writing Exercise.* The sum will be positive when the positive number is greater than the sum of the absolute values of the negative numbers.

93. Starting with the final value, we "undo" the deposit and original amount by adding their opposites. The result is the amount of the check.

Rewording: $\underbrace{\text{Final value}}$ plus $\underbrace{\text{opposite of deposit}}$ plus

$\qquad\qquad \downarrow \qquad \downarrow \qquad\quad \downarrow \qquad\qquad \downarrow$

Translating: $\quad -\$42.37 \;+\quad (-\$152)\quad +$

$\underbrace{\text{opposite of original amount}}$ is check amount.

$\qquad\qquad \downarrow \qquad\quad \downarrow \qquad\quad \downarrow$

$\qquad -\$257.33\;\; = \text{check amount.}$

Since $-42.37 + (-152) + (-257.33) = (-194.37) + (-257.33)$
$$= -451.70,$$

So the amount of the check was \$451.70.

95. $\quad 4x + \underline{} + (-9x) + (-2y)$

$\quad = 4x + (-9x) + \underline{} + (-2y)$

$\quad = [4 + (-9)]x + \underline{} + (-2y)$

$\quad = -5x + \underline{} + (-2y)$

This expression is equivalent to $-5x - 7y$, so the missing term is the term which yields $-7y$ when added to $-2y$. Since $-5y + (-2y) = -7y$, the missing term is $-5y$.

97. $\quad 3m + 2n + \underline{} + (-2m)$

$\quad = 2n + \underline{} + (-2m) + 3m$

$\quad = 2n + \underline{} + (-2 + 3)m$

$\quad = 2n + \underline{} + m$

This expression is equivalent to $2n + (-6m)$, so the missing term is the term which yields $-6m$ when added to m. Since $-7m + m = -6m$, the missing term is $-7m$.

99. Note that, in order for the sum to be 0, the two missing terms must be the opposites of the given terms. Thus, the missing terms are $-7t$ and -23.

101. $-3 + (-3) + 2 + (-2) + 1 = -5$

Since the total is 5 under par after the five rounds and $-5 = -1 + (-1) + (-1) + (-1) + (-1)$, the golfer was 1 under par on average.

Exercise Set 1.6

1. $-x$ is read "the opposite of x," so choice (d) is correct.

3. $12 - (-x)$ is read "twelve minus the opposite of x," so choice (f) is correct.

5. $x - (-12)$ is read "x minus negative twelve," so choice (a) is correct.

7. $-x - x$ is read "the opposite of x minus x," so choice (b) is correct.

9. $6 - 10$ is read "six minus ten."

11. $2 - (-12)$ is read "two minus negative twelve."

13. $9 - (-t)$ is read "nine minus the opposite of t."

15. $-x - y$ is read "the opposite of x minus y."

17. $-3 - (-n)$ is read "negative three minus the opposite of n."

19. The opposite of 51 is -51 because $51 + (-51) = 0$.

21. The opposite of $-\dfrac{11}{3}$ is $\dfrac{11}{3}$ because $-\dfrac{11}{3} + \dfrac{11}{3} = 0$.

23. The opposite of -3.14 is 3.14 because $-3.14 + 3.14 = 0$.

25. If $x = -45$, then $-x = -(-45) = 45$. (The opposite of 45 is -45.)

27. If $x = -\dfrac{14}{3}$, then $-x = -\left(-\dfrac{14}{3}\right) = \dfrac{14}{3}$.

$\left(\text{The opposite of } -\dfrac{14}{3} \text{ is } \dfrac{14}{3}.\right)$

29. If $x = 0.101$, then $-x = -(0.101) = -0.101$.
(The opposite of 0.101 is -0.101.)

31. If $x = 37$, then $-(-x) = -(-37) = 37$
(The opposite of the opposite of 37 is 37.)

33. If $x = -\dfrac{2}{5}$, then $-(-x) = -\left[-\left(-\dfrac{2}{5}\right)\right] = -\dfrac{2}{5}$.

$\left(\text{The opposite of the opposite of } -\dfrac{2}{5} \text{ is } -\dfrac{2}{5}.\right)$

35. When we change the sign of -1 we obtain 1.

37. When we change the sign we obtain -15.

39. $7 - 10 = 7 + (-10) = -3$

41. $0 - 6 = 0 + (-6) = -6$

43. $2 - 5 = 2 + (-5) = -3$

45. $-4 - 3 = -4 + (-3) = -7$

47. $-9 - (-3) = -9 + 3 = -6$

49. Note that we are subtracting a number from itself. The result is 0. We could also do this exercise as follows:
$$-8 - (-8) = -8 + 8 = 0$$

51. $14 - 19 = 14 + (-19) = -5$

53. $30 - 40 = 30 + (-40) = -10$

55. $0 - 11 = 0 + (-11) = -11$

57. $-9 - (-9) = -9 + 9 = 0$
(See Exercise 49.)

59. $5 - 5 = 5 + (-5) = 0$
(See Exercise 49.)

61. $4 - (-4) = 4 + 4 = 8$

63. $-7 - 4 = -7 + (-4) = -11$

65. $6 - (-10) = 6 + 10 = 16$

67. $-4 - 15 = -4 + (-15) = -19$

69. $-6 - (-7) = -6 + 7 = 1$

71. $5 - (-12) = 5 + 12 = 17$

73. $0 - (-3) = 0 + 3 = 3$

75. $-5 - (-2) = -5 + 2 = -3$

77. $-7 - 14 = -7 + (-14) = -21$

79. $0 - (-10) = 0 + 10 = 10$

81. $-8 - 0 = -8 + 0 = -8$

83. $-52 - 8 = -52 + (-8) = -60$

85. $2 - 25 = 2 + (-25) = -23$

87. $-4.2 - 3.1 = -4.2 + (-3.1) = -7.3$

89. $-1.3 - (-2.4) = -1.3 + 2.4 = 1.1$

91. $3.2 - 8.7 = 3.2 + (-8.7) = -5.5$

93. $0.072 - 1 = 0.072 + (-1) = -0.928$

95. $\dfrac{2}{11} - \dfrac{9}{11} = \dfrac{2}{11} + \left(-\dfrac{9}{11}\right) = -\dfrac{7}{11}$

97. $\dfrac{-1}{5} - \dfrac{3}{5} = \dfrac{-1}{5} + \left(\dfrac{-3}{5}\right) = \dfrac{-4}{5}$, or $-\dfrac{4}{5}$

99. $-\dfrac{4}{17} - \left(-\dfrac{9}{17}\right) = -\dfrac{4}{17} + \dfrac{9}{17} = \dfrac{5}{17}$

101. We subtract the smaller number from the larger.

Translate: $3.8 - (-5.2)$

Simplify: $3.8 - (-5.2) = 3.8 + 5.2 = 9$

103. We subtract the smaller number from the larger.

Translate: $114 - (-79)$

Simplify: $114 - (-79) = 114 + 79 = 193$

105. $-8 - 32 = -8 + (-32) = -40$

107. $18 - (-25) = 18 + 25 = 43$

109. $16 - (-12) - 1 - (-2) + 3 = 16 + 12 + (-1) + 2 + 3 = 32$

111. $-31 + (-28) - (-14) - 17 = (-31) + (-28) + 14 + (-17)$
$= -62$

113. $-34 - 28 + (-33) - 44 = (-34) + (-28) + (-33) + (-44)$
$= -139$

115. $-93 + (-84) - (-93) - (-84)$

Note that we are subtracting -93 from -93 and -84 from -84. Thus, the result will be 0. We could also do this exercise as follows:
$-93 + (-84) - (-93) - (-84) = -93 + (-84) + 93 + 84 = 0$

117. $-3y - 8x = -3y + (-8x)$, so the terms are $-3y$ and $-8x$.

119. $9 - 5t - 3st = 9 + (-5t) + (-3st)$, so the terms are 9, $-5t$, and $-3st$.

121. $10x - 13x$
$= 10x + (-13x)$ Adding the opposite
$= (10 + (-13))x$ Using the distributive law
$= -3x$

123. $7a - 12a + 4$
$= 7a + (-12a) + 4$ Adding the opposite
$= (7 + (-12))a + 4$ Using the distributive law
$= -5a + 4$

125. $-8n - 9 + 7n$
$= -8n + (-9) + 7n$ Adding the opposite
$= -8n + 7n + (-9)$ Using the commutative law of addition
$= -n - 9$ Adding like terms

127. $5 - 3x - 11$
$= 5 + (-3x) + (-11)$
$= -3x + 5 + (-11)$
$= -3x - 6$

129. $2 - 6t - 9 - 2t$
$= 2 + (-6t) + (-9) + (-2t)$
$= 2 + (-9) + (-6t) + (-2t)$
$= -7 - 8t$

131. $5y + (-3x) - 9x + 1 - 2y + 8$
$= 5y + (-3x) + (-9x) + 1 + (-2y) + 8$
$= 5y + (-2y) + (-3x) + (-9x) + 1 + 8$
$= 3y - 12x + 9$

133. $13x - (-2x) + 45 - (-21) - 7x$
$= 13x + 2x + 45 + 21 + (-7x)$
$= 13x + 2x + (-7x) + 45 + 21$
$= 8x + 66$

135. We subtract the lower temperature from the higher temperature:
$134 - (-80) = 134 + 80 = 214$
The temperature range is $214°$F.

137. We subtract the lower elevation from the higher elevation:
$29,035 - (-1312) = 29,035 + 1312 = 30,347$
The difference in elevation is 30,347 ft.

139. We subtract the lower elevation from the higher elevation:
$-40 - (-156) = -40 + 156 = 116$
Lake Assal is 116 m lower than the Valdes Peninsula.

141. *Writing Exercise.* Yes; rewrite subtraction as addition of the opposite.

143. Area $= lw = (36$ ft$)(12$ ft$) = 432$ ft^2

145. Answers will vary. The symbol "$-$" can represent the opposite, or a negative, or subtraction.

147. If the clock reads 8:00 A.M. on the day following the black-out when the actual time is 3:00 P.M., then the clock is 7 hr behind the actual time. This indicates that the power outage lasted 7 hr, so power was restored 7 hr after 4:00 P.M., or at 11:00 P.M. on August 14.

149. False. For example, let $m = -3$ and $n = -5$. Then $-3 > -5$, but $-3 + (-5) = -8 \not> 0$.

151. True. For example, for $m = 4$ and $n = -4$, $4 = -(-4)$ and $4 + (-4) = 0$; for $m = -3$ and $n = 3$, $-3 = -3$ and $-3 + 3 = 0$.

153. $\boxed{(-)}\,\boxed{9}\,\boxed{-}\,\boxed{(-)}\,\boxed{7}\,\boxed{\text{ENTER}}$

Exercise Set 1.7

---

1. The product of two reciprocals is 1.

3. The sum of a pair of additive inverses is 0.

5. The number 0 has no reciprocal.

7. The number 1 is the multiplicative identity.

9. A nonzero number divided by itself is 1.

11. $-4 \cdot 10 = -40$ Think: $4 \cdot 10 = 40$, make the answer negative.

13. $-8 \cdot 7 = -56$ Think: $8 \cdot 7 = 56$, make the answer negative.

15. $4 \cdot (-10) = -40$

17. $-9 \cdot (-8) = 72$ Multiplying absolute values; the answer is positive.

19. $-6 \cdot 7 = -42$

21. $-5 \cdot (-9) = 45$ Multiplying absolute values; the answer is positive.

23. $-19 \cdot (-10) = 190$

25. $11 \cdot (-12) = -132$

27. $-25 \cdot (-48) = 1200$

29. $4.5 \cdot (-28) = -126$

31. $-5 \cdot (-2.3) = 11.5$

33. $-(25) \cdot 0 = 0$ The product of 0 and any real number is 0.

35. $\dfrac{2}{5} \cdot \left(-\dfrac{5}{7}\right) = -\left(\dfrac{2 \cdot 5}{5 \cdot 7}\right) = -\left(\dfrac{2}{7} \cdot \dfrac{5}{5}\right) = \dfrac{-2}{7}$

37. $-\dfrac{3}{8} \cdot \left(-\dfrac{2}{9}\right) = \dfrac{\not3 \cdot \not2 \cdot 1}{4 \cdot \not2 \cdot \not3 \cdot 3} = \dfrac{1}{12}$

39. $(-5.3)(2.1) = -11.13$

41. $-\dfrac{5}{9} \cdot \dfrac{3}{4} = -\dfrac{5 \cdot \not3}{\not3 \cdot 3 \cdot 4} = -\dfrac{5}{12}$

43. $3 \cdot (-7) \cdot (-2) \cdot 6$
$= -21 \cdot (-12)$ Multiplying the first two numbers and the last two numbers
$= 252$

45. 0, The product of 0 and any real number is 0.

47. $-\dfrac{1}{3} \cdot \dfrac{1}{4} \cdot \left(-\dfrac{3}{7}\right) = -\dfrac{1}{12} \cdot \left(-\dfrac{3}{7}\right) = \dfrac{3}{12 \cdot 7}$
$= \dfrac{\not3 \cdot 1}{\not3 \cdot 4 \cdot 7} = \dfrac{1}{28}$

49. $-2 \cdot (-5) \cdot (-3) \cdot (-5) = 10 \cdot 15 = 150$

51. 0, The product of 0 and any real number is 0.

53. $(-8)(-9)(-10) = 72(-10) = -720$

55. $(-6)(-7)(-8)(-9)(-10) = 42 \cdot 72 \cdot (-10)$
$= 3024 \cdot (-10) = -30,240$

57. $18 \div (-2) = -9$ Check: $-9 \cdot (-2) = 18$

59. $\dfrac{36}{-9} = -4$ Check: $-4 \cdot (-9) = 36$

61. $\dfrac{-56}{8} = -7$ Check: $-7 \cdot 8 = -56$

63. $\dfrac{-48}{-12} = 4$ Check: $4(-12) = -48$

65. $72 \div 8 = -9$ Check: $-9 \cdot 8 = -72$

67. $-10.2 \div (-2) = 5.1$ Check: $5.1(-2) = -10.2$

69. $-100 \div (-11) = \dfrac{100}{11}$

71. $\dfrac{400}{-50} = -8$ Check: $-8 \cdot (-50) = 400$

73. Undefined

75. $-4.8 \div 1.2 = -4$ Check: $-4(1.2) = -4.8$

77. $\dfrac{0}{-9} = 0$

79. $\dfrac{9.7(-2.8)0}{4.3}$
Since the numerator has a factor of 0, the product in the numerator is 0. The denominator is nonzero, so the quotient is 0.

81. $\dfrac{-8}{3} = \dfrac{8}{-3}$ and $\dfrac{-8}{3} = -\dfrac{8}{3}$

83. $\dfrac{29}{-35} = \dfrac{-29}{35}$ and $\dfrac{29}{-35} = -\dfrac{29}{35}$

85. $-\dfrac{7}{3} = \dfrac{-7}{3}$ and $-\dfrac{7}{3} = \dfrac{7}{-3}$

87. $\dfrac{-x}{2} = \dfrac{x}{-2}$ and $\dfrac{-x}{2} = -\dfrac{x}{2}$

89. The reciprocal of $\dfrac{4}{-5}$ is $\dfrac{-5}{4}$ $\left(\text{or equivalently, } -\dfrac{5}{4}\right)$ because $\dfrac{4}{-5} \cdot \dfrac{-5}{4} = 1$.

91. The reciprocal of $\dfrac{51}{-10}$ is $-\dfrac{10}{51}$ because
$$\dfrac{51}{-10} \cdot \left(-\dfrac{10}{51}\right) = 1.$$

93. The reciprocal of -10 is $\dfrac{1}{-10}$ $\left(\text{or equivalently, } -\dfrac{1}{10}\right)$ because $-10\left(\dfrac{1}{-10}\right) = 1.$

95. The reciprocal of 4.3 is $\dfrac{1}{4.3}$ because $4.3\left(\dfrac{1}{4.3}\right) = 1.$

Since $\dfrac{1}{4.3} = \dfrac{1}{4.3} \cdot \dfrac{10}{10} = \dfrac{10}{43}$, the reciprocal can also be expressed as $\dfrac{10}{43}$.

97. The reciprocal of $\dfrac{-9}{4}$ is $\dfrac{4}{-9}$ $\left(\text{or equivalently, } -\dfrac{4}{9}\right)$ because
$$\dfrac{-9}{4} \cdot \dfrac{4}{-9} = 1.$$

99. The reciprocal of 0 does not exist. (There is no number n for which $0 \cdot n = 1$.)

101. $\left(\dfrac{-7}{4}\right)\left(-\dfrac{3}{5}\right)$

$\qquad = \left(-\dfrac{7}{4}\right)\left(-\dfrac{3}{5}\right)$ Rewriting $\dfrac{-7}{4}$ as $-\dfrac{7}{4}$

$\qquad = \dfrac{21}{20}$

103. $\dfrac{-3}{8} + \dfrac{-5}{8} = \dfrac{-8}{8} = -1$

105. $\left(\dfrac{-9}{5}\right)\left(\dfrac{5}{-9}\right)$

Note that this is the product of reciprocals. Thus, the result is 1.

107. $\left(-\dfrac{3}{11}\right) - \left(-\dfrac{6}{11}\right) = \dfrac{-3}{11} + \dfrac{6}{11} = \dfrac{3}{11}$

109. $\dfrac{7}{8} \div \left(-\dfrac{1}{2}\right) = \dfrac{7}{8} \cdot \left(-\dfrac{2}{1}\right) = -\dfrac{14}{8} = -\dfrac{7 \cdot 2}{2 \cdot 4 \cdot 1} = -\dfrac{7}{4}$

111. $-\dfrac{5}{9} \div \left(-\dfrac{5}{9}\right)$

Note that we have a number divided by itself. Thus, the result is 1.

113. $\dfrac{-3}{10} + \dfrac{2}{5} = \dfrac{-3}{10} + \dfrac{2}{5} \cdot \dfrac{2}{2} = \dfrac{-3}{10} + \dfrac{4}{10} = \dfrac{1}{10}$

115. $\dfrac{7}{10} \div \left(\dfrac{-3}{5}\right) = \dfrac{7}{10} \div \left(-\dfrac{3}{5}\right) = \dfrac{7}{10} \cdot \left(-\dfrac{5}{3}\right) = -\dfrac{35}{30}$

$\qquad = -\dfrac{7 \cdot 5}{2 \cdot 5 \cdot 3} = -\dfrac{7}{6}$

117. $\dfrac{14}{-9} \div \dfrac{0}{3} = \dfrac{14}{-9} \cdot \dfrac{3}{0}$ Undefined

119. $\dfrac{-4}{15} + \dfrac{2}{-3} = \dfrac{-4}{15} + \dfrac{-2}{3} = \dfrac{-4}{15} + \dfrac{-2}{3} \cdot \dfrac{5}{5} = \dfrac{-4}{15} + \dfrac{-10}{15}$

$\qquad = \dfrac{-14}{15}, \text{ or } -\dfrac{14}{15}$

121. *Writing Exercise.* You get the original number. The reciprocal of the reciprocal of a number is the original number.

123. $\dfrac{264}{468} = \dfrac{2 \cdot 2 \cdot 2 \cdot 3 \cdot 11}{2 \cdot 2 \cdot 3 \cdot 3 \cdot 13} = \dfrac{22}{39}$

125. *Writing Exercise.* Yes; consider n $(n \neq 0)$ and its opposite $-n$. The reciprocals of these numbers are $\dfrac{1}{n}$ and $\dfrac{1}{-n}$. Now $\dfrac{1}{n} + \dfrac{1}{-n} = \dfrac{1}{n} + \dfrac{-1}{n} = 0$, so the reciprocals are also opposites.

127. Let a and b represent the numbers. The $a + b$ is the sum and $\dfrac{1}{a+b}$ is the reciprocal of the sum.

129. Let a and b represent the numbers. Then $a + b$ is the sum and $-(a + b)$ is the opposite of the sum.

131. Let x represent a real number. $x = -x$

133. For 2 and 3, the reciprocal of the sum is $\dfrac{1}{(2+3)}$ or $\dfrac{1}{5}$. But $\dfrac{1}{5} \neq \dfrac{1}{2} + \dfrac{1}{3}$.

135. The starting temperature is $-3°\text{F}$.

Rewording:	starting temp.	rise	$2°$	for	3hr	rise	$3°$	for	6hrs
	↓	↓	↓	↓	↓	↓	↓	↓	↓
Translating:	-3	$+$	2	$\cdot$	3	$+$	3	$\cdot$	6

fall	$2°$	for	3hr	fall	$5°$	for	2hr
↓	↓	↓	↓	↓	↓	↓	
$-$	2	$\cdot$	3	$-$	5	$\cdot$	2

Since $-3 + 2 \cdot 3 + 3 \cdot 6 - 2 \cdot 3 - 5 \cdot 2$

$\qquad = -3 + 6 + 18 - 6 - 10$

$\qquad = 5.$

The temperature forecast at 8PM is $5°\text{F}$

137. $-n$ and $-m$ are both positive, so $\dfrac{-n}{-m}$ is positive.

139. $-m$ is positive, so $\dfrac{n}{-m}$ is negative and $-\left(\dfrac{n}{-m}\right)$ is positive.

141. $-n$ and $-m$ are both positive, so $-n - m$, or $-n + (-m)$ is positive; $\dfrac{n}{m}$ is also positive, so $(-n - m)\dfrac{n}{m}$ is positive.

143. $a(-b) + ab = a[-b + b]$ Distributive law

$\qquad\qquad\qquad = a(0)$ Law of opposites

$\qquad\qquad\qquad = 0$ Multiplicative property of 0

Therefore, $a(-b)$ is the opposite of ab by the law of opposites.

Chapter 1 Connecting the Concepts

1. $-8 + (-2) = -10$

3. $-8 \div (-2) = \frac{-8}{-2} = 4$

5. $12 \cdot (-10) = -120$

7. $-5 - 18 = -23$

9. $\frac{3}{5} - \frac{8}{5} = -\frac{5}{5} = -1$

11. $-5.6 + 4.8 = -0.8$

13. $-44.1 \div 6.3 = -7$

15. $\frac{9}{5} \cdot \left(-\frac{20}{3}\right) = -\frac{3 \cdot 3 \cdot 4 \cdot 5}{5 \cdot 3} = -12$

17. $38 - (-62) = 38 + 62 = 100$

19. $(-15) \cdot (-12) = 180$

Exercise Set 1.8

1. a) $4 + 8 \div 2 \cdot 2$

There are no grouping symbols or exponential expressions, so we multiply and divide from left to right. This means that we divide first.

b) $7 - 9 + 15$

There are no grouping symbols, exponential expressions, multiplications, or divisions, so we add and subtract from left to right. This means that we subtract first.

c) $5 - 2(3 + 4)$

We perform the operation in the parentheses first. This means that we add first.

d) $6 + 7 \cdot 3$

There are no grouping symbols or exponential expressions, so we multiply and divide from left to right. This means that we multiply first.

e) $18 - 2[4 + (3 - 2)]$

We perform the operation in the innermost grouping symbols first. This means that we perform the subtraction in the parentheses first.

f) $\dfrac{5 - 6 \cdot 7}{2}$

Since the denominator does not need to be simplified, we consider the numerator. There are no grouping symbols or exponential expressions, so we multiply and divide from left to right. This means that we multiply first.

3. $\underbrace{x \cdot x \cdot x \cdot x \cdot x \cdot x}_{6 \text{ factors}} = x^6$

5. $\underbrace{(-5)(-5)(-5)}_{3 \text{ factors}} = (-5)^3$

7. $3t \cdot 3t \cdot 3t \cdot 3t \cdot 3t = (3t)^5$

9. $2 \cdot \underbrace{n \cdot n \cdot n \cdot n}_{4 \text{ factors}} = 2n^4$

11. $4^2 = 4 \cdot 4 = 16$

13. $(-3)^2 = (-3)(-3) = 9$

15. $-3^2 = -(3 \cdot 3) = -9$

17. $4^3 = 4 \cdot 4 \cdot 4 = 16 \cdot 4 = 64$

19. $(-5)^4 = (-5)(-5)(-5)(-5) = 25 \cdot 25 = 625$

21. $7^1 = 7 \qquad$ (1 factor)

23. $(-2)^5 = (-2)(-2)(-2)(-2)(-2) = -32$

25. $(3t)^4 = (3t)(3t)(3t)(3t)$
$= 3 \cdot 3 \cdot 3 \cdot 3 \cdot t \cdot t \cdot t \cdot t = 81t^4$

27. $(-7x)^3 = (-7x)(-7x)(-7x)$
$= (-7)(-7)(-7)(x)(x)(x) = -343x^3$

29. $\begin{aligned}5 + 3 \cdot 7 &= 5 + 21 &&\text{Multiplying} \\ &= 26 &&\text{Adding}\end{aligned}$

31. $\begin{aligned}10 \cdot 5 + 1 \cdot 1 &= 50 + 1 &&\text{Multiplying} \\ &= 51 &&\text{Adding}\end{aligned}$

33. $\begin{aligned}6 - 70 \div 7 - 2 &= 6 - 10 - 2 &&\text{Multiplying} \\ &= -4 - 2 &&\text{Adding} \\ &= -6\end{aligned}$

35. $14 \cdot 19 \div (19 \cdot 14)$

Since $14 \cdot 19$ and $19 \cdot 14$ are equivalent, we are dividing the product $14 \cdot 19$ by itself. Thus the result is 1.

37. $\begin{aligned}&3(-10)^2 - 8 \div 2^2 \\ &= 3(100) - 8 \div 4 &&\text{Simplifying the} \\ &&&\text{exponential expressions} \\ &= 300 - 8 \div 4 &&\text{Multiplying and} \\ &= 300 - 2 &&\text{dividing from left to right} \\ &= 298 &&\text{Subtracting}\end{aligned}$

39. $\begin{aligned}&8 - (2 \cdot 3 - 9) \\ &= 8 - (6 - 9) &&\text{Multiplying inside the} \\ &&&\text{parentheses} \\ &= 8 - (-3) &&\text{Subtracting inside the} \\ &&&\text{parentheses} \\ &= 8 + 3 &&\text{Removing parentheses} \\ &= 11 &&\text{Adding}\end{aligned}$

41. $\begin{aligned}&(8 - 2)(3 - 9) \\ &= 6(-6) &&\text{Subtracting inside the} \\ &&&\text{parentheses} \\ &= -36 &&\text{Multiplying}\end{aligned}$

43. $13(-10)^2 + 45 \div (-5)$

$\quad = 13(100) + 45 \div (-5)$ Simplifying the exponential expression

$\quad = 1300 + 45 \div (-5)$ Multiplying and

$\quad = 1300 - 9$ dividing from left to right

$\quad = 1291$ Subtracting

45. $5 + 3(2-9)^2 = 5 + 3(-7)^2 = 5 + 3 \cdot 49 = 5 + 147 = 152$

47. $[2 \cdot (5-8)]^2 = [2 \cdot (-3)]^2 = (-6)^2 = 36$

49. $\dfrac{7+2}{5^2 - 4^2} = \dfrac{9}{25-16} = \dfrac{9}{9} = 1$

51. $8(-7) + |3(-4)| = -56 + |-12| = -56 + 12 = -44$

53. $36 \div (-2)^2 + 4[5 - 3(8-9)^5]$

$\quad = 36 \div 4 + 4[5 - 3(-1)^5]$

$\quad = 9 + 4[5 + 3]$

$\quad = 9 + 32$

$\quad = 41$

55. $\dfrac{(7)^2 - (-1)^7}{5 \cdot 7 - 4 \cdot 3^2 - 2^2} = \dfrac{49 - (-1)}{35 - 4 \cdot 9 - 4} = \dfrac{49 + 1}{35 - 36 - 4} =$

$\dfrac{50}{-5} = -10$

57. $\dfrac{-3^3 - 2 \cdot 3^2}{8 \div 2^2 - (6 - |2 - 15|)} = \dfrac{-27 - 2 \cdot 9}{8 \div 4 - (6 - |-13|)}$

$\quad = \dfrac{-27 - 18}{2 - (6 - 13)} = \dfrac{-45}{2 - (-7)} = \dfrac{-45}{9} = -5$

59. $9 - 4x$

$\quad = 9 - 4 \cdot 7$ Substituting 7 for x

$\quad = 9 - 28$ Multiplying

$\quad = -19$

61. $24 \div t^3$

$\quad = 24 \div (-2)^3$ Substituting -2 for t

$\quad = 24 \div (-8)$ Simplifying the exponential expression

$\quad = -3$ Dividing

63. $45 \div a \cdot 5 = 45 \div (-3) \cdot 5$ Substituting -3 for a

$\quad = -15 \cdot 5$ Multiplying and dividing in order from left to right

$\quad = -75$

65. $5x \div 15x^2$

$\quad = 5 \cdot 3 \div 15(3)^2$ Substituting 3 for x

$\quad = 5 \cdot 3 \div 15 \cdot 9$ Simplifying the exponential expression

$\quad = 15 \div 15 \cdot 9$ Multiplying and dividing

$\quad = 1 \cdot 9$ in order from

$\quad = 9$ left to right

67. $45 \div 3^2 x(x-1)$

$\quad = 45 \div 3^2 \cdot 3(3-1)$ Substituting 3 for x

$\quad = 45 \div 3^2 \cdot 3(2)$ Subtracting inside the parentheses

$\quad = 45 \div 9 \cdot 3(2)$ Evaluating the exponential expression

$\quad = 5 \cdot 3(2)$ Dividing and

$\quad = 15(2)$ multiplying

$\quad = 30$ from left to right

69. $-x^2 - 5x = -(-3)^2 - 5(-3) = -9 - 5(-3)$

$\quad = -9 + 15 = 6$

71. $\dfrac{3a - 4a^2}{a^2 - 20} = \dfrac{3 \cdot 5 - 4(5)^2}{(5)^2 - 20} = \dfrac{3 \cdot 5 - 4 \cdot 25}{25 - 20}$

$\quad = \dfrac{15 - 100}{5} = \dfrac{-85}{5} = -17$

73. $-(9x + 1) = -9x - 1$ Removing parentheses and changing the sign of each term

75. $-[7n + 8] = 7n - 8$ Removing grouping symbols and changing the sign of each term

77. $-(4a - 3b + 7c) = -4a + 3b - 7c$

79. $-(3x^2 + 5x - 1) = -3x^2 - 5x + 1$

81. $8x - (6x + 7)$

$\quad = 8x - 6x - 7$ Removing parentheses and changing the sign of each term

$\quad = 2x - 7$ Collecting like terms

83. $2x - 7x - (4x - 6) = 2x - 7x - 4x + 6 = -9x + 6$

85. $9t - 7r + 2(3r + 6t) = 9t - 7r + 6r + 12t = 21t - r$

87. $15x - y - 5(3x - 2y + 5z)$

$\quad = 15x - y - 15x + 10y - 25z$ Multiplying each term in parentheses by -5

$\quad = 9y - 25z$

89. $3x^2 + 11 - (2x^2 + 5) = 3x^2 + 11 - 2x^2 - 5$

$\quad = x^2 + 6$

91. $5t^3 + t + 3(t - 2t^3) = 5t^3 + t + 3t - 6t^3$

$\quad = -t^3 + 4t$

93. $12a^2 - 3ab + 5b^2 - 5(-5a^2 + 4ab - 6b^2)$

$\quad = 12a^2 - 3ab + 5b^2 + 25a^2 - 20ab + 30b^2$

$\quad = 37a^2 - 23ab + 35b^2$

95. $-7t^3 - t^2 - 3(5t^3 - 3t)$

$\quad = -7t^3 - t^2 - 15t^3 + 9t$

$\quad = -22t^3 - t^2 + 9t$

97. $5(2x - 7) - [4(2x - 3) + 2]$

$\quad = 5(2x - 7) - [8x - 12 + 2]$

$\quad = 5(2x - 7) - [8x - 10]$

$\quad = 10x - 35 - 8x + 10$

$\quad = 2x - 25$

99. *Writing Exercise.* Operations should be performed in the following order.

1. **P**arentheses: Perform all calculations within parentheses (and other grouping symbols) first.

2. **E**xponents: Evaluate all exponential expressions.

3. **M**ultiply and **D**ivide in order from left to right.

4. **A**dd and **S**ubtract in order from left to right.

This is a valid approach because it describes the rules for the order of operations.

101. Let n represent "a number." Then we have $2n - 9$.

103. *Writing Exercise.* Finding the opposite of a number and then squaring it is not equivalent to squaring the number and then finding the opposite of the result.
$$(-x)^2 = (-1 \cdot x)^2 = (-1 \cdot x)(-1 \cdot x)$$
$$= (-1)(-1)(x)(x) = x^2 \neq -x^2 \text{ for } x \neq 0.$$

105. $5t - \{7t - [4r - 3(t - 7)] + 6r\} - 4r$
$$= 5t - \{7t - [4r - 3t + 21] + 6r\} - 4r$$
$$= 5t - \{7t - 4r + 3t - 21 + 6r\} - 4r$$
$$= 5t - \{10t + 2r - 21\} - 4r$$
$$= 5t - 10t - 2r + 21 - 4r$$
$$= -5t - 6r + 21$$

107. $\{x - [f - (f - x)] + [x - f]\} - 3x$
$$= \{x - [f - f + x] + [x - f]\} - 3x$$
$$= \{x - [x] + [x - f]\} - 3x$$
$$= \{x - x + x - f\} - 3x$$
$$= x - f - 3x$$
$$= -2x - f$$

109. *Writing Exercise.* No; let $a = 2$, $b = -1$, and $c = 3$. Then $a|b - c| = 2|-1 - 3| = 2|-4| = 2 \cdot 4 = 8$, but $ab - ac = 2(-1) - 2 \cdot 3 = -2 - 6 = -8$.

111. True; $m - n = -n + m = -(n - m)$

113. False; let $m = 2$ and $n = 1$. Then $-2(1 - 2) = -2(-1) = 2$, but $-(2 \cdot 1 + 2^2) = -(2 + 4) = -6$.

115. $[x + 3(2 - 5x) \div 7 + x](x - 3)$

When $x = 3$, the factor $x - 3$ is 0, so the product is 0.

117. $\dfrac{x^2 + 2^x}{x^2 - 2^x} = \dfrac{3^2 + 2^3}{3^2 - 2^3} = \dfrac{9 + 8}{9 - 8} = \dfrac{17}{1} = 17$

119. $4 \cdot 20^3 + 17 \cdot 20^2 + 10 \cdot 20 + 0 \cdot 1$
$$= 4 \cdot 8000 + 17 \cdot 400 + 10 \cdot 20 + 0 \cdot 1$$
$$= 32,000 + 6800 + 200 + 0$$
$$= 39,000$$

121. The tower is composed of cubes with sides of length x. The volume of each cube is $x \cdot x \cdot x$, or x^3. Now we count the number of cubes in the tower. The two lowest levels each contain 3×3, or 9 cubes. The next level contains one cube less than the two lowest levels, so it has $9 - 1$, or 8 cubes. The fourth level from the bottom contains one cube less than the level below it, so it has $8 - 1$, or 7 cubes. The fifth level from the bottom contains one cube less than the level below it, so it has $7 - 1$, or 6 cubes. Finally, the top level contains one cube less than the level below it, so it has $6 - 1$, or 5 cubes. All together there are $9 + 9 + 8 + 7 + 6 + 5$, or 44 cubes, each with volume x^3, so the volume of the tower is $44x^3$.

Chapter Review Exercises

1. True

3. False

5. False

7. True

9. False

11. $8t = 8 \cdot 3$ substitute 3 for t
$$= 24$$

13. $9 - y^2 = 9 - (-5)^2$ Substitute -5 for y
$$= 9 - 25$$
$$= -16$$

15. $y - 7$

17. Let b represent Brandt's speed and w represent the wind speed; $15(b - w)$

19. Let d represent the number of digital prints made in 2006, in billions.

Translating: 14.1   is   3.2   more than d.
$$14.1 = 3.2 + d$$
or $\qquad\qquad 14.1 = d + 3.2$

21. $(2x + y) + z = 2x + (y + z)$

23. $6(3x + 5y) = 6 \cdot 3x + 6 \cdot 5y$
$$= 18x + 30y$$

25. $21x + 15y = 3 \cdot 7x + 3 \cdot 5y$
$$= 3(7x + 5y)$$

27. $56 = 7 \cdot 8 = 7 \cdot 2 \cdot 2 \cdot 2$ or $2 \cdot 2 \cdot 2 \cdot 7$

29. $\dfrac{18}{8} = \dfrac{2 \cdot 3 \cdot 3}{2 \cdot 2 \cdot 2} = \dfrac{9}{4}$

31. $\dfrac{9}{16} \div 3 = \dfrac{9}{16} \cdot \dfrac{1}{3}$ Multiply by the reciprocal of the divisor
$$= \dfrac{3 \cdot 3 \cdot 1}{16 \cdot 3}$$
$$= \dfrac{3}{16}$$

33. $\dfrac{9}{10} \cdot \dfrac{6}{5} = \dfrac{9 \cdot 6}{10 \cdot 5} = \dfrac{9 \cdot \cancel{2} \cdot 3}{\cancel{2} \cdot 5 \cdot 5} = \dfrac{27}{25}$

35.

37. $2 \geq -8$ True, since 2 is to the right of -8.

39. $-\dfrac{4}{9} = -\left(\dfrac{4}{9}\right) = -\left(4 \div 9\right)$, so we divide.

$$
\begin{array}{r}
0.4\ 4\ 4 \\
9\,)\,\overline{4.0\ 0\ 0} \\
\underline{3\ 6} \\
4\ 0 \\
\underline{3\ 6} \\
4\ 0 \\
\underline{3\ 6} \\
4
\end{array}
$$

$\dfrac{4}{9} = 0.\overline{4}$, so $-\dfrac{4}{9} = -0.\overline{4}$

41. $-(-x) = -(-(-12))$ Subsitute -12 for x

$= -(12) = -12$

43. $-\dfrac{2}{3} + \dfrac{1}{12} = \dfrac{-8}{12} + \dfrac{1}{12}$ The absolute values are $\dfrac{8}{12}$ and $\dfrac{1}{12}$. The difference is $\dfrac{8}{12} - \dfrac{1}{12}$, or $\dfrac{8-1}{12} = \dfrac{7}{12}$. The negative number has the greater absolute value, so the answer is negative. $-\dfrac{2}{3} + \dfrac{1}{12} = -\dfrac{7}{12}$

45. $-3.8 + 5.1 + (-12) + (-4.3) + 10$

$= (5.1 + 10) + [(-3.8) + (-12) + (-4.3)]$

$= 15.1 + [-20.1]$ Adding positive nos. and adding negative nos.

$= -5$ Adding a positive and a negative no.

47. $\dfrac{-9}{10} - \dfrac{1}{2} = \dfrac{-9}{10} + \left(\dfrac{-1}{2}\right)$

$= \dfrac{-9}{10} + \left(\dfrac{-5}{10}\right) = \dfrac{-14}{10} = \dfrac{-7}{5}$

49. $-9 \cdot (-7) = 63$

51. $\dfrac{2}{3} \cdot \left(-\dfrac{3}{7}\right) = -\left(\dfrac{2}{3} \cdot \dfrac{3}{7}\right) = -\left(\dfrac{2 \cdot 3}{3 \cdot 7}\right) = -\dfrac{2}{7}$

53. $35 \div (-5) = -7$

55. $-\dfrac{3}{5} \div \left(-\dfrac{4}{15}\right) = \dfrac{3}{5} \cdot \dfrac{15}{4} = \dfrac{3 \cdot 3 \cdot 5}{5 \cdot 4} = \dfrac{9}{4}$

57. $\left(120 - 6^2\right) \div 4 \cdot 8 = (120 - 36) \div 4 \cdot 8$

$= 84 \div 4 \cdot 8$

$= 21 \cdot 8$

$= 168$

59. $16 \div (-2)^3 - 5[3 - 1 + 2(4 - 7)]$

$= 16 \div (-8)^3 - 5[3 - 1 + 2(4 - 7)]$

$= 16 \div (-2)^3 - 5[3 - 1 + 2(4 - 7)]$

$= -2^3 - 5[3 - 1 + 2(-3)]$

$= -2 - 5[-4]$

$= -2 + 20$

$= 18$

61. $\dfrac{4\,(18 - 8) + 7 \cdot 9}{9^2 - 8^2}$

$= \dfrac{4\,(18 - 8) + 7 \cdot 9}{81 - 64}$

$= \dfrac{4\,(10) + 7 \cdot 9}{81 - 64}$

$= \dfrac{40 + 63}{81 - 64}$

$= \dfrac{103}{17}$

63. $7x - 3y - 11x + 8y$

$= 7x + (-11x) + (-3y) + 8y$

$= [7 + (-11)]x + (-3 + 8)y$

$= -4x + 5y$

65. The reciprocal of -7 is $-\dfrac{1}{7}$, since $-\dfrac{1}{7} \cdot -7 = 1$

67. $(-5x)^3$

$= -5x \cdot (-5x) \cdot (-5x)$

$= -5 \cdot (-5) \cdot (-5) \cdot x \cdot x \cdot x$

$= -125x^3$

69. $5b + 3(2b - 9)$

$= 5b + 3 \cdot 2b - 3 \cdot 9$

$= 5b + 6b - 27$

$= 11b - 27$

71. $2n^2 - 5(-3n^2 + m^2 - 4mn) + 6m^2$

$= 2n^2 + 15n^2 - 5m^2 + 20mn + 6m^2$

$= (2 + 15)n^2 + (-5 + 6)m^2 + 20mn$

$= 17n^2 + m^2 + 20mn$

73. *Writing Exercise.* The value of a constant never varies. A variable can represent a variety of numbers.

75. *Writing Exercise.* The distributive law is used in factoring algebraic expressions, multiplying algebraic expressions, combining like terms, finding the opposite of a sum, and subtracting algebraic expressions.

77. Substitute 1 for a, 2 for b, and evaluate: $a^{50} - 20a^{25}b^4 + 100b^8$

$= 1^{50} - 20 \cdot 1^{25} 2^4 + 100 \cdot 2^8$

$= 1 - 20 \cdot 1 \cdot 16 + 100 \cdot 256$

$= 1 - 320 + 25,600$

$= 25,281$

79. $-\left|\dfrac{7}{8} - \left(-\dfrac{1}{2}\right) - \dfrac{3}{4}\right|$ Use 8: the common denominator

$-\left|\dfrac{7}{8} - \left(-\dfrac{4}{8}\right) - \dfrac{6}{8}\right| = -\left|\dfrac{7}{8} + \dfrac{4}{8} + \dfrac{-6}{8}\right|$

$= -\left|\dfrac{5}{8}\right| = -\dfrac{5}{8}$

81. i

83. a

85. k

87. c

89. d

91. g

Chapter 1 Test

1. Substitute 10 for x and 5 for y.

$$\frac{2x}{y} = \frac{2 \cdot 10}{5} = \frac{2 \cdot 2 \cdot \cancel{5}}{\cancel{5}} = 4$$

3. $A = \frac{1}{2} \cdot b \cdot h$

$$= \frac{1}{2} \cdot (16 \text{ ft}) \cdot (30 \text{ ft})$$

$$= \frac{1}{2} \cdot 16 \cdot 30 \cdot \text{ft} \cdot \text{ft}$$

$$= 8 \cdot 30 \text{ ft} \cdot \text{ft}$$

$$= 240 \text{ ft}^2, \text{ or } 240 \text{ square feet}$$

5. $x \cdot (4 \cdot y) = (x \cdot 4) \cdot y$

7. 45,950 is 4250 less than the maximum

Rewording: $\underbrace{45,950}_{\downarrow}$ is $\underbrace{\text{maximum}}_{\downarrow}$ less $\underbrace{4250}_{\downarrow}$

Translating: 45,950   = p −   4250

where p represents the maximum production capability.

9. $-5(y-2) = -5 \cdot y - 5(-2) = -5y + 10$

11. $7x + 7 + 49y$
$$= 7 \cdot x + 7 \cdot 1 + 7 \cdot 2y$$
$$= 7(x + 1 + 7y)$$

13. $\dfrac{24}{56} = \dfrac{3 \cdot \cancel{8}}{7 \cdot \cancel{8}} = \dfrac{3}{7}$

15. $-3 > -8$, since -3 is right of -8.

17. $|-3.8| = 3.8$, since -3.8 is 3.8 units from 0.

19. $\dfrac{-7}{4}$, since $\dfrac{-7}{4} \cdot \dfrac{-4}{7} = 1$

21. $-5 \geq x$ has the same meaning as $x \leq -5$.

23. $-8 + 4 + (-7) + 3$
$$= [-8 + (-7)] + [4 + 3]$$
$$= -15 + 7 = -8$$

25. $-\dfrac{1}{8} - \dfrac{3}{4}$

$$= -\dfrac{1}{8} - \dfrac{2}{2} \cdot \dfrac{3}{4}$$

$$= -\dfrac{1}{8} - \dfrac{6}{8} = \dfrac{-1-6}{8} = -\dfrac{7}{8}$$

27. $-\dfrac{1}{2} \cdot \left(-\dfrac{4}{9}\right) = \dfrac{1}{2} \cdot \dfrac{4}{9} = \dfrac{1 \cdot \cancel{2} \cdot 2}{\cancel{2} \cdot 9} = \dfrac{2}{9}$

29. $-\dfrac{3}{5} \div \left(-\dfrac{4}{5}\right) = \dfrac{3}{5} \cdot \dfrac{5}{4} = \dfrac{3 \cdot \cancel{5}}{\cancel{5} \cdot 4} = \dfrac{3}{4}$

31. $-2(16) - |2(-8) - 5^3|$
$$= -2(16) - |2(-8) - 125|$$
$$= -2(16) - |-16 - 125|$$
$$= -2(16) - |-141|$$
$$= -2(16) - 141$$
$$= -32 - 141$$
$$= -32 + (-141) = -173$$

33. $256 \cdot (-16) \cdot 4 = -16 \cdot 4 = -64$

35. $18y + 30a - 9a + 4y$
$$= 30a + (-9a) + 18y + 4y$$
$$= [30 + (-9)]a + (18 + 4)y$$
$$= 21a + 22y$$

37. $4x - (3x - 7)$
$$= 4x - 3x + 7$$
$$= (4 - 3)x + 7 = x + 7$$

39. $3[5(y - 3) + 9] - 2(8y - 1)$
$$= 3[5y - 15 + 9] - 16y + 2$$
$$= 3[5y - 6] - 16y + 2$$
$$= 15y - 18 - 16y + 2$$
$$= -y - 16$$

41. $9 - (3 - 4) + 5 = 15$

43. $a - \{3a - [4a - (2a - 4a)]\}$
$$= a - \{3a - [4a - (-2a)]\}$$
$$= a - \{3a - [4a + 2a]\}$$
$$= a - \{3a - 6a\}$$
$$= a - \{-3a\}$$
$$= a + 3a = 4a$$

Chapter 2

Equations, Inequalities, and Problem Solving

Exercise Set 2.1

1. The equations $x + 3 = 7$ and $6x = 24$ are <u>equivalent</u> <u>equations</u>.

3. A <u>solution</u> is a replacement that makes an equation true.

5. The <u>multiplication principle</u> is used to solve $\frac{2}{3} \cdot x = -4$.

7. For $6x = 30$, the next step is (d) divide both sides by 6.

9. For $\frac{1}{6}x = 30$, the next step is (c) multiply both sides by 6.

11.
$$x + 10 = 21$$
$$x + 10 - 10 = 21 - 10$$
$$x = 11$$

Check:
$$\frac{x + 10 = 21}{11 + 10 \mid 21}$$
$$21 \overset{?}{=} 21 \quad \text{TRUE}$$

The solution is 11.

13.
$$y + 7 = -18$$
$$y + 7 - 7 = -18 - 7$$
$$y = -25$$

Check:
$$\frac{y + 7 = -18}{-25 + 7 \mid 18}$$
$$-18 \overset{?}{=} -18 \quad \text{TRUE}$$

The solution is -25.

15.
$$-6 = y + 25$$
$$-6 - 25 = y + 25 - 25$$
$$-31 = y$$

Check:
$$\frac{-6 = y + 25}{-6 \mid -31 + 25}$$
$$-6 \overset{?}{=} -6 \quad \text{TRUE}$$

The solution is -31.

17.
$$x - 18 = 23$$
$$x - 18 + 18 = 23 + 18$$
$$x = 41$$

Check:
$$\frac{x - 18 = 23}{41 - 18 \mid 23}$$
$$23 \overset{?}{=} 23 \quad \text{TRUE}$$

The solution is 41.

19.
$$12 = -7 + y$$
$$7 + 12 = 7 + (-7) + y$$
$$19 = y$$

Check:
$$\frac{12 = -7 + y}{12 \mid -7 + 19}$$
$$12 \overset{?}{=} 12 \quad \text{TRUE}$$

The solution is 19.

21.
$$-5 + t = -11$$
$$5 + (-5) + t = 5 + (-11)$$
$$t = -6$$

Check:
$$\frac{-5 + t = -11}{-5 + (-6) \mid -11}$$
$$-11 \overset{?}{=} -11 \quad \text{TRUE}$$

The solution is -6.

23.
$$r + \frac{1}{3} = \frac{8}{3}$$
$$r + \frac{1}{3} - \frac{1}{3} = \frac{8}{3} - \frac{1}{3}$$
$$r = \frac{7}{3}$$

Check:
$$\frac{r + \frac{1}{3} = \frac{8}{3}}{\frac{7}{3} + \frac{1}{3} \mid \frac{8}{3}}$$
$$\frac{8}{3} \overset{?}{=} \frac{8}{3} \quad \text{TRUE}$$

The solution is $\frac{7}{3}$.

25.
$$x - \frac{3}{5} = -\frac{7}{10}$$
$$x - \frac{3}{5} + \frac{3}{5} = -\frac{7}{10} + \frac{3}{5}$$
$$x = -\frac{7}{10} + \frac{3}{5} \cdot \frac{2}{2}$$
$$x = -\frac{7}{10} + \frac{6}{10}$$
$$x = -\frac{1}{10}$$

Check:
$$\frac{x - \frac{3}{5} = -\frac{7}{10}}{-\frac{1}{10} - \frac{3}{5} \mid -\frac{7}{10}}$$
$$-\frac{1}{10} - \frac{6}{10} \mid$$
$$-\frac{7}{10} \overset{?}{=} -\frac{7}{10} \quad \text{TRUE}$$

The solution is $-\frac{1}{10}$.

27.
$$x - \frac{5}{6} = \frac{7}{8}$$
$$x - \frac{5}{6} + \frac{5}{6} = \frac{7}{8} + \frac{5}{6}$$
$$x = \frac{7}{8} \cdot \frac{3}{3} + \frac{5}{6} \cdot \frac{4}{4}$$
$$x = \frac{21}{24} + \frac{20}{24}$$
$$x = \frac{41}{24}$$

Check:

$$\begin{array}{c|c} x - \frac{5}{6} = \frac{7}{8} \\ \hline \frac{41}{24} - \frac{5}{6} & \frac{7}{8} \\ \frac{41}{24} - \frac{20}{24} & \frac{21}{24} \\ \frac{21}{24} \stackrel{?}{=} \frac{21}{24} & \text{TRUE} \end{array}$$

The solution is $\frac{41}{24}$.

29.
$$-\frac{1}{5} + z = -\frac{1}{4}$$
$$\frac{1}{5} - \frac{1}{5} + z = \frac{1}{5} - \frac{1}{4}$$
$$z = \frac{1}{5} \cdot \frac{4}{4} - \frac{1}{4} \cdot \frac{5}{5}$$
$$z = \frac{4}{20} - \frac{5}{20}$$
$$z = -\frac{1}{20}$$

Check:

$$\begin{array}{c|c} -\frac{1}{5} + z = -\frac{1}{4} \\ \hline -\frac{1}{5} + \left(-\frac{1}{20}\right) & -\frac{1}{4} \\ -\frac{4}{20} + \left(-\frac{1}{20}\right) & -\frac{5}{20} \\ -\frac{5}{20} \stackrel{?}{=} -\frac{5}{20} & \text{TRUE} \end{array}$$

The solution is $-\frac{1}{20}$.

31.
$$m - 2.8 = 6.3$$
$$m - 2.8 + 2.8 = 6.3 + 2.8$$
$$m = 9.1$$

Check:

$$\begin{array}{c|c} m - 2.8 = 6.3 \\ \hline 9.1 - 2.8 & 6.3 \\ 6.3 \stackrel{?}{=} 6.3 & \text{TRUE} \end{array}$$

The solution is 9.1.

33.
$$-9.7 = -4.7 + y$$
$$4.7 + (-9.7) = 4.7 + (-4.7) + y$$
$$-5 = y$$

Check:

$$\begin{array}{c|c} -9.7 = -4.7 + y \\ \hline -9.7 & -4.7 + (-5) \\ -9.7 \stackrel{?}{=} -9.7 & \text{TRUE} \end{array}$$

The solution is -5.

35.
$$8a = 56$$
$$\frac{8a}{8} = \frac{56}{8} \qquad \text{Dividing both sides by 8}$$
$$1 \cdot a = 7 \qquad \text{Simplifying}$$
$$a = 7 \qquad \text{Identity property of 1}$$

Check:

$$\begin{array}{c|c} 8a = 56 \\ \hline 8 \cdot 7 & 56 \\ 56 \stackrel{?}{=} 56 & \text{TRUE} \end{array}$$

The solution is 7.

37.
$$84 = 7x$$
$$\frac{84}{7} = \frac{7x}{7} \qquad \text{Dividing both sides by 7}$$
$$12 = 1 \cdot x$$
$$12 = x$$

Check:

$$\begin{array}{c|c} 84 = 7x \\ \hline 84 & 7 \cdot 12 \\ 84 \stackrel{?}{=} 84 & \text{TRUE} \end{array}$$

The solution is 12.

39.
$$-x = 38$$
$$-1 \cdot x = 38$$
$$-1 \cdot (-1 \cdot x) = -1 \cdot 38$$
$$1 \cdot x = -38$$
$$x = -38$$

Check:

$$\begin{array}{c|c} -x = 38 \\ \hline -(-38) & 38 \\ 38 \stackrel{?}{=} 38 & \text{TRUE} \end{array}$$

The solution is -38.

41. $-t = -8$

The equation states that the opposite of t is the opposite of 8. Thus, $t = 8$. We could also do this exercise as follows.

$$-t = -8$$
$$-1(-t) = -1(-8) \qquad \text{Multiplying both sides by } -1$$
$$t = 8$$

Check:

$$\begin{array}{c|c} -t = -8 \\ \hline -(8) & -8 \\ -8 \stackrel{?}{=} -8 & \text{TRUE} \end{array}$$

The solution is 8.

43.
$$-7x = 49$$
$$\frac{-7x}{-7} = \frac{49}{-7}$$
$$1 \cdot x = -7$$
$$x = -7$$

Check:

$$\begin{array}{c|c} -7x = 49 \\ \hline -7(-7) & 49 \\ 49 \stackrel{?}{=} 49 & \text{TRUE} \end{array}$$

The solution is -7.

45. $-1.3a = -10.4$

$$\frac{-1.3a}{-1.3} = \frac{-10.4}{-1.3}$$

$$a = 8$$

Check: $\dfrac{-1.3a = -10.4}{-1.3(8) \;\big|\; -10.4}$

$$-10.4 \overset{?}{=} -10.4 \qquad \text{TRUE}$$

The solution is 8.

47. $\dfrac{y}{8} = 11$

$$\frac{1}{8} \cdot y = 11$$

$$8\left(\frac{1}{8}\right) \cdot y = 8 \cdot 11$$

$$y = 88$$

Check: $\dfrac{y}{8} = 11$

$$\dfrac{88}{8} \;\bigg|\; 11$$

$$11 \overset{?}{=} 11 \qquad \text{TRUE}$$

The solution is 88.

49. $\dfrac{4}{5}x = 16$

$$\frac{5}{4} \cdot \frac{4}{5}x = \frac{5}{4} \cdot 16$$

$$x = \frac{5 \cdot \cancel{4} \cdot 4}{\cancel{4} \cdot 1}$$

$$x = 20$$

Check: $\dfrac{4}{5}x = 16$

$$\dfrac{4}{5} \cdot 20 \;\bigg|\; 16$$

$$16 \overset{?}{=} 16 \qquad \text{TRUE}$$

The solution is 20.

51. $\dfrac{-x}{6} = 9$

$$-\frac{1}{6} \cdot x = 9$$

$$-6\left(-\frac{1}{6}\right) \cdot x = -6 \cdot 9$$

$$x = -54$$

Check: $\dfrac{-x}{6} = 9$

$$\dfrac{-(-54)}{6} \;\bigg|\; 9$$

$$\dfrac{54}{6} \;\bigg|\;$$

$$9 \overset{?}{=} 9 \qquad \text{TRUE}$$

The solution is -54.

53. $\dfrac{1}{9} = \dfrac{z}{-5}$

$$\frac{1}{9} = -\frac{1}{5} \cdot z$$

$$-5 \cdot \frac{1}{9} = -5 \cdot \left(-\frac{1}{5} \cdot z\right)$$

$$-\frac{5}{9} = z$$

Check: $\dfrac{1}{9} = \dfrac{z}{-5}$

$$\dfrac{1}{9} \;\bigg|\; \dfrac{-5/9}{-5}$$

$$\;\bigg|\; -\frac{5}{9} \cdot \frac{1}{-5}$$

$$\frac{1}{9} \overset{?}{=} \frac{1}{9} \qquad \text{TRUE}$$

The solution is $\dfrac{-5}{9}$.

55. $-\dfrac{3}{5}r = -\dfrac{3}{5}$

The solution of the equation is the number that is multiplied by $-\dfrac{3}{5}$ to get $-\dfrac{3}{5}$. That number is 1. We could also do this exercise as follows:

$$-\frac{3}{5}r = -\frac{3}{5}$$

$$-\frac{5}{3} \cdot \left(-\frac{3}{5}r\right) = -\frac{5}{3}\left(-\frac{3}{5}\right)$$

$$r = 1$$

Check: $-\dfrac{3}{5}r = -\dfrac{3}{5}$

$$-\dfrac{3}{5} \cdot 1 \;\bigg|\; -\dfrac{3}{5}$$

$$-\frac{3}{5} \overset{?}{=} -\frac{3}{5} \qquad \text{TRUE}$$

The solution is 1.

57. $\dfrac{-3r}{2} = -\dfrac{27}{4}$

$$-\frac{3}{2}r = -\frac{27}{4}$$

$$-\frac{2}{3} \cdot \left(-\frac{3}{2}r\right) = -\frac{2}{3} \cdot \left(-\frac{27}{4}\right)$$

$$r = \frac{\cancel{2} \cdot \cancel{3} \cdot 3 \cdot 3}{\cancel{3} \cdot \cancel{2} \cdot 2}$$

$$r = \frac{9}{2}$$

Check: $\dfrac{-3r}{2} = -\dfrac{27}{4}$

$$-\dfrac{3}{2} \cdot \dfrac{9}{2} \;\bigg|\; -\dfrac{27}{4}$$

$$-\frac{27}{4} \overset{?}{=} -\frac{27}{4} \qquad \text{TRUE}$$

The solution is $\dfrac{9}{2}$.

59.
$$4.5 + t = -3.1$$
$$4.5 + t - 4.5 = -3.1 - 4.5$$
$$t = -7.6$$
The solution is -7.6.

61.
$$-8.2x = 20.5$$
$$\frac{-8.2x}{-8.2} = \frac{20.5}{-8.2}$$
$$x = -2.5$$
The solution is -2.5.

63.
$$x - 4 = -19$$
$$x - 4 + 4 = -19 + 4$$
$$x = -15$$
The solution is -15.

65.
$$t - 3 = 8$$
$$t - 3 + 3 = -8 + 3$$
$$t = -5$$
The solution is -5.

67.
$$-12x = 14$$
$$\frac{-12x}{-12} = \frac{14}{-12}$$
$$1 \cdot x = -\frac{7}{6}$$
$$x = -\frac{7}{6}$$
The solution is $-\frac{7}{6}$.

69.
$$48 = -\frac{3}{8}y$$
$$-\frac{8}{3} \cdot 48 = -\frac{8}{3}\left(-\frac{3}{8}y\right)$$
$$-\frac{8 \cdot \cancel{3} \cdot 16}{\cancel{3}} = y$$
$$-128 = y$$
The solution is -128.

71.
$$a - \frac{1}{6} = -\frac{2}{3}$$
$$a - \frac{1}{6} + \frac{1}{6} = -\frac{2}{3} + \frac{1}{6}$$
$$a = -\frac{4}{6} + \frac{1}{6}$$
$$a = -\frac{3}{6}$$
$$a = -\frac{1}{2}$$
The solution is $-\frac{1}{2}$.

73.
$$-24 = \frac{8x}{5}$$
$$-24 = \frac{8}{5}x$$
$$\frac{5}{8}(-24) = \frac{5}{8} \cdot \frac{8}{5}x$$
$$-\frac{5 \cdot \cancel{8} \cdot 3}{\cancel{8} \cdot 1} = x$$
$$-15 = x$$
The solution is -15.

75.
$$-\frac{4}{3}t = -12$$
$$-\frac{3}{4}\left(-\frac{4}{3}t\right) = -\frac{3}{4}(-12)$$
$$t = \frac{3 \cdot \cancel{4} \cdot 3}{\cancel{4}}$$
$$t = 9$$
The solution is 9.

77.
$$-483.297 = -794.053 + t$$
$$-483.297 + 794.053 = -794.053 + t + 794.053$$
$$310.756 = t \qquad \text{Using a calculator}$$
The solution is 310.756.

79. *Writing Exercise.* For an equation $x + a = b$, add the opposite of a (or subtract a) on both sides of the equation. For an equation $ax = b$, multiply by $1/a$ (or divide by a) on both sides of the equation.

81.
$$3 \cdot 4 - 18$$
$$= 12 - 18 \qquad \text{Multiplying}$$
$$= -6 \qquad \text{Subtracting}$$

83.
$$16 \div (2 - 3 \cdot 2) + 5$$
$$= 16 \div (2 - 6) + 5 \qquad \text{Simplifying inside}$$
$$= 16 \div (-4) + 5 \qquad \text{the parentheses}$$
$$= -4 + 5 \qquad \text{Dividing}$$
$$= 1 \qquad \text{Adding}$$

85. *Writing Exercise.* Yes, it will form an equivalent equation by the addition principle. It will not help to solve the equation, however. The multiplication principle should be used to solve the equation.

87.
$$2x = x + x$$
$$2x = 2x \qquad \text{Adding on the right side}$$
This is an identity.

89.
$$9x = 0$$
$$\frac{9x}{9} = \frac{0}{9}$$
$$x = 0$$
The solution is 0.

91.
$$x + 8 = 3 + x + 7$$
$$x + 8 = 10 + x \qquad \text{Adding on the right side}$$
$$x + 8 - x = 10 + x - x$$
$$8 = 10$$
This is a contradiction.

93. $2|x| = -14$

$$\frac{2|x|}{2} = -\frac{14}{2}$$

$$|x| = -7$$

Since the absolute value of a number is always nonnegative, this is a contradiction.

95. $mx = 11.6m$

$$\frac{mx}{m} = \frac{11.6m}{m}$$

$$x = 11.6$$

The solution is 11.6.

97. $cx + 5c = 7c$

$$cx + 5c - 5c = 7c - 5c$$

$$cx = 2c$$

$$\frac{cx}{c} = \frac{2c}{c}$$

$$x = 2$$

The solution is 2.

99. $7 + |x| = 30$

$$-7 + 7 + |x| = -7 + 30$$

$$|x| = 23$$

x represents a number whose distance from 0 is 23. Thus $x = -23$ or $x = 23$.

101. $t - 3590 = 1820$

$$t - 3590 + 3590 = 1820 + 3590$$

$$t = 5410$$

$$t + 3590 = 5410 + 3590$$

$$t + 3590 = 9000$$

103. To "undo" the last step, divide 22.5 by 0.3.

$$22.5 \div 0.3 = 75$$

Now divide 75 by 0.3.

$$75 \div 0.3 = 250$$

The answer should be 250 not 22.5.

Exercise Set 2.2

1. $3x - 1 = 7$

$3x - 1 + 1 = 7 + 1$ Adding 1 to both sides

$3x = 7 + 1$

Choice (c) is correct.

3. $6(x - 1) = 2$

$6x - 6 = 2$ Using the distributive law

Choice (a) is correct.

5. $4x = 3 - 2x$

$4x + 2x = 3 - 2x + 2x$ Adding $2x$ to both sides

$4x + 2x = 3$

Choice (b) is correct.

7. $2x + 9 = 25$

$2x + 9 - 9 = 25 - 9$ Subtracting 9 from both sides

$2x = 16$ Simplifying

$$\frac{2x}{2} = \frac{16}{2}$$ Dividing both sides by 2

$x = 8$ Simplifying

Check:

$$\begin{array}{c|c} 2x + 9 = 25 & \\ \hline 2 \cdot 8 + 9 & 25 \\ 16 + 9 & \\ 25 \overset{?}{=} 25 & \text{TRUE} \end{array}$$

The solution is 8.

9. $6z + 5 = 47$

$6z + 5 - 5 = 47 - 5$ Subtracting 5 from both sides

$6z = 42$ Simplifying

$$\frac{6z}{6} = \frac{42}{6}$$ Dividing both sides by 6

$z = 7$ Simplifying

Check:

$$\begin{array}{c|c} 6z + 5 = 47 & \\ \hline 6 \cdot 7 + 5 & 47 \\ 42 + 5 & \\ 47 \overset{?}{=} 47 & \text{TRUE} \end{array}$$

The solution is 7.

11. $7t - 8 = 27$

$7t - 8 + 8 = 27 + 8$ Adding 8 to both sides

$7t = 35$

$$\frac{7t}{7} = \frac{35}{7}$$ Dividing both sides by 7

$t = 5$

Check:

$$\begin{array}{c|c} 7t - 8 = 27 & \\ \hline 7 \cdot 5 - 8 & 27 \\ 35 - 8 & \\ 27 \overset{?}{=} 27 & \text{TRUE} \end{array}$$

The solution is 5.

13. $3x - 9 = 1$

$3x - 9 + 9 = 1 + 9$

$3x = 10$

$$\frac{3x}{3} = \frac{10}{3}$$

$$x = \frac{10}{3}$$

Check:

$$\begin{array}{c|c} 3x - 9 = 1 & \\ \hline 3 \cdot \frac{10}{3} - 9 & 1 \\ 10 - 9 & \\ 1 \overset{?}{=} 1 & \text{TRUE} \end{array}$$

The solution is $\frac{10}{3}$.

15.
$$8z + 2 = -54$$
$$8z + 2 - 2 = -54 - 2$$
$$8z = -56$$
$$\frac{8z}{8} = \frac{-56}{8}$$
$$z = -7$$

Check:

$$\begin{array}{c|c} 8z + 2 = -54 \\ \hline 8(-7) + 2 & -54 \\ -56 + 2 & \\ & -54 \stackrel{?}{=} -54 \quad \text{TRUE} \end{array}$$

The solution is -7.

17.
$$-37 = 9t + 8$$
$$-37 - 8 = 9t + 8 - 8$$
$$-45 = 9t$$
$$\frac{-45}{9} = \frac{9t}{9}$$
$$-5 = t$$

Check:

$$\begin{array}{c|c} -37 = 9t + 8 \\ \hline -37 & 9 \cdot (-5) + 8 \\ & -45 + 8 \\ -37 \stackrel{?}{=} -37 & \quad \text{TRUE} \end{array}$$

The solution is -5.

19.
$$12 - t = 16$$
$$-12 + 12 - t = -12 + 16$$
$$-t = 4$$
$$\frac{-t}{-1} = \frac{4}{-1}$$
$$t = -4$$

Check:

$$\begin{array}{c|c} 12 - t = 16 \\ \hline 12 - (-4) & 16 \\ 12 + 4 & \\ & 16 \stackrel{?}{=} 16 \quad \text{TRUE} \end{array}$$

The solution is -4.

21.
$$-6z - 18 = -132$$
$$-6z - 18 + 18 = -132 + 18$$
$$-6z = -114$$
$$\frac{-6z}{-6} = \frac{-114}{-6}$$
$$z = 19$$

Check:

$$\begin{array}{c|c} -6z - 18 = -132 \\ \hline -6 \cdot 19 - 18 & -132 \\ -114 - 18 & \\ & -132 \stackrel{?}{=} -132 \quad \text{TRUE} \end{array}$$

The solution is 19.

23.
$$5.3 + 1.2n = 1.94$$
$$1.2n = -3.36$$
$$\frac{1.2n}{1.2} = \frac{-3.36}{1.2}$$
$$n = -2.8$$

Check:

$$\begin{array}{c|c} 5.31 + 1.2n = 1.94 \\ \hline 5.3 + 1.2(-2.8) & 1.94 \\ 5.3 + (-3.36) & \\ & 1.94 \stackrel{?}{=} 1.94 \quad \text{TRUE} \end{array}$$

The solution is -2.8.

25.
$$32 - 7x = 11$$
$$-32 + 32 - 7x = -32 + 11$$
$$-7x = -21$$
$$\frac{-7x}{-7} = \frac{-21}{-7}$$
$$x = 3$$

Check:

$$\begin{array}{c|c} 32 - 7x = 11 \\ \hline 32 - 7 \cdot 3 & 11 \\ 32 - 21 & \\ & 11 \stackrel{?}{=} 11 \quad \text{TRUE} \end{array}$$

The solution is 3.

27.
$$\frac{3}{5}t - 1 = 8$$
$$\frac{3}{5}t - 1 + 1 = 8 + 1$$
$$\frac{3}{5}t = 9$$
$$\frac{5}{3} \cdot \frac{3}{5}t = \frac{5}{3} \cdot 9$$
$$t = \frac{5 \cdot \cancel{3} \cdot 3}{\cancel{3} \cdot 1}$$
$$t = 15$$

Check:

$$\begin{array}{c|c} \frac{3}{5}t - 1 = 8 \\ \hline \frac{3}{5} \cdot 15 - 1 & 8 \\ 9 - 1 & \\ & 8 \stackrel{?}{=} 8 \quad \text{TRUE} \end{array}$$

The solution is 15.

29.
$$6 + \frac{7}{2}x = -15$$
$$-6 + 6 + \frac{7}{2}x = -6 - 15$$
$$\frac{7}{2}x = -21$$
$$\frac{2}{7} \cdot \frac{7}{2}x = \frac{2}{7}(-21)$$
$$x = -\frac{2 \cdot 3 \cdot 7}{7 \cdot 1}$$
$$x = -6$$

Check:
$$\frac{6 + \frac{7}{2}x = -15}{\begin{array}{c|c} 6 + \frac{7}{2}(-6) & -15 \\ 6 + (-21) & \\ & -15 \overset{?}{=} -15 \end{array}} \quad \text{TRUE}$$

The solution is -6.

31.
$$-\frac{4a}{5} - 8 = 2$$
$$-\frac{4a}{5} - 8 + 8 = 2 + 8$$
$$-\frac{4a}{5} = 10$$
$$-\frac{5}{4}\left(-\frac{4a}{5}\right) = -\frac{5}{4} \cdot 10$$
$$a = -\frac{5 \cdot 5 \cdot \cancel{2}}{2 \cdot \cancel{2}}$$
$$a = -\frac{25}{2}$$

Check:
$$\frac{-\frac{4a}{5} - 8 = 2}{\begin{array}{c|c} -\frac{4}{5}\left(-\frac{25}{2}\right) - 8 & 2 \\ 10 - 8 & \\ & 2 \overset{?}{=} 2 \end{array}} \quad \text{TRUE}$$

The solution is $-\frac{25}{2}$.

33.
$$-5z - 6z = -44$$
$$-11z = -44 \quad \text{Combining like terms}$$
$$\frac{-11z}{-11} = \frac{-44}{-11}$$
$$z = 4$$

Check:
$$\frac{-5z - 6z = -44}{\begin{array}{c|c} -5 \cdot 4 - 6 \cdot 4 & -44 \\ -20 - 24 & \\ & -44 \overset{?}{=} -44 \end{array}} \quad \text{TRUE}$$

The solution is 4.

35.
$$4x - 6 = 6x$$
$$-6 = 6x - 4x \quad \text{Subtracting } 4x \text{ from both sides}$$
$$-6 = 2x \quad \text{Simplifying}$$
$$\frac{-6}{2} = \frac{2x}{2} \quad \text{Dividing both sides by 2}$$
$$-3 = x$$

Check:
$$\frac{4x - 6 = 6x}{\begin{array}{c|c} 4(-3) - 6 & 6(-3) \\ -12 - 6 & -18 \\ & -18 \overset{?}{=} -18 \end{array}} \quad \text{TRUE}$$

The solution is -3.

37.
$$2 - 5y = 26 - y$$
$$2 - 5y + y = 26 - y + y \quad \text{Adding } y \text{ to both sides}$$
$$2 - 4y = 26 \quad \text{Simplifying}$$
$$-2 + 2 - 4y = -2 + 26 \quad \text{Adding -2 to both sides}$$
$$-4y = 24 \quad \text{Simplifying}$$
$$\frac{-4y}{-4} = \frac{24}{-4} \quad \text{Dividing both sides by -4}$$
$$y = -6$$

Check:
$$\frac{2 - 5y = 26 - y}{\begin{array}{c|c} 2 - 5(-6) & 26 - (-6) \\ 2 + 30 & 26 + 6 \\ & 32 \overset{?}{=} 32 \end{array}} \quad \text{TRUE}$$

The solution is -6.

39.
$$7(2a - 1) = 21$$
$$14a - 7 = 21 \quad \text{Using the distributive law}$$
$$14a = 21 + 7 \quad \text{Adding 7}$$
$$14a = 28$$
$$a = 2 \quad \text{Dividing by 14}$$

Check:
$$\frac{7(2a - 1) = 21}{\begin{array}{c|c} 7(2 \cdot 2 - 1) & 21 \\ 7(4 - 1) & \\ 7 \cdot 3 & \\ & 21 \overset{?}{=} 21 \end{array}} \quad \text{TRUE}$$

The solution is 2.

41. We can write $11 = 11(x + 1)$ as $11 \cdot 1 = 11(x + 1)$. Then $1 = x + 1$, or $x = 0$. The solution is 0.

43.
$$2(3 + 4m) - 6 = 48$$
$$6 + 8m - 6 = 48$$
$$8m = 48 \quad \text{Combining like terms}$$
$$m = 6$$

Check:
$$\frac{2(3 + 4m) - 6 = 48}{\begin{array}{c|c} 2(3 + 4 \cdot 6) - 6 & 48 \\ 2(3 + 24) - 6 & \\ 2 \cdot 27 - 6 & \\ 54 - 6 & \\ & 48 \overset{?}{=} 48 \end{array}} \quad \text{TRUE}$$

The solution is 6.

45.
$$2r + 8 = 6r + 10$$
$$2r + 8 - 10 = 6r + 10 - 10$$
$$2r - 2 = 6r \qquad \text{Combining like terms}$$
$$-2r + 2r - 2 = -2r + 6r$$
$$-2 = 4r$$
$$\frac{-2}{4} = \frac{4r}{4}$$
$$-\frac{1}{2} = r$$

Check:

$$\begin{array}{c|c}
\multicolumn{2}{c}{2r + 8 = 6r + 10} \\
\hline
2\left(-\dfrac{1}{2}\right) + 8 & 6\left(-\dfrac{1}{2}\right) + 10 \\
-1 + 8 & -3 + 10 \\
\end{array}$$
$$7 \overset{?}{=} 7 \qquad \text{TRUE}$$

The solution is $-\dfrac{1}{2}$.

47.
$$6x + 3 = 2x + 3$$
$$6x - 2x = 3 - 3$$
$$4x = 0$$
$$\frac{4x}{4} = \frac{0}{4}$$
$$x = 0$$

Check:

$$\begin{array}{c|c}
\multicolumn{2}{c}{6x + 3 = 2x + 3} \\
\hline
6 \cdot 0 + 3 & 2 \cdot 0 + 3 \\
0 + 3 & 0 + 3 \\
\end{array}$$
$$3 \overset{?}{=} 3 \qquad \text{TRUE}$$

The solution is 0.

49.
$$5 - 2x = 3x - 7x + 25$$
$$5 - 2x = -4x + 25$$
$$4x - 2x = 25 - 5$$
$$2x = 20$$
$$\frac{2x}{2} = \frac{20}{2}$$
$$x = 10$$

Check:

$$\begin{array}{c|c}
\multicolumn{2}{c}{5 - 2x = 3x - 7x + 25} \\
\hline
5 - 2 \cdot 10 & 3 \cdot 10 - 7 \cdot 10 + 25 \\
5 - 20 & 30 - 70 + 25 \\
-15 & -40 + 25 \\
\end{array}$$
$$-15 \overset{?}{=} -15 \qquad \text{TRUE}$$

The solution is 10.

51.
$$7 + 3x - 6 = 3x + 5 - x$$
$$3x + 1 = 2x + 5 \qquad \text{Combining like terms}$$
$$\text{on each side}$$
$$3x - 2x = 5 - 1$$
$$x = 4$$

Check:

$$\begin{array}{c|c}
\multicolumn{2}{c}{7 + 3x - 6 = 3x + 5 - x} \\
\hline
7 + 3 \cdot 4 - 6 & 3 \cdot 4 + 5 - 4 \\
7 + 12 - 6 & 12 + 5 - 4 \\
19 - 6 & 17 - 4 \\
\end{array}$$
$$13 \overset{?}{=} 13 \qquad \text{TRUE}$$

The solution is 4.

53.
$$4y - 4 + y + 24 = 6y + 20 - 4y$$
$$5y + 20 = 2y + 20$$
$$5y - 2y = 20 - 20$$
$$3y = 0$$
$$y = 0$$

Check:

$$\begin{array}{c|c}
\multicolumn{2}{c}{4y - 4 + y + 24 = 6y + 20 - 4y} \\
\hline
4 \cdot 0 - 4 + 0 + 24 & 6 \cdot 0 + 20 - 4 \cdot 0 \\
0 - 4 + 0 + 24 & 0 + 20 - 0 \\
\end{array}$$
$$20 \overset{?}{=} 20 \qquad \text{TRUE}$$

The solution is 0.

55.
$$19 - 3(2x - 1) = 7$$
$$19 - 6x + 3 = 7$$
$$22 - 6x = 7$$
$$-6x = 7 - 22$$
$$-6x = -15$$
$$x = \frac{15}{6}$$
$$x = \frac{5}{2}$$

Check:

$$\begin{array}{c|c}
\multicolumn{2}{c}{19 - 3(2x - 1) = 7} \\
\hline
19 - 3(2 \cdot \frac{5}{2} - 1) & 7 \\
19 - 3(5 - 1) & \\
19 - 3(4) & \\
19 - 12 & \\
\end{array}$$
$$7 \overset{?}{=} 7 \qquad \text{TRUE}$$

The solution is $\frac{5}{2}$.

57.
$$7(5x - 2) = 6(6x - 1)$$
$$35x - 14 = 36x - 6$$
$$-14 + 6 = 36x - 35x$$
$$-8 = x$$

Check:

$$\begin{array}{c|c}
\multicolumn{2}{c}{7(5x - 2) = 6(6x - 1)} \\
\hline
7(5(-8) - 2) & 6(6(-8) - 1) \\
7(-40 - 2) & 6(-48 - 1) \\
7(-42) & 6(-49) \\
\end{array}$$
$$-294 \overset{?}{=} -294 \qquad \text{TRUE}$$

The solution is -8.

59. $2(3t+1)-5 = t-(t+2)$
$6t+2-5 = t-t-2$
$6t-3 = -2$
$6t = -2+3$
$6t = 1$
$t = \frac{1}{6}$

Check:

$$\begin{array}{c|c} 2(3t+1)-5 & = t-(t+2) \\ \hline 2\left(3\cdot\frac{1}{6}+1\right)-5 & \frac{1}{6}-\left(\frac{1}{6}+2\right) \\ 2\left(\frac{1}{2}+1\right)-5 & \frac{1}{6}-2\frac{1}{6} \\ 2\cdot\frac{3}{2}-5 & -2 \\ & -2 \overset{?}{=} -2 \quad \text{TRUE} \end{array}$$

The solution is $\frac{1}{6}$.

61. $19-(2x+3) = 2(x+3)+x$
$19-2x-3 = 2x+6+x$
$16-2x = 3x+6$
$16-6 = 3x+2x$
$10 = 5x$
$2 = x$

Check:

$$\begin{array}{c|c} 19-(2x+3) & = 2(x+3)+x \\ \hline 19-(2\cdot2+3) & 2(2+3)+2 \\ 19-(4+3) & 2\cdot5+2 \\ 19-7 & 10+2 \\ 12 & \overset{?}{=} 12 \quad \text{TRUE} \end{array}$$

The solution is 2.

63. $\frac{2}{3}+\frac{1}{4}t = 2$

The number 12 is the least common denominator, so we multiply by 12 on both sides.

$$12\left(\frac{2}{3}+\frac{1}{4}t\right) = 12\cdot2$$
$$12\cdot\frac{2}{3}+12\cdot\frac{1}{4}t = 24$$
$$8+3t = 24$$
$$3t = 24-8$$
$$3t = 16$$
$$t = \frac{16}{3}$$

Check:

$$\begin{array}{c|c} \frac{2}{3}+\frac{1}{4}t & = 2 \\ \hline \frac{2}{3}+\frac{1}{4}\left(\frac{16}{3}\right) & 2 \\ \frac{2}{3}+\frac{4}{3} & \\ & 2 \overset{?}{=} 2 \quad \text{TRUE} \end{array}$$

The solution is $\frac{16}{3}$.

65. $\frac{2}{3}+4t = 6t-\frac{2}{15}$

The number 15 is the least common denominator, so we multiply by 15 on both sides.

$$15\left(\frac{2}{3}+4t\right) = 15\left(6t-\frac{2}{15}\right)$$
$$15\cdot\frac{2}{3}+15\cdot4t = 15\cdot6t-15\cdot\frac{2}{15}$$
$$10+60t = 90t-2$$
$$10+2 = 90t-60t$$
$$12 = 30t$$
$$\frac{12}{30} = t$$
$$\frac{2}{5} = t$$

Check:

$$\begin{array}{c|c} \frac{2}{3}+4t & = 6t-\frac{2}{15} \\ \hline \frac{2}{3}+4\cdot\frac{2}{5} & 6\cdot\frac{2}{5}-\frac{2}{15} \\ \frac{2}{3}+\frac{8}{5} & \frac{12}{5}-\frac{2}{15} \\ \frac{10}{15}+\frac{24}{15} & \frac{36}{15}-\frac{2}{15} \\ \frac{34}{15} & \overset{?}{=} \frac{34}{15} \quad \text{TRUE} \end{array}$$

The solution is $\frac{2}{5}$.

67. $\frac{1}{3}x+\frac{2}{5} = \frac{4}{5}+\frac{3}{5}x-\frac{2}{3}$

The number 15 is the least common denominator, so we multiply by 15 on both sides.

$$15\left(\frac{1}{3}x+\frac{2}{5}\right) = 15\left(\frac{4}{5}+\frac{3}{5}x-\frac{2}{3}\right)$$
$$15\cdot\frac{1}{3}x+15\cdot\frac{2}{5} = 15\cdot\frac{4}{5}+15\cdot\frac{3}{5}x-15\cdot\frac{2}{3}$$
$$5x+6 = 12+9x-10$$
$$5x+6 = 2+9x$$
$$5x-9x = 2-6$$
$$-4x = -4$$
$$\frac{-4x}{-4} = \frac{-4}{-4}$$
$$x = 1$$

Check:

$$\begin{array}{c|c} \frac{1}{3}x+\frac{2}{5} & = \frac{4}{5}+\frac{3}{5}x-\frac{2}{3} \\ \hline \frac{1}{3}\cdot1+\frac{2}{5} & \frac{4}{5}+\frac{3}{5}\cdot1-\frac{2}{3} \\ \frac{1}{3}+\frac{2}{5} & \frac{4}{5}+\frac{3}{5}-\frac{2}{3} \\ \frac{5}{15}+\frac{6}{15} & \frac{12}{15}+\frac{9}{15}-\frac{10}{15} \\ \frac{11}{15} & \frac{11}{15} \\ \frac{11}{15} & \overset{?}{=} \frac{11}{15} \quad \text{TRUE} \end{array}$$

The solution is 1.

69.
$$2.1x + 45.2 = 3.2 - 8.4x$$

Greatest number of decimal places is 1

$$10(2.1x + 45.2) = 10(3.2 - 8.4x)$$

Multiplying by 10 to clear decimals

$$10(2.1x) + 10(45.2) = 10(3.2) - 10(8.4x)$$

$$21x + 452 = 32 - 84x$$

$$21x + 84x = 32 - 452$$

$$105x = -420$$

$$x = \frac{-420}{105}$$

$$x = -4$$

Check:

$$\begin{array}{c|c} \multicolumn{2}{c}{2.1x + 45.2 = 3.2 - 8.4x} \\ \hline 2.1(-4) + 45.2 & 3.2 - 8.4(-4) \\ -8.4 + 45.2 & 3.2 + 33.6 \\ & \\ 36.8 \overset{?}{=} 36.8 & \text{TRUE} \end{array}$$

The solution is -4.

71.
$$0.76 + 0.21t = 0.96t - 0.49$$

Greatest number of decimal places is 2

$$100(0.76 + 0.21t) = 100(0.96t - 0.49)$$

Multiplying by 100 to clear decimals

$$100(0.76) + 100(0.21t) = 100(0.96t) - 100(0.49)$$

$$76 + 21t = 96t - 49$$

$$76 + 49 = 96t - 21t$$

$$125 = 75t$$

$$\frac{125}{75} = t$$

$$\frac{5}{3} = t, \text{ or}$$

$$1.\overline{6} = t$$

The answer checks. The solution is $\frac{5}{3}$, or $1.\overline{6}$.

73.
$$\frac{2}{5}x - \frac{3}{2}x = \frac{3}{4}x + 3$$

The least common denominator is 20.

$$20\left(\frac{2}{5}x - \frac{3}{2}x\right) = 20\left(\frac{3}{4}x + 3\right)$$

$$20 \cdot \frac{2}{5}x - 20 \cdot \frac{3}{2}x = 20 \cdot \frac{3}{4}x + 20 \cdot 3$$

$$8x - 30x = 15x + 60$$

$$-22x = 15x + 60$$

$$-22x - 15x = 60$$

$$-37x = 60$$

$$x = -\frac{60}{37}$$

Check:

$$\begin{array}{c|c} \multicolumn{2}{c}{\frac{2}{5}x - \frac{3}{2}x = \frac{3}{4}x + 3} \\ \hline \frac{2}{5}\left(-\frac{60}{37}\right) - \frac{3}{2}\left(-\frac{60}{37}\right) & \frac{3}{4}\left(-\frac{60}{37}\right) + 3 \\ -\frac{24}{37} + \frac{90}{37} & -\frac{45}{37} + \frac{111}{37} \\ \frac{66}{37} \overset{?}{=} \frac{66}{37} & \text{TRUE} \end{array}$$

The solution is $-\frac{60}{37}$.

75.
$$\frac{1}{3}(2x - 1) = 7$$

$$3 \cdot \frac{1}{3}(2x - 1) = 3 \cdot 7$$

$$2x - 1 = 21$$

$$2x = 22$$

$$x = 11$$

Check:

$$\begin{array}{c|c} \multicolumn{2}{c}{\frac{1}{3}(2x - 1) = 7} \\ \hline \frac{1}{3}(2 \cdot 11 - 1) & 7 \\ \frac{1}{3} \cdot 21 & \\ & \\ 7 \overset{?}{=} 7 & \text{TRUE} \end{array}$$

The solution is 11.

77.
$$\frac{3}{4}(3t - 4) = 15$$

$$\frac{4}{3} \cdot \frac{3}{4}(3t - 4) = \frac{4}{3} \cdot 15$$

$$3t - 4 = 20$$

$$3t = 24$$

$$t = 8$$

Check:

$$\begin{array}{c|c} \multicolumn{2}{c}{\frac{3}{4}(3t - 4) = 15} \\ \hline \frac{3}{4}(3 \cdot 8 - 4) & 15 \\ \frac{3}{4} \cdot (24 - 4) & \\ \frac{3}{4} \cdot 20 & \\ 15 \overset{?}{=} 15 & \text{TRUE} \end{array}$$

The solution is 8.

79.
$$\frac{1}{6}\left(\frac{3}{4}x - 2\right) = -\frac{1}{5}$$

$$30 \cdot \frac{1}{6}\left(\frac{3}{4}x - 2\right) = 30\left(-\frac{1}{5}\right)$$

$$5\left(\frac{3}{4}x - 2\right) = -6$$

$$\frac{15}{4}x - 10 = -6$$

$$\frac{15}{4}x = 4$$

$$4 \cdot \frac{15}{4}x = 4 \cdot 4$$

$$15x = 16$$

$$x = \frac{16}{15}$$

Check: $\dfrac{1}{6}\left(\dfrac{3}{4}x - 2\right) = -\dfrac{1}{5}$

$$\begin{array}{c|c} \dfrac{1}{6}\left(\dfrac{3}{4}\cdot\dfrac{16}{15} - 2\right) & -\dfrac{1}{5} \\[2mm] \dfrac{1}{6}\left(\dfrac{4}{5} - 2\right) & \\[2mm] \dfrac{1}{6}\left(-\dfrac{6}{5}\right) & \end{array}$$

$$-\dfrac{1}{5} \overset{?}{=} -\dfrac{1}{5} \qquad \text{TRUE}$$

The solution is $\dfrac{16}{15}$.

81. $\quad 0.7(3x + 6) = 1.1 - (x - 3)$

$\quad\quad 2.1x + 4.2 = 1.1 - x + 3$

$\quad\quad 2.1x + 4.2 = -x + 4.1$

$\quad 10(2.1x + 4.2) = 10(-x + 4.1) \qquad$ Clearing decimals

$\quad\quad\quad 21x + 42 = -10x + 41$

$\quad\quad\quad\quad\quad 21x = -10x + 41 - 42$

$\quad\quad\quad\quad\quad 21x = -10x - 1$

$\quad\quad\quad\quad\quad 31x = -1$

$$x = -\dfrac{1}{31}$$

The check is left to the student. The solution is $-\dfrac{1}{31}$.

83. $\quad a + (a - 3) = (a + 2) - (a + 1)$

$\quad\quad a + a - 3 = a + 2 - a - 1$

$\quad\quad\quad 2a - 3 = 1$

$\quad\quad\quad\quad 2a = 1 + 3$

$\quad\quad\quad\quad 2a = 4$

$\quad\quad\quad\quad a = 2$

Check: $\quad\dfrac{a + (a - 3) = (a + 2) - (a + 1)}{\begin{array}{c|c} 2 + (2 - 3) & (2 + 2) - (2 + 1) \\ 2 - 1 & 4 - 3 \end{array}}$

$$1 \overset{?}{=} 1 \qquad\qquad \text{TRUE}$$

The solution is 2.

85. *Writing Exercise.* By adding $t - 13$ to both sides of $45 - t = 13$ we have $32 = t$. This approach is preferable since we found the solution in just one step.

87. $3 - 5a = 3 - 5 \cdot 2 = 3 - 10 = -7$

89. $7x - 2x = 7(-3) - 2(-3) = -21 + 6 = -15$

91. *Writing Exercise.* Multiply by 100 to clear decimals. Next multiply by 12 to clear fractions. (These steps could be reversed.) Then proceed as usual. The procedure could be streamlined by multiplying by 1200 to clear decimals and fractions in one step.

93. $\quad 8.43x - 2.5(3.2 - 0.7x) = -3.455x + 9.04$

$\quad\quad 8.43x - 8 + 1.75x = -3.455x + 9.04$

$\quad\quad\quad 10.18x - 8 = -3.455x + 9.04$

$\quad 10.18x + 3.455x = 9.04 + 8$

$\quad\quad\quad 13.635x = 17.04$

$$x = 1.\overline{2497}, \text{ or } \dfrac{1136}{909}$$

The solution is $1.\overline{2497}$, or $\dfrac{1136}{909}$.

95. $\quad -2[3(x - 2) + 4] = 4(5 - x) - 2x$

$\quad\quad -2[3x - 6 + 4] = 20 - 4x - 2x$

$\quad\quad\quad -2[3x - 2] = 20 - 6x$

$\quad\quad\quad -6x + 4 = 20 - 6x$

$\quad\quad\quad\quad 4 = 20 \qquad$ Adding $6x$ to both sides

This is a contradiction.

97. $\quad 3(x + 5) = 3(5 + x)$

$\quad\quad 3x + 15 = 15 + 3x$

$\quad 3x + 15 - 15 = 15 - 15 + 3x$

$\quad\quad\quad 3x = 3x$

This is an identity.

99. $\quad 2x(x + 5) - 3(x^2 + 2x - 1) = 9 - 5x - x^2$

$\quad 2x^2 + 10x - 3x^2 - 6x + 3 = 9 - 5x - x^2$

$\quad\quad\quad -x^2 + 4x + 3 = 9 - 5x - x^2$

$\quad\quad\quad\quad 4x + 3 = 9 - 5x \qquad$ Adding x^2

$\quad\quad\quad\quad 4x + 5x = 9 - 3$

$\quad\quad\quad\quad 9x = 6$

$$x = \dfrac{2}{3}$$

The solution is $\dfrac{2}{3}$.

101. $\quad 9 - 3x = 2(5 - 2x) - (1 - 5x)$

$\quad\quad 9 - 3x = 10 - 4x - 1 + 5x$

$\quad\quad 9 - 3x = 9 + x$

$\quad\quad 9 - 9 = x + 3x$

$\quad\quad\quad 0 = 4x$

$\quad\quad\quad 0 = x$

The solution is 0.

103. $[7 - 2(8 \div (-2))]x = 0$

Since $7 - 2(8 \div (-2)) \neq 0$ and the product on the left side of the equation is 0, then x must be 0.

105. $$\dfrac{5x + 3}{4} + \dfrac{25}{12} = \dfrac{5 + 2x}{3}$$

$$12\left(\dfrac{5x + 3}{4} + \dfrac{25}{12}\right) = 12\left(\dfrac{5 + 2x}{3}\right)$$

$$12\left(\dfrac{5x + 3}{4}\right) + 12 \cdot \dfrac{25}{12} = 4(5 + 2x)$$

$$3(5x + 3) + 25 = 4(5 + 2x)$$

$$15x + 9 + 25 = 20 + 8x$$

$$15x + 34 = 20 + 8x$$

$$7x = -14$$

$$x = -2$$

The solution is -2.

Exercise Set 2.3

1. We substitute 0.9 for t and calculate d.

$$d = 344t = 344 \cdot 0.9 = 309.6$$

The fans were 309.6m from the stage.

3. We substitute 21,345 for n and calculate f.

$$f = \frac{n}{15} = \frac{21,345}{15} = 1423$$

There are 1423 full-time equivalent students.

5. We substitute 0.025 for I and 0.044 for U and calculate f.

$$f = 8.5 + 1.4(I - U)$$
$$= 8.5 + 1.4(0.025 - 0.044)$$
$$= 8.5 + 1.4(-0.019)$$
$$= 8.5 - 0.0266$$
$$= 8.4734$$

The federal funds rate should be 8.4734.

7. Substitute 1 for t and calculate n.

$$n = 0.5t^4 + 3.45t^3 - 96.65t^2 + 347.7t$$
$$= 0.5(1)^4 + 3.45(1)^3 - 96.65(1)^2 + 347.7(1)$$
$$= 0.5 + 3.45 - 96.65 + 347.7$$
$$= 255$$

255 mg of ibuprofen remains in the bloodstream.

9. $A = bh$

$\frac{A}{h} = \frac{bh}{h}$ Dividing both sides by h

$\frac{A}{h} = b$

11. $d = rt$

$\frac{d}{t} = \frac{rt}{t}$ Dividing both sides by t

$\frac{d}{t} = r$

13. $I = Prt$

$\frac{I}{rt} = \frac{Prt}{rt}$ Dividing both sides by rt

$\frac{I}{rt} = P$

15. $\quad H = 65 - m$

$H + m = 65$ Adding m to both sides

$\quad\quad m = 65 - H$ Subtracting H from both sides

17. $\quad\quad P = 2l + 2w$

$P - 2w = 2l + 2w - 2w$ Subtracting $2w$ from both sides

$P - 2w = 2l$

$\frac{P - 2w}{2} = \frac{2l}{2}$ Dividing both sides by 2

$\frac{P - 2w}{2} = l$, or

$\frac{P}{2} - w = l$

19. $A = \pi r^2$

$\frac{A}{r^2} = \frac{\pi r^2}{r^2}$

$\frac{A}{r^2} = \pi$

21. $A = \frac{1}{2}bh$

$2A = 2 \cdot \frac{1}{2}bh$ Multiplying both sides by 2

$2A = bh$

$\frac{2A}{b} = \frac{bh}{b}$ Dividing both sides by h

$\frac{2A}{b} = h$

23. $E = mc^2$

$\frac{E}{m} = \frac{mc^2}{m}$ Dividing both sides by m

$\frac{E}{m} = c^2$

25. $\quad Q = \frac{c + d}{2}$

$2Q = 2 \cdot \frac{c + d}{2}$ Multiplying both sides by 2

$2Q = c + d$

$2Q - c = c + d - c$ Subtracting c from both sides

$2Q - c = d$

27. $\quad\quad A = \frac{a + b + c}{3}$

$3A = 3 \cdot \frac{a + b + c}{3}$ Multiplying both sides by 3

$3A = a + b + c$

$3A - a - c = a + b + c - a - c$ Subtracting a and c from both sides

$3A - a - c = b$

29. $w = \frac{r}{f}$

$f \cdot w = f \cdot \frac{r}{f}$ Multiplying both sides by f

$fw = r$

31. $\quad\quad F = \frac{9}{5}C + 32$

$F - 32 = \frac{9}{5}C$

$\frac{5}{9}(F - 32) = \frac{5}{9} \cdot \frac{9}{5}C$

$\frac{5}{9}(F - 32) = C$

33. $\quad\quad 2x - y = 1$

$2x - y + y - 1 = 1 + y - 1$ Adding $y - 1$ to both sides

$2x - 1 = y$

35. $2x + 5y = 10$
$5y = -2x + 10$
$y = \dfrac{-2x + 10}{5}$
$y = -\dfrac{2}{5}x + 2$

37. $4x - 3y = 6$
$-3y = -4x + 6$
$y = \dfrac{-4x + 6}{-3}$
$y = \dfrac{4}{3}x - 2$

39. $9x + 8y = 4$
$8y = -9x + 4$
$y = \dfrac{-9x + 4}{8}$
$y = -\dfrac{9}{8}x + \dfrac{1}{2}$

41. $3x - 5y = 8$
$-5y = -3x + 8$
$y = \dfrac{-3x + 8}{-5}$
$y = \dfrac{3}{5}x - \dfrac{8}{5}$

43. $z = 13 + 2(x + y)$
$z - 13 = 2(x + y)$
$z - 13 = 2x + 2y$
$z - 13 - 2y = 2x$
$\dfrac{z - 13 - 2y}{2} = x$
$\dfrac{1}{2}z - \dfrac{13}{2} - y = x$

45. $t = 27 - \dfrac{1}{4}(w - l)$
$t - 27 = -\dfrac{1}{4}(w - l)$
$-4(t - 27) = w - l$ Multiplying by -4
$-4t + 108 = w - l$
$-4t + 108 - w = -l$
$4t - 108 + w = l$ Multiplying by -1

47. $A = at + bt$
$A = t(a + b)$ Factoring
$\dfrac{A}{a + b} = t$ Dividing both sides by $a+b$

49. $A = \dfrac{1}{2}ah + \dfrac{1}{2}bh$
$2A = 2\left(\dfrac{1}{2}ah + \dfrac{1}{2}bh\right)$
$2A = ah + bh$
$2A = h(a + b)$
$\dfrac{2A}{a + b} = h$

51. $R = r + \dfrac{400(W - L)}{N}$
$N \cdot R = N\left(r + \dfrac{400(W - L)}{N}\right)$
       Multiplying both sides by N
$NR = Nr + 400(W - L)$
$NR = Nr + 400W - 400L$
$NR + 400L = Nr + 400W$ Adding $400L$ to
                        both sides
$400L = Nr + 400W - NR$ Adding
                       $-NR$ to both sides
$L = \dfrac{Nr + 400W - NR}{400}$

53. *Writing Exercise.* Given the formula for converting Celsius temperature C to Fahrenheit temperature F, solve for C. This yields a formula for converting Fahrenheit temperature to Celsius temperature.

55. $-2 + 5 - (-4) - 17$
$= -2 + 5 + 4 - 17$
$= 3 + 4 - 17$
$= 7 - 17$
$= -10$

57. $4.2(-11.75)(0) = 0$

59. $20 \div (-4) \cdot 2 - 3$
$= -5 \cdot 2 - 3$ Dividing and
$= -10 - 3$ multiplying from left to
               right
$= -13$ Subtracting

61. *Writing Exercise.* Answers may vary. A decorator wants to have a carpet cut for a bedroom. The perimeter of the room is 54 ft and its length is 15 ft. How wide should the carpet be?

63. $K = 21.235w + 7.75h - 10.54a + 102.3$
$2852 = 21.235(80) + 7.75(190) - 10.54a + 102.3$
$2852 = 1698.8 + 1472.5 - 10.54a + 102.3$
$2852 = 3273.6 - 10.54a$
$-421.6 = -10.54a$
$40 = a$

The man is 40 years old.

65. First we substitute 54 for A and solve for s to find the length of a side of the cube.
$A = 6s^2$
$54 = 6s^2$
$9 = s^2$
$3 = s$ Taking the positive square root

Now we substitute 3 for s in the formula for the volume of a cube and compute the volume.
$V = s^3 = 3^3 = 27$

The volume of the cube is 27 in^3.

67. $c = \dfrac{w}{a} \cdot d$

$ac = a \cdot \dfrac{w}{a} \cdot d$

$ac = wd$

$a = \dfrac{wd}{c}$

69. $ac = bc + d$

$ac - bc = d$

$c(a - b) = d$

$c = \dfrac{d}{a - b}$

71. $3a = c - a(b + d)$

$3a = c - ab - ad$

$3a + ab + ad = c$

$a(3 + b + d) = c$

$a = \dfrac{c}{3 + b + d}$

73. $K = 21.235w + 7.75h - 10.54a + 102.3$

$K = 21.235\left(\dfrac{w}{2.2046}\right) + 7.75\left(\dfrac{h}{0.3937}\right) - 10.54a + 102.3$

$K = 9.632w + 19.685h - 10.54a + 102.3$

Exercise Set 2.4

1. "What percent of 57 is 23?" can be translated as $n \cdot 57 = 23$, so choice (d) is correct.

3. "23 is 57% of what number?" can be translated as $23 = 0.57y$, so choice (e) is correct.

5. "57 is what percent of 23?" can be translated as $n \cdot 23 = 57$, so choice (c) is correct.

7. "What is 23% of 57?" can be translated as $a = (0.23)57$, so choice (f) is correct.

9. "23% of what number is 57?" can be translated as $57 = 0.23y$, so choice (b) is correct.

11. $49\% = 49.0\%$

$49\%\qquad 0.49.0$

Move the decimal point 2 places to the left.

$49\% = 0.49$

13. $1\% = 1.0\%$

$1\%\qquad 0.01.0$

Move the decimal point 2 places to the left.

$1\% = 0.01$

15. $4.1\% = 4.10\%$

$4.1\%\qquad 0.04.10$

Move the decimal point 2 places to the left.

$4.1\% = 0.041$

17. $20\% = 20.0\%$

$20\%\qquad 0.20.0$

Move the decimal point 2 places to the left.

$20\% = 0.20$, or 0.2

19. $62.5\%\qquad 0.62.5$

Move the decimal point 2 places to the left.

$62.5\% = 0.625$

21. $0.2\%\qquad 0.00.2$

Move the decimal point 2 places to the left.

$0.2\% = 0.002$

23. $175\% = 175.0\%\qquad 1.75.0$

Move the decimal point 2 places to the left.

$175\% = 1.75$

25. 0.38

First move the decimal point $0.38.$
two places to the right;
then write a % symbol: 38%

27. 0.039

First move the decimal point $0.03.9$
two places to the right;
then write a % symbol: 3.9%

29. 0.45

First move the decimal point $0.45.$
two places to the right;
then write a % symbol: 45%

31. 0.7

First move the decimal point $0.70.$
two places to the right;
then write a % symbol: 70%

33. 0.0009

First move the decimal point $0.00.09$
two places to the right;
then write a % symbol: 0.09%

35. 1.06

First move the decimal point $1.06.$
two places to the right;
then write a % symbol: 106%

37. 1.8

First move the decimal point $1.80.$
two places to the right;
then write a % symbol: 180%

39. $\dfrac{3}{5}$ $\left(\text{Note: } \dfrac{3}{5} = 0.6\right)$

Move the decimal point $0.60.$
two places to the right;
then write a % symbol: 60%

41. $\dfrac{8}{25}$ $\left(\text{Note: } \dfrac{8}{25} = 0.32\right)$

First move the decimal point   0.32.
two places to the right;
then write a % symbol: 32%

43. *Translate*.

$$\underbrace{\text{What percent}}_{y} \text{ of } \underset{\cdot}{76} \text{ is } \underset{19}{19}?$$

We solve the equation and then convert to percent notation.

$$y \cdot 76 = 19$$
$$y = \frac{19}{76}$$
$$y = 0.25 = 25\%$$

The answer is 25%.

45. *Translate*.

$$\underbrace{\text{What percent}}_{y} \text{ of } \underset{150}{150} \text{ is } \underset{39}{39}?$$

We solve the equation and then convert to percent notation.

$$y \cdot 150 = 39$$
$$y = \frac{39}{150}$$
$$y = 0.26 = 26\%$$

The answer is 26%.

47. *Translate*.

$$14 \text{ is } 30\% \text{ of } \underbrace{\text{what number}}_{y}?$$
$$14 = 30\% \cdot y$$

We solve the equation.

$$14 = 0.3y \qquad (30\% = 0.3)$$
$$\frac{14}{0.3} = y$$
$$46.\overline{6} = y$$

The answer is $46.\overline{6}$, or $46\dfrac{2}{3}$, or $\dfrac{140}{3}$.

49. *Translate*.

$$0.3 \text{ is } 12\% \text{ of } \underbrace{\text{what number}}_{y}?$$
$$0.3 = 12\% \cdot y$$

We solve the equation.

$$0.3 = 0.12y \qquad (12\% = 0.12)$$
$$\frac{0.3}{0.12} = y$$
$$2.5 = y$$

The answer is 2.5.

51. *Translate*.

$$\underbrace{\text{What number}}_{y} \text{ is } 1\% \text{ of one million}?$$
$$y = 1\% \cdot 1,000,000$$

We solve the equation.

$$y = 0.01 \cdot 1,000,000 \qquad (1\% = 0.01)$$
$$y = 10,000 \qquad \text{Multiplying}$$

The answer is 10,000.

53. *Translate*.

$$\underbrace{\text{What percent}}_{y} \text{ of } \underset{60}{60} \text{ is } \underset{75}{75}?$$

We solve the equation and then convert to percent notation.

$$y \cdot 60 = 75$$
$$y = \frac{75}{60}$$
$$y = 1.25 = 125\%$$

The answer is 125%.

55. *Translate*.

$$\text{What is } 2\% \text{ of } 40?$$
$$x = 2\% \cdot 40$$

We solve the equation.

$$x = 0.02 \cdot 40 \qquad (2\% = 0.02)$$
$$x = 0.8 \qquad \text{Multiplying}$$

The answer is 0.8.

57. Observe that 25 is half of 50. Thus, the answer is 0.5, or 50%. We could also do this exercise by translating to an equation.

Translate.

$$25 \text{ is } \underbrace{\text{what percent}}_{y} \text{ of } 50?$$
$$25 = y \cdot 50$$

We solve the equation and convert to percent notation.

$$25 = y \cdot 50$$
$$\frac{25}{50} = y$$
$$0.5 = y, \text{ or } 50\% = y$$

The answer is 50%.

59. *Translate*.

$$\text{What percent of } 69 \text{ is } \$23?$$
$$y \cdot 69 = 23$$

We solve the equation and convert to percent notation.

$$y \cdot 69 = 23 \quad y = \frac{23}{69} \quad y = 0.33\overline{3} = 33.\overline{3}\% \text{ or } 33\frac{1}{3}\%$$

The answer is $33.\overline{3}\%$ or 33 1/3%.

61. First we reword and translate, letting c represent Americans who commute to work, in millions.

$$\text{What is } 5\% \text{ of } 57?$$
$$c = 0.05 \cdot 57$$
$$c = 0.05 \cdot 57 = 2.85$$

There are 2.85 million Americans who bicycle to commute to school or work.

63. First we reword and translate, letting h represent Americans who bicycle to exercise for health.

What is 41% of 57%

$\downarrow \quad \downarrow \quad \downarrow \quad \downarrow \quad \downarrow$

$h \quad = 0.41 \quad \cdot \quad 57$

$h = 0.41 \cdot 57 = 23.37$

There are 23.37 million Americans who bicycle to exercise for health.

65. First we reword and translate, letting c represent the number of credits Cody has completed.

What is 60% of 125?

$\downarrow \quad \downarrow \quad \downarrow \quad \downarrow \quad \downarrow$

$c \quad = 0.6 \quad \cdot \quad 125$

$c = 0.6 \cdot 125 = 75$

Cody has completed 75 credits.

67. First we reword and translate, letting b represent the number of at-bats.

216 is 36.3% of what number?

$\downarrow \quad \downarrow \quad \downarrow \quad \downarrow \quad\quad \downarrow$

$216 \quad = \quad 0.363 \quad \cdot \quad\quad b$

$\dfrac{216}{0.363} = b$

$595 \approx b$

Magglio Ordonez had 595 at-bats.

69. a) First we reword and translate, letting p represent the unknown percent.

What percent of \$25 is \$4?

$\downarrow \qquad\qquad \downarrow \quad \downarrow \quad \downarrow \quad \downarrow$

$p \qquad\qquad \cdot \quad 25 \ = \ 4$

$\dfrac{p \cdot 25}{25} = \dfrac{4}{25}$

$p = 0.16 = 16\%$

The tip was 16% of the cost of the meal.

b) We add to find the total cost of the meal, including tip:

$\$25 + \$4 = \$29$

71. To find the percent of crude oil came from Canada and Mexico, we first reword and translate, letting p represent the unknown percent.

3.4 million is what percent of 10.2 million?

$\downarrow \qquad\quad \downarrow \qquad \downarrow \qquad \downarrow \qquad \downarrow$

$3.4 \qquad = \qquad p \qquad \cdot \qquad 10.2$

$\dfrac{3.4}{10.2} = p$

$0.33\overline{3} = p$

$33.\overline{3}\% = p$ or 33 1/3%

About $33.\overline{3}\%$ or 33 1/3% of crude oil came from Canada and Mexico.

To find the percent of crude oil that came from the rest of the world, we subtract:

$100\% - 33\ 1/3\% = 66\ 2/3\%$ or $66.\overline{6}\%$.

About 66 2/3% or $66.\overline{6}\%$ of crude oil came from the rest of the world.

73. Let $I =$ the amount of interest Glenn will pay. Then we have:

I is 7% of \$2400.

$\downarrow\downarrow \quad \downarrow \quad \downarrow \quad\quad \downarrow$

$I = 0.07 \quad \cdot \quad \$2400$

$I = \$168$

Glenn will pay \$168 interest.

75. If $n =$ the number of women who had babies in good or excellent health, we have:

n is 95% of 300.

$\downarrow\downarrow \quad \downarrow \quad \downarrow \quad\quad \downarrow$

$n = 0.95 \quad \cdot \quad 300$

$n = 285$

285 women had babies in good or excellent health.

77. A self-employed person must earn 120% as much as a non-self-employed person. Let $a =$ the amount Tia would need to earn, in dollars per hour, on her own for a comparable income.

a is 120% of \$16.

$\downarrow\downarrow \quad \downarrow \quad \downarrow \quad\quad \downarrow$

$a = 1.2 \quad \cdot \quad 16$

$a = 19.20$

Tia would need to earn \$19.20 per hour on her own.

79. We reword and translate.

What percent of 103 is 45?

$\downarrow \qquad\qquad \downarrow \quad \downarrow \quad \downarrow \quad \downarrow$

$p \qquad\qquad \cdot \quad 103 \ = \ 45$

$p \cdot 103 = 45$

$p \approx 0.437 = 43.7\%$

The actual cost exceeds initial estimate by about 43.7%.

81. When the sales tax is 6%, the total amount paid is 106% of the cost of the merchandise. Let $c =$ the cost of the merchandise. Then we have:

\$47.70 is 106% of c.

$\downarrow \quad\quad \downarrow \quad \downarrow \quad \downarrow\downarrow$

$47.70 = 1.06 \quad \cdot \quad c$

$\dfrac{47.70}{1.06} = c$

$45 = c$

The price of the merchandise was \$45.

83. When the sales tax is 6%, the total amount paid is 106% of the cost of the merchandise. Let c = the amount the school group owes, or the cost of the software without tax. Then we have:

$157.41 is 106% of c.
$$\downarrow \quad \downarrow \quad \downarrow \quad \downarrow \downarrow$$
$$157.41 \;=\; 1.06 \;\cdot\; c$$
$$\frac{157.41}{1.06} = c$$
$$148.5 = c$$

The school group owes $148.50.

85. First we reword and translate.

What is 16.5% of 191?
$$\downarrow \quad \downarrow \quad \downarrow \quad \downarrow \quad \downarrow$$
$$a \;\;=\; 0.165 \;\cdot\; 191$$

Solve. We convert 16.5% to decimal notation and multiply.

$$a = 0.165 \cdot 191$$
$$a = 31.515 \approx 31.5$$

About 31.5 lb of the author's body weight is fat.

87. Let m = the number of mailed ads that led to a sale or response from customers. Then we have:

m is 2.15% of 114.
$$\downarrow \downarrow \quad \downarrow \quad \downarrow \quad \downarrow$$
$$m = 0.0215 \;\cdot\; 114$$
$$m \approx \;\; 2.45$$

About 2.45 billion pieces of mail led to a response.

89. The number of calories in a serving of Light Style Bread is 85% of the number of calories in a serving of regular bread. Let c = the number of calories in a serving of regular bread. Then we have:

$\underbrace{140 \text{ calories}}$ is 85% of c.
$$\downarrow \qquad\quad \downarrow \quad \downarrow \quad \downarrow \downarrow$$
$$140 \quad\;\; = 0.85 \;\cdot\; c$$
$$\frac{140}{0.85} = c$$
$$165 \approx c$$

There are about 165 calories in a serving of regular bread.

91. *Writing Exercise.* The book is marked up $30. Since Campus Bookbuyers paid $30 for the book, this is a 100% markup.

93. Let l represent represent the length and w the width. Then twice the length plus twice the width is $2l + 2w$.

95. Let p represent the number of points Tino scored. Then $p - 5$ is five fewer than p.

97. Half of a is $\frac{1}{2}a$. So the product of 10 and half of a is $10\left(\frac{1}{2}a\right)$.

99. Let l represent the length and w the width. Then, the width is 2 in. less than the length which is $w = l - 2$.

101. (a) In the survey report, 40% of all sick days on Monday or Friday sounds excessive. However, for a traditional 5-day business week, 40% is the same as $\frac{2}{5}$. That is, just 2 days out of 5.
(b) In the FBI statistics, 26% of home burglaries occurring between Memorial Day and Labor Day sounds excessive. However, 26% of a 365-day year is 73 days, For the months of June, July, and August there are at least 90 days. So 26% is less than one home burglary per day.

103. Let p = the population of Bardville. Then we have:

1332 is 15% of 48% of $\underbrace{\text{the population.}}$
$$\downarrow \downarrow \downarrow \downarrow \downarrow \downarrow \qquad\quad \downarrow$$
$$1332 = 0.15 \;\cdot\; 0.48 \;\cdot\; \qquad p$$
$$\frac{1332}{0.15(0.48)} = p$$
$$18,500 = p$$

The population of Bardville is 18,500.

105. Since 6 ft = 6×1 ft = 6×12 in. = 72 in., we can express 6 ft 4 in. as 72 in.+4 in., or 76 in. We reword and translate. Let a = Jaraan's final adult height.

$\underbrace{76\text{in.}}$ is 96.1% of $\underbrace{\text{adult height}}$
$$\downarrow \quad \downarrow \quad \downarrow \quad \downarrow \qquad\quad \downarrow$$
$$76 \;\; = 0.961 \;\cdot\; \qquad a$$
$$\frac{76}{0.961} = a$$
$$79 \approx a$$

Note that 79 in. = 72 in. + 7 in. = 6 ft 7 in.

Jaraan's final adult height will be about 6 ft 7 in.

107. Using the formula for the area A of a rectangle with length l and width w, $A = l \cdot w$, we first find the area of the photo.

$$A = 8 \text{ in.} \times 6 \text{ in.} = 48 \text{ in}^2$$

Next we find the area of the photo that will be visible using a mat intended for a 5-in. by 7-in. photo.

$$A = 7 \text{ in.} \times 5 \text{ in.} = 35 \text{ in}^2$$

Then the area of the photo that will be hidden by the mat is 48 in^2 − 35 in^2, or 13 in^2.

We find what percentage of the area of the photo this represents.

$\underbrace{\text{What percent}}$ of $\underbrace{48 \text{ in}^2}$ is $\underbrace{13 \text{ in}^2}$?
$$\downarrow \qquad\qquad \downarrow \quad \downarrow \quad \downarrow \quad \downarrow$$
$$p \qquad\;\; \cdot \quad 48 \;\; = \;\; 13$$
$$\frac{p \cdot 48}{48} = \frac{13}{48}$$
$$p \approx 0.27$$
$$p \approx 27\%$$

The mat will hide about 27% of the photo.

109. *Writing Exercise.* Suppose Jorge has x dollars of taxable income. If he makes a $50 tax-deductible contribution, then he pays tax of $0.3(x - \$50)$, or $0.3x - \$15$ and his assets are reduced by $0.3x - \$15 + \$50$, or $0.3x + \$35$. If he makes a $40 non-tax-deductible contribution, he pays tax of $0.3x$ and his assets are reduced by $0.3x + \$40$. Thus, it costs him less to make a $50 tax-deductible contribution.

<hr>

Exercise Set 2.5

<hr>

1. **Familiarize.** Let n = the number. Then three less than two times the number is $2n - 3$.

   Translate.

   $\underbrace{\text{Three less than twice a number}}$ is 19.

   $\downarrow$
   $2n - 3$　　　$\begin{array}{cc}\downarrow & \downarrow \\ = & 19\end{array}$

   Carry out. We solve the equation.

   $$2n - 3 = 19$$
   $$2n = 22 \quad \text{Adding 3}$$
   $$n = 11 \quad \text{Dividing by 2}$$

   Check. Twice 11 is 22 and three fewer than 19. The answer checks.

   State. The number is 11.

3. **Familiarize.** Let a = the number. Then "five times the sum of 3 and twice some number" translates to $5(2a + 3)$.

   Translate.

   $\underbrace{\begin{array}{c}\text{Five times the sum of} \\ \text{3 and twice some number}\end{array}}$ is 70.

   $\downarrow$
   $5(2a + 3)$　　$\begin{array}{cc}\downarrow & \downarrow \\ = & 70\end{array}$

   Carry out. We solve the equation.

   $$5(2a + 3) = 70$$
   $$10a + 15 = 70 \quad \text{Using the distributive law}$$
   $$10a = 55 \quad \text{Subtracting 15}$$
   $$a = \tfrac{11}{2} \quad \text{Dividing by 10}$$

   Check. The sum of $2 \cdot \tfrac{11}{2}$ and 3 is 14, and $5 \cdot 14 = 70$. The answer checks.

   State. The number is $\tfrac{11}{2}$.

5. **Familiarize.** Let p = the regular price of the iPod. At 20% off, Kyle paid $(100 - 20)\%$, or 80% of the regular price.

   Translate.

   \$120 is 80% of $\underbrace{\text{the regular price.}}$

   $\begin{array}{cccc}\downarrow & \downarrow & \downarrow & \downarrow \\ 120 & = & 0.80 & \cdot\end{array}$　　$\begin{array}{c}\downarrow \\ p\end{array}$

   Carry out. We solve the equation.

   $$120 = 0.80p$$
   $$150 = p$$

   Check. 80% of \$150, or 0.80(\$150), is \$120. The answer checks.

   State. The regular price was \$150.

7. **Familiarize.** Let c = the price of the graphing calculator itself. When the sales tax rate is 6%, the tax paid on the calculator is 6% of c, or $0.06c$.

   Translate.

   $\underbrace{\text{Price of calculator}}$ plus $\underbrace{\text{sales tax}}$ is \$137.80.

   $\begin{array}{ccccc}\downarrow & \downarrow & \downarrow & \downarrow & \downarrow \\ c & + & 0.06c & = & 137.80\end{array}$

   Carry out. We solve the equation.

   $$c + 0.06c = 137.80$$
   $$1.06c = 137.80$$
   $$c = 130$$

   Check. 6% of \$130, or 0.06(\$130), is \$7.80 and \$7.80 + \$130 is \$137.80, the total cost. The answer checks.

   State. The graphing calculator itself cost \$130.

9. **Familiarize.** Let d = Looi's distance, in miles, from the start after 8 hr. Then the distance from the finish line is $2d$.

   Translate.

   $\underbrace{\begin{array}{c}\text{Distance} \\ \text{from start}\end{array}}$ plus $\underbrace{\begin{array}{c}\text{distance} \\ \text{from finish}\end{array}}$ is 235.3 mi.

   $\begin{array}{ccccc}\downarrow & \downarrow & \downarrow & \downarrow & \downarrow \\ d & + & 2d & = & 235.3\end{array}$

   Carry out. We solve the equation.

   $$d + 2d = 235.3$$
   $$3d = 235.3$$
   $$d \approx 78.4$$

   Check. If Looi is 78.4 mi from the start, then he is $2 \cdot (78.4)$, or 156.8 mi from the finish. Since $78.4 + 156.8$ is approximately 235.3, the total distance, the answer checks.

   State. Looi had traveled approximately 78.4 mi.

11. **Familiarize.** Let d = the distance, in miles, that Danica had traveled to the given point after the start. Then the distance from the finish line was $300d$ miles.

    Translate.

    $\underbrace{\begin{array}{c}\text{Distance} \\ \text{to finish}\end{array}}$ plus $\underbrace{\begin{array}{c}\text{20 mi} \\ \text{more}\end{array}}$ was $\underbrace{\begin{array}{c}\text{distance} \\ \text{to start.}\end{array}}$

    $\begin{array}{ccccc}\downarrow & \downarrow & \downarrow & \downarrow & \downarrow \\ 300 - d & + & 20 & = & d\end{array}$

    Carry out. We solve the equation.

    $$300 - d + 20 = d$$
    $$320 - d = d$$
    $$320 = 2d$$
    $$160 = d$$

    Check. If Danica was 160 mi from the start, she was 300160, or 140 mi from the finish. Since 160 is 20 more than 140, the answer checks.

    State. Danica had traveled 160 mi at the given point.

13. **Familiarize.** Let n = the number of Erica's apartment. Then $n + 1$ = the number of her next-door neighbor's apartment.

    Translate.

    $\underbrace{\text{Erica's number}}$ plus $\underbrace{\text{neighbor's number}}$ is 2409.

    $\begin{array}{ccccc}\downarrow & \downarrow & \downarrow & \downarrow & \downarrow \\ n & + & (n + 1) & = & 2409\end{array}$

Carry out. We solve the equation.

$$n + (n + 1) = 2409$$
$$2n + 1 = 2409$$
$$2n = 2408$$
$$n = 1204$$

If Erica's apartment number is 1204, then her next-door neighbor's number is $1204 + 1$, or 1205.

Check. 1204 and 1205 are consecutive numbers whose sum is 2409. The answer checks.

State. The apartment numbers are 1204 and 1205.

15. Familiarize. Let $n =$ the smaller house number. Then $n + 2 =$ the larger number.

Translate.

$$\underbrace{\text{Smaller number}}_{n} \quad \text{plus} \atop + \quad \underbrace{\text{larger number}}_{(n+2)} \quad \text{is} \atop = \quad 572.$$

Carry out. We solve the equation.

$$n + (n + 2) = 572$$
$$2n + 2 = 572$$
$$2n = 570$$
$$n = 285$$

If the smaller number is 285, then the larger number is $285 + 2$, or 287.

Check. 285 and 287 are consecutive odd numbers and $285 + 287 = 572$. The answer checks.

State. The house numbers are 285 and 287.

17. Familiarize. Let $x =$ the first page number. Then $x + 1 =$ the second page number, and $x + 2 =$ the third page number.

Translate.

$$\underbrace{\text{The sum of three} \atop \text{consecutive page numbers}}_{x + (x+1) + (x+2)} \quad \text{is} \atop = \quad 99.$$

Carry out. We solve the equation.

$$x + (x + 1) + (x + 2) = 99$$
$$3x + 3 = 99$$
$$3x = 96$$
$$x = 32$$

If x is 32, then $x + 1$ is 33 and $x + 2 = 34$.

Check. 32, 33, and 34 are consecutive integers, and $32 + 33 + 34 = 99$. The result checks.

State. The page numbers are 32, 33, and 34.

19. Familiarize. Let $m =$ the man's age. Then $m - 2 =$ the woman's age.

Translate.

$$\underbrace{\text{Man's age}}_{m} \quad \text{plus} \atop + \quad \underbrace{\text{Woman's age}}_{(m-2)} \quad \text{is} \atop = \quad 204.$$

Carry out. We solve the equation.

$$m + (m - 2) = 204$$
$$2m - 2 = 204$$
$$2m = 206$$
$$m = 103$$

If m is 103, then $m - 2$ is 101.

Check. 103 is 2 more than 101, and $103 + 101 = 204$. The answer checks.

State. The man was 103 yr old, and the woman was 101 yr old.

21. Familiarize. Familiarize. Let $m =$ the number non-spam messages, in billions. Then $4m$ is the number of spam messages.

Translate.

$$\underbrace{\text{spam}}_{4m} \quad \text{plus} \atop + \quad \underbrace{\text{non-spam}}_{m} \quad \text{is} \atop = \quad 125$$

Carry out. We solve the equation.

$$4m + m = 125$$
$$5m = 125$$
$$m = 25$$

If m is 25, then $4m$ is 100.

Check. 100 is four times 25, and $25 + 100 = 125$. The answer checks.

State. There were 100 billion spam messages and 25 billion non-spam messages sent each day in 2006.

23. Familiarize. The page numbers are consecutive integers. If we let $x =$ the smaller number, then $x + 1 =$ the larger number.

Translate. We reword the problem.

$$\underbrace{\text{First integer}}_{x} \quad + \atop + \quad \underbrace{\text{Second integer}}_{(x+1)} \quad = \atop = \quad 281$$

Carry out. We solve the equation.

$$x + (x + 1) = 281$$
$$2x + 1 = 281 \qquad \text{Combining like terms}$$
$$2x = 280 \qquad \text{Adding } -1 \text{ on both sides}$$
$$x = 140 \qquad \text{Dividing on both sides by 2}$$

Check. If $x = 140$, then $x + 1 = 141$. These are consecutive integers, and $140 + 141 = 281$. The answer checks.

State. The page numbers are 140 and 141.

25. Familiarize. We draw a picture. Let $w =$ the width of the rectangle, in feet. Then $w + 60 =$ the length.

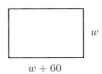

The perimeter is twice the length plus twice the width, and the area is the product of the length and the width.

Translate.

$$\underbrace{\text{Twice the length}}_{2(w+60)} \quad \underset{+}{\text{plus}} \quad \underbrace{\text{twice the width}}_{2w} \quad \underset{=}{\text{is}} \quad \underbrace{520 \text{ ft}}_{520}.$$

Carry out. We solve the equation.

$$2(w + 60) + 2w = 520$$
$$2w + 120 + 2w = 520$$
$$4w + 120 = 520$$
$$4w = 400$$
$$w = 100$$

Then $w + 60 = 100 + 60 = 160$, and the area is 160 ft · 100 ft $= 16,000$ ft^2.

Check. The length, 160 ft, is 60 ft more than the width, 100 ft. The perimeter is $2 \cdot 160$ ft $+ 2 \cdot 100$ ft, or 320 ft $+$ 200 ft, or 520 ft. We can check the area by doing the calculation again. The answer checks.

State. The length is 160 ft, the width is 100 ft, and the area is 16,000 ft^2.

27. **Familiarize**. Let $w =$ the width, in meters. Then $w + 4$ is the length. The perimeter is twice the length plus twice the width.

Translate.

$$\underbrace{\text{Twice the width}}_{2w} \quad \underset{+}{\text{plus}} \quad \underbrace{\text{twice the length}}_{2(w+4)} \quad \underset{=}{\text{is 92}}_{92}.$$

Carry out. We solve the equation.

$$2w + 2(w + 4) = 92$$
$$2w + 2w + 8 = 92$$
$$4w = 84$$
$$w = 21$$

Then $w + 4 = 21 + 4 = 25$.

Check. The length, 25 m is 4 more than the width, 21 m. The perimeter is $2 \cdot 21$ m $+ 2 \cdot 25$ m $= 42$ m $+ 50$ m $= 92$ m. The answer checks.

State. The length of the garden is 25 m and the width is 21 m.

29. **Familiarize**. Let $w =$ the width, in inches. Then $2w =$ the length. The perimeter is twice the length plus twice the width. We express $10\frac{1}{2}$ as 10.5.

Translate.

$$\underbrace{\text{Twice the length}}_{2 \cdot 2w} \quad \underset{+}{\text{plus}} \quad \underbrace{\text{twice the width}}_{2w} \quad \underset{=}{\text{is}} \quad \underbrace{10.5 \text{ in}}_{10.5}.$$

Carry out. We solve the equation.

$$2 \cdot 2w + 2w = 10.5$$
$$4w + 2w = 10.5$$
$$6w = 10.5$$
$$w = 1.75, \text{ or } 1\frac{3}{4}$$

Then $2w = 2(1.75) = 3.5$, or $3\frac{1}{2}$.

Check. The length, $3\frac{1}{2}$ in., is twice the width, $1\frac{3}{4}$ in. The perimeter is $2\left(3\frac{1}{2} \text{ in.}\right) + 2\left(1\frac{3}{4} \text{ in.}\right) =$ 7 in. $+ 3\frac{1}{2}$ in. $= 10\frac{1}{2}$ in. The answer checks.

State. The actual dimensions are $3\frac{1}{2}$ in. by $1\frac{3}{4}$ in.

31. **Familiarize.** We draw a picture. We let $x =$ the measure of the first angle. Then $3x =$ the measure of the second angle, and $x + 30 =$ the measure of the third angle.

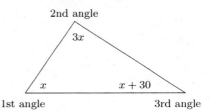

Recall that the measures of the angles of any triangle add up to 180°.

Translate.

$$\underbrace{\text{Measure of first angle}}_{x} \; \underset{+}{+} \; \underbrace{\text{measure of second angle}}_{3x} \; \underset{+}{+}$$

$$\underbrace{\text{measure of third angle}}_{x+30} \; \underset{=}{\text{is 180°}}_{180}.$$

Carry out. We solve the equation.

$$x + 3x + (x + 30) = 180$$
$$5x + 30 = 180$$
$$5x = 150$$
$$x = 30$$

Possible answers for the angle measures are as follows:

First angle: $x = 30°$

Second angle: $3x = 3(30)° = 90°$

Third angle: $x + 30° = 30° + 30° = 60°$

Check. Consider 30°, 90°, and 60°. The second angle is three times the first, and the third is 30° more than the first. The sum of the measures of the angles is 180°. These numbers check.

State. The measure of the first angle is 30°, the measure of the second angle is 90°, and the measure of the third angle is 60°.

33. Familiarize. Let $x =$ the measure of the first angle. Then $4x =$ the measure of the second angle, and $x + 4x + 5 = 5x + 5$ is the measure of the third angle.

Translate.

Measure of first angle	+	measure of second angle	+
↓	↓	↓	↓
x	+	$4x$	+

measure of third angle	is 180°.
↓	↓ ↓
$(5x + 5)$	= 180

Carry out. We solve the equation.

$$x + 4x + (5x + 5) = 180$$
$$10x + 5 = 180$$
$$10x = 175$$
$$x = 17.5$$

If $x = 17.5$, then $4x = 4(17.5) = 70$, and $5x + 5 = 5(17.5) + 5 = 87.5 + 5 = 92.5$.

Check. Consider 17.5°, 70°, and 92.5°. The second is four times the first, and the third is 5° more than the sum of the other two. The sum of the measures of the angles is 180°. These numbers check.

State. The measure of the second angle is 70°.

35. Familiarize. Let $b =$ the length of the bottom section of the rocket, in feet. Then $\frac{1}{6}b =$ the length of the top section, and $\frac{1}{2}b =$ the length of the middle section.

Translate.

Length of top section	+	length of middle section	+	length of bottom section	is	240 ft.
↓	↓	↓	↓	↓		
$\frac{1}{6}b$	+	$\frac{1}{2}b$	+	b	=	240

Carry out. We solve the equation. First we multiply by 6 on both sides to clear the fractions.

$$\frac{1}{6}b + \frac{1}{2}b + b = 240$$
$$6\left(\frac{1}{6}b + \frac{1}{2}b + b\right) = 6 \cdot 240$$
$$6 \cdot \frac{1}{6}b + 6 \cdot \frac{1}{2}b + 6 \cdot b = 1440$$
$$b + 3b + 6b = 1440$$
$$10b = 1440$$
$$b = 144$$

Then $\frac{1}{6}b = \frac{1}{6} \cdot 144 = 24$ and $\frac{1}{2}b = \frac{1}{2} \cdot 144 = 72$.

Check. 24 ft is $\frac{1}{6}$ of 144 ft, and 72 ft is $\frac{1}{2}$ of 144 ft. The sum of the lengths of the sections is 24 ft + 72 ft + 144 ft = 240 ft. The answer checks.

State. The length of the top section is 24 ft, the length of the middle section is 72 ft, and the length of the bottom section is 144 ft.

37. Familiarize. Let $m =$ the number of miles that can be traveled on a $18 budget. Then the total cost of the taxi ride, in dollars, is $2.250 + 1.80m$, or $2.25 + 1.8m$.

Translate.

Cost of taxi ride	is	$18.
↓	↓	↓
$2.25 + 1.8m$	=	18

Carry out. We solve the equation.

$$2.25 + 1.8m = 18$$
$$1.8m = 15.75$$
$$m = \frac{15.75}{1.8} = 8.75 = 8\frac{3}{4}$$

Check. The mileage charge is $1.80(8.75), or $15.75, and the total cost of the ride is $2.25 + $15.75 = $18. The answer checks.

State. Debbie can travel 8.75 mi on her budget.

39. Familiarize. The total cost is the daily charge plus the mileage charge. Let $d =$ the distance that can be traveled, in miles, in one day for $100. The mileage charge is the cost per mile times the number of miles traveled, or $0.39d$.

Translate.

Daily rate	plus	mileage charge	is	$100.
↓	↓	↓	↓	↓
49.95	+	$0.39d$	=	100

Carry out. We solve the equation.

$$49.95 + 0.39d = 100$$
$$0.39d = 50.05$$
$$d = 128.\overline{3}, \ or \ 128\frac{1}{3}$$

Check. For a trip of $128\frac{1}{3}$ mi, the mileage charge is $0.39\left(128\frac{1}{3}\right)$, or $50.05, and $49.95 + $50.05 = $100. The answer checks.

State. Concert Productions can travel $128\frac{1}{3}$ mi in one day and stay within their budget.

41. Familiarize. Let $x =$ the measure of one angle. Then $90 - x =$ the measure of its complement.

Translate.

Measure of one angle	is	15°	more than	twice the measure of its complement.
↓	↓	↓	↓	↓
x	=	15	+	$2(90 - x)$

Carry out. We solve the equation.

$$x = 15 + 2(90 - x)$$
$$x = 15 + 180 - 2x$$
$$x = 195 - 2x$$
$$3x = 195$$
$$x = 65$$

If x is 65, then $90 - x$ is 25.

Check. The sum of the angle measures is 90°. Also, 65° is 15° more than twice its complement, 25°. The answer checks.

State. The angle measures are 65° and 25°.

43. Familiarize. Let x = the measure of one angle. Then $180 - x$ = the measure of its supplement.

Translate.

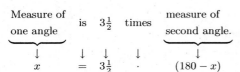

Measure of one angle	is	$3\frac{1}{2}$	times	measure of second angle.
↓	↓	↓	↓	↓
x	$=$	$3\frac{1}{2}$	$\cdot$	$(180 - x)$

Carry out. We solve the equation.
$$x = 3\tfrac{1}{2}(180 - x)$$
$$x = 630 - 3.5x$$
$$4.5x = 630$$
$$x = 140$$

If $x = 140$, then $180 - 140 = 40°$.

Check. The sum of the angles is 180°. Also 140° is three and a half times 40°. The answer checks.

State. The angles are 40° and 140°.

45. **Familiarize**. Let l = the length of the paper, in cm. Then $l - 6.3$ = the width. The perimeter is twice the length plus twice the width.

Translate.

Twice the length	plus	twice the width	is	99 cm.
↓	↓	↓	↓	↓
$2l$	$+$	$2(l - 6.3)$	$=$	99

Carry out. We solve the equation.
$$2l + 2(l - 6.3) = 99$$
$$2l + 2l - 12.6 = 99$$
$$4l - 12.6 = 99$$
$$4l = 111.6$$
$$l = 27.9$$

Then $l - 6.3 = 27.9 - 6.3 = 21.6$.

Check. The width, 21.6 cm, is 6.3 cm less than the length, 27.9 cm. The perimeter is $2(27.9 \text{ cm}) + 2(21.6 \text{ cm}) = 55.8 \text{ cm} + 43.2 \text{ cm} = 99 \text{ cm}$. The answer checks.

State. The length of the paper is 27.9 cm, and the width is 21.6 cm.

47. **Familiarize**. Let a = the amount Janeka invested. Then the simple interest for one year is $6\% \cdot a$, or $0.06a$.

Translate.

Amount invested	plus	interest	is	$6996.
↓	↓	↓	↓	
a	$+$	$0.06a$	$=$	6996

Carry out. We solve the equation.
$$a + 0.06a = 6996$$
$$1.06a = 6996$$
$$a = 6600$$

Check. An investment of $6600 at 6% simple interest earns 0.06($6600), or $396, in one year. Since $6600 + $396 = $6996, the answer checks.

State. Janeka invested $6600.

49. **Familiarize**. Let w = the winning score. Then $w - 340$ = the losing score.

Translate.

Winning score	plus	losing score	was	1320 points.
↓	↓	↓	↓	↓
w	$+$	$w - 340$	$=$	1320

Carry out. We solve the equation.
$$w + (w - 340) = 1320$$
$$2w - 340 = 1320$$
$$2w = 1660$$
$$w = 830$$

Then $w - 340 = 830 - 340 = 490$.

Check. The winning score, 830, is 340 points more than the losing score, 490. The total of the two scores is $830 + 490 = 1320$ points. The answer checks.

State. The winning score was 830 points.

51. **Familiarize**. Let a = the selling price of the house. Then the commission on the selling price is 6% times a, or $0.06a$.

Translate.

Selling price	minus	commission	is	$ 117,500.
↓	↓	↓	↓	↓
a	$-$	$0.06a$	$=$	$117,500$

Carry out. We solve the equation.
$$a - 0.06a = 117,500$$
$$0.94a = 117,500$$
$$a = 125,000$$

Check. A selling price of $125,000 gives a commission of $7500. Since $125,000 - $7500 = $117,500, the answer checks.

State. They must sell the house for $125,000.

53. **Familiarize**. We will use the equation
$$T = \frac{1}{4}N + 40.$$

Translate. We substitute 80 for T.
$$80 = \frac{1}{4}N + 40$$

Carry out. We solve the equation.
$$80 = \frac{1}{4}N + 40$$
$$40 = \frac{1}{4}N$$
$$160 = N \qquad \text{Multiplying by 4 on both sides}$$

Check. When $N = 160$, we have $T = \frac{1}{4} \cdot 160 + 40 = 40 + 40 = 80$. The answer checks.

State. A cricket chirps 160 times per minute when the temperature is $80°$F.

55. *Writing Exercise.* Although many of the problems in this section might be solved by guessing, using the five-step problem-solving process to solve them would give the student practice is using a technique that can be used to solve other problems whose answers are not so readily guessed.

57. Since -8 is to the left of $1, -8 < 1$.

59. Since $\frac{1}{2}$ is to the right of $0, \frac{1}{2} > 0$.

61. $-4 \le x$

63. $y < 5$

65. *Writing Exercise.* Answers may vary.

The sum of three consecutive odd integers is 375. What are the integers?

67. *Familiarize.* Let c = the amount the meal originally cost. The 15% tip is calculated on the original cost of the meal, so the tip is $0.15c$.

Translate.

Original cost plus tip less \$10 is \$32.55.

$$c + 0.15c - 10 = 32.55$$

Carry out. We solve the equation.

$$c + 0.15c - 10 = 32.55$$
$$1.15c - 10 = 32.55$$
$$1.15c = 42.55$$
$$c = 37$$

Check. If the meal originally cost \$37, the tip was 15% of \$37, or 0.15(\$37), or \$5.55. Since $\$37 + \$5.55 - \$10 = \$32.55$, the answer checks.

State. The meal originally cost \$37.

69. *Familiarize.* Let s = one score. Then four score = $4s$ and four score and seven = $4s + 7$.

Translate. We reword .

1776 plus four score and seven is 1863

$$1776 + (4s + 7) = 1863$$

Carry out. We solve the equation.

$$1776 + (4s + 7) = 1863$$
$$4s + 1783 = 1863$$
$$4s = 80$$
$$s = 20$$

Check. If a score is 20 years, then four score and seven represents 87 years. Adding 87 to 1776 we get 1863. This checks.

State. A score is 20.

71. *Familiarize.* Let n = the number of half dollars. Then the number of quarters is $2n$; the number of dimes is $2 \cdot 2n$, or $4n$; and the number of nickels is $3 \cdot 4n$, or $12n$. The total value of each type of coin, in dollars, is as follows.

Half dollars: $0.5n$

Quarters: $0.25(2n)$, or $0.5n$

Dimes: $0.1(4n)$, or $0.4n$

Nickels: $0.05(12n)$, or $0.6n$

Then the sum of these amounts is $0.5n + 0.5n + 0.4n + 0.6n$, or $2n$.

Translate.

Total amount of change is \$10.

$$2n = 10$$

Carry out. We solve the equation.

$$2n = 10$$
$$n = 5$$

Then $2n = 2 \cdot 5 = 10$, $4n = 4 \cdot 5 = 20$, and $12n = 12 \cdot 5 = 60$.

Check. If there are 5 half dollars, 10 quarters, 20 dimes, and 60 nickels, then there are twice as many quarters as half dollars, twice as many dimes as quarters, and 3 times as many nickels as dimes. The total value of the coins is $\$0.5(5) + \$0.25(10) + \$0.1(20) + \$0.05(60) = \$2.50 + \$2.50 + \$2 + \$3 = \$10$. The answer checks.

State. The shopkeeper got 5 half dollars, 10 quarters, 20 dimes, and 60 nickels.

73. Familiarize. Let p = the price before the two discounts. With the first 10% discount, the price becomes 90% of p, or $0.9p$. With the second 10% discount, the final price is 90% of $0.9p$, or $0.9(0.9p)$.

Translate.

10% discount and 10% discount of price is \$ 77.75.

$$0.9 \cdot 0.9p = 77.75$$

Carry out. We solve the equation.

$$0.9(0.9p) = 77.75$$
$$0.81p = 77.75$$
$$p = 95.99$$

Check. Since 90% of \$95.99 is \$86.39, and 90% of \$86.39 is \$77.75, the answer checks.

State. The original price before discounts was \$95.99.

75. *Familiarize.* Let n = the number of DVDs purchased. Assume that two more DVDs were purchased. Then the first DVD costs \$9.99 and the total cost of the remaining $(n - 1)$ DVDs is \$6.99$(n - 1)$. The shipping and handling costs are \$3 for the first DVD, \$1.50 for the second (half of \$3), and a total of \$1$(n - 2)$ for the remaining $n - 2$ DVDs.

Translate.

1st DVD	plus	remaining DVDs	plus	1stS&H charges
↓	↓	↓	↓	↓
9.99	+	$6.99(n-1)$	+	3

... plus	2ndS&H charges	plus	remaining S&Hcharges	is	$45.45.
↓	↓	↓	↓	↓	↓
+	1.50	+	$1(n-2)$	=	45.45

Carry out. We solve the equation.

$$9.99 + 6.99(n-1) + 3 + 1.5 + (n-2) = 45.45$$
$$9.99 + 6.99n - 6.99 + 4.5 + n - 2 = 45.45$$
$$7.99n + 5.5 = 45.45$$
$$7.99n = 39.95$$
$$n = 5$$

Check. If there are 5 DVDs, the cost of the DVDs is $9.99 + $6.99(5 − 1), or $9.99 + $27.96, or $37.95. The cost for shipping and handling is $3 + $1.50 + $1(5 − 2) = $7.50. The total cost is $37.95 + $7.50, or $45.45. The answer checks.

State. There were 5 DVDs in the shipment.

77. Familiarize. Let d = the distance, in miles, that Glenda traveled. At $0.40 per $\frac{1}{5}$ mile, the mileage charge can also be given as 5($0.40), or $2 per mile. Since it took 20 min to complete what is usually a 10-min drive, the taxi was stopped in traffic for 20 − 10, or 10 min.

Translate.

Initial charge	plus	$2 per mile	plus	stopped in traffic charge	is	$18.50.
↓	↓	↓	↓	↓	↓	↓
2.50	+	$2d$	+	0.40(10)	=	18.50

Carry out. We solve the equation.

$$2.5 + 2d + 0.4(10) = 18.5$$
$$2.5 + 2d + 4 = 18.5$$
$$2d + 6.5 = 18.5$$
$$2d = 12$$
$$d = 6$$

Check. Since $2(6) = $12, and $0.40(10) = $4, and $12 + $4 + $2.50 = $18.50, the answer checks.

State. Glenda traveled 6 mi.

79. *Writing Exercise.* If the school can invest the $2000 so that it earns at least 7.5% and thus grows to at least $2150 by the end of the year, the second option should be selected. If not, the first option is preferable.

81. Familiarize. Let w = the width of the rectangle, in cm. Then $w + 4.25$ = the length.

Translate.

The perimeter	is	101.74 cm.
↓	↓	↓
$2(w + 4.25) + 2w$ =		101.74

Carry out. We solve the equation.

$$2(w + 4.25) + 2w = 101.74$$
$$2w + 8.5 + 2w = 101.74$$
$$4w + 8.5 = 101.74$$
$$4w = 93.24$$
$$w = 23.31$$

Then $w + 4.25 = 23.31 + 4.25 = 27.56$.

Check. The length, 27.56 cm, is 4.25 cm more than the width, 23.31 cm. The perimeter is 2(27.56) cm + 2(23.31 cm) = 55.12 cm + 46.62 cm = 101.74 cm. The answer checks.

State. The length of the rectangle is 27.56 cm, and the width is 23.31 cm.

Exercise Set 2.6

1. $-5x \leq 30$

    $x \geq -6$ Dividing by −5 and reversing the inequality symbol

3. $-2t > -14$

    $t < 7$ Dividing by −2 and reversing the inequality symbol

5. $x < -2$ and $-2 > x$ are equivalent.

7. If we add 1 to both sides of $-4x - 1 \leq 15$, we get $-4x \leq 16$. The two given inequalities are equivalent.

9. $x > -4$

  a) Since $4 > -4$ is true, 4 is a solution.

  b) Since $-6 > -4$ is false, −6 is not a solution.

  c) Since $-4 > -4$ is false, −4 is not a solution.

11. $y \leq 19$

  a) Since $18.99 \leq 19$ is true, 18.99 is a solution.

  b) Since $19.07 \leq 19$ is false, 19.01 is not a solution.

  c) Since $19 \leq 19$ is true, 19 is a solution.

13. $c \geq -7$

  a) Since $0 \geq -7$ is true, 0 is a solution.

  b) Since $-5.4 \geq -7$ is true, −5.4 is a solution.

  c) Since $7.1 \geq -7$ is true, 7.1 is a solution.

15. $z < -3$

  a) Since $0 < -3$ is false, 0 is not a solution.

  b) Since $-3\frac{1}{3} < -3$ is true, $-3\frac{1}{3}$ is a solution.

  c) Since $1 < -3$ is false, 1 is not a solution.

17. The solutions of $y < 2$ are those numbers less than 2. They are shown on the graph by shading all points to the left of 2. The open circle at 2 indicates that 2 is not part of the graph.

$y < 2$

19. The solutions of $x \geq -1$ are those numbers greater than or equal to -1. They are shown on the graph by shading all points to the right of -1. The closed circle at -1 indicates that the point -1 is part of the graph.

$$x \geq -1$$

$$\xleftarrow{\;\;}\!+\!+\!+\!+\!\bullet\!+\!+\!+\!+\!+\!\xrightarrow{\;\;}$$
$$-4\;-2\;\;\;0\;\;\;2\;\;\;4$$

21. The solutions of $0 \leq t$, or $t \geq 0$, are those numbers greater than or equal to zero. They are shown on the graph by shading the point 0 and all points to the right of 0. The closed circle at 0 indicates that 0 is part of the graph.

$$0 \leq t$$

$$\xleftarrow{\;\;}\!+\!+\!+\!+\!\bullet\!+\!+\!+\!+\!+\!\xrightarrow{\;\;}$$
$$-4\;-2\;\;\;0\;\;\;2\;\;\;4$$

23. In order to be solution of the inequality $-5 \leq x < 2$, a number must be a solution of both $-5 \leq x$ and $x < 2$. The solution set is graphed as follows:

$$-5 \leq x < 2$$

$$\xleftarrow{\;\;}\!+\!\bullet\!+\!+\!+\!+\!+\!\circ\!+\!+\!\xrightarrow{\;\;}$$
$$-6\;-4\;-2\;\;\;0\;\;\;2\;\;\;4$$

The closed circle at -5 means that -5 is part of the graph. The open circle at 2 means that 2 is not part of the graph.

25. In order to be a solution of the inequality $-4 < x < 0$, a number must be a solution of both $-4 < x$ and $x < 0$. The solution set is graphed as follows:

$$-4 < x < 0$$

$$\xleftarrow{\;\;}\!+\!+\!+\!\circ\!+\!+\!+\!+\!\circ\!+\!+\!+\!\xrightarrow{\;\;}$$
$$-6\;-4\;-2\;\;\;0\;\;\;2$$

The open circles at -4 and 0 mean that -4 and 0 are both part of the graph.

27. All points to the right of -4 are shaded. The open circle at -4 indicates that -4 is not part of the graph. Using set-builder notation we have $\{x | x > -4\}$.

29. The point 2 and all points to the left of 2 are shaded. Using set-builder notation we have $\{x | x \leq 2\}$.

31. All points to the left of -1 are shaded. The open circle at -1 indicates that -1 is not part of the graph. Using set-builder notation we have $\{x | x < -1\}$.

33. The point 0 and all points to the right of 0 are shaded. Using set-builder notation we have $\{x | x \geq 0\}$.

35.
$$y + 6 > 9$$
$$y + 6 - 6 > 9 - 6 \quad \text{Adding } -6 \text{ to both sides}$$
$$y > 3 \quad \text{Simplifying}$$

The solution set is $\{y | y > 3\}$. The graph is as follows:

$$0\qquad 3$$

37.
$$x + 9 \leq -12$$
$$x + 9 - 9 \leq -12 - 9 \quad \text{Adding } -9 \text{ to both sides}$$
$$x \leq -21 \quad \text{Simplifying}$$

The solution set is $\{x | x \leq -21\}$. The graph is as follows:

$$-21 \qquad\qquad 0$$

39.
$$n - 6 < 11$$
$$n - 6 + 6 < 11 + 6 \quad \text{Adding 6 to both sides}$$
$$n < 17 \quad \text{Simplifying}$$

The solution set is $\{n | n < 17\}$. The graph is as follows:

$$\xleftarrow{\;\;}\!\!\!=\!\!=\!\!=\!\!+\!\!=\!\!=\!\!\circ\!\xrightarrow{\;\;}$$
$$0\qquad 17$$

41.
$$2x \leq x - 9$$
$$2x - x \leq x - 9 - x$$
$$x \leq -9$$

The solution set is $\{x | x \leq -9\}$. The graph is as follows:

$$\xleftarrow{\;\;}\!\!=\!\!=\!\!=\!\!\bullet\!\!+\!\!+\!\!\xrightarrow{\;\;}$$
$$-9\quad 0$$

43.
$$y + \frac{1}{3} \leq \frac{5}{6}$$
$$y + \frac{1}{3} - \frac{1}{3} \leq \frac{5}{6} - \frac{1}{3}$$
$$y \leq \frac{5}{6} - \frac{2}{6}$$
$$y \leq \frac{3}{6}$$
$$y \leq \frac{1}{2}$$

The solution set is $\left\{ y \middle| y \leq \frac{1}{2} \right\}$. The graph is as follows:

$$\xleftarrow{\;\;}\!\!=\!\!=\!\!=\!\!\bullet\!\!+\!\!+\!\!\xrightarrow{\;\;}$$
$$0\;\tfrac{1}{2}$$

45.
$$t - \frac{1}{8} > \frac{1}{2}$$
$$t - \frac{1}{8} + \frac{1}{8} > \frac{1}{2} + \frac{1}{8}$$
$$t > \frac{4}{8} + \frac{1}{8}$$
$$t > \frac{5}{8}$$

The solution set is $\left\{ t \middle| t > \frac{5}{8} \right\}$. The graph is as follows:

$$\xleftarrow{\;\;}\!\!+\!\!\circ\!\!=\!\!=\!\!=\!\!\xrightarrow{\;\;}$$
$$0\;\tfrac{5}{8}$$

47.
$$-9x + 17 > 17 - 8x$$
$$-9x + 17 - 17 > 17 - 8x - 17 \quad \text{Adding } -17$$
$$-9x > -8x$$
$$-9x + 9x > -8x + 9x \qquad \text{Adding } 9x$$
$$0 > x$$

The solution set is $\{x | x < 0\}$. The graph is as follows:

49. $-23 < -t$

The inequality states that the opposite of 23 is less than the opposite of t. Thus, t must be less than 23, so the solution set is $\{t | t < 23\}$. To solve this inequality using the addition principle, we would proceed as follows:
$$-23 < -t$$
$$t - 23 < 0 \quad \text{Adding } t \text{ to both sides}$$
$$t < 23 \quad \text{Adding 23 to both sides}$$

The solution set is $\{t | t < 23\}$. The graph is as follows:

51.
$$10 - y \leq -12$$
$$-10 + 10 - y \leq -10 - 12 \quad \text{Adding -10}$$
$$-y \leq -22$$
$$\qquad \qquad \qquad \text{The symbol has to be reversed.}$$
$$-1(-y) \geq -1(-22)$$
$$y \geq 22$$

The solution set is $\{y | y \geq 22\}$.

53.
$$4x < 28$$
$$\frac{1}{4} \cdot 4x < \frac{1}{4} \cdot 28 \quad \text{Multiplying by } \frac{1}{4}$$
$$x < 7$$

The solution set is $\{x | x < 7\}$. The graph is as follows:

55.
$$-7x < 13$$
$$-\frac{1}{7} \cdot (-7x) > -\frac{1}{7} \cdot 13 \quad \text{Multiplying by } -\frac{1}{7}$$
$$\qquad \qquad \text{The symbol has to be reversed.}$$
$$x > -\frac{13}{7} \quad \text{Simplifying}$$

The solution set is $\left\{ x \middle| x > -\frac{13}{7} \right\}$.

57.
$$-24 > 8t$$
$$-3 > t$$

The solution set is $\{t | t < -3\}$.

59.
$$1.8 \geq -1.2n$$
$$\frac{-1}{1.2} \cdot 1.8 \leq \frac{-1}{1.2}(-1.2n) \quad \text{Multiplying by } \frac{1}{7}$$
$$\qquad \qquad \qquad \text{The symbol has to be reversed.}$$
$$-1.5 \leq n$$

The solution set is $\{n | n \geq -1.5\}$.

61.
$$-2y \leq \frac{1}{5}$$
$$-\frac{1}{2} \cdot (-2y) \geq -\frac{1}{2} \cdot \frac{1}{5} \quad \text{The symbol has to be reversed.}$$
$$y \geq -\frac{1}{10}$$

The solution set is $\left\{ y \middle| y \geq -\frac{1}{10} \right\}$.

63.
$$-\frac{8}{5} > -2x$$
$$-\frac{1}{2} \cdot \left(-\frac{8}{5} \right) < -\frac{1}{2} \cdot (-2x)$$
$$\frac{8}{10} < x$$
$$\frac{4}{5} < x, \text{ or } x > \frac{4}{5}$$

The solution set is $\left\{ x \middle| \frac{4}{5} < x \right\}$, or $\left\{ x \middle| x > \frac{4}{5} \right\}$.

65.
$$2 + 3x < 20$$
$$2 + 3x - 2 < 20 - 2 \quad \text{Adding } -2 \text{ to both sides}$$
$$3x < 18 \qquad \text{Simplifying}$$
$$x < 6 \qquad \text{Multiplying both sides by } \frac{1}{3}$$

The solution set is $\{x | x < 6\}$.

67.
$$4t - 5 \leq 23$$
$$4t - 5 + 5 \leq 23 + 5 \quad \text{Adding 5 to both sides}$$
$$4t \leq 28$$
$$\frac{1}{4} \cdot 4t \leq \frac{1}{4} \cdot 28 \quad \text{Multiplying both sides}$$
$$\text{by } \frac{1}{4}$$
$$t \leq 7$$

The solution set is $\{t | t \leq 7\}$.

69.
$$16 \leq 6 - 10y$$
$$16 - 6 \leq 6 - 10y - 6 \quad \text{Adding } -4 \text{ to both sides}$$
$$10 \leq -10y$$
$$-\frac{1}{10} \cdot 10 \geq -\frac{1}{10} \cdot (-10y) \quad \text{Multiplying by } -\frac{1}{10}$$
$$\underline{} \quad \text{The symbol has to be reversed.}$$
$$-1 \geq y$$

The solution set is $\{y | y \leq -1\}$.

71.
$$39 > 3 - 9x$$
$$39 - 3 > 3 - 9x - 3 \quad \text{Adding } -3$$
$$36 > -9x$$
$$-\frac{1}{9} \cdot 36 < -\frac{1}{9} \cdot (-9x) \quad \text{Multiplying by } -\frac{1}{9}$$
$$\underline{} \quad \text{The symbol has to be reversed.}$$
$$-4 < x$$

The solution set is $\{x | -4 < x\}$, or $\{x | x > -4\}$.

73.
$$5 - 6y > 25$$
$$-5 + 5 - 6y > -5 + 25$$
$$-6y > 20$$
$$-\frac{1}{6} \cdot (-6y) < -\frac{1}{6} \cdot 20$$
$$\underline{} \quad \text{The symbol has to be}$$
$$\text{reversed.}$$
$$y < -\frac{20}{6}$$
$$y < -\frac{10}{3}$$

The solution set is $\left\{y \Big| y < -\frac{10}{3}\right\}$.

75.
$$-3 < 8x + 7 - 7x$$
$$-3 < x + 7 \quad \text{Collecting like terms}$$
$$-3 - 7 < x + 7 - 7$$
$$-10 < x$$

The solution set is $\{x | -10 < x\}$, or $\{x | x > -10\}$.

77.
$$6 - 4y > 6 - 3y$$
$$6 - 4y + 4y > 6 - 3y + 4y \quad \text{Adding } 4y$$
$$6 > 6 + y$$
$$-6 + 6 > -6 + 6 + y \quad \text{Adding } -4$$
$$0 > y, \text{ or } y < 0$$

The solution set is $\{y | 0 > y\}$, or $\{y | y < 0\}$.

79.
$$7 - 9y \leq 4 - 7y$$
$$7 - 9y + 9y \leq 4 - 7y + 9y$$
$$7 \leq 4 + 2y$$
$$-4 + 7 \leq -4 + 4 + 2y$$
$$3 \leq 2y$$
$$\frac{3}{2} \leq y, \text{ or } y \geq \frac{3}{2}$$

The solution set is $\{y | y \geq \frac{3}{2}\}$.

81.
$$33 - 12x < 4x + 97$$
$$33 - 12x - 97 < 4x + 97 - 97$$
$$-64 - 12x < 4x$$
$$-64 - 12x + 12x < 4x + 12x$$
$$-64 < 16x$$
$$-4 < x$$

The solution set is $\{x | -4 < x\}$, or $\{x | x > -4\}$.

83.
$$2.1x + 43.2 > 1.2 - 8.4x$$
$$10(2.1x + 43.2) > 10(1.2 - 8.4x) \quad \text{Multiplying by}$$
$$10 \text{ to clear decimals}$$
$$21x + 432 > 12 - 84x$$
$$21x + 84x > 12 - 432 \quad \text{Adding } 84x \text{ and}$$
$$-432$$
$$105x > -420$$
$$x > -4 \quad \text{Multiplying by } \frac{1}{105}$$

The solution set is $\{x | x > -4\}$.

85.
$$1.7t + 8 - 1.62t < 0.4t - 0.32 + 8$$
$$0.08t + 8 < 0.4t + 7.68 \quad \text{Collecting like tern}$$
$$100(0.08t + 8) < 100(0.4t + 7.68) \quad \text{Multiplying by 100}$$
$$8t + 800 < 40t + 768$$
$$-8t - 768 + 8t + 800 < 40t + 768 - 8t - 768$$
$$32 < 32t$$
$$1 < t$$

The solution set is $\{t | t > 1\}$.

87.
$$\frac{x}{3} + 4 \leq 1$$
$$3\left(\frac{x}{3} + 4\right) \leq 3 \cdot 1 \quad \text{Multiplying by 3 to}$$
$$\text{to clear the fraction}$$
$$x + 12 \leq 3$$
$$x \leq -9$$

The solution set is $\{x | x \leq -9\}$.

89.
$$3 < 5 - \frac{t}{7}$$
$$-2 < -\frac{t}{7}$$
$$-7(-2) > -7\left(-\frac{t}{7}\right)$$
$$14 > t$$

The solution set is $\{t | t < 14\}$.

91.
$$4(2y - 3) \leq -44$$
$$8y - 12 \leq -44 \quad \text{Removing parentheses}$$
$$8y \leq -32 \quad \text{Adding 12}$$
$$y \leq -4 \quad \text{Multiplying by } \frac{1}{8}$$

The solution set is $\{y | y \leq -4\}$.

93.
$$8(2t + 1) > 4(7t + 7)$$
$$16t + 8 > 28t + 28$$
$$-12t + 8 > 28$$
$$-12t > 20$$
$$t < -\frac{5}{3} \quad \text{Multiplying by}$$
$$-\frac{1}{12} \text{ and}$$
$$\text{reversing the symbol}$$

The solution set is $\left\{t | t < -\frac{5}{3}\right\}$

95. $3(r-6)+2 < 4(r+2)-21$

$3r-18+2 < 4r+8-21$

$3r-16 < 4r-13$

$-16+13 < 4r-3r$

$-3 < r,$ or $r > -3$

The solution set is $\{r|r > -3\}$.

97. $\dfrac{4}{5}(3x-4) \le 20$

$\dfrac{5}{4} \cdot \dfrac{4}{5}(3x+4) \le \dfrac{5}{4} \cdot 20$

$3x+4 \le 25$

$3x \le 21$

$x \le 7$

The solution set is $\{x|x \le 7\}$.

99. $\dfrac{2}{3}\left(\dfrac{7}{8}-4x\right) - \dfrac{5}{8} < \dfrac{3}{8}$

$\dfrac{2}{3}\left(\dfrac{7}{8}-4x\right) < 1$ \qquad Adding $\frac{5}{8}$

$\dfrac{7}{12} - \dfrac{8}{3}x < 1$ \qquad Removing parentheses

$12\left(\dfrac{7}{12} - \dfrac{8}{3}x\right) < 12 \cdot 1$ \qquad Clearing fractions

$7 - 32x < 12$

$-32x < 5$

$x > -\dfrac{5}{32}$

The solution is $\left\{x\middle|x > -\dfrac{5}{32}\right\}$.

101. *Writing Exercise.* The inequalities $x > -3$ and $x \ge -2$ are not equivalent because they do not have the same solution set. For example, -2.5 is a solution of $x > -3$, but it is not a solution of $x \ge -2$.

103. $5x - 2(3-6x) = 5x - 6 + 12x = 17x - 6$

105. $x - 2[4y + 3(8-x) - 1]$

$= x - 2[4y + 24 - 3x - 1]$

$= x - 2[4y - 3x + 23]$

$= x - 8y + 6x - 46$

$= 7x - 8y - 46$

107. $3[5(2a-b)+1] - 5[4-(a-b)]$

$= 3[10a - 5b + 1] - 5[4 - a + b]$

$= 30a - 15b + 3 - 20 + 5a - 5b$

$= 35a - 20b - 17$

109. *Writing Exercise.* The graph of an inequality of the form $a \le x \le a$ consists of just one number, a.

111. $x < x + 1$

When any real number is increased by 1, the result is greater than the original number. Thus the solution set is $\{x|x$ is a real number$\}$.

113. $27 - 4[2(4x-3)+7] \ge 2[4-2(3-x)] - 3$

$27 - 4[8x - 6 + 7] \ge 2[4 - 6 + 2x] - 3$

$27 - 4[8x + 1] \ge 2[-2 + 2x] - 3$

$27 - 32x - 4 \ge -4 + 4x - 3$

$23 - 32x \ge -7 + 4x$

$23 + 7 = 4x + 32x$

$30 \ge 36x$

$\dfrac{5}{6} \ge x$

The solution set is $\left\{x\middle|x \le \dfrac{5}{6}\right\}$.

115. $-(x+5) \ge 4a - 5$

$-x - 5 \ge 4a - 5$

$-x \ge 4a - 5 + 5$

$-x \ge 4a$

$-1(-x) \le -1 \cdot 4a$

$x \le -4a$

The solution set is $\{x|x \le -4a\}$.

117. $y < ax + b$ \qquad Assume $a > 0$.

$y - b < ax$

$\dfrac{y-b}{a} < x$ \qquad Since $a > 0$, the inequality symbol stays the same.

The solution set is $\left\{x\middle|x > \dfrac{y-b}{a}\right\}$.

119. $|x| > -3$

Since absolute value is always nonnegative, the absolute value of any real number will be greater than -3. Thus, the solution set is $\{x|x$ is a real number$\}$.

Chapter 2 Connecting the Concepts

1. $x - 6 = 15$

$x = 21$ \qquad Adding 6 to both sides

The solution is 21.

3. $3x = -18$

$x = -6$ \qquad Dividing both sides by 3

The solution is -6.

5. $-3x > -18$

$x < 6$ \qquad Dividing both sides by -3 and reversing the direction of the inequality symbol

The solution is $\{x < 6\}$.

7. $7 - 3x = 8$

$-3x = 1$ \qquad Subtracting 7 from both sides

$x = \dfrac{-1}{3}$ \qquad Dividing both sides by -3

9. $3 - t \geq 19$

$\qquad -t \geq 16 \qquad$ Subtracting 3 from both sides

$\qquad t \leq -16 \qquad$ Dividing both sides by -1 and reversing the direction of the inequality symblol

The solution is $\{t | t \leq -16\}$.

11. $3 - 5a > a + 9$

$\qquad -5a > a + 6 \qquad$ Subtracting 3 from both sides

$\qquad -6a > 6 \qquad$ Subtracting a from both sides

$\qquad a < -1 \qquad$ Dividing both sides by -6 and reversing the direction of the inequality symbol

The solution is $\{a | a < -1\}$.

13. $\dfrac{2}{3}(x + 5) \geq -4$

$\qquad x + 5 \geq -6 \qquad$ Multiplying both sides by $\frac{3}{2}$

$\qquad x \geq -11 \qquad$ Subracting 5 from both sides

The solution is $\{x | x \geq -11\}$.

15. $0.5x - 2.7 = 3x + 7.9$

$\qquad 0.5x = 3x + 10.6 \qquad$ Adding 2.7 to both sides

$\qquad -2.5x = 10.6 \qquad$ Subtracting $3x$ from both sides

$\qquad x = -4.24 \qquad$ Dividing both sides by -2.5

The solution is -4.24.

17. $8 - \dfrac{y}{3} \leq 7$

$\qquad \dfrac{-y}{3} \leq -1 \qquad$ Subtracting 8 from both sides

$\qquad y \geq 3 \qquad$ Multiplying both sides by -3 and reversing the direction of the inequality symbol

The solution is $\{y | y \geq 3\}$.

19. $-15 > 7 - 5x$

$\qquad -22 > -5x \qquad$ Subtracting 7 from both sides

$\qquad \dfrac{22}{5} < x, \text{ or } x > \dfrac{22}{5} \qquad$ Dividing both sides by -5 and reversing the direction of the inequality symbol.

The solution is $\left\{ x | x > \dfrac{22}{5} \right\}$

Exercise Set 2.7

---

1. a is at least b can be translated as $b \leq a$.

3. a is at most b can be translated as $a \leq b$.

5. b is no more than a can be translated as $b \leq a$.

7. b is less than a can be translated as $b < a$.

9. Let n represent the number. Then we have $n < 10$.

11. Let t represent the temperature. Then we have $t \leq -3$.

13. Let d represent the number of years of driving experience. Then we have $d \geq 5$.

15. Let a represent the age of the Mayan altar. Then we have $a > 1200$.

17. Let h represent Tania's hourly wage. Then we have $12 < h < 15$.

19. Let w represent the wind speed. Then we have $w > 50$.

21. Let c represent the cost of a room at Pine Tree Bed and Breakfast. Then we have $c \leq 120$.

23. *Familiarize*. Let s = the length of the service call, in hours. The total charge is \$55 plus \$40 times the number of hours RJ's was there.

Translate.

\$55 charge	plus	hourly rate	times	number of hours	is greater than	\$150 .
↓	↓	↓	↓	↓	↓	↓
55	+	40	·	s	>	150

Carry out. We solve the inequality.

$$55 + 40s > 150$$
$$40s > 95$$
$$s > 2.375$$

Check. As a partial check, we show that the cost of a 2.375 hour service call is \$150.

$$\$55 + \$30(2.375) = \$55 + \$95 = \$150$$

State. The length of the service call was more than 2.375 hr.

25. *Familiarize*. Let q = Robbin's undergraduate grade point average. Unconditional acceptance is 500 plus 200 times the grade point average.

Translate.

GMAT score of 500	plus	200 times	grade point average	is at least	950 .
↓	↓	↓	↓	↓	↓
500	+	200 ·	q	≥	950

Carry out. We solve the inequality.

$$500 + 200q \geq 950$$
$$200q \geq 450$$
$$q \geq 2.25$$

Check. As a partial check we show that the acceptance score is 950.

$500 + 200(2.25) = 500 + 450 = 950.$

State.For unconditional acceptance, Robbin must have a gpa of at least 2.25.

27. **Familiarize.** The average of the five scores is their sum divided by the number of tests, 5. We let s represent Rod's score on the last test.

    Translate. The average of the five scores is given by

    $$\frac{73 + 75 + 89 + 91 + s}{5}.$$

    Since this average must be at least 85, this means that it must be greater than or equal to 85. Thus, we can translate the problem to the inequality

    $$\frac{73 + 75 + 89 + 91 + s}{5} \geq 85.$$

    Carry out. We first multiply by 5 to clear the fraction.

    $$5\left(\frac{73 + 75 + 89 + 91 + s}{5}\right) \geq 5 \cdot 85$$
    $$73 + 75 + 89 + 91 + s \geq 425$$
    $$328 + s \geq 425$$
    $$s \geq 97$$

    Check. As a partial check, we show that Rod can get a score of 97 on the fifth test and have an average of at least 85:

    $$\frac{73 + 75 + 89 + 91 + 97}{5} = \frac{425}{5} = 85.$$

    State. Scores of 97 and higher will earn Rod an average quiz grade of at least 85.

29. **Familiarize.** Let $c =$ the number of credits Millie must complete in the fourth quarter.

    Translate.

    $$\underbrace{\text{Average number of credits}}_{\substack{\downarrow \\ \dfrac{5 + 7 + 8 + c}{4}}} \underbrace{\text{is at least}}_{\substack{\downarrow \\ \geq}} \underbrace{7.}_{\substack{\downarrow \\ 7}}$$

    Carry out. We solve the inequality.

    $$\frac{5 + 7 + 8 + c}{4} \geq 7$$
    $$4\left(\frac{5 + 7 + 8 + c}{4}\right) \geq 4 \cdot 7$$
    $$5 + 7 + 8 + c \geq 28$$
    $$20 + c \geq 28$$
    $$c \geq 8$$

    Check. As a partial check, we show that Millie can complete 8 credits in the fourth quarter and average 7 credits per quarter.

    $$\frac{5 + 7 + 8 + 8}{4} = \frac{28}{4} = 7$$

    State. Millie must complete 8 credits or more in the fourth quarter.

31. **Familiarize**. The average number of plate appearances for 10 days is the sum of the number of appearance per

day divided by the number of days, 10. We let p represent the number of plate appearances on the tenth day.

Translate. The average for 10 days is given by

$$\frac{5 + 1 + 4 + 2 + 3 + 4 + 4 + 3 + 2 + p}{10}.$$

Since the average must be at least 3.1, this means that it must be greater than or equal to 3.1. Thus, we can translate the problem to the inequality

$$\frac{5 + 1 + 4 + 2 + 3 + 4 + 4 + 3 + 2 + p}{10} \geq 3.1.$$

Carry out. We first multiply by 10 to clear the fraction.

$$10\left(\frac{5 + 1 + 4 + 2 + 3 + 4 + 4 + 3 + 2 + p}{10}\right) \geq 10 \cdot 3.1$$
$$5 + 1 + 4 + 2 + 3 + 4 + 4 + 3 + 2 + p \geq 31$$
$$28 + p \geq 31$$
$$p \geq 3$$

Check. As a partial check, we show that 3 plate appearances in the 10th game will average 3.1

$$\frac{5 + 1 + 4 + 2 + 3 + 4 + 4 + 3 + 2 + 3}{10} = \frac{31}{10} = 3.1$$

State. On the tenth day, 3 or more plate appearances will give an average of at least 3.1.

33. **Familiarize**. We first make a drawing. We let b represent the length of the base. Then the lengths of the other sides are $b - 2$ and $b + 3$.

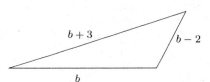

The perimeter is the sum of the lengths of the sides or $b + b - 2 + b + 3$, or $3b + 1$.

Translate.

$$\underbrace{\text{The perimeter}}_{\substack{\downarrow \\ 3b + 1}} \underbrace{\text{is greater than}}_{\substack{\downarrow \\ >}} \underbrace{\text{19 cm.}}_{\substack{\downarrow \\ 19}}$$

Carry out.

$$3b + 1 > 19$$
$$3b > 18$$
$$b > 6$$

Check. We check to see if the solution seems reasonable.

When $b = 5$, the perimeter is $3 \cdot 5 + 1$, or 16 cm.

When $b = 6$, the perimeter is $3 \cdot 6 + 1$, or 19 cm.

When $b = 7$, the perimeter is $3 \cdot 7 + 1$, or 22 cm.

From these calculations, it would appear that the solution is correct.

State. For lengths of the base greater than 6 cm the perimeter will be greater than 19 cm.

35. **Familiarize**. Let $d =$ the depth of the well, in feet. Then the cost on the pay-as-you-go plan is $\$500 + \$8d$. The cost of the guaranteed-water plan is \$4000. We want to find

the values of d for which the pay-as-you-go plan costs less than the guaranteed-water plan.

Translate.

Cost of pay-as-you-go plan is less than cost of guaranteed-water plan

$$500 + 8d < 4000$$

Carry out.

$$500 + 8d < 4000$$
$$8d < 3500$$
$$d < 437.5$$

Check. We check to see that the solution is reasonable.

When $d = 437$, $\$500 + \$8 \cdot 437 = \$3996 < \$4000$

When $d = 437.5$, $\$500 + \$8(437.5) = \$4000$

When $d = 438$, $\$500 + \$8(438) = \$4004 > \$4000$

From these calculations, it appears that the solution is correct.

State. It would save a customer money to use the pay-as-you-go plan for a well of less than 437.5 ft.

37. Familiarize. Let $v =$ the blue book value of the car. Since the car was repaired, we know that $\$8500$ does not exceed $0.8v$ or, in other words, $0.8v$ is at least $\$8500$.

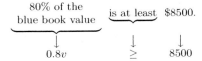

Translate.

80% of the blue book value is at least $8500.

$$0.8v \geq 8500$$

Carry out.

$$0.8v \geq 8500$$
$$v \geq \frac{8500}{0.8}$$
$$v \geq 10,625$$

Check. As a partial check, we show that 80% of $\$10,625$ is at least $\$8500$:

$$0.8(\$10,625) = \$8500$$

State. The blue book value of the car was at least $\$10,625$.

39. Familiarize. Let $L =$ the length of the package.

Translate.

Length and girth is less than 84 in

$$L + 29 < 84$$

Carry out.

$$L + 29 < 84$$
$$L < 55$$

Check. We check to see if the solution seems reasonable.

When $L = 60$ $60 + 29 = 89$ in.
When $L = 55$ $55 + 29 = 84$ in.
When $L = 50$ $50 + 29 = 79$ in.

From these calculations, it would appear that the solution is correct.

State. For lengths less than 55 in, the box is considered a "package."

41. Familiarize. We will use the formula $F = \frac{9}{5}C + 32$.

Translate.

Fahrenheit temperature is above $98.6°$.

$$F > 98.6$$

Substituting $\frac{9}{5}C + 32$ for F, we have

$$\frac{9}{5}C + 32 > 98.6.$$

Carry out. We solve the inequality.

$$\frac{9}{5}C + 32 > 98.6$$
$$\frac{9}{5}C > 66.6$$
$$C > \frac{333}{9}$$
$$C > 37$$

Check. We check to see if the solution seems reasonable.

When $C = 36$, $\frac{9}{5} \cdot 36 + 32 = 96.8$.

When $C = 37$, $\frac{9}{5} \cdot 37 + 32 = 98.6$.

When $C = 38$, $\frac{9}{5} \cdot 38 + 32 = 100.4$.

It would appear that the solution is correct, considering that rounding occurred.

State. The human body is feverish for Celsius temperatures greater than $37°$.

43. Familiarize. Let $h =$ the height of the triangle, in ft. Recall that the formula for the area of a triangle with base b and height h is $A = \frac{1}{2}bh$.

Translate.

Area less than or equal to 12 ft².

$$\frac{1}{2}(8)h \leq 12$$

Carry out. We solve the inequality.

$$\frac{1}{2}(8)h \leq 12$$
$$4h \leq 12$$
$$h \leq 3$$

Check. As a partial check, we show that a length of 3 ft will result in an area of 12 ft².

$$\frac{1}{2}(8)(3) = 12$$

State. The height should be no more than 3 ft.

45. Familiarize. Let r = the amount of fat in a serving of the regular peanut butter, in grams. If reduced fat peanut butter has at least 25% less fat than regular peanut butter, then it has at most 75% as much fat as the regular peanut butter.

Translate.

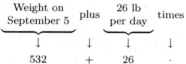

$$
\begin{array}{ccccc}
12 & \leq & 0.75 & \cdot & r
\end{array}
$$

Carry out.

$$12 \leq 0.75r$$
$$16 \leq r$$

Check. As a partial check, we show that 12 g of fat does not exceed 75% of 16 g of fat:

$$0.75(16) = 12$$

State. Regular peanut butter contains at least 16 g of fat per serving.

47. Familiarize. Let d = the number of days after September 5.

Translate.

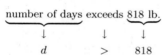

$$
\begin{array}{ccc}
d & > & 818
\end{array}
$$

Carry out. We solve the inequality.

$$532 + 26d > 818$$
$$26d > 286$$
$$d > 11$$

Check. As a partial check, we can show that the weight of the pumpkin is 818 lb 11 days after September 5.

$$532 + 26 \cdot 11 = 532 + 286 = 818 \text{ lb}$$

State. The pumpkin's weight will exceed 818 lb more than 11 days after September 5, or on dates after September 16.

49. Familiarize. Let n = the number of text messages. The total cost is the monthly fee of \$39.95 plus taxes of \$6.65 plus .10 times the number of text messages, or $.10n$.

Translate.

Monthly fee	plus	taxes	plus	text messages.	cannot exceed	\$60
↓	↓	↓	↓	↓	↓	↓
39.95	+	6.65	+	.10n	≤	60

Carry out. We solve the inequality.

$$39.95 + 6.65 + .10n \leq 60$$
$$46.60 + .10n \leq 60$$
$$.10n \leq 13.4$$
$$n \leq 134$$

Check. As a partial check, if the number of text messages is 134, the budget of \$60 will not be exceeded.

State. Liam can send or receive 134 text messages and stay within his budget.

51. Familiarize. We will use the formula $R = -0.0065t + 4.3259$.

Translate.

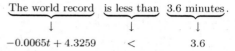

$$
\begin{array}{ccc}
-0.0065t + 4.3259 & < & 3.6
\end{array}
$$

Carry out. We solve the inequality.

$$-0.0065t + 4.3259 < 3.6$$
$$-0.0065t < -0.7259$$
$$t > 111.68$$

Check. As a partial check, we can show that the record is more than 3.6 min 111 yr after 1900 and is less than 3.6 min 112 yr after 1900.

For $t = 111$, $R = -0.0065(111) + 4.3259 = 3.7709$.

For $t = 112$, $R = -0.0065(112) + 4.3259 = 3.5979$.

State. The world record in the mile run is less than 3.6 min more than 112 yr after 1900, or in years after 2012.

53. Familiarize. We will use the equation $y = 0.06x + 0.50$.

Translate.

The cost	is at most	\$14.
↓	↓	↓
0.06x + 0.50	≤	14

Carry out. We solve the inequality.

$$0.06x + 0.50 \leq 14$$
$$0.06x \leq 13.50$$
$$x \leq 225$$

Check. As a partial check, we show that the cost for driving 225 mi is \$14.

$$0.06(225) + 0.50 = 14$$

State. The cost will be at most \$14 for mileages less than or equal to 225 mi.

55. Writing Exercise. Answers may vary. Fran is more than 3 years older than Todd.

57. $-2 + (-5) - 7 = -2 + (-5) + (-7) = -14$

59. $3 \cdot (-10) \cdot (-1) \cdot (-2) = (-30) \cdot (-1) \cdot (-2)$
$= (30) \cdot (-2) = -60$

61. $(3 - 7) - (4 - 8) = (-4) - (-4) = (-4) + (4) = 0$

63. $\dfrac{-2 - (-6)}{8 - 10} = \dfrac{-2 + 6}{8 + (-10)} = \dfrac{4}{-2} = -2$

65. Writing Exercise. Answers may vary.

A boat has a capacity of 2800 lb. How many passengers can go on the boat if each passenger is considered to weigh 150 lb.

67. *Familiarize.* We use the formula $F = \frac{9}{5}C + 32$.

Translate. We are interested in temperatures such that $5° < F < 15°$. Substituting for F, we have:

$$5 < \frac{9}{5}C + 32 < 15$$

Solve.

$$5 < \frac{9}{5}C + 32 < 15$$

$$5 \cdot 5 < 5\left(\frac{9}{5}C + 32\right) < 5 \cdot 15$$

$$25 < 9C + 160 < 75$$

$$-135 < 9C < -85$$

$$-15 < C < -9\frac{4}{9}$$

Check. The check is left to the student.

State. Green ski wax works best for temperatures between $-15°C$ and $-9\frac{4}{9}°C$.

69. Since $8^2 = 64$, the length of a side must be less than or equal to 8 cm (and greater than 0 cm, of course). We can also use the five-step problem-solving procedure.

Familiarize. Let s represent the length of a side of the square. The area s is the square of the length of a side, or s^2.

Translate.

$$\underbrace{\text{The area}}_{s^2} \underbrace{\text{is no more than}}_{\leq} \underbrace{64 \text{ cm}^2.}_{64}$$

Carry out.

$$s^2 \leq 64$$

$$s^2 - 64 \leq 0$$

$$(s + 8)(s - 8) \leq 0$$

We know that $(s+8)(s-8) = 0$ for $s = -8$ or $s = 8$. Now $(s+8)(s-8) < 0$ when the two factors have opposite signs. That is:

$s+8>0$ *and* $s-8<0$ *or* $s+8<0$ *and* $s-8>0$

$s>-8$ *and* $s<8$ *or* $s<-8$ *and* $s>8$

This can be expressed   This is not possible.

as $-8 < s < 8$.

Then $(s + 8)(s - 8) \leq 0$ for $-8 \leq s \leq 8$.

Check. Since the length of a side cannot be negative we only consider positive values of s, or $0 < s \leq 8$. We check to see if this solution seems reasonable.

When $s = 7$, the area is 7^2, or 49 cm^2.

When $s = 8$, the area is 8^2, or 64 cm^2.

When $s = 9$, the area is 9^2, or 81 cm^2.

From these calculations, it appears that the solution is correct.

State. Sides of length 8 cm or less will allow an area of no more than 64 cm^2. (Of course, the length of a side must be greater than 0 also.)

71. *Familiarize.* Let $f =$ the fat content of a serving of regular tortilla chips, in grams. A product that contains 60% less fat than another product has 40% of the fat content of that product. If Reduced Fat Tortilla Pops cannot be labeled lowfat, then they contain at least 3 g of fat.

Translate.

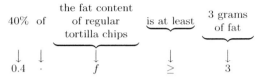

Carry out.

$$0.4f \geq 3$$

$$f \geq 7.5$$

Check. As a partial check, we show that 40% of 7.5 g is not less than 3 g.

$$0.4(7.5) = 3$$

State. A serving of regular tortilla chips contains at least 7.5 g of fat.

73. *Familiarize.* Let $p =$ the price of Neoma's tenth book. If the average price of each of the first 9 books is $12, then the total price of the 9 books is $9 \cdot \$12$, or $108. The average price of the first 10 books will be $\frac{\$108 + p}{10}$.

Translate.

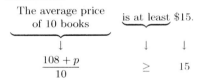

Carry out. We solve the inequality.

$$\frac{108 + p}{10} \geq 15$$

$$108 + p \geq 150$$

$$p \geq 42$$

Check. As a partial check, we show that the average price of the 10 books is $15 when the price of the tenth book is $42.

$$\frac{\$108 + \$42}{10} = \frac{\$150}{10} = \$15$$

State. Neoma's tenth book should cost at least $42 if she wants to select a $15 book for her free book.

75. Let $b =$ the total purchases of bestsellers, $h =$ the total purchases of hardcovers, $p =$ the total purchases of other items at Barnes & Noble.

(1) Solving $0.40b > 25$, we get $62.50

(2) Solving $0.20h > 25$, we get $125

(3) Solving $0.10p > 25$, we get $250

Thus when a customer's bestseller purchases are more than $62.50, or hardcover purchases are more than $125, or other purchases are more than $250, the customer saves money by purchasing the card.

Chapter Review Exercises

1. True

3. True

5. True

7. True

9. $x + 9 = -16$
$x + 9 - 9 = -16 - 9$ Adding -9
$x = -25$ Simplifying

The solution is -25.

11. $-\frac{x}{5} = 13$
$-5\left(-\frac{x}{5}\right) = -5\,(13)$ Multiplying by -5
$x = -65$ Simplifying

The solution is -65.

13. $\frac{2}{5}t = -8$

$\frac{5}{2} \cdot \frac{2}{5}t = \frac{5}{2}(-8)$ Multiplying by $\frac{5}{2}$

$t = -20$

The solution is -20.

15. $-\frac{2}{3} + x = -\frac{1}{6}$
$6\left(-\frac{2}{3} + x\right) = 6\left(-\frac{1}{6}\right)$ Multiplying by 6
$-4 + 6x = -1$ Simplifying
$-4 + 6x + 4 = -1 + 4$ Adding 4
$6x = 3$ Simplifying
$x = \frac{1}{2}$ Multiplying by $\frac{1}{6}$

The solution is $\frac{1}{2}$.

17. $5 - x = 13$
$5 - x - 5 = 13 - 5$ Adding -5
$-x = 8$ Simplifying
$x = -8$ Multiplying by -1

The solution is -8.

19. $7x - 6 = 25x$
$7x - 6 - 7x = 25x - 7x$ Adding $-7x$
$-6 = 18x$ Simplifying
$-\frac{1}{3} = x$ Multiplying by $\frac{1}{18}$

The solution is $-\frac{1}{3}$.

21. $14y = 23y - 17 - 10$
$14y = 23y - 27$ Simplifying
$14y - 14y = 23y - 27 - 14y$ Adding $-14y$
$0 = 9y - 27$ Simplifying
$0 + 27 = 9y - 27 + 27$ Adding 27
$27 = 9y$ Simplifying
$3 = y$ Multiplying by $\frac{1}{2}$

The solution is 3.

23. $\frac{1}{4}x - \frac{1}{8}x = 3 - \frac{1}{16}x$
$16\left(\frac{1}{4}x - \frac{1}{8}x\right) = 16\left(3 - \frac{1}{16}x\right)$ Multiplying by 16
$4x - 2x = 48 - x$ Distributive Law
$2x = 48 - x$ Simplifying
$2x + x = 48 - x + x$ Adding x
$3x = 48$ Simplifying
$x = 16$ Multiplying by $\frac{1}{3}$

The solution is 16.

25. $4\,(5x - 7) = -56$
$20x - 28 = -56$ Distributive Law
$20x - 28 + 28 = -56 + 28$ Adding 28
$20x = -28$ Simplifying
$x = -\frac{28}{20}$ Multiplying by $\frac{1}{20}$
$x = -\frac{7}{5}$ Simplifying

The solution is $-\frac{7}{5}$.

27. $-5x + 3\,(x + 8) = 16$
$-5x + 3x + 24 = 16$ Distributive Law
$-2x + 24 = 16$ Simplifying
$-2x + 24 - 24 = 16 - 24$ Adding -24
$-2x = -8$ Simplifying
$x = 4$ Multiplying by $-\frac{1}{2}$

The solution is 4.

29. $V = \frac{1}{3}Bh$
$3 \cdot V = 3\left(\frac{1}{3}Bh\right)$ Multiplying by 3
$3V = Bh$ Simplifying
$\frac{1}{h}\,(3V) = \frac{1}{h}\,(Bh)$ Multiplying by $\frac{1}{h}$
$\frac{3V}{h} = B$ Simplifying

31. $tx = ax + b$
$tx - ax = ax + b - ax$ Adding $-ax$
$tx - ax = b$ Simplifying
$x(t - a) = b$ Factoring x
$x = \dfrac{b}{t - a}$ Multiplying by $\dfrac{1}{t - a}$

33. $\frac{11}{25} = \frac{4}{4} \cdot \frac{11}{25} = \frac{44}{100} = 0.44$ $0.44.$
$\quad\quad\quad\quad\quad\quad\quad\quad\quad\quad\quad\quad\quad\quad\quad\quad\quad\quad\;\; 44\%$

First, move the decimal point two places to the right; then write a % symbol: The answer is 44%.

35. Translate.

49   is   35%   of   What number?

↓   ↓   ↓   ↓   ↓
49   =   0.35   · y

We solve the equation and then convert to percent notation.

$49 = 0.35y$

$\dfrac{49}{0.35} = y$

$140 = y$

The answer is 140.

37. $x \le -5$

We substitute -7 for x giving $-7 \le -5$, which is a true statement since -7 is to the left of -5 on the number line, so -7 is a solution of the inequality $x \le -5$.

39.
$$5x - 6 < 2x + 3$$
$$5x - 6 + 6 < 2x + 3 + 6 \qquad \text{Adding 6}$$
$$5x < 2x + 9 \qquad \text{Simplifying}$$
$$5x - 2x < 2x + 9 - 2x \qquad \text{Adding } -2x$$
$$3x < 9 \qquad \text{Simplifying}$$
$$x < 3 \qquad \text{Multiplying by } \tfrac{1}{3}$$

The solution set is $\{x | x < 3\}$. The graph is as follows:
$$5x - 6 < 2x + 3$$

41. $t > 0$

The solution set is $\{t | t > 0\}$. The graph is as follows:
$$t > 0$$

43.
$$9x \ge 63$$
$$\tfrac{1}{9}(9x) \ge \tfrac{1}{9} \cdot 63 \qquad \text{Multiplying by } \tfrac{1}{9}$$
$$x \ge 7 \qquad \text{Simplifying}$$

The solution set is $\{x | x \ge 7\}$.

45.
$$7 - 3y \ge 27 + 2y$$
$$7 - 3y - 7 \ge 27 + 2y - 7 \qquad \text{Adding } -7$$
$$-3y \ge 20 + 2y \qquad \text{Simplifying}$$
$$-3y - 2y \ge 20 + 2y - 2y \qquad \text{Adding } -2y$$
$$-5y \ge 20 \qquad \text{Simplifying}$$
$$y \le -4 \qquad \text{Multiplying by } -\tfrac{1}{5}$$
$$\qquad \text{and reversing}$$
$$\qquad \text{the inequality}$$
$$\qquad \text{symbol}$$

The solution set is $\{y | y \le -4\}$.

47.
$$-4y < 28$$
$$-\tfrac{1}{4}(-4y) > -\tfrac{1}{4} \cdot 28 \qquad \text{Multiplying by } -\tfrac{1}{4} \text{ and re-}$$
$$\qquad \text{versing the inequality sym-}$$
$$\qquad \text{bol}$$
$$y > -7 \qquad \text{Simplifying}$$

The solution set is $\{y | y > -7\}$.

49.
$$4 - 8x < 13 + 3x$$
$$4 - 8x - 4 < 13 + 3x - 4 \qquad \text{Adding } -4$$
$$-8x < 9 + 3x \qquad \text{Simplifying}$$
$$-8x - 3x < 9 + 3x - 3x \qquad \text{Adding } -3x$$
$$-11x < 9 \qquad \text{Simplifying}$$
$$-\tfrac{1}{11}(-11x) > -\tfrac{1}{11} \cdot 9 \qquad \text{Multiplying by } -\tfrac{1}{11}$$
$$x > -\tfrac{9}{11} \qquad \text{Simplifying}$$

The solution set is $\left\{ x \left| x > -\tfrac{9}{11} \right. \right\}$.

51.
$$7 \le 1 - \tfrac{3}{4}x$$
$$7 - 1 \le 1 - \tfrac{3}{4}x - 1 \qquad \text{Adding - 1}$$
$$6 \le -\tfrac{3}{4}x \qquad \text{Simplifying}$$
$$-\tfrac{4}{3} \cdot 6 \ge -\tfrac{4}{3} \left(-\tfrac{3}{4}x \right) \qquad \text{Multiplying by } -\tfrac{4}{3}$$
$$-8 \ge x \qquad \text{Simplifying}$$

The solution set is $\{x | -8 \ge x\}$, or $\{x | x \le -8\}$.

53. *Familiarize.* Let x = the length of the first piece, in ft. Since the second piece is 2 ft longer than the first piece, it must be $x + 2$ ft.

Translate.

$$\underbrace{\text{The sum of the lengths of the two pieces}}_{x + (x + 2)} \underset{=}{\text{ is }} \underset{18}{\text{18 ft.}}$$

Carry out. We solve the equation.
$$x + (x + 2) = 18$$
$$2x + 2 = 18$$
$$2x = 16$$
$$x = 8$$

Check. If the first piece is 8 ft long, then the second piece must be 8+2, or 10 ft long. The sum of the lengths of the two pieces is 8 ft+10 ft, or 18 ft. The answer checks.

State. The lengths of the two pieces are 8 ft and 10 ft.

55. *Familiarize.* Let x = the first odd integer and let $x + 2$ = the next consecutive odd integer.

Translate.

$$\underbrace{\text{The sum of the two consecutive odd integers}}_{x + (x + 2)} \underset{=}{\text{ is 116}} \underset{116}{}$$

Carry out. We solve the equation.
$$x + (x + 2) = 116$$
$$2x + 2 = 116$$
$$2x = 114$$
$$x = 57$$

Check. If the first odd integer is 57, then the next consecutive odd integer would be 57+2, or 59. The sum of these two integers is 57+59, or 116. This result checks.

State. The integers are 57 and 59.

57. *Familiarize.* Let x = the regular price of the picnic table. Since the picnic table was reduced by 25%, it actually sold for 75% of its original price.

Translate.

$$\underset{0.75}{75\%} \underset{\cdot}{\text{ of }} \underbrace{\text{the original price}}_{x} \underset{=}{\text{ is \$120?}} \underset{120}{}$$

Carry out. We solve the equation.
$$0.75x = 120$$
$$x = \frac{120}{0.75}$$
$$x = 160$$

Check. If the original price was \$160 with a 25% discount, then the purchaser would have paid 75% of \$160, or $0.75 \cdot$ \$160, or \$120. This result checks.

State. The original price was \$160.

59. *Familiarize.* Let $x =$ the measure of the first angle. The measure of the second angle is $x+50°$, and the measure of the third angle is $2x-10°$. The sum of the measures of the angles of a triangle is $180°$.

Translate.

$$\underbrace{\text{The sum of the measures of the angles}} \quad \text{is } 180°$$

$$\downarrow \qquad\qquad\qquad \downarrow \quad \downarrow$$

$$x + (x + 50) + (2x - 10) \quad = \quad 180$$

Carry out. We solve the equation.

$$
\begin{aligned}
x + (x + 50) + (2x - 10) &= 180 \\
4x + 40 &= 180 \\
4x &= 140 \\
x &= 35
\end{aligned}
$$

Check. If the measure of the first angle is $35°$, then the measure of the second angle is $35°+50°$, or $85°$, and the measure of the third angle is $2 \cdot 35° - 10°$, or $60°$. The sum of the measures of the first, second, and third angles is $35°+85°+60°$, or $180°$. These results check.

State. The measures of the angles are $35°$, $85°$, and $60°$.

61. *Familiarize.* Let $n =$ the number of copies. The total cost is the setuup fee of $6 plus $4 per copy, or $4n$.

Translate.

$$\underset{6}{\underbrace{\overset{\text{Set up}}{\underset{\text{fee}}{}}}} \;\; \underset{+}{\overset{\text{plus}}{}} \;\; \underset{4n}{\underbrace{\overset{\text{cost per}}{\underset{\text{copy}}{}}}} \;\; \underset{\leq}{\overset{\text{cannot}}{\underset{\text{exceed}}{}}} \;\; \underset{65}{\$65}$$

Carry out. We solve the inequality.

$$
\begin{aligned}
6 + 4n &\leq 65 \\
4n &\leq 59 \\
n &\leq \frac{59}{4} \\
n &\leq 14.75
\end{aligned}
$$

Check. As a partial check, if the number of copies is 14, the total cost $6 + \$4 \cdot 14$, ir $62 does not exceed the budget of $65.

State. Myra can make 14 or fewer copies.

63. *Writing Exercise.* The solutions of an equation can usually each be checked. The solutions of an inequality are normally too numerous to check. Checking a few numbers from the solution set found cannot guarantee that the answer is correct, although if any number does not check, the answer found is incorrect.

65. *Familiarize.* Let $x =$ the length of the Nile River, in mi. Let $x+65$ represent the length of the Amazon River, in mi.

Translate.

$$\underbrace{\overset{\text{The combined length}}{\underset{\text{of both rivers}}{}}} \quad \text{is 8385 mi}$$

$$\downarrow \qquad\qquad\qquad \downarrow \quad \downarrow$$

$$x + (x + 65) \qquad = \quad 8385$$

Carry out. We solve the equation.

$$
\begin{aligned}
x + (x + 65) &= 8385 \\
2x + 65 &= 8385 \\
2x &= 8320 \\
x &= 4160
\end{aligned}
$$

Check. If the Nile River is 4160 mi long, then the Amazon River is 4160 mi+65 mi, or 4225 mi. The combined length of both rivers is then 4160 mi+4225 mi, or 8385 mi. These results check..

State. The Amazon River is 4225 mi long, and the Nile River is 4160 mi long.

67.
$$
\begin{aligned}
2\,|n| + 4 &= 50 \\
2\,|n| &= 46 \\
|n| &= 23
\end{aligned}
$$

The distance from some number n and the origin is 23 units. The solution is $n = 23$, or $n = -23$.

69.
$$
\begin{aligned}
y &= 2a - ab + 3 \\
y &= a\,(2 - b) + 3 \\
y - 3 &= a\,(2 - b) \\
\frac{y-3}{2-b} &= a
\end{aligned}
$$

The solution is $a = \frac{y-3}{2-b}$.

Chapter 2 Test

1.
$$
\begin{array}{ll}
t + 7 = 16 & \\
t + 7 - 7 = 16 - 7 & \text{Adding } -7 \\
t = 9 & \text{Simplifying}
\end{array}
$$
The solution is 9.

3.
$$
\begin{array}{ll}
6x = -18 & \\
\frac{1}{6}(6x) = \frac{1}{6}(-18) & \text{Multiplying by } \frac{1}{6} \\
x = -3 & \text{Simplifying}
\end{array}
$$
The solution is -3.

5.
$$
\begin{array}{ll}
3t + 7 = 2t - 5 & \\
3t + 7 - 7 = 2t - 5 - 7 & \text{Adding -7} \\
3t = 2t - 12 & \text{Simplifying} \\
3t - 2t = 2t - 12 - 2t & \text{Adding } -2 \\
t = -12 & \text{Simplifying}
\end{array}
$$
The solution is -12.

7.
$$8 - y = 16$$
$$8 - y - 8 = 16 - 8 \qquad \text{Adding } -8$$
$$-y = 8 \qquad \text{Simplifying}$$
$$y = -8 \qquad \text{Multiply by } -1$$
The solution is -8.

9.
$$4(x + 2) = 36$$
$$4x + 8 = 36 \qquad \text{Distributive Law}$$
$$4x + 8 - 8 = 36 - 8 \qquad \text{Adding } -8$$
$$4x = 28 \qquad \text{Simplifying}$$
$$\tfrac{1}{4}(4x) = \tfrac{1}{4}(28) \qquad \text{Multiplying by } \tfrac{1}{4}$$
$$x = 7 \qquad \text{Simplifying}$$
The solution is 7.

11.
$$\tfrac{5}{6}(3x + 1) = 20$$
$$\tfrac{6}{5}[\tfrac{5}{6}(3x + 1)] = \tfrac{6}{5} \cdot 20 \qquad \text{Multiplying by } \tfrac{6}{5}$$
$$3x + 1 = 24 \qquad \text{Simplifying}$$
$$3x + 1 - 1 = 24 - 1 \qquad \text{Adding } -1$$
$$3x = 23 \qquad \text{Simplifying}$$
$$\tfrac{1}{3}(3x) = \tfrac{1}{3}(23) \qquad \text{Multiplying by } \tfrac{1}{3}$$
$$x = \tfrac{23}{3} \qquad \text{Simplifying}$$
The solution is $\tfrac{23}{3}$.

13.
$$14x + 9 > 13x - 4$$
$$14x + 9 - 9 > 13x - 4 - 9 \qquad \text{Adding } -9$$
$$14x > 13x - 13 \qquad \text{Simplifying}$$
$$14x - 13x > 13x - 13 - 13x \qquad \text{Adding } -13x$$
$$x > -13 \qquad \text{Simplifying}$$
The solution set is $\{x | x > -13\}$.

15.
$$4y \leq -30$$
$$\tfrac{1}{4}(4y) \leq \tfrac{1}{4}(-30) \qquad \text{Multiplying by } \tfrac{1}{4}$$
$$y \leq -\tfrac{15}{2} \qquad \text{Simplifying}$$
The solution set is $\{y | y \leq -\tfrac{15}{2}\}$.

17.
$$3 - 5x > 38$$
$$3 - 5x - 3 > 38 - 3 \qquad \text{Adding -3}$$
$$-5x > 35 \qquad \text{Simplifying}$$
$$-\tfrac{1}{5}(-5x) < -\tfrac{1}{5}(35) \qquad \text{Multiplying by } -\tfrac{1}{5} \text{ and reversing the inequality symbol}$$
$$x < -7 \qquad \text{Simplifying}$$
The solution set is $\{x | x < -7\}$.

19.
$$5 - 9x \geq 19 + 5x$$
$$5 - 9x - 5 \geq 19 + 5x - 5 \qquad \text{Adding } -5$$
$$-9x \geq 14 + 5x \qquad \text{Simplifying}$$
$$-9x - 5x \geq 14 + 5x - 5x \qquad \text{Adding } -5x$$
$$-14x \geq 14 \qquad \text{Simplifying}$$
$$-\tfrac{1}{14}(-14x) \leq -\tfrac{1}{14}(14) \qquad \text{Multiplying by } -\tfrac{1}{14} \text{ and reversing the inequality symbol}$$
$$x \leq -1 \qquad \text{Simplifying}$$
The solution set is $\{x | x \leq -1\}$.

21.
$$w = \tfrac{P+l}{2}$$
$$2 \cdot w = 2\left(\tfrac{P+l}{2}\right) \qquad \text{Multiplying by 2}$$
$$2w = P + l \qquad \text{Simplifying}$$
$$2w - P = P + l - P \qquad \text{Adding } -P$$
$$2w - P = l \qquad \text{Simplifying}$$
The solution is $l = 2w - P$.

23. 0.003 First move the decimal point two places to the right; then write a % symbol. The answer is 0.3%.

25. *Translate.*

$$\underbrace{\text{What percent}}_{y} \quad \text{of} \quad 75 \quad \text{is} \quad 33?$$
$$y \quad \cdot \quad 75 \quad = \quad 33$$

We solve the equation and then convert to percent notation.
$$y \cdot 75 = 33$$
$$y = \frac{33}{75}$$
$$y = 0.44 = 44\%$$
The solution is 44%.

27.
$$-2 \leq x \leq 2$$
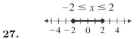

29. *Familiarize.* Let $x =$ the distance from Springer Mountain in miles. then $3 \times$ mi is the distance from Mt. Katahdin.

Translate.

Southern Distance	and	northern distance	is	Appalachian trail.
↓	↓	↓	↓	↓
x	$+$	$3x$	$\cdot$	2100

Carry out. We solve the equation.
$$x + 3x = 2100$$
$$4x = 2100$$
$$x = 525$$
$$3x = 1575$$
Check.
$$525 + 1575 = 2100.$$

State. The distance is 525 mi from Springer Moutain and 1575 mi from Mt. Katahdin.

31. *Familiarize.* Let $x =$ the electric bill before the temperature of the water heater was lowered. If the bill dropped by 7%, then the Kellys paid 93% of their original bill.
Translate.

93%	of	the original bill	is	$60.45.
↓	↓	↓	↓	↓
0.93	$\cdot$	x	$=$	60.45

Carry out. We solve the equation.
$$0.93x = 60.45$$
$$x = \frac{60.45}{0.93}$$
$$x = 65$$
Check. If the original bill was $65, and the bill was reduced by 7%, or $0.07 \cdot \$65$, or \$4.55, the new bill would be $65−\$4.55$, or \$60.45. This result checks. *State.* The original bill was $65.

33.

$$c = \frac{2cd}{a-d}$$

$(a-d)c = (a-d)(\frac{2cd}{a-d})$ Multiplying   by $a-d$

$$ac - dc = 2cd$$ Simplifying

$$ac - dc + dc = 2cd + dc$$ Adding dc

$$ac = 3cd$$ Simplifying

$$\frac{1}{3c}(ac) = \frac{1}{3c}(3cd)$$ Multiplying by $\frac{1}{3c}$

$$\frac{a}{3} = d$$ Simplifying

The solution is $d = \frac{a}{3}$.

35. Let $h =$ the number of hours of sun each day. Then we have $4 \le h \le 6$.

Chapter 3

Introduction to Graphing

1. The x-values extend from -9 to 4 and the y-values range from -1 to 5, so (a) is the best choice.

3. The x-values extend from -2 to 4 and the y-values range from -9 to 1, so (b) is the best choice.

5. We go to the top of the bar that is above the body weight 100 lb. Then we move horizontally from the top of the bar to the vertical scale listing numbers of drinks. It appears that consuming approximately 2 drinks in one hour will give a 100 lb person a blood-alcohol level of 0.08%.

7. For 3 drinks in one hour, we use the horizontal line at 3. For persons weighing 140 lb or less, their blood-alcohol level is 0.08% or more For persons weighing more than 140 lbs. their blood-alcohol level is under 0.08%. Therefore the person weighs more than 140 lbs.

9. Familiarize. From the pie chart we see that 51% of student aid is Federal loans. The average aid is the total aid distributed of $134.8 billion divided by the total number of full-time students, 13,334,170, or

$$\frac{\$134.8\text{billion}}{13,334,170}$$

Let f = the average federal loan per full-time student.

Translate. We reword and translate the problem.

What is 51% of $\frac{\$134.8}{13.334.170}$ billion?

$$f \;=\; 51\% \;\cdot\; \frac{134.8}{13,334,170}$$

Carry out. We solve the equation.

$$f = 0.51 \cdot \frac{134.8}{13,334,170} = \$5156$$

Check. We repeat the calculations. The answer checks.

State. The average federal loan per full-time equivalent student is $5156.

11. Familiarize. From the pie chart we see that 2% of the total aid is Federal campus-based, or 2% of $134.8 billion = $2.696 billion. Let t = the amount given to students in two-year public institutions.

Translate. We reword the problem.

What is 8.6% of 2.696?

$$t \;=\; 8.6\% \;\cdot\; 2.696$$

Carry out.

$$t = 0.086 \cdot 2.696 = 0.231856 \text{ billion}$$

Check. We repeat the calculations.

State. In 2006 the campus-based aid given to students at two-year institutions is $231,856,000.

13. Familiarize. From the pie chart we see that 11.9% of solid waste is plastic. We let p = the amount of plastic, in millions of tons, in the waste generated in 2005.

Translate. We reword the problem.

What is 11.9% of 245?

$$x \;=\; 11.9\% \;\cdot\; 245$$

Carry out.

$$p = 0.119 \cdot 245 \approx 29.2$$

Check. We can repeat the calculations.

State. In 2005, about 29.2 million tons of waste was plastic.

15. Familiarize. From the pie chart we see that 5.2% of solid waste is glass. From Exercise 13 we know that Americans generated 245 million tons of waste in 2005. Then the amount of this that is glass is

$$0.052(245), \text{ or } 12.74 \text{ million tons}$$

We let g = the amount of glass, in millions of tons, that Americans recycled in 2005.

Translate. We reword the problem.

What is 25.3% of 12.74 million tons?

$$g \;=\; 25.3\% \;\cdot\; 12.74$$

Carry out.

$$x = 0.253(12.74) \approx 3.2$$

Check. The result checks.

State. Americans recycled about 3.2 million tons of glass in 2005.

17. We locate 2002 on the horizontal axis and then move up to the line. From there we move left to the vertical axis and read the value of home videos, in billions. We estimate that about $12 billion was spent on home videos in 2002.

19. We locate 10.5 on the vertical axis and move right to the line. From there we move down to the horizontal scale and read the year. We see that approximately $10.5 billion was spent on home videos in 2001.

21. Starting at the origin:

$(1,2)$ is 1 unit right and 2 units up;

$(-2,3)$ is 2 units left and 3 units up;

$(4,-1)$ is 4 units right and 1 unit down;

$(-5,-3)$ is 5 units left and 3 units down;

$(4,0)$ is 4 units right and 0 units up or down;

$(0, -2)$ is 0 units right or left and 2 units down.

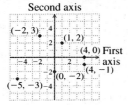

23. Starting at the origin:

$(4,4)$ is 4 units right and 4 units up;

$(-2, 4)$ is 2 units left and 4 units up;

$(5, -3)$ is 5 units right and 3 units down;

$(-5, -5)$ is 5 units left and 5 units down;

$(0,4)$ is 0 units right or left and 4 units up;

$(0, -4)$ is 0 units right or left and 4 units down;

$(0,0)$ is 0 units right and 0 units up or down;

$(-4, 0)$ is 4 units left and 0 units up or down.

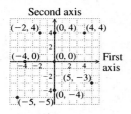

25. We plot the points $(2000, 12)$, $(2001, 34)$, $(2002, 931)$, $(2003, 1221)$, $(2004, 2862)$, $(2005, 7253)$ and $(2006, 8000)$ and connect adjacent points with line segments.

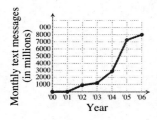

27.

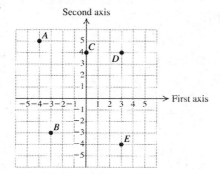

Point A is 4 units left and 5 units up. The coordinates of A are $(-4, 5)$.

Point B is 3 units left and 3 units down. The coordinates of B are $(-3, -3)$.

Point C is 0 units right or left and 4 units up. The coordinates of C are $(0,4)$.

Point D is 3 units right and 4 units up. The coordinates of D are $(3,4)$.

Point E is 3 units right and 4 units down. The coordinates of E are $(3, -4)$.

29.

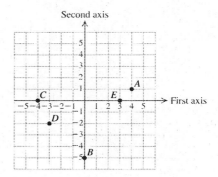

Point A is 4 units right and 1 unit up. The coordinates of A are $(4,1)$.

Point B is 0 units right or left and 5 units down. The coordinates of B are $(0, -5)$.

Point C is 4 units left and 0 units up or down. The coordinates of C are $(-4, 0)$.

Point D is 3 units left and 2 units down. The coordinates of D are $(-3, -2)$.

Point E is 3 units right and 0 units up or down. The coordinates of E are $(3,0)$.

31. Since the x-values range from -75 to 9, the 10 horizontal squares must span $9 - (-75)$, or 84 units. Since 84 is close to 100 and it is convenient to count by 10's, we can count backward from 0 eight squares to -80 and forward from 0 two squares to 20 for a total of $8 + 2$, or 10 squares.

Since the y-values range from -4 to 5, the 10 vertical squares must span $5 - (-4)$, or 9 units. It will be convenient to count by 2's in this case. We count down from 0 five squares to -10 and up from 0 five squares to 10 for a total of $5 + 5$, or 10 squares. (Instead, we might have chosen to count by 1's from -5 to 5.)

Then we plot the points $(-75, 5)$, $(-18, -2)$, and $(9, -4)$.

33. Since the x-values range from -5 to 5, the 10 horizontal squares must span $5 - (-5)$, or 10 units. It will be convenient to count by 2's in this case. We count backward from 0 five squares to -10 and forward from 0 five squares to 10 for a total of $5 + 5$, or 10 squares.

Since the y-values range from -14 to 83, the 10 vertical squares must span $83 - (-14)$, or 97 units. To include both -14 and 83, the squares should extend from about -20 to 90, or $90 - (-20)$, or 110 units. We cannot do this counting by 10's, so we use 20's instead. We count down from 0 four units to -80 and up from 0 six units to 120 for a total of $4 + 6$, or 10 units. There are other ways to cover the values from -14 to 83 as well.

Then we plot the points $(-1, 83)$, $(-5, -14)$, and $(5, 37)$.

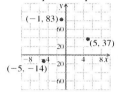

35. Since the x-values range from -16 to 3, the 10 horizontal squares must span $3 - (-16)$, or 19 units. We could number by 2's or 3's. We number by 3's, going backward from 0 eight squares to -24 and forward from 0 two squares to 6 for a total of $8 + 2$, or 10 squares.

Since the y-values range from -4 to 15, the 10 vertical squares must span $15 - (-4)$, or 19 units. We will number the vertical axis by 3's as we did the horizontal axis. We go down from 0 four squares to -12 and up from 0 six squares to 18 for a total of $4 + 6$, or 10 squares.

Then we plot the points $(-10, -4)$, $(-16, 7)$, and $(3, 15)$.

37. Since the x-values range from -100 and 800, the 10 horizontal squares must span $800 - (-100)$, or 900 units. Since 900 is close to 1000 we can number by 100's. We go backward from 0 two squares to -200 and forward from 0 eight squares to 800 for a total of $2 + 8$, or 10 squares. (We could have numbered from -100 to 900 instead.)

Since the y-values range from -5 to 37, the 10 vertical squares must span $37 - (-5)$, or 42 units. Since 42 is close to 50, we can count by 5's. We go down from 0 two squares to -10 and up from 0 eight squares to 40 for a total of $2 + 8$, or 10 squares.

Then we plot the points $(-100, -5)$, $(350, 20)$, and $(800, 37)$.

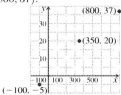

39. Since the x-values range from -124 to 54, the 10 horizontal squares must span $54 - (-124)$, or 178 units. We can number by 25's. We go backward from 0 six squares to

-150 and forward from 0 four squares to 100 for a total of $6 + 4$, or 10 squares.

Since the y-values range from -238 to 491, the 10 vertical squares must span $491 - (-238)$, or 729 units. We can number by 100's. We go down from 0 four squares to -400 and up from 0 six squares to 600 for a total of $4 + 6$, or 10 squares.

Then we plot the points $(-83, 491)$, $(-124, -95)$, and $(54, -238)$.

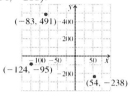

41. Since the first coordinate is positive and the second coordinate negative, the point $(7, -2)$ is located in quadrant IV.

43. Since both coordinates are negative, the point $(-4, -3)$ is in quadrant III.

45. Since both coordinates are positive, the point $(2, 1)$ is in quadrant I.

47. Since the first coordinate is negative and the second coordinate is positive, the point $(-4.9, 8.3)$ is in quadrant II.

49. First coordinates are positive in the quadrants that lie to the right of the origin, or in quadrants I and IV.

51. Points for which both coordinates are positive lie in quadrant I, and points for which both coordinates are negative life in quadrant III. Thus, both coordinates have the same sign in quadrants I and III.

53. *Writing Exercise.* The vertical scale above $80\cancel{c}$ is not labeled. The actual years in question are not given either.

55. $5y = 2x$
   $y = \frac{2}{5}x$ Divide both sides by 5

57. $x - y = 8$
   $-y = -x + 8$ Add $-x$ to both sides
   $y = x - 8$

59. $2x + 3y = 5$
   $3y = -2x + 5$
   $y = \frac{-2}{3}x + \frac{5}{3}$

61. *Writing Exercise.* As time passes from 2004-6, the line graph is almost horizontal. The indicates that there is no new business involving home videos.

63. The coordinates have opposite signs, so the point could be in quadrant II or quadrant IV.

65.

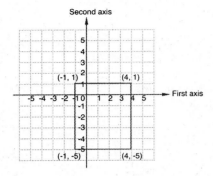

The coordinates of the fourth vertex are $(-1, -5)$.

67. Answers may vary.

We select eight points such that the sum of the coordinates for each point is 7.

$$
\begin{array}{ll}
(0,7) & 0+7=7 \\
(1,6) & 1+6=7 \\
(2,5) & 2+5=7 \\
(3,4) & 3+4=7 \\
(4,3) & 4+3=7 \\
(5,2) & 5+2=7 \\
(6,1) & 6+1=7 \\
(7,0) & 7+0=7
\end{array}
$$

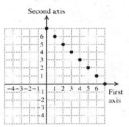

69.

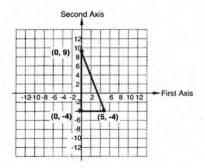

The base is 5 units and the height is 13 units.

$$A = \frac{1}{2}bh = \frac{1}{2} \cdot 5 \cdot 13 = \frac{65}{2} \text{ sq units, or } 32\frac{1}{2} \text{ sq units}$$

71. Latitude 27° North,

Longitude 81° West

73. *Writing Exercise.* Eight "quadrants" will exist. Think of the coordinate system being formed by the intersection of one coordinate plane with another plane perpendicular to one of its axes such that the origins of the two planes coincide. Then there are four quadrants "above" the x,y-plane and four "below" it.

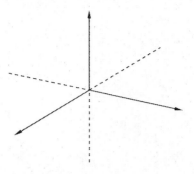

Exercise Set 3.2

1. False. A linear equation in two variables has infinitely many ordered pairs that are solutions.

3. True. All of the points on the graph of the line are solutions to the equation.

5. True. A solution may be found by selecting a value for x and solving for y. The ordered pair is a solution to the equation.

7. We substitute 2 for x and 1 for y.

$$
\begin{array}{c|c}
y = 4x - 7 & \\
\hline
1 & 4(2) - 7 \\
& 8 - 7 \\
\hline
1 \stackrel{?}{=} 1 & \text{TRUE}
\end{array}
$$

Since $1 = 1$ is true, the pair $(2,1)$ is a solution.

9. We substitute 5 for x and 1 for y.

$$
\begin{array}{c|c}
3y + 4x = 19 & \\
\hline
3(1) + 4(5) & 19 \\
3 + 20 & \\
\hline
23 \stackrel{?}{=} 19 & \text{FALSE}
\end{array}
$$

Since $23 = 19$ is false, the pair $(5, 1)$ is not a solution.

11. We substitute 3 for m and -1 for n.

$$
\begin{array}{c|c}
4m - 5n = 7 & \\
\hline
4(3) - 5(-1) & 7 \\
12 + 5 & \\
\hline
17 \stackrel{?}{=} 7 & \text{FALSE}
\end{array}
$$

Since $17 = 7$ is false, the pair $(3, -1)$ is not a solution.

13. To show that a pair is a solution, we substitute, replacing x with the first coordinate and y with the second coordinate in each pair.

$$
\begin{array}{c|c}
y = x + 3 \\
\hline
2 & -1 + 3 \\
\end{array}
$$
$$2 \stackrel{?}{=} 2 \qquad \text{TRUE}$$

$$
\begin{array}{c|c}
y = x + 3 \\
\hline
7 & 4 + 3 \\
\end{array}
$$
$$7 \stackrel{?}{=} 7 \qquad \text{TRUE}$$

In each case the substitution results in a true equation. Thus, $(-1, 2)$ and $(4, 7)$ are both solutions of $y = x + 3$. We graph these points and sketch the line passing through them.

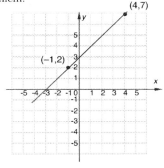

The line appears to pass through $(0, 3)$ also. We check to determine if $(0, 3)$ is a solution of $y = x + 3$.

$$
\begin{array}{c|c}
y = x + 3 \\
\hline
3 & 0 + 3 \\
\end{array}
$$
$$3 \stackrel{?}{=} 3 \qquad \text{TRUE}$$

Thus, $(0, 3)$ is another solution. There are other correct answers, including $(-5, -2)$, $(-4, -1)$, $(-3, 0)$, $(-2, 1)$, $(1, 4)$, $(2, 5)$, and $(3, 6)$.

15. To show that a pair is a solution, we substitute, replacing x with the first coordinate and y with the second coordinate in each pair.

$$
\begin{array}{c|c}
y = \frac{1}{2}x + 3 \\
\hline
5 & \frac{1}{2} \cdot 4 + 3 \\
& 2 + 3 \\
\end{array}
$$
$$5 \stackrel{?}{=} 5 \qquad \text{TRUE}$$

$$
\begin{array}{c|c}
y = \frac{1}{2}x + 3 \\
\hline
2 & \frac{1}{2}(-2) + 3 \\
& -1 + 3 \\
\end{array}
$$
$$2 \stackrel{?}{=} 2 \qquad \text{TRUE}$$

In each case the substitution results in a true equation. Thus, $(4, 5)$ and $(-2, 2)$ are both solutions of $y = \frac{1}{2}x + 3$. We graph these points and sketch the line passing through them.

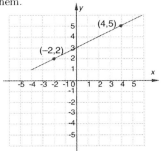

The line appears to pass through $(0, 3)$ also. We check to determine if $(0, 3)$ is a solution of $y = \frac{1}{2}x + 3$.

$$
\begin{array}{c|c}
y = \frac{1}{2}x + 3 \\
\hline
3 & \frac{1}{2} \cdot 0 + 3 \\
\end{array}
$$
$$3 \stackrel{?}{=} 3 \qquad \text{TRUE}$$

Thus, $(0, 3)$ is another solution. There are other correct answers, including $(-6, 0)$, $(-4, 1)$, $(2, 4)$, and $(6, 6)$.

17. To show that a pair is a solution, we substitute, replacing x with the first coordinate and y with the second coordinate in each pair.

$$
\begin{array}{c|c}
y + 3x = 7 \\
\hline
1 + 3 \cdot 2 & 7 \\
1 + 6 & \\
\end{array}
$$
$$7 \stackrel{?}{=} 7 \quad \text{TRUE}$$

$$
\begin{array}{c|c}
y + 3x = 7 \\
\hline
-5 + 3 \cdot 4 & 7 \\
-5 + 12 & \\
\end{array}
$$
$$7 \stackrel{?}{=} 7 \quad \text{TRUE}$$

In each case the substitution results in a true equation. Thus, $(2, 1)$ and $(4, -5)$ are both solutions of $y + 3x = 7$. We graph these points and sketch the line passing through them.

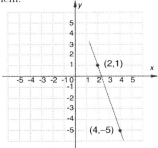

The line appears to pass through $(1, 4)$ also. We check to determine if $(1, 4)$ is a solution of $y + 3x = 7$.

$$
\begin{array}{c|c}
y + 3x = 7 \\
\hline
4 + 3 \cdot 1 & 7 \\
4 + 3 & \\
\end{array}
$$
$$7 \stackrel{?}{=} 7 \quad \text{TRUE}$$

Thus, $(1, 4)$ is another solution. There are other correct answers, including $(3, -2)$.

19. To show that a pair is a solution, we substitute, replacing x with the first coordinate and y with the second coordinate in each pair.

$$
\begin{array}{c|c}
4x - 2y = 10 \\
\hline
4 \cdot 0 - 2(-5) & 10 \\
\end{array}
$$
$$10 \stackrel{?}{=} 10 \quad \text{TRUE}$$

$$
\begin{array}{c|c}
4x - 2y = 10 \\
\hline
4 \cdot 4 - 2 \cdot 3 & 10 \\
16 - 6 & \\
\end{array}
$$
$$10 \stackrel{?}{=} 10 \quad \text{TRUE}$$

In each case the substitution results in a true equation. Thus, $(0, -5)$ and $(4, 3)$ are both solutions of $4x - 2y = 10$.

We graph these points and sketch the line passing through them.

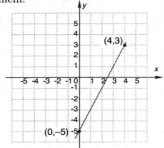

The line appears to pass through $(2, -1)$ also. We check to determine if $(2, -1)$ is a solution of $4x - 2y = 10$.

$$
\begin{array}{c|c}
4x - 2y = 10 \\
\hline
4 \cdot 2 - 2(-1) & 10 \\
8 + 2 & \\
 & \overset{?}{10 \overset{?}{=} 10} \text{ TRUE}
\end{array}
$$

Thus, $(2, -1)$ is another solution. There are other correct answers, including $(1, -3)$, $(2, -1)$, $(3, 1)$, and $(5, 5)$.

21. $y = x + 1$

The equation is in the form $y = mx + b$. The y-intercept is $(0, 1)$. We find two other pairs.

When $x = 3$, $y = 3 + 1 = 4$.
When $x = -5$, $y = -5 + 1 = -4$.

x	y
0	1
3	4
-5	-4

Plot these points, draw the line they determine, and label the graph $y = x + 1$.

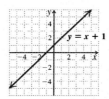

23. $y = -x$

The equation is equivalent to $y = -x + 0$. The y-intercept is $(0, 0)$. We find two other points.

When $x = -2$, $y = -(-2) = 2$.
When $x = 3$, $y = -3$.

x	y
0	0
-2	2
3	-3

Plot these points, draw the line they determine, and label the graph $y = -x$.

25. $y = 2x$

The y-intercept is $(0, 0)$. We find two other points.

When $x = 1$, $y = 2(1) = 2$.

When $x = -1$, $y = 2(-1) = -2$.

x	y
0	0
1	2
-1	-2

Plot these points, draw the line they determine, and label the graph $y = 2x$.

27. $y = 2x + 2$

The y-intercept is $(0, 2)$. We find two other points.

When $x = -3$, $y = 2(-3) + 2 = -6 + 2 = -4$.

When $x = 1$, $y = 2 \cdot 1 + 2 = 2 + 2 = 4$.

x	y
0	2
-3	-4
1	4

Plot these points, draw the line they determine, and label the graph $y = 2x + 2$.

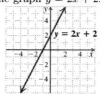

29. $y = -\frac{1}{2}x = -\frac{1}{2}x + 0$

The y-intercept is $(0, 0)$. We find two other points.

When $x = 2$, $y = -\frac{1}{2}(2) = -1$.

When $x = -2$, $y = -\frac{1}{2}(-2) = 1$.

x	y
0	0
2	-1
-2	1

Plot these points, draw the line they determine, and label the graph $y = -\frac{1}{2}x$.

31. $y = \frac{1}{3}x - 4 = \frac{1}{3}x + (-4)$

The y-intercept is $(0, -4)$. We find two other points, using multiples of 3 for x to avoid fractions.

When $x = -3$, $y = \frac{1}{3}(-3) - 4 = -1 - 4 = -5$.

When $x = 3$, $y = \frac{1}{3} \cdot 3 - 4 = 1 - 4 = -3$.

x	y
0	-4
-3	-5
3	-3

Plot these points, draw the line they determine, and label the graph $y = \frac{1}{3}x - 4$.

33. $x + y = 4$

$y = -x + 4$

The y-intercept is $(0, 4)$. We find two other points.

When $x = -1$, $y = -(-1) + 4 = 1 + 4 = 5$.

When $x = 2$, $y = -2 + 4 = 2$.

x	y
0	4
-1	5
2	2

Plot these points, draw the line they determine, and label the graph $x + y = 4$.

35. $x - y = -2$

$y = x + 2$

The y-intercept is $(0, 2)$. We find two other points.

When $x = 1$, $y = 1 + 2 = 3$.

When $x = -1$, $y = -1 + 2 = 1$.

x	y
0	2
1	3
-1	1

Plot these points, draw the line they determine, and label the graph $x - y = -2$.

37. $x + 2y = -6$

$2y = -x - 6$

$y = -\frac{1}{2}x - 3$

$y = -\frac{1}{2}x + (-3)$

The y-intercept is $(0, -3)$. We find two other points, using multiples of 2 for x to avoid fractions.

When $x = -4$, $y = -\frac{1}{2}(-4) - 3 = 2 - 3 = -1$.

When $x = 2$, $y = -\frac{1}{2} \cdot 2 - 3 = -1 - 3 = -4$.

x	y
0	-3
-4	-1
2	-4

Plot these points, draw the line they determine, and label the graph $x + 2y = -6$.

39. $y = -\frac{2}{3}x + 4$

The y-intercept is $(0, 4)$. We find two other points, using multiples of 3 for x to avoid fractions.

When $x = 3$, $y = -\frac{2}{3} \cdot 3 + 4 = -2 + 4 = 2$.

When $x = 6$, $y = -\frac{2}{3} \cdot 6 + 4 = -4 + 4 = 0$.

x	y
0	4
3	2
6	0

Plot these points, draw the line they determine, and label the graph $y = -\frac{2}{3}x + 4$.

41. $4x = 3y$

$y = \frac{4}{3}x = \frac{4}{3}x + 0$

The y-intercept is $(0,0)$. We find two other points.

When $x = 3$, $y = \frac{4}{3}(3) = 4$.

When $x = -3$, $y = \frac{4}{3}(-3) = -4$.

x	y
0	0
3	4
-3	-4

Plot these points, draw the line they determine, and label the graph $4x = 3y$.

43. $5x - y = 0$

$y = 5x = 5x + 0$

The y-intercept is $(0,0)$. We find two other points.

When $x = 1$, $y = 5(1) = 5$.

When $x = -1$, $y = 5(-1) = -5$.

x	y
0	0
1	5
-1	-5

Plot these points, draw the line they determine, and label the graph $5x - y = 0$.

45. $6x - 3y = 9$

$-3y = -6x + 9$

$y = 2x - 3$

$y = 2x + (-3)$

The y-intercept is $(0,-3)$. We find two other points.

When $x = -1$, $y = 2(-1) - 3 = -2 - 3 = -5$.

When $x = 3$, $y = 2 \cdot 3 - 3 = 6 - 3 = 3$.

x	y
0	-3
-1	-5
3	3

Plot these points, draw the line they determine, and label the graph $6x - 3y = 9$.

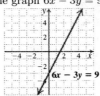

47. $6y + 2x = 8$

$6y = -2x + 8$

$y = -\frac{1}{3}x + \frac{4}{3}$

The y-intercept is $\left(0, \frac{4}{3}\right)$. We find two other points.

When $x = -2$, $y = -\frac{1}{3}(-2) + \frac{4}{3} = \frac{2}{3} + \frac{4}{3} = 2$.

When $x = 1$, $y = -\frac{1}{3} \cdot 1 + \frac{4}{3} = -\frac{1}{3} + \frac{4}{3} = 1$.

x	y
0	$\frac{4}{3}$
-2	2
1	1

Plot these points, draw the line they determine, and label the graph $6y + 2x = 8$.

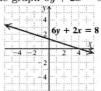

49. We graph $a = 0.08t + 2.5$ by selecting values for t and then calculating the associated values for a.

If $t = 0$, $a = 0.08(0) + 2.5 = 2.5$.

If $t = 10$, $a = 0.08(10) + 2.5 = 3.3$.

If $t = 20$, $a = 0.08(20) + 2.5 = 4.1$.

t	a
0	2.5
10	3.3
20	4.1

We plot the points and draw the graph.

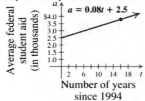

Since $2010 - 1994 = 16$, the year 2010 is 16 years after 1994. Thus, to estimate the average amount of federal student aid per student in 2010, we find the second coordinate associated with 16. Locate the point on the line that is above 16 and then find a value on the vertical axes that corresponds to that point. That value is about 3.8, so we estimate that the average amount of federal student aid per student in 2010 is $3.8 thousand or $3,800.

51. We graph $c = 3.1w + 29.07$ by selecting values for w and then calculating the associated values for c.

If $w = 1$, $c = 3.1(1) + 29.07 = 32.17$.

If $w = 5$, $c = 3.1(5) + 29.07 = 44.57$.

If $w = 10$, $c = 3.1(10) + 29.07 = 60.07$.

w	c
1	32.17
5	44.57
10	60.07

We plot the points and draw the graph.

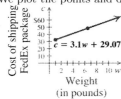

Locate the point on the line that is above $6\frac{1}{2}$ and then find the value on the vertical axes that corresponds to that point. That value is 49, so we estimate the cost of shipping a $6\frac{1}{2}$-lb package to be $49.

53. We graph $p = 3.5n + 9$ by selecting values for n and then calculating the associated values for p.

If $n = 10$, $p = 3.5(10) + 9 = 44$.

If $n = 20$, $p = 3.5(20) + 9 = 79$.

If $n = 30$, $p = 3.5(30) + 9 = 114$.

n	p
10	44
20	79
30	114

We plot the points and draw the graph.

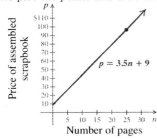

Locate the point on the line that is above 25 and then find the value on the vertical axes that corresponds to that point. That value is 97, so we estimate the price of a scrapbook containing 25 pages as $97.

55. We graph $w = 1.6t + 16.7$ by selecting values for t and then calculating the associated values for w.

If $t = 5$, $w = 1.6(5) + 16.7 = 24.7$.

If $t = 7$, $w = 1.6(7) + 16.7 = 27.9$.

If $t = 10$, $w = 1.6(10) + 16.7 = 32.7$.

We plot the points and draw the graph.

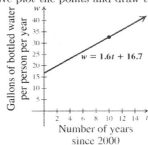

To predict the number of gallons consumed per person in 2010 we find the second coordinate associated with 10. (2010 is 10 years after 2000.) The value is 32.7, so we predict that about 33 gallons of bottled water will be consumed per person in 2010.

57. We graph $T = \frac{5}{4}c + 2$. Since number of credits cannot be negative, we select only nonnegative values for c.

If $c = 4$, $T = \frac{5}{4}(4) + 2 = 7$.

If $c = 8$, $T = \frac{5}{4}(8) + 2 = 12$.

If $c = 12$, $T = \frac{5}{4}(12) + 2 = 17$.

We plot the points and draw the graph.

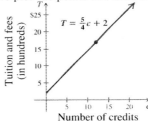

Four three-credit courses total $4 \cdot 3$ or 12, credits. Locate the point in the graph, the value is 17. So tuition and fees will cost $17 hundred, or $1700.

59. *Writing Exercise.* Most would probably say that the second equation would be easier to graph because it has been solved for y. This makes it more efficient to find the y-value that corresponds to a given x-value.

61.
$$5x + 3 \cdot 0 = 12$$
$$5x + 0 = 12$$
$$5x = 12$$
$$x = \frac{12}{5}$$

Check:
$$\begin{array}{c|c} 5x + 3 \cdot 0 = 12 & \\ \hline 5 \cdot \dfrac{12}{5} + 3 \cdot 0 & 12 \\ 12 + 0 & \\ & 12 \overset{?}{=} 12 \quad \text{TRUE} \end{array}$$

The solution is $\frac{12}{5}$.

63. $5x + 3(2 - x) = 12$
$$5x + 6 - 3x = 12$$
$$2x + 6 = 12$$
$$2x = 6$$
$$x = 3$$

Check: $\dfrac{5x + 3(2 - x) = 12}{\begin{array}{c|c} 5(3) + 3(2 - 3) & 12 \\ 15 + 3(-1) & \\ 15 - 3 & \\ 12 & \\ & \overset{?}{12 = 12} \quad \text{TRUE} \end{array}}$

The solution is 3.

65.
$$A = \frac{T + Q}{2}$$
$$2A = T + Q$$
$$2A - T = Q$$

67. $Ax + By = C$
$$By = C - Ax \quad \text{Subtracting } Ax$$
$$y = \frac{C - Ax}{B} \quad \text{Dividing by } B$$

69. *Writing Exercise.* Her graph will be a reflection of the correct graph across the line $y = x$.

71. Let s represent the gear that Laura uses on the southbound portion of her ride and n represent the gear she uses on the northbound portion. Then we have $s + n = 24$. We graph this equation, using only positive integer values for s and n.

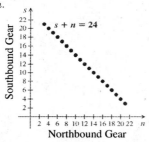

73. Note that the sum of the coordinates of each point on the graph is 5. Thus, we have $x + y = 5$, or $y = -x + 5$.

75. Note that each y-coordinate is 2 more than the corresponding x-coordinate. Thus, we have $y = x + 2$.

77. The equation is $25d + 5l = 225$.

Since the number of dinners cannot be negative, we choose only nonnegative values of d when graphing the equation. The graph stops at the horizontal axis since the number of lunches cannot be negative.

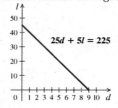

We see that three points on the graph are $(1, 40)$, $(5, 20)$, and $(8, 5)$. Thus, three combinations of dinners and lunches that total \$225 are

1 dinner, 40 lunches,

5 dinners, 20 lunches,

8 dinners, 5 lunches.

79. $y = -|x|$

x	y
-3	-3
-2	-2
-1	-1
0	0
1	-1
2	-2
3	-3

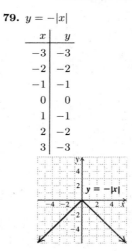

81. $y = x^2$

x	y
-3	9
-2	4
-1	1
0	0
1	1
2	4
3	9

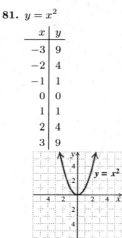

83.

$$y = -2.8x + 3.5$$

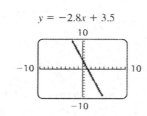

85.

$$y = 2.8x - 3.5$$

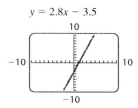

87.

$$y = x^2 + 4x + 1$$

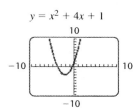

89. $t = -0.1s + 13.1$

s	t
55	7.6
70	6.1

At 55 mph, the efficiency is 7.6 mpg, so a 500 mile trip will use $\frac{500}{7.6} \approx 65.79$ gal at a cost of \$230.27. At 70 mph, the efficiency is 6.1 mpg, so a 500 mile trip will use $\frac{500}{6.1} \approx 81.97$ gal at a cost of \$286.89. Driving at 55 mph instead of 70 mph will save \$56.62 and 16.2 gal.

Exercise Set 3.3

1. The graph of $x = -4$ is a vertical line, so (f) is the most appropriate choice.

3. The point $(0, 2)$ lies on the y-axis, so (d) is the most appropriate choice.

5. The point $(3, -2)$ does not lie on an axis, so it could be used as a check when we graph using intercepts. Thus (b) is the most appropriate choice.

7. (a) The graph crosses the y-axis at $(0, 5)$, so the y-intercept is $(0, 5)$.

(b) The graph crosses the x-axis at $(2, 0)$, so the x-intercept is $(2, 0)$.

9. (a) The graph crosses the y-axis at $(0, -4)$, so the y-intercept is $(0, -4)$.

(b) The graph crosses the x-axis at $(3, 0)$, so the x-intercept is $(3, 0)$.

11. (a) The graph crosses the y-axis at $(0, -2)$, so the y-intercept is $(0, -2)$.

(b) The graph crosses the x-axis at $(-3, 0)$ and $(3, 0)$, so the x-intercepts are $(-3, 0)$ and $(3, 0)$.

13. (a) The graph crosses the y-axis at $(0, 0)$, so the y-intercept is $(0, 0)$.

(b) The graph crosses the x-axis at $(-2, 0), (0, 0)$ and $(5, 0)$, so the x-intercepts are $(-2, 0), (0, 0)$ and $(5, 0)$.

15. $3x + 5y = 15$

(a) To find the y-intercept, let $x = 0$. This is the same as temporarily ignoring the x-term and then solving.

$$5y = 15$$
$$y = 3$$

The $y-$intercept is $(0, 3)$.

(b) To find the x-intercept, let $y = 0$. This is the same as temporarily ignoring the y-term and then solving.

$$3x = 15$$
$$x = 5$$

The x-intercept is $(5, 0)$.

17. $9x - 2y = 36$

(a) To find the y-intercept, let $x = 0$. This is the same as temporarily ignoring the x-term and then solving.

$$-2y = 36$$
$$y = -18$$

The y-intercept is $(0, -18)$.

(b) To find the x-intercept, let $y = 0$. This is the same as temporarily ignoring the y-term and then solving.

$$9x = 36$$
$$x = 4$$

The x-intercept is $(4, 0)$.

19. $-4x + 5y = 80$

(a) To find the y-intercept, let $x = 0$. This is the same as temporarily ignoring the x-term and then solving.

$$5y = 80$$
$$y = 16$$

The y-intercept is $(0, 16)$.

(b) To find the x-intercept, let $y = 0$. This is the same as temporarily ignoring the y-term and then solving.

$$-4x = 80$$
$$x = -20$$

The x-intercept is $(-20, 0)$.

21. $x = 12$

Observe that this is the equation of a vertical line 12 units to the right of the y-axis. Thus, (a) there is no y-intercept and (b) the x-intercept is $(12, 0)$.

23. $y = -9$

Observe that this is the equation of a horizontal line 9 units below the x-axis. Thus, (a) the y-intercept is $(0, -9)$ and (b) there is no x-intercept.

25. $3x + 5y = 15$

Find the y-intercept:

$$5y = 15 \quad \text{Ignoring the } x\text{-term}$$
$$y = 3$$

The y-intercept is $(0, 3)$.

Find the x-intercept:

$$3x = 15 \quad \text{Ignoring the } y\text{-term}$$
$$x = 5$$

The x-intercept is $(5, 0)$.

To find a third point we replace x with -5 and solve for y.

$$3(-5) + 5y = 15$$
$$-15 + 5y = 15$$
$$5y = 30$$
$$y = 6$$

The point $(-5, 6)$ appears to line up with the intercepts, so we draw the graph.

27. $x + 2y = 4$

Find the y-intercept:

$$2y = 4 \quad \text{Ignoring the } x\text{-term}$$
$$y = 2$$

The y-intercept is $(0, 2)$.

Find the x-intercept:

$$x = 4 \quad \text{Ignoring the } y\text{-term}$$

The x-intercept is $(4, 0)$.

To find a third point we replace x with 2 and solve for y.

$$2 + 2y = 4$$
$$2y = 2$$
$$y = 1$$

The point $(2, 1)$ appears to line up with the intercepts, so we draw the graph.

29. $-x + 2y = 8$

Find the y-intercept:

$$2y = 8 \quad \text{Ignoring the } x\text{-term}$$
$$y = 4$$

The y-intercept is $(0, 4)$.

Find the x-intercept:

$$-x = 8 \quad \text{Ignoring the } y\text{-term}$$
$$x = -8$$

The x-intercept is $(-8, 0)$.

To find a third point we replace x with 4 and solve for y.

$$-4 + 2y = 8$$
$$2y = 12$$
$$y = 6$$

The point $(4, 6)$ appears to line up with the intercepts, so we draw the graph.

31. $3x + y = 9$

Find the y-intercept:

$$y = 9 \quad \text{Ignoring the } x\text{-term}$$

The y-intercept is $(0, 9)$.

Find the x-intercept:

$$3x = 9 \quad \text{Ignoring the } y\text{-term}$$
$$x = 3$$

The x-intercept is $(3, 0)$.

To find a third point we replace x with 2 and solve for y.

$$3 \cdot 2 + y = 9$$
$$6 + y = 9$$
$$y = 3$$

The point $(2, 3)$ appears to line up with the intercepts, so we draw the graph.

33. $y = 2x - 6$

Find the y-intercept:

$$y = -6 \quad \text{Ignoring the } x\text{-term}$$

The y-intercept is $(0, -6)$.

Find the x-intercept:

$$0 = 2x - 6 \quad \text{Replacing } y \text{ with } 0$$
$$6 = 2x$$
$$3 = x$$

The x-intercept is $(3, 0)$.

To find a third point we replace x with 2 and find y.

$$y = 2 \cdot 2 - 6 = 4 - 6 = -2$$

The point $(2, -2)$ appears to line up with the intercepts, so we draw the graph.

35. $5x - 10 = 5y$

We can leave the equation in the given form or rewrite it in the form $Ax + By = C$. We will use the given form.

Find the y-intercept:

$$-10 = 5y \quad \text{Ignoring the } x\text{-term}$$
$$-2 = y$$

The y-intercept is $(0, -2)$.

To find the x-intercept, let $y = 0$.

$$5x - 10 = 5 \cdot 0$$
$$5x - 10 = 0$$
$$5x = 10$$
$$x = 2$$

The x-intercept is $(2, 0)$.

To find a third point we replace x with 5 and solve for y.

$$5 \cdot 5 - 10 = 5y$$
$$25 - 10 = 5y$$
$$15 = 5y$$
$$3 = y$$

The point $(5, 3)$ appears to line up with the intercepts, so we draw the graph.

37. $2x - 5y = 10$

Find the y-intercept:

$$-5y = 10 \quad \text{Ignoring the } x\text{-term}$$
$$y = -2$$

The y-intercept is $(0, -2)$.

Find the x-intercept:

$$2x = 10 \quad \text{Ignoring the } y\text{-term}$$
$$x = 5$$

The x-intercept is $(5, 0)$.

To find a third point we replace x with -5 and solve for y.

$$2(-5) - 5y = 10$$
$$-10 - 5y = 10$$
$$-5y = 20$$
$$y = -4$$

The point $(-5, -4)$ appears to line up with the intercepts, so we draw the graph.

39. $6x + 2y = 12$

Find the y-intercept:

$$2y = 12 \quad \text{Ignoring the } x\text{-term}$$
$$y = 6$$

The y-intercept is $(0, 6)$.

Find the x-intercept:

$$6x = 12 \quad \text{Ignoring the } y\text{-term}$$
$$x = 2$$

The x-intercept is $(2, 0)$.

To find a third point we replace x with 3 and solve for y.

$$6 \cdot 3 + 2y = 12$$
$$18 + 2y = 12$$
$$2y = -6$$
$$y = -3$$

The point $(3, -3)$ appears to line up with the intercepts, so we draw the graph.

41. $4x + 3y = 16$

Find the y-intercept:

$$3y = 16 \quad \text{Ignoring the } x\text{-term}$$
$$y = \frac{16}{3}$$

The y-intercept is $(0, \frac{16}{3})$.

Find the x-intercept:

$$4x = 16 \quad \text{Ignoring the } y\text{-term}$$
$$x = 4$$

The x-intercept is $(4, 0)$.

To find a third point we replace x with -2 and solve for y.

$$4(-2) + 3y = 16$$
$$-8 + 3y = 16$$
$$3y = 24$$
$$y = 8$$

The point $(-2, 8)$ appears to line up with the intercepts, so we draw the graph.

43. $2x + 4y = 1$

Find the y-intercept:

$$4y = 1 \quad \text{Ignoring the } x\text{-term}$$
$$y = \frac{1}{4}$$

The y-intercept is $\left(0, \frac{1}{4}\right)$.

Find the x-intercept:

$$2x = 1 \quad \text{Ignoring the } y\text{-term}$$
$$x = \frac{1}{2}$$

The x-intercept is $\left(\frac{1}{2}, 0\right)$.

To find a third point we replace x with $-\frac{3}{2}$ and solve for y.

$$2\left(-\frac{3}{2}\right) + 4y = 1$$
$$-3 + 4y = 1$$
$$4y = 4$$
$$y = 1$$

The point $\left(-\frac{3}{2}, 1\right)$ appears to line up with the intercepts, so we draw the graph.

45. $5x - 3y = 180$

Find the y-intercept:

$$-3y = 180 \quad \text{Ignoring the } x\text{-term}$$
$$y = -60$$

The y-intercept is $(0, -60)$.

Find the x-intercept:
$$5x = 180 \quad \text{Ignoring the } y\text{-term}$$
$$x = 36$$

The x-intercept is $(36, 0)$.

To find a third point we replace x with 6 and solve for y.
$$5 \cdot 6 - 3y = 180$$
$$30 - 3y = 180$$
$$-3y = 150$$
$$y = -50$$

This means that $(6, -50)$ is on the graph.

To graph all three points, the y-axis must go to at least 60 and the x-axis must go to at least 36. Using a scale of 10 units per square allows us to display both intercepts and $(6, -50)$ as well as the origin.

The point $(6, -50)$ appears to line up with the intercepts, so we draw the graph.

47. $y = -30 + 3x$

Find the y-intercept:
$$y = -30 \quad \text{Ignoring the } x\text{-term}$$

The y-intercept is $(0, -30)$.

To find the x-intercept, let $y = 0$.
$$0 = -30 + 3x$$
$$30 = 3x$$
$$10 = x$$

The x-intercept is $(10, 0)$.

To find a third point we replace x with 5 and solve for y.
$$y = -30 + 3 \cdot 5$$
$$y = -30 + 15$$
$$y = -15$$

This means that $(5, -15)$ is on the graph.

To graph all three points, the y-axis must go to at least -30 and the x-axis must go to at least 10. Using a scale of 5 units per square allows us to display both intercepts and $(5, -15)$ as well as the origin.

The point $(5, -15)$ appears to line up with the intercepts, so we draw the graph.

49. $-4x = 20y + 80$

To find the y-intercept, we let $x = 0$.
$$-4 \cdot 0 = 20y + 80$$
$$0 = 20y + 80$$
$$-80 = 20y$$
$$-4 = y$$

The y-intercept is $(0, -4)$.

Find the x-intercept:
$$-4x = 80 \quad \text{Ignoring the } y\text{-term}$$
$$x = -20$$

The x-intercept is $(-20, 0)$.

To find a third point we replace x with -40 and solve for y.
$$-4(-40) = 20y + 80$$
$$160 = 20y + 80$$
$$80 = 20y$$
$$4 = y$$

This means that $(-40, 4)$ is on the graph.

To graph all three points, the y-axis must go at least from -4 to 4 and the x-axis must go at least from -40 to -20. Since we also want to include the origin we can use a scale of 10 units per square on the x-axis and 1 unit per square on the y-axis.

The point $(-40, 4)$ appears to line up with the intercepts, so we draw the graph.

51. $y - 3x = 0$

Find the y-intercept:
$$y = 0 \quad \text{Ignoring the } x\text{-term}$$

The y-intercept is $(0, 0)$. Note that this is also the x-intercept.

In order to graph the line, we will find a second point.

When $x = 1$, $y - 3 \cdot 1 = 0$
$$y - 3 = 0$$
$$y = 3$$

To find a third point we replace $x = -1$ and solve for y.
$$y - 3(-1) = 0$$
$$y + 3 = 0$$
$$y = -3$$

The point $(-1, -3)$ appears to line up with the other two points, so we draw the graph.

53. $y = 1$

Any ordered pair $(x, 1)$ is a solution. The variable y must be 1, but the x variable can be any number we choose. A few solutions are listed below. Plot these points and draw the line.

x	y
-3	1
0	1
2	1

55. $x = 3$

Any ordered pair $(3, y)$ is a solution. The variable x must be 3, but the y variable can be any number we choose. A few solutions are listed below. Plot these points and draw the line.

x	y
3	-2
3	0
3	4

57. $y = -2$

Any ordered pair $(x, -2)$ is a solution. The variable y must be -2, but the x variable can be any number we choose. A few solutions are listed below. Plot these points and draw the line.

x	y
-3	-2
0	-2
4	-2

59. $x = -1$

Any ordered pair $(-1, y)$ is a solution. The variable x must be -1, but the y variable can be any number we choose. A few solutions are listed below. Plot these points and draw the line.

x	y
-1	-3
-1	0
-1	2

61. $y = -15$

Any ordered pair $(x, -15)$ is a solution. The variable y must be -15, but the x variable can be any number we choose. A few solutions are listed below. Plot these points and draw the line.

x	y
-1	-15
0	-15
3	-15

63. $y = 0$

Any ordered pair $(x, 0)$ is a solution. A few solutions are listed below. Plot these points and draw the line.

x	y
-4	0
0	0
2	0

65. $x = -\dfrac{5}{2}$

Any ordered pair $\left(-\dfrac{5}{2}, y \right)$ is a solution. A few solutions are listed below. Plot these points and draw the line.

x	y
$-\dfrac{5}{2}$	-3
$-\dfrac{5}{2}$	0
$-\dfrac{5}{2}$	5

67.
$$-4x = -100$$
$$x = 25 \qquad \text{Dividing by } -4$$

The graph is a vertical line 25 units to the right of the y-axis.

69. $35 + 7y = 0$

$$7y = -35$$

$$y = -5$$

The graph is a horizontal line 5 units below the x-axis.

71. Note that every point on the horizontal line passing through $(0, -1)$ has -1 as the y-coordinate. Thus, the equation of the line is $y = -1$.

73. Note that every point on the vertical line passing through $(4, 0)$ has 4 as the x-coordinate. Thus, the equation of the line is $x = 4$.

75. Note that every point on the vertical line passing through $(0, 0)$ has 0 as the x-coordinate. Thus, the equation of the line is $x = 0$.

77. *Writing Exercise.* Any solution of $y = 8$ is an ordered pair $(x, 8)$. Thus, all points on the graph of $y = 8$ are 8 units above the x-axis, so they lie on a horizontal line.

79. $d - 7$

81. Let n represent the number. Then we have $7 + 4n$.

83. Let x and y represent the numbers. Then we have $2(x + y)$.

85. *Writing Exercise.* The graph will be a line with y-intercept $(0, C)$ and x-intercept $(C, 0)$.

87. The x-axis is a horizontal line, so it is of the form $y = b$. All points on the x-axis are of the form $(x, 0)$, so b must be 0 and the equation is $y = 0$.

89. A line parallel to the y-axis has an equation of the form $x = a$. Since the x-coordinate of one point on the line is -2, then $a = -2$ and the equation is $x = -2$.

91. Since the x-coordinate of the point of intersection must be -3 and y must equal 4, the point of intersection is $(-3, 4)$.

93. The y-intercept is $(0, 5)$, so we have $y = mx + 5$. Another point on the line is $(-3, 0)$ so we have

$$0 = m(-3) + 5$$

$$-5 = -3m$$

$$\frac{5}{3} = m$$

The equation is $y = \frac{5}{3}x + 5$, or $5x - 3y = -15$, or $-5x + 3y = 15$.

95. Substitute 0 for x and -8 for y.

$$4 \cdot 0 = C - 3(-8)$$

$$0 = C + 24$$

$$-24 = C$$

97. $Ax + D = C$

$$Ax = C - D$$

$$x = \frac{C - D}{A}$$

The x-intercept is $\left(\dfrac{C - D}{A}, 0 \right)$.

99. Find the y-intercept:

$$-7y = 80 \qquad \text{Covering the } x\text{-term}$$

$$y = -\frac{80}{7} = -11.\overline{428571}$$

The y-intercept is $\left(0, -\dfrac{80}{7} \right)$, or $(0, -11.\overline{428571})$.

Find the x-intercept:

$$2x = 80 \quad \text{Covering the } y\text{-term}$$

$$x = 40$$

The x-intercept is $(40, 0)$.

101. From the equation we see that the y-intercept is $(0, -9)$.

To find the x-intercept, let $y = 0$.

$$0 = 0.2x - 9$$

$$9 = 0.2x$$

$$45 = x$$

The x-intercept is $(45, 0)$.

103. Find the y-intercept.

$$25y = 1 \qquad \text{Covering the } x\text{-term}$$

$$y = \frac{1}{25}, \text{ or } 0.04$$

The y-intercept is $\left(0, \dfrac{1}{25} \right)$, or $(0, 0.04)$.

Find the x-intercept:

$$50x = 1 \qquad \text{Covering the } y\text{-term}$$

$$x = \frac{1}{50}, \text{ or } 0.02$$

The x-intercept is $\left(\dfrac{1}{50}, 0 \right)$, or $(0.02, 0)$.

Exercise Set 3.4

1. $\dfrac{100 \text{ miles}}{5 \text{ hours}} = 20$ miles per hour, or miles/hour

3. $\dfrac{300 \text{ dollars}}{150 \text{ miles}} = 2$ dollars per mile, or dollars/mile

5. $\dfrac{40 \text{ minutes}}{8 \text{ errands}} = 5$ minutes per errand, or minutes/errand

7. a) We divide the number of miles traveled by the number of gallons of gas used for distance traveled.

Rate, in miles per gallon

$$= \frac{14,131 \text{ mi} - 13,741 \text{ mi}}{13 \text{ gal}}$$

$$= \frac{390 \text{ mi}}{13 \text{ gal}}$$

$$= 30 \text{ mi/gal}$$

$$= 30 \text{ miles per gallon}$$

b) We divide the cost of the rental by the number of days. From June 5 to June 8 is $8 - 5$, or 3 days.

Average cost, in dollars per day

$$= \frac{118 \text{ dollars}}{3 \text{ days}}$$

$$\approx 39.33 \text{ dollars/day}$$

$$\approx \$39.33 \text{ per day}$$

c) We divide the number of miles traveled by the number of days. The car was driven 390 miles, and was rented for 3 days.

Rate, in miles per day

$$= \frac{390 \text{ mi}}{3 \text{ days}}$$

$$= 130 \text{ mi/day}$$

$$= 130 \text{ mi per day}$$

d) Note that $\$118 = 11,800\cent$. The car was driven 390 miles.

Rate, in cents per mile $= \dfrac{11,800\cent}{390 \text{ mi}}$

$$\approx 30\cent \text{ per mi}$$

9. a) From 9:00 to 11:00 is $11 - 9$, or 2 hr.

Average speed, in miles per hour $= \dfrac{14 \text{ mi}}{2 \text{ hr}}$

$$= 7 \text{ mph}$$

b) From part (a) we know that the bike was rented for 2 hr.

Rate, in dollars per hour $= \dfrac{\$15}{2 \text{ hr}}$

$$= \$7.50 \text{ per hr}$$

c) Rate, in dollars per mile $= \dfrac{\$15}{14 \text{ mi}}$

$$\approx \$1.07 \text{ per mi}$$

11. a) It is 3 hr from 9:00 A.M. to noon and 2 more hours from noon to 2:00 P.M., so the proofreader worked $3 + 2$, or 5 hr.

Rate, in dollars per hour $= \dfrac{\$110}{5 \text{ hr}}$

$$= \$22 \text{ per hr}$$

b) The number of pages proofread is $195 - 92$, or 103.

Rate, in pages per hour $= \dfrac{103 \text{ pages}}{5 \text{ hr}}$

$$= 20.6 \text{ pages per hr}$$

c) Rate, in dollars per page $= \dfrac{\$110}{103 \text{ pages}}$

$$\approx \$1.07 \text{ per page}$$

13. Increase in debt: 8612 billion -5770 billion or 2842 billion.

Change in time $2006 - 2001 = 5$ year

$$\text{Rate of increase} = \frac{\text{Change in debt}}{\text{Change in time}}$$

$$= \frac{\$2842 \text{ billion}}{5 \text{ yr}}$$

$$\approx \$568.4 \text{ billion/yr}$$

15. a) The elevator traveled $34 - 5$, or 29 floors in $2{:}40 - 2{:}38$, or 2 min.

Average rate of travel $= \dfrac{29 \text{ floors}}{2 \text{ min}}$

$$= 14.5 \text{ floors per min}$$

b) In part (a) we found that the elevator traveled 29 floors in 2 min. Note that 2 min $= 2 \times 1$ min $= 2 \times 60$ sec $= 120$ sec.

Average rate of travel $= \dfrac{120 \text{ sec}}{29 \text{ floors}}$

$$\approx 4.14 \text{ sec per floor}$$

17. Ascended $29,028 \text{ ft} - 17,552 \text{ ft} = 11,476 \text{ ft}$. The time of ascent: 8 hr, 10 min, or 8 hr $+$ 10 min $= 480$ min $+$ 10 min $= 490$ min. a)

Rate, in feet per minute $= \dfrac{11,476 \text{ ft}}{490 \text{ min}}$

$$\approx 23.42 \text{ ft/min}$$

b) Rate, in minutes per foot $= \dfrac{490 \text{ min}}{11,476 \text{ ft}}$

$$\approx 0.04 \text{ min/ft}$$

19. The rate of decrease is given in tons per year, so we list the number of tons in the vertical axis and the year on the the horizontal axis. If we count by increments of 10 million on the vertical axis we can easily reach 35.7 million and beyond. We label the units on the vertical axis in millions of tons. We list the years on the horizontal axis, beginning with 2006. We plot the point $(2006, 35.7 \text{ million})$. Then, to display the decreased rate we move from that point to a point that represents a decrease of 700,000 tons one year later. The coordinates of this point are $(2006 + 1, 35.7 - 0.7 \text{ million})$, or $(2007, 35 \text{ million})$. Finally, we draw a line through the two points.

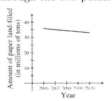

21. The rate is given in dollars per year, so we list the amount in sales of asthma drug products on the vertical axis and year on the horizontal axis. We can count by increments of 2 billion on the vertical axis. We plot the point $(2006, \$11 \text{ billion})$. Then to display the rate of growth, we move from that point to a point that represents \$1.2 billion more a year later. The coordinates of this point are $(2006+1, \$11+1.2 \text{ billion})$ or $(2007, \$12.2 \text{ billion})$. Finally, we draw a line through the two points.

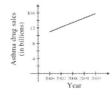

23. The rate is given in miles per hour, so we list the number of miles traveled on the vertical axis and the time of day

on the horizontal axis. If we count by 100's of miles on the vertical axis we can easily reach 230 without needing a terribly large graph. We plot the point (3:00, 230). Then to display the rate of travel, we move from that point to a point that represents 90 more miles traveled 1 hour later. The coordinates of this point are (3:00 + 1 hr, 230 + 90), or (4:00, 320). Finally, we draw a line through the two points.

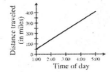

25. The rate is given in dollars per hour so we list money earned on the vertical axis and the time of day on the horizontal axis. We can count by $20 on the vertical axis and reach $50 without needing a terribly large graph. Next we plot the point (2:00 P.M., $50). To display the rate we move from that point to a point that represents $15 more 1 hour later. The coordinates of this point are (2 + 1, $50 + $15), or (3:00 P.M., $65). Finally, we draw a line through the two points.

27. The rate is given in cost per minute so we list the amount of the telephone bill on the vertical axis and the number of additional minutes on the horizontal axis. We begin with $7.50 on the vertical axis and count by $0.50. A jagged line at the base of the axis indicates that we are not showing amounts smaller than $7.50. We begin with 0 additional minutes on the horizontal axis and plot the point (0, $7.50). We move from there to a point that represents $0.10 more 1 minute later. The coordinates of this point are (0 + 1 min, $7.50 + $0.10), or (1 min, $7.60). Then we draw a line through the two points.

29. The points (10:00, 30 calls) and (1:00, 90 calls) are on the graph. This tells us that in the 3 hr between 10:00 and 1:00 there were 90 − 30 = 60 calls completed. The rate is
$$\frac{60 \text{ calls}}{3 \text{ hr}} = 20 \text{ calls/hour.}$$

31. The points (12:00, 100 mi) and (2:00, 250 mi) are on the graph. This tells us that in the 2 hr between 12:00 and 2:00 the train traveled 250 − 100 = 150 mi. The rate is
$$\frac{150 \text{ mi}}{2\text{hr}} = 75 \text{ mi per hr.}$$

33. The points $(5 \text{ min}, 60¢)$ and $(10 \text{ min}, 120¢)$ are on the graph. This tells us that in $10 - 5 = 5$ min the cost of the call increased $120¢ - 60¢ = 60¢$. The rate is
$$\frac{60¢}{5 \text{ min}} = 12¢ \text{ per min.}$$

35. The points (1970, 140 thousand) and (2000, 80 thousand) are on the graph. This tells us that in $2000 - 1970 = 30$yrs the population decreases $80 - 140 = -60$thousand. The rate is
$$\frac{-60 \text{ thousand}}{30 \text{ yr}} = -2000 \text{ people/year.}$$

37. The points (90 mi, 2 gal) and (225 mi, 5 gal) are on the graph. This tells us that when driven $225 - 90 = 135$ mi the vehicle consumed $5 - 2 = 3$ gal of gas. The rate is
$$\frac{3 \text{ gal}}{135 \text{ mi}} = 0.02 \text{ gal/mi.}$$

39. Since swimming is the slowest of the three sports and biking is the fastest, the slope of the line representing swimming speed will be the least steep of the three and that representing biking speed will be the steepest. The second segment of graph (e) rises most steeply and the third segment is the least steep of the three segments. Thus this graph represents running followed by biking and then swimming.

41. Since swimming is the slowest of the three sports and biking is the fastest, the slope of the line representing swimming speed will be the least steep of the three and that representing biking speed will be the steepest. The first segment of graph (d) is the least steep and the second segment is the steepest of the three segments. Thus this graph represents swimming followed by biking and then running.

43. Since swimming is the slowest of the three sports and biking is the fastest, the slope of the line representing swimming speed will be the least steep of the three and that representing biking speed will be the steepest. The first segment of graph (b) is the steepest and the second segment is the least steep of the three segments. Thus this graph represents biking followed by swimming and then running.

45. *Writing Exercise.* A negative rate of travel indicates that an object is moving backwards.

47. $-2 - (-7) = -2 + 7 = 5$

49. $\dfrac{5 - (-4)}{-2 - 7} = \dfrac{9}{-9} = -1$

51. $\dfrac{-4 - 8}{11 - 2} = \dfrac{-12}{9} = \dfrac{-4}{3}$

53. $\dfrac{-6 - (-6)}{-2 - 7} = \dfrac{-6 + 6}{-2 - 7} = \dfrac{0}{9} = 0$

55. *Writing Exercise.* a) The graph of Jon's total earnings would be above Jenny's total earnings, with Jon's rate or slope steeper than Jenny's.

b) Jenny's graph is above Jon's but the slope or rate is the same.

c) The final result (total earnings) can be compared, but not the rate.

57. Let $t =$ flight time and $a =$ altitude. While the plane is climbing at a rate of 6300 ft/min, the equation $a = 6300t$ describes the situation. Solving $31,500 = 6300t$, we find that the cruising altitude of 31,500 ft is reached after about 5 min. Thus we graph $a = 6300t$ for $0 \le t \le 5$.

The plane cruises at 31,500 ft for 3 min, so we graph $a = 31,500$ for $5 < t \le 8$. After 8 min the plane descends at a rate of 3500 ft/min and lands. The equation $a = 31,500 - 3500(t - 8)$, or $a = -3500t + 59,500$, describes this situation. Solving $0 = -3500t + 59,500$, we find that the plane lands after about 17 min. Thus we graph $a = -3500t + 59,500$ for $8 < t \le 17$. The entire graph is show below.

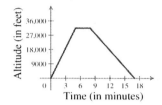

59. Let the horizontal axis represent the distance traveled, in miles, and let the vertical axis represent the fare, in dollars. Use increments of 1/5, or 0.2 mi, on the horizontal axis and of \$1 on the vertical axis. The fare for traveling 0.2 mi is $2 + $0.50 \cdot 1$, or \$2.50 and for 0.4 mi, or 0.2 mi $\times$ 2, we have $2 + $0.50(2)$, or \$3. Plot the points (0.2 mi, \$2.50) and (0.4 mi, \$3) and draw the line through them.

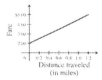

61. 95 mph + 39 mph = 134 mph

$\dfrac{134 \text{ mi}}{1 \text{ hr}}$ gives us $\dfrac{1 \text{ hr}}{134 \text{ mi}}$.

$\dfrac{1 \text{ hr}}{134 \text{ min}} = \dfrac{1 \text{ hr}}{134 \text{ mi}} \cdot \dfrac{60 \text{ min}}{1 \text{ hr}} \approx 0.45 \text{ min per mi}$

63. First we find Anne's speed in minutes per kilometer.

$\text{Speed} = \dfrac{15.5 \text{ min}}{7 \text{ km} - 4 \text{ km}} = \dfrac{15.5}{3} \dfrac{\text{min}}{\text{km}}$

Now we convert min/km to min/mi.

$\dfrac{15.5}{3} \dfrac{\text{min}}{\text{km}} \approx \dfrac{15.5}{3} \dfrac{\text{min}}{\text{km}} \cdot \dfrac{1 \text{ km}}{0.621 \text{ mi}} \approx \dfrac{15.5}{1.863} \dfrac{\text{min}}{\text{mi}}$

At a rate of $\dfrac{15.5}{1.863} \dfrac{\text{min}}{\text{mi}}$, to run a 5-mi race it would take $\dfrac{15.5}{1.863} \dfrac{\text{min}}{\text{mi}} \cdot 5 \text{ mi} \approx 41.6 \text{ min}$.

(Answers may vary slightly depending on the conversion factor used.)

65. First we find Ryan's rate. Then we double it to find Alex's rate. Note that 50 minutes $= \dfrac{50}{60}$ hr $= \dfrac{5}{6}$ hr.

$\text{Ryan's rate} = \dfrac{\text{change in number of bushels picked}}{\text{corresponding change in time}}$

$= \dfrac{5\frac{1}{2} - 4 \text{ bushels}}{\frac{5}{6} \text{ hr}}$

$= \dfrac{1\frac{1}{2} \text{ bushels}}{\frac{5}{6} \text{ hr}}$

$= \dfrac{3}{2} \cdot \dfrac{6}{5} \dfrac{\text{bushels}}{\text{hr}}$

$= \dfrac{9}{5} \text{ bushels per hour, or}$

$1.8 \text{ bushels per hour}$

Then Alex's rate is $2(1.8) = 3.6$ bushels per hour.

Exercise Set 3.5

1. A teenager's height increases over time, so the rate is positive.

3. The water level decreases during a drought, so the rate is negative.

5. The distance from the starting point increases during a race, so the rate is positive.

7. The number of U.S. senators does not change, so the rate is zero.

9. The number of people present decreases in the moments following the final buzzer, so the rate is negative.

11. The rate can be found using the coordinates of any two points on the line. We use (10, \$600) and (25, \$1500).

$\text{Rate} = \dfrac{\text{change in compensation}}{\text{corresponding change in number of blogs}}$

$= \dfrac{\$1500 - \$600}{25 - 10}$

$= \dfrac{\$900}{15 \text{ blogs}}$

$= \$60 \text{ per blog}$

13. The rate can be found using the coordinates of any two points on the line. We use (May 2005, \$380) and (Mar 2006, \$320).

$\text{Rate} = \dfrac{\text{change in price}}{\text{corresponding change in time}}$

$= \dfrac{\$320 - \$380}{\text{Mar 2006} - \text{May 2005}}$

$= \dfrac{-\$60}{15 - 5 \text{ months}}$

$= \dfrac{-60}{10}$

$= -\$6 \text{ per month}$

15. The rate can be found using the coordinates of any two points on the line. We use (35, 480) and (65, 510), where 35 and 65 are in \$1000's.

$$\text{Rate} = \frac{\text{change in score}}{\text{corresponding change in income}}$$
$$= \frac{510 - 480 \text{ points}}{65 - 35}$$
$$= \frac{30 \text{ points}}{30}$$
$$= 1 \text{ point per } \$1000 \text{ income}$$

17. The rate can be found using the coordinates of any two points on the line. We use $(0 \text{ min}, 54°)$ and $(27 \text{ min}, -4°)$.

$$\text{Rate} = \frac{\text{change in temperature}}{\text{corresponding change in time}}$$
$$= \frac{-4° - 54°}{27 \text{ min} - 0 \text{ min}}$$
$$= \frac{-58°}{27 \text{ min}}$$
$$\approx -2.1° \text{per min}$$

19. We can use any two points on the line, such as $(0,1)$ and $(3,5)$.
$$m = \frac{\text{change in } y}{\text{change in } x}$$
$$= \frac{5-1}{3-0} = \frac{4}{3}$$

21. We can use any two points on the line, such as $(1,0)$ and $(3,3)$.
$$m = \frac{\text{change in } y}{\text{change in } x}$$
$$= \frac{3-0}{3-1} = \frac{3}{2}$$

23. We can use any two points on the line, such as $(2,2)$ and $(4,6)$.
$$m = \frac{\text{change in } y}{\text{change in } x}$$
$$= \frac{6-2}{4-2}$$
$$= \frac{4}{2} = 2$$

25. We can use any two points on the line, such as $(0,2)$ and $(2,0)$.
$$m = \frac{\text{change in } y}{\text{change in } x}$$
$$= \frac{2-0}{0-2} = \frac{2}{-2} = -1$$

27. This is the graph of a horizontal line. Thus, the slope is 0.

29. We can use any two points on the line, such as $(0,2)$ and $(3,1)$.
$$m = \frac{\text{change in } y}{\text{change in } x}$$
$$= \frac{1-2}{3-0} = -\frac{1}{3}$$

31. This is the graph of a vertical line. Thus, the slope is undefined.

33. We can use any two points on the line, such as $(-2,1)$ and $(2,-2)$.
$$m = \frac{\text{change in } y}{\text{change in } x}$$
$$= \frac{-2-1}{2-(-2)} = -\frac{3}{4}$$

35. We can use any two points on the line, such as $(-2,0)$ and $(2,1)$.
$$m = \frac{\text{change in } y}{\text{change in } x}$$
$$= \frac{1-0}{2-(-2)} = \frac{1}{4}$$

37. This is the graph of a horizontal line, so the slope is 0.

39. $(1,3)$ and $(5,8)$
$$m = \frac{8-3}{5-1} = \frac{5}{4}$$

41. $(-2,4)$ and $(3,0)$
$$m = \frac{4-0}{-2-3} = \frac{4}{-5} = -\frac{4}{5}$$

43. $(-4,0)$ and $(5,6)$
$$m = \frac{6-0}{5-(-4)} = \frac{6}{9} = \frac{2}{3}$$

45. $(0,7)$ and $(-3,10)$
$$m = \frac{10-7}{-3-0} = \frac{3}{0-3} = -1$$

47. $(-2,3)$ and $(-6,5)$
$$m = \frac{5-3}{-6-(-2)} = \frac{2}{-6+2} = \frac{2}{-4} = -\frac{1}{2}$$

49. $\left(-2, \frac{1}{2}\right)$ and $\left(-5, \frac{1}{2}\right)$
Observe that the points have the same y-coordinate. Thus, they lie on a horizontal line and its slope is 0. We could also compute the slope.
$$m = \frac{\frac{1}{2} - \frac{1}{2}}{-2-(-5)} = \frac{\frac{1}{2} - \frac{1}{2}}{-2+5} = \frac{0}{3} = 0$$

51. $(5,-4)$ and $(2,-7)$
$$m = \frac{-7-(-4)}{2-5} = \frac{-3}{-3} = 1$$

53. $(6,-4)$ and $(6,5)$
Observe that the points have the same x-coordinate. Thus, they lie on a vertical line and its slope is undefined. We could also compute the slope.
$$m = \frac{-4-5}{6-6} = \frac{-9}{0}, \text{ undefined}$$

55. The line $y = 5$ is a horizontal line. A horizontal line has slope 0.

57. The line $x = -8$ is a vertical line. Slope is undefined.

59. The line $x = 9$ is a vertical line. The slope is undefined.

61. The line $y = -10$ is a horizontal line. A horizontal line has slope 0.

63. The grade is expressed as a percent.
$$m = \frac{792}{5280} = 0.15 = 15\%$$

65. The slope is expressed as a percent.
$$m = \frac{28}{80} = 0.35 = 35\%.$$

67. 2 ft 5 in. $= 2 \cdot 12$ in. $+ 5$ in. $= 24$ in. $+ 5$ in. $= 29$ in.
8 ft 2 in. $= 8 \cdot 12$ in. $+ 2$ in. $= 96$ in. $+ 2$ in. $= 98$ in.
$$m = \frac{29}{98}, \text{ or about } 30\%$$

69. Dooley Mountain rises $5400 - 3500 = 1900$ ft.
$$m = \frac{1900}{37000} \approx 0.051 \approx 5.1\%$$
Yes, it qualifies as part of the Tour de France.

71. *Writing Exercise.*
$$\frac{y_2 - y_1}{x_2 - x_1} = \frac{-(y_1 - y_2)}{-(x_1 - x_2)} = \frac{y_1 - y_2}{x_1 - x_2}$$

73. $ax + by = c$
$$by = c - ax \quad \text{Adding } -ax \text{ to both sides}$$
$$y = \frac{c - ax}{b} \quad \text{Dividing both sides by } b$$

75. $ax - by = c$
$$-by = c - ax \quad \text{Adding } -ax \text{ to both sides}$$
$$y = \frac{c - ax}{-b} \quad \text{Dividing both sides by } -b$$
We could also express this result as $y = \dfrac{ax - c}{b}$.

77. $8x + 6y = 24$
$$6y = -8x + 24$$
$$y = \frac{-4}{3}x + 4$$

x	y
0	4
3	0
6	-4

79. *Writing Exercise.*

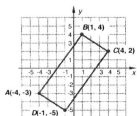

We find the slope of each side of the quadrilateral.

For side $\overline{AB}$, $m = \dfrac{4 - (-3)}{1 - (-4)} = \dfrac{7}{5}$.

For side $\overline{BC}$, $m = \dfrac{2 - 4}{4 - 1} = -\dfrac{2}{3}$.

For side $\overline{CD}$, $m = \dfrac{2 - (-5)}{4 - (-1)} = \dfrac{7}{5}$.

For side $\overline{DA}$, $m = \dfrac{-5 - (-3)}{-1 - (-4)} = -\dfrac{2}{3}$.

Since the opposite sides of the quadrilateral have the same slopes but lie on different lines, the lines on which they lie never intersect so they are parallel. Thus the quadrilateral is a parallelogram.

81. From the dimensions on the drawing, we see that the ramps labeled A have a rise of 61 cm and a run of 167.6 cm.
$$m = \frac{61 \text{ cm}}{167.6 \text{ cm}} \approx 0.364, \text{ or } 36.4\%$$

83. If the line passes through $(2, 5)$ and never enters the second quadrant, then it slants up from left to right or is vertical. This means that its slope is positive. The line slants least steeply if it passes through $(0, 0)$. In this case, $m = \dfrac{5 - 0}{2 - 0} = \dfrac{5}{2}$. Thus, the numbers the line could have for it slope are $\left\{ m \middle| m \geq \dfrac{5}{2} \right\}$.

85. Let $t =$ the number of units each tick mark on the vertical axis represents. Note that the graph drops 4 units for every 3 units of horizontal change. Then we have:
$$\frac{-4t}{3} = -\frac{2}{3}$$
$$-4t = -2 \quad \text{Multiplying by 3}$$
$$t = \frac{1}{2} \quad \text{Dividing by } -4$$
Each tick mark on the vertical axis represents $\dfrac{1}{2}$ unit.

Exercise Set 3.6

1. We can read the slope, 3, directly from the equation. Choice (f) is correct.

3. We can read the slope, $\dfrac{2}{3}$, directly from the equation. Choice (d) is correct.

5. $y = 3x - 2 = 3x + (-2)$
The y-intercept is $(0, -2)$, so choice (e) is correct.

7. Slope $\dfrac{2}{3}$; y-intercept $(0, 1)$

We plot $(0, 1)$ and from there move up 2 units and right 3 units. This locates the point $(3, 3)$. We plot $(3, 3)$ and draw a line passing through $(0, 1)$ and $(3, 3)$.

9. Slope $\frac{5}{3}$; y-intercept $(0, -2)$

We plot $(0, -2)$ and from there move up 5 units and right 3 units. This locates the point $(3, 3)$. We plot $(3, 3)$ and draw a line passing through $(0, -2)$ and $(3, 3)$.

11. Slope $-\frac{1}{3}$; y-intercept $(0, 5)$

We plot $(0, 5)$. We can think of the slope as $\frac{-1}{3}$, so from $(0, 5)$ we move down 1 unit and right 3 units. This locates the point $(3, 4)$. We plot $(3, 4)$ and draw a line passing through $(0, 5)$ and $(3, 4)$.

13. Slope 2; y-intercept $(0, 0)$

We plot $(0, 0)$. We can think of the slope as $\frac{2}{1}$, so from $(0, 0)$ we move up 2 units and right 1 unit. This locates the point $(1, 2)$. We plot $(1, 2)$ and draw a line passing through $(0, 0)$ and $(1, 2)$.

15. Slope -3; y-intercept $(0, 2)$

We plot $(0, 2)$. We can think of the slope as $\frac{-3}{1}$, so from $(0, 2)$ we move down 3 units and right 1 unit. This locates the point $(1, -1)$. We plot $(1, -1)$ and draw a line passing through $(0, 2)$ and $(1, -1)$.

17. Slope 0; y-intercept $(0, -5)$

Since the slope is 0, we know the line is horizontal, so from $(0, -5)$ we move right 1 unit. This locates the point $(1, -5)$. We plot $(1, -5)$ and draw a line passing through $(0, -5)$ and $(1, -5)$.

19. We read the slope and y-intercept from the equation.

$$y = -\frac{2}{7}x + 5$$

The slope is $-\frac{2}{7}$.   The y-intercept is $(0, 5)$.

21. We read the slope and y-intercept from the equation.

$$y = \frac{1}{3}x + 7$$

The slope is $\frac{1}{3}$.   The y-intercept is $(0, 7)$.

23. $y = \frac{9}{5}x - 4$

$y = \frac{9}{5}x + (-4)$

The slope is $\frac{9}{5}$, and the y-intercept is $(0, -4)$.

25. We solve for y to rewrite the equation in the form $y = mx + b$.

$-3x + y = 7$

$y = 3x + 7$

The slope is 3, and the y-intercept is $(0, 7)$.

27. $4x + 2y = 8$

$2y = -4x + 8$

$y = \frac{1}{2}(-4x + 8)$

$y = -2x + 4$

The slope is -2, and the y-intercept is $(0, 4)$.

29. Observe that this is the equation of a horizontal line that lies 3 units above the x-axis. Thus, the slope is 0, and the y-intercept is $(0, 3)$. We could also write the equation in slope-intercept form.

$y = 3$

$y = 0x + 3$

The slope is 0, and the y-intercept is $(0, 3)$.

31. $2x - 5y = -8$

$-5y = -2x - 8$

$y = -\frac{1}{5}(-2x - 8)$

$y = \frac{2}{5}x + \frac{8}{5}$

The slope is $\frac{2}{5}$, and the y-intercept is $\left(0, \frac{8}{5}\right)$.

33. $9x - 8y = 0$

$-8y = -9x$

$y = \frac{9}{8}x$ or $y = \frac{9}{8}x + 0$

Slope: $\frac{9}{8}$, y-intercept: $(0, 0)$

35. We use the slope-intercept equation, substituting 5 for m and 7 for b:
$$y = mx + b$$
$$y = 5x + 7$$

37. We use the slope-intercept equation, substituting $\frac{7}{8}$ for m and -1 for b:
$$y = mx + b$$
$$y = \frac{7}{8}x - 1$$

39. We use the slope-intercept equation, substituting $-\frac{5}{3}$ for m and -8 for b:
$$y = mx + b$$
$$y = -\frac{5}{3}x - 8$$

41. We use the slope-intercept equation, substituting 0 for m and $\frac{1}{3}$ for b.
$$y = mx + b$$
$$y = 0x + \frac{1}{3}$$
$$y = \frac{1}{3}$$

43. From the graph we see that the y-intercept is $(0, 17)$. We also see that the point $(4, 23)$ is on the graph. We find the slope:
$$m = \frac{23 - 17}{4 - 0} = \frac{6}{4} = \frac{3}{2}$$
Substituting $\frac{3}{2}$ for m and 17 for b in the slope-intercept equation $y = mx + b$, we have
$$y = \frac{3}{2}x + 17,$$
where y is the number of gallons of bottled water consumed per person and x is the number of years since 2000.

45. From the graph we see that the y-intercept is $(0, 15)$. We also see that the point $(5, 17)$ is on the graph. We find the slope:
$$m = \frac{17 - 15}{5 - 0} = \frac{2}{5}$$
Substituting $\frac{2}{5}$ for m and 15 for b in the slope-intercept equation $y = mx + b$, we have
$$y = \frac{2}{5}x + 15,$$
where y is the number of jobs in millions, and x is the number of years since 2000.

47. $y = \frac{2}{3}x + 2$

First we plot the y-intercept $(0, 2)$. We can start at the y-intercept and use the slope, $\frac{2}{3}$, to find another point. We move up 2 units and right 3 units to get a new point $(3, 4)$. Thinking of the slope as $\frac{-2}{-3}$ we can start at $(0, 2)$ and move down 2 units and left 3 units to get another point $(-3, 0)$.

49. $y = -\frac{2}{3}x + 3$

First we plot the y-intercept $(0, 3)$. We can start at the y-intercept and, thinking of the slope as $\frac{-2}{3}$, find another point by moving down 2 units and right 3 units to the point $(3, 1)$. Thinking of the slope as $\frac{2}{-3}$ we can start at $(0, 3)$ and move up 2 units and left 3 units to get another point $(-3, 5)$.

51. $y = \frac{3}{2}x + 3$

First we plot the y-intercept $(0, 3)$. We can start at the y-intercept and use the slope, $\frac{3}{2}$, to find another point. We move up 3 units and right 2 units to get a new point $(2, 6)$. Thinking of the slope as $\frac{-3}{-2}$ we can start at $(0, 3)$ and move down 3 units and left 2 units to get another point $(-2, 0)$.

53. $y = \frac{-4}{3}x + 3$

First we plot the y-intercept $(0, 3)$. We can start at the y-intercept and, thinking of the slope as $\frac{-4}{3}$, find another point by moving down 4 units and right 3 units to the point $(3, -1)$. Thinking of the slope as $\frac{4}{-3}$ we can start at $(0, 3)$ and move up 4 units and left 3 units to get another point $(-3, 7)$.

55. We first rewrite the equation in slope-intercept form.
$$2x + y = 1$$
$$y = -2x + 1$$
Now we plot the y-intercept $(0, 1)$. We can start at the y-intercept and, thinking of the slope as $\frac{-2}{1}$, find another point by moving down 2 units and right 1 unit to the point $(1, -1)$. In a similar manner, we can move from the point $(1, -1)$ to find a third point $(2, -3)$.

57. We first rewrite the equation in slope-intercept form.
$$3x + y = 0$$
$$y = -3x, \text{ or } y = -3x + 0$$
Now we plot the y-intercept $(0,0)$. We can start at the y-intercept and, thinking of the slope as $\dfrac{-3}{1}$, find another point by moving down 3 units and right 1 unit to the point $(1, -3)$. Thinking of the slope as $\dfrac{3}{-1}$ we can start at $(0,0)$ and move up 3 units and left 1 unit to get another point $(-1, 3)$.

59. We first rewrite the equation in slope-intercept form.
$$4x + 5y = 15$$
$$5y = -4x + 15$$
$$y = \frac{1}{5}(-4x + 15)$$
$$y = -\frac{4}{5}x + 3$$
Now we plot the y-intercept $(0,3)$. We can start at the y-intercept and, thinking of the slope as $\dfrac{-4}{5}$, find another point by moving down 4 units and right 5 units to the point $(5, -1)$. Thinking of the slope as $\dfrac{4}{-5}$ we can start at $(0,3)$ and move up 4 units and left 5 units to get another point $(-5, 7)$.

61. We first rewrite the equation in slope-intercept form.
$$x - 4y = 12$$
$$-4y = -x + 12$$
$$y = -\frac{1}{4}(-x + 12)$$
$$y = \frac{1}{4}x - 3$$
Now we plot the y-intercept $(0, -3)$. We can start at the y-intercept and use the slope, $\dfrac{1}{4}$, to find another point. We move up 1 unit and right 4 units to the point $(4, -2)$. Thinking of the slope as $\dfrac{-1}{-4}$ we can start at $(0, -3)$ and move down 1 unit and left 4 units to get another point $(-4, -4)$.

63. The equation $y = \dfrac{3}{4}x + 6$ represents a line with slope $\dfrac{3}{4}$, and the y-intercept is $(0, 6)$.

The equation $y = \dfrac{3}{4}x - 2$ represents a line with slope $\dfrac{3}{4}$, and the y-intercept is $(0, -2)$.

Since both lines have slope $\dfrac{3}{4}$ but different y-intercepts, their graphs are parallel.

65. The equation $y = 2x - 5$ represents a line with slope 2 and y-intercept $(0, -5)$. We rewrite the second equation in slope-intercept form.
$$4x + 2y = 9$$
$$2y = -4x + 9$$
$$y = \frac{1}{2}(-4x + 9)$$
$$y = -2x + \frac{9}{2}$$
The slope is -2 and the y-intercept is $\left(0, \dfrac{9}{2}\right)$. Since the lines have different slopes, their graphs are not parallel.

67. Rewrite each equation in slope-intercept form.
$$3x + 4y = 8$$
$$4y = -3x + 8$$
$$y = \frac{1}{4}(-3x + 8)$$
$$y = -\frac{3}{4}x + 2$$
The slope is $-\dfrac{3}{4}$, and the y-intercept is $(0, 2)$.
$$7 - 12y = 9x$$
$$-12y = 9x - 7$$
$$y = -\frac{1}{12}(9x - 7)$$
$$y = -\frac{3}{4}x + \frac{7}{12}$$
The slope is $-\dfrac{3}{4}$, and the y-intercept is $\left(0, \dfrac{7}{12}\right)$.

Since both lines have slope $-\dfrac{3}{4}$ but different y-intercepts, their graphs are parallel.

69. *Writing Exercise.* Yes; think of the slope as $\dfrac{0}{a}$ for any nonzero value of a.

71. $y - k = m(x - h)$
$\quad\quad y = m(x - h) + k$ Adding k to both sides

73. $-10 - (-3) = -10 + 3 = -7$

75. $-4 - 5 = -4 + (-5) = -9$

77. *Writing Exercise.* Some such circumstances include using an incorrect slope and/or y-intercept when drawing the graph.

79. When $x = 0$, $y = b$, so $(0, b)$ is on the line. When $x = 1$, $y = m + b$, so $(1, m + b)$ is on the line. Then,

slope $= \dfrac{(m + b) - b}{1 - 0} = m$.

81. Rewrite each equation in slope-intercept form.

$$2x - 6y = 10$$
$$-6y = -2x + 10$$
$$y = \frac{1}{3}x - \frac{5}{3}$$

The slope of the line is $\dfrac{1}{3}$.

$$9x + 6y = 18$$
$$6y = -9x + 18$$
$$y = -\frac{3}{2}x + 3$$

The y-intercept of the line is $(0, 3)$.

The equation of the line is $y = \dfrac{1}{3}x + 3$.

83. Rewrite each equation in slope-intercept form.

$$4x + 5y = 9$$
$$5y = -4x + 9$$
$$y = \frac{1}{5}(-4x + 9)$$
$$y = -\frac{4}{5}x + \frac{9}{5}$$

The slope of the line is $-\dfrac{4}{5}$.

$$2x + 3y = 12$$
$$3y = -2x + 12$$
$$y = \frac{1}{3}(-2x + 12)$$
$$y = -\frac{2}{3}x + 4$$

The y-intercept of the line is $(0, 4)$.

The equation of the line is $y = -\dfrac{4}{5}x + 4$.

85. Rewrite each equation in slope-intercept form.

$$-4x + 8y = 5$$
$$8y = 4x + 5$$
$$y = \frac{1}{8}(4x + 5)$$
$$y = \frac{1}{2}x + \frac{5}{8}$$

The slope is $\dfrac{1}{2}$.

$$4x - 3y = 0$$
$$-3y = -4x$$
$$y = -\frac{1}{3}(-4x)$$
$$y = \frac{4}{3}x, \text{ or } y = \frac{4}{3}x + 0$$

The y-intercept is $(0, 0)$.

The equation of the line is $y = \dfrac{1}{2}x + 0$, or $y = \dfrac{1}{2}x$.

87. Rewrite each equation in slope-intercept form.

$$y + 3x = 10$$
$$y = -3x + 10$$

The slope is -3.

$$2x - 6y = 18$$
$$-6y = -2x + 18$$
$$y = -\frac{1}{6}(-2x + 18)$$
$$y = \frac{1}{3}x - 3$$

The slope is $\dfrac{1}{3}$.

Since $-3 \cdot \dfrac{1}{3} = -1$, the graphs of the equations are perpendicular.

89. Rewrite each equation in slope-intercept form.

$$10 - 4y = 7x$$
$$-4y = 7x - 10$$
$$y = -\frac{1}{4}(7x - 10)$$
$$y = -\frac{7}{4}x + \frac{5}{2}$$

The slope is $-\dfrac{7}{4}$.

$$7y + 21 = 4x$$
$$7y = 4x - 21$$
$$y = \frac{1}{7}(4x - 21)$$
$$y = \frac{4}{7}x - 3$$

The slope is $\dfrac{4}{7}$.

Since $-\dfrac{7}{4} \cdot \dfrac{4}{7} = -1$, the graphs of the equations are perpendicular.

91. The slope of $y = -2x$ is -2.

The slope of $x = \dfrac{1}{2}$ is undefined.

Since the product of the slopes is not -1, the graphs of the equations are not perpendicular.

93. Rewrite the first equation in slope-intercept form.

$$3x - 5y = 8$$
$$-5y = -3x + 8$$
$$y = -\frac{1}{5}(-3x + 8)$$
$$y = \frac{3}{5}x - \frac{8}{5}$$

The slope is $\frac{3}{5}$.

The slope of a line perpendicular to this line is a number m such that

$$\frac{3}{5}m = -1, \text{ or}$$
$$m = -\frac{5}{3}.$$

Now rewrite the second equation in slope-intercept form.

$$2x + 4y = 12$$
$$4y = -2x + 12$$
$$y = \frac{1}{4}(-2x + 12)$$
$$y = -\frac{1}{2}x + 3$$

The y-intercept of the line is $(0, 3)$.

The equation of the line is $y = -\frac{5}{3}x + 3$.

95. Rewrite $2x + 5y = 6$ in slope-intercept form.

$$2x + 5y = 6$$
$$5y = -2x + 6$$
$$y = \frac{1}{5}(-2x + 6)$$
$$y = -\frac{2}{5}x + \frac{6}{5}$$

The slope is $-\frac{2}{5}$.

The slope of a line perpendicular to this line is a number m such that

$$-\frac{2}{5}m = -1, \text{ or}$$
$$m = \frac{5}{2}.$$

We graph the line whose equation we want to find. First we plot the given point $(2, 6)$. Now think of the slope as $\frac{-5}{-2}$. From the point $(2, 6)$ go down 5 units and left 2 units to the point $(0, 1)$. Plot this point and draw the graph.

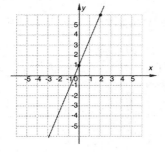

We see that the y-intercept is $(0, 1)$, so the desired equation is $y = \frac{5}{2}x + 1$.

Chapter 3 Connecting the Concepts

1. a) $x = 3$ is linear.
   b) Draw a vertical line through $(3, 0)$.

3. a) $y = \frac{1}{2}x + 3$ is linear.
   b) The y-intercept is $(0, 3)$. Another point is $(2, 4)$.

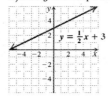

5. a) $y - 5 = x$ is linear.

   b) Rewriting in point-slope form $y = x + 5$. The y-intercept is $(0, 5)$. Another point is $(1, 6)$.

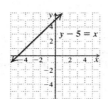

7. $3xy = 6$ is not linear.

9. a) $3 - y = 4$ is linear.

   b) Solving for y,
$$3 - y = 4$$
$$-1 = y$$
   Draw a horizontal line through $(0, -1)$.

11. a) $2y = 9x - 10$ is linear.

b) Rewriting in slope-intercept form $y = \frac{9}{2}x - 5$. The y-intercept is $(0, -5)$. Another point is $(2, 4)$.

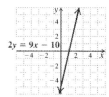

13.   **a)** $2x - 6 = 3y$ is linear.

     b) Solving for y,
$$3y = 2x - 6$$
$$y = \frac{2}{3}x - 2$$
The y-intercept is $(0, -2)$. Another point is $(3, 0)$.

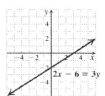

15.   **a)** $2y - x = 4$ is linear.

     b) When $x = 0$,
$$2y = 4$$
$$y = 2$$
the y-intercept is $(0, 2)$.
When $y = 0$,
$$-x = 4$$
$$x = -4$$
the x-intercept is $(-4, 0)$.

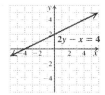

17.   **a)** $x - 2y = 0$ is linear.

     b) When $x = 0$,
$$-2y = 0$$
$$y = 0$$
the y-intercept is $(0, 0)$, also the x-intercept.

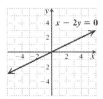

19.   **a)** $y = 4 + x$ is linear.

     b) The y-intercept is $(0, 4)$. Another point is $(1, 5)$.

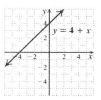

Exercise Set 3.7

1. Substituting 5 for m, 2 for x_1, and 3 for y_1 in the point-slope equation $y - y_1 = m(x - x_1)$, we have $y - 3 = 5(x - 2)$. Choice (g) is correct.

3. Substituting -5 for m, 2 for x_1, and 3 for y_1 in the point-slope equation $y - y_1 = m(x - x_1)$, we have $y - 3 = -5(x - 2)$. Choice (d) is correct.

5. Substituting -5 for m, -2 for x_1, and -3 for y_1 in the point-slope equation $y - y_1 = m(x - x_1)$, we have $y - (-3) = -5(x - (-2))$, or $y + 3 = -5(x + 2)$. Choice (e) is correct.

7. Substituting -5 for m, -3 for x_1, and -2 for y_1 in the point-slope equation $y - y_1 = m(x - x_1)$, we have $y - (-2) = -5(x - (-3))$, or $y + 2 = -5(x + 3)$. Choice (f) is correct.

9. We see that the points $(1, -4)$ and $(-3, 2)$ are on the line. To go from $(1, -4)$ to $(-3, 2)$ we go up 6 units and left 4 units so the slope of the line is $\frac{6}{-4}$, or $-\frac{3}{2}$. Then, substituting $-\frac{3}{2}$ for m, 1 for x_1, and -4 for y_1 in the point-slope equation $y - y_1 = m(x - x_1)$, we have $y - (-4) = -\frac{3}{2}(x - 1)$, or $y + 4 = -\frac{3}{2}(x - 1)$. Choice (c) is correct.

11. We see that the points $(1, -4)$ and $(5, 2)$ are on the line. To go from $(1, -4)$ to $(5, 2)$ we go up 6 units and right 4 units so the slope of the line is $\frac{6}{4}$, or $\frac{3}{2}$. Then, substituting $\frac{3}{2}$ for m, 1 for x_1, and -4 for y_1 in the point-slope equation $y - y_1 = m(x - x_1)$, we have $y - (-4) = \frac{3}{2}(x - 1)$, or $y + 4 = \frac{3}{2}(x - 1)$. Choice (d) is correct.

13. $y - y_1 = m(x - x_1)$
We substitute 3 for m, 1 for x_1, and 6 for y_1.
$$y - 6 = 3(x - 1)$$

15. $y - y_1 = m(x - x_1)$
We substitute $\frac{3}{5}$ for m, 2 for x_1, and 8 for y_1.
$$y - 8 = \frac{3}{5}(x - 2)$$

17. $y - y_1 = m(x - x_1)$
We substitute -4 for m, 3 for x_1, and 1 for y_1.
$$y - 1 = -4(x - 3)$$

19. $y - y_1 = m(x - x_1)$

We substitute $\frac{3}{2}$ for m, 5 for x_1, and -4 for y_1.

$$y - (-4) = \frac{3}{2}(x - 5)$$

21. $y - y_1 = m(x - x_1)$

We substitute $\frac{-5}{4}$ for m, -2 for x_1, and 6 for y_1.

$$y - 6 = \frac{-5}{4}(x - (-2))$$

23. $y - y_1 = m(x - x_1)$

We substitute -2 for m, -4 for x_1, and -1 for y_1.

$$y - (-1) = -2(x - (-4))$$

25. $y - y_1 = m(x - x_1)$

We substitute 1 for m, -2 for x_1, and 8 for y_1.

$$y - 8 = 1(x - (-2))$$

27. First we write the equation in point-slope form.

$$y - y_1 = m(x - x_1)$$
$$y - 5 = 4(x - 3) \quad \text{Substituting}$$

Next we find an equivalent equation of the form $y = mx + b$.

$$y - 5 = 4(x - 3)$$
$$y - 5 = 4x - 12$$
$$y = 4x - 7$$

29. First we write the equation in point-slope form.

$$y - y_1 = m(x - x_1)$$
$$y - (-2) = \frac{7}{4}(x - 4) \quad \text{Substituting}$$

Next we find an equivalent equation of the form $y = mx + b$.

$$y - (-2) = \frac{7}{4}(x - 4)$$
$$y + 2 = \frac{7}{4}x - 7$$
$$y = \frac{7}{4}x - 9$$

31. First we write the equation in point-slope form.

$$y - y_1 = m(x - x_1)$$
$$y - 7 = -2(x - (-3))$$

Next we find an equivalent equation of the form $y = mx + b$.

$$y - 7 = -2(x - (-3))$$
$$y - 7 = -2(x + 3)$$
$$y - 7 = -2x - 6$$
$$y = -2x + 1$$

33. First we write the equation in point-slope form.

$$y - y_1 = m(x - x_1)$$
$$y - (-1) = -4(x - (-2))$$

Next we find an equivalent equation of the form $y = mx + b$.

$$y - (-1) = -4(x - (-2))$$
$$y + 1 = -4(x + 2)$$
$$y + 1 = -4x - 8$$
$$y = -4x - 9$$

35. First we write the equation in point-slope form.

$$y - y_1 = m(x - x_1)$$
$$y - 6 = \frac{2}{3}(x - 5)$$

Next we find an equivalent equation of the form $y = mx + b$.

$$y - 6 = \frac{2}{3}x - \frac{10}{3}$$
$$y = \frac{2}{3}x + \frac{8}{3}$$

37. The slope is $-\frac{5}{6}$ and the y-intercept is $(0, 4)$. Substituting $-\frac{5}{6}$ for m and 4 for b in the slope-intercept equation $y = mx + b$, we have $y = -\frac{5}{6}x + 4$.

39. We plot $(1, 2)$, move up 4 and to the right 3 to $(4, 6)$ and draw the line.

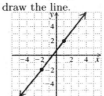

41. We plot $(2, 5)$, move down 3 and to the right 4 to $(6, 2)$ $\left(\text{since } -\frac{3}{4} = \frac{-3}{4}\right)$, and draw the line. We could also think of $-\frac{3}{4}$ and $\frac{3}{-4}$ and move up 3 and to the left 4 from the point $(2, 5)$ to $(-2, 8)$.

43. $y - 5 = \frac{1}{3}(x - 2)$ Point-slope form

The line has slope $\frac{1}{3}$ and passes through $(2, 5)$. We plot $(2, 5)$ and then find a second point by moving up 1 unit and right 3 units to $(5, 6)$. We draw the line through these points.

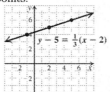

45. $y - 1 = -\frac{1}{4}(x - 3)$ Point-slope form

The line has slope $-\frac{1}{4}$, or $\frac{1}{-4}$ passes through $(3, 1)$. We plot $(3, 1)$ and then find a second point by moving up 1 unit and left 4 units to $(-1, 2)$. We draw the line through these points.

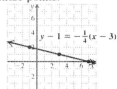

47. $y + 2 = \frac{2}{3}(x - 1)$, or $y - (-2) = \frac{2}{3}(x - 1)$

The line has slope $\frac{2}{3}$ and passes through $(1, -2)$. We plot $(1, -2)$ and then find a second point by moving up 2 units and right 3 units to $(4, 0)$. We draw the line through these points.

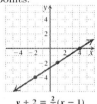

48. $y - 1 = \frac{3}{4}(x + 5)$, or $y - 1 = \frac{3}{4}(x - (-5))$; $m = \frac{3}{4}$

Plot $(-5, 1)$, move up 3 units and right 4 units to $(-1, 4)$, and draw the line.

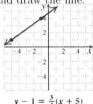

49. $y + 4 = 3(x + 1)$, or $y - (-4) = 3(x - (-1))$

The line has slope 3, or $\frac{3}{1}$, and passes through $(-1, -4)$. We plot $(-1, -4)$ and then find a second point by moving up 3 units and right 1 unit to $(0, -1)$. We draw the line through these points.

51. $y - 4 = -2(x + 1)$, or $y - 4 = -2(x - (-1))$

The line has slope -2, or $\frac{-2}{1}$, and passes through $(-1, 4)$. We plot $(-1, 4)$ and then find a second point by moving

down 2 units and right 1 unit to $(0, 2)$. We draw the line through these points.

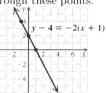

53. $y + 4 = 3(x + 2)$, or $y - (-4) = 3(x - (-2))$

The line has slope 3, or $\frac{3}{1}$, and passes through $(-2, -4)$. We plot $(-2, -4)$ and then find a second point by moving up 3 units and right 1 unit to $(-1, -1)$. We draw the line through these points.

55. $y + 1 = -\frac{3}{5}(x - 2)$, or $y - (-1) = -\frac{3}{5}(x - (-2))$

The line has slope $-\frac{3}{5}$, or $\frac{-3}{5}$ and passes through $(-2, -1)$. We plot $(2, -1)$ and then find a second point by moving up 3 units and left 5 units to $(-3, 2)$, and draw the line.

57. First find the slope of the line passing through the points $(1, 62.1)$ and $(17, 41.1)$.

$$m = \frac{41.1 - 62.1}{17 - 1} = \frac{-21}{16} = -1.3125$$

Now write an equation of the line. We use $(1, 62.1)$ in the point-slope equation and then write an equivalent slope-intercept equation.

$$y - y_1 = m(x - x_1)$$
$$y - 62.1 = -1.3125(x - 1)$$
$$y - 62.1 = -1.3125x + 1.3125$$
$$y = -1.3125x + 63.4125$$

a) Since 1999 is 9 yr after 1990, we substitute 9 for x to calculate the birth rate in 1999.

$$y = -1.3125(9) + 63.4125 = -11.8125 + 63.4125 = 51.6$$

In 1999, there were 51.6 births per 1000 females age 15 to 19.

b) 2008 is 18 yr after 1990 (2008 − 1990 = 18), so we substitute 18 for x.

$y = -1.3125(18) + 63.4125 = -23.625 + 63.4125 = 39.7875 \approx 39.8$

We predict that the birth rate among teenagers will be 39.8 births per 1000 females in 2008.

59. First find the slope of the line passing through the points $(0, 14.2)$ and $(3, 10.8)$. In each case, we let the first coordinate represent the number of years after 2000.

$$m = \frac{10.8 - 14.2}{3 - 0} = \frac{-3.4}{3} \approx -1.13$$

The y-intercept of the line is $(0, 14.2)$. We write the slope-intercept equation: $y = -1.13x + 14.2$.

a) Since 2002−2001 = 1, we substitute 1 for x to calculate the percentage in 2002.

$y = -1.13(1) + 14.2 = -1.131 + 14.2 \approx 13.1\%$

b) Since 2008 − 2001 = 7, we substitute 7 for x to find the percentage in 2008.

$y = -1.13(7) + 14.2 = -10.17 + 14.2 \approx 4\%$

(Answers will vary depending on how the slope was rounded in part (a).)

61. First find the slope of the line passing through $(0, 14.3)$ and $(10, 17.4)$. In each case, we let the first coordinate represent the number of years after 1995.

$$m = \frac{17.4 - 14.3}{10 - 0} = \frac{3.1}{10} = 0.31$$

The y-intercept is $(0, 14.3)$. We write the slope-intercept equation: $y = 0.31x + 14.3$.

a) Since 2002 is 7 yr after 1995, we substitute 7 for x to calculate the college enrollment in 2002.

$y = 0.31(7) + 14.3 = 2.17 + 14.3 = 16.47$ million students

b) Since 2010 − 1995 = 15 yr after 1990, we substitute 15 for x to find the enrollment in 2010.

$y = 0.31(15) + 14.3 = 4.65 + 14.3 = 18.9$ million students

63. First find the slope of the line through $(0, 31)$ and $(12, 36.3)$. In each case, we let the first coordinate represent the number of years after 1990 and the second millions of residents.

$$m = \frac{36.3 - 31}{14 - 0} = \frac{5.3}{14} \approx 0.38$$

The y-intercept is $(0, 31)$. We write the slope-intercept equation: $y = 0.38x + 31$.

a) Since 1997 is 7 yr after 1990, we substitute 7 for x to find the number of U.S. residents over the age of 65 in 1997.

$y = 0.38(7) + 31 = 33.6$ million residents

b) Since 2010 is 20 yr after 1990, we substitute 20 for x to find the number of U.S. residents over the age of 65 in 2010.

$y = 0.38(20) + 31 = 38.6$ million residents

(Answers will vary depending on how the slope is rounded.)

65. $(2, 3)$ and $(4, 1)$

First we find the slope.

$$m = \frac{1 - 3}{4 - 2} = \frac{-2}{2} = -1$$

Then we write an equation of the line in point-slope form using either of the points above.

$$y - 3 = -1(x - 2)$$

Finally, we find an equivalent equation in slope-intercept form.

$$y - 3 = -1(x - 2)$$
$$y - 3 = -x + 2$$
$$y = -x + 5$$

67. $(-3, 1)$ and $(3, 5)$

First we find the slope.

$$m = \frac{1 - 5}{-3 - 3} = \frac{-4}{-6} = \frac{2}{3}$$

Then we write an equation of the line in point-slope form using either of the points above.

$$y - 5 = \frac{2}{3}(x - 3)$$

Finally, we find an equivalent equation in slope-intercept form.

$$y - 5 = \frac{2}{3}(x - 3)$$
$$y - 5 = \frac{2}{3}x - 2$$
$$y = \frac{2}{3}x + 3$$

69. $(5, 0)$ and $(0, -2)$

First we find the slope.

$$m = \frac{0 - (-2)}{5 - 0} = \frac{2}{5}$$

Then we write an equation of the line in point-slope form using either of the points above.

$$y - 0 = \frac{2}{5}(x - 5)$$

Finally, we find an equivalent equation in slope-intercept form.

$$y - 0 = \frac{2}{5}(x - 5)$$
$$y = \frac{2}{5}x - 2$$

71. $(-4, -1)$ and $(1, 9)$

First we find the slope.

$$m = \frac{9 - (-1)}{1 - (-4)} = \frac{9 + 1}{1 + 4} = 2$$

Then we write an equation of the line in point-slope form using either of the points above.

$$y - 9 = 2(x - 1)$$

Finally, we find an equivalent equation in slope-intercept form.

$$y - 9 = 2(x - 1)$$
$$y - 9 = 2x - 2$$
$$y = 2x + 7$$

73. *Writing Exercise.* The equation of a horizontal line $y = b$ can be written in point-slope form:

$$y - b = 0(x - x_1)$$

The equation of a vertical line cannot be written in point-slope form because the slope of a vertical line is undefined.

75. $(-5)^3 = (-5)(-5)(-5) = -125$

77. $-2^6 = -2 \cdot 2 \cdot 2 \cdot 2 \cdot 2 \cdot 2 = -64$

79. $2 - (3 - 2^2) + 10 \div 2 \cdot 5 = 2 - (3 - 4) + 10 \div 2 \cdot 5$
$= 2 - (-1) + 10 \div 2 \cdot 5 = 2 + 1 + 10 \div 2 \cdot 5$
$= 2 + 1 + 5 \cdot 5 = 2 + 1 + 25 = 28$

81. *Writing Exercise.*

(1) Find the slope of the line using $m = \dfrac{y_2 - y_1}{x_2 - x_1}$.

(2) Substitute in the point-slope equation, $y - y_1 = m(x - x_1)$.

(3) Solve for y.

83. $y - 3 = 0(x - 52)$

Observe that the slope is 0. Then this is the equation of a horizontal line that passes through $(52, 3)$. Thus, its graph is a horizontal line 3 units above the x-axis.

$y - 3 = 0(x - 52)$

85. First we find the slope of the line using any two points on the line. We will use $(3, -3)$ and $(4, -1)$.

$$m = \frac{-3 - (-1)}{3 - 4} = \frac{-2}{-1} = 2$$

Then we write an equation of the line in point-slope form using either of the points above.

$$y - (-3) = 2(x - 3)$$

Finally, we find an equivalent equation in slope-intercept form.

$$y - (-3) = 2(x - 3)$$
$$y + 3 = 2x - 6$$
$$y = 2x - 9$$

87. First we find the slope of the line using any two points on the line. We will use $(2, 5)$ and $(5, 1)$.

$$m = \frac{5 - 1}{2 - 5} = \frac{4}{-3} = -\frac{4}{3}$$

Then we write an equation of the line in point-slope form using either of the points above.

$$y - 5 = -\frac{4}{3}(x - 2)$$

Finally, we find an equivalent equation in slope-intercept form.

$$y - 5 = -\frac{4}{3}(x - 2)$$
$$y - 5 = -\frac{4}{3}x + \frac{8}{3}$$
$$y = -\frac{4}{3}x + \frac{23}{3}$$

89. First find the slope of $2x + 3y = 11$.

$$2x + 3y = 11$$
$$3y = -2x + 11$$
$$y = -\frac{2}{3}x + \frac{11}{3}$$

The slope is $-\dfrac{2}{3}$.

Then write a point-slope equation of the line containing $(-4, 7)$ and having slope $-\dfrac{2}{3}$.

$$y - 7 = -\frac{2}{3}(x - (-4))$$

91. The slope of $y = 3 - 4x$ is -4. We are given the y-intercept of the line, so we use slope-intercept form. The equation is $y = -4x + 7$.

93. First find the slope of the line passing through $(2, 7)$ and $(-1, -3)$.

$$m = \frac{-3 - 7}{-1 - 2} = \frac{-10}{-3} = \frac{10}{3}$$

Now find an equation of the line containing the point $(-1, 5)$ and having slope $\dfrac{10}{3}$.

$$y - 5 = \frac{10}{3}(x - (-1))$$
$$y - 5 = \frac{10}{3}(x + 1)$$
$$y - 5 = \frac{10}{3}x + \frac{10}{3}$$
$$y = \frac{10}{3}x + \frac{25}{3}$$

95. $\dfrac{x}{2} + \dfrac{y}{5} = 1$

Using the form $\dfrac{x}{a} + \dfrac{y}{b} = 1$

The x-intercept is $(2, 0)$.
The y-intercept is $(0, 5)$.

97. $4y - 3x = 12$

$$\frac{1}{12}(4y - 3x) = 12 \cdot \frac{1}{12}$$
$$\frac{y}{3} - \frac{x}{4} = 1$$
$$\frac{x}{-4} + \frac{y}{3} = 1$$

The x-intercept is $(-4, 0)$.
The y-intercept is $(0, 3)$.

99. *Writing Exercise.* Equations are entered on most graphing calculators in slope-intercept form. Writing point-slope form in the modified form $y = m(x - x_1) + y_1$ better accommodates graphing calculators.

Chapter 3 Connecting the Concepts

1. $y = -\frac{1}{2}x - 7$ is in slope-intercept form.

3. $x = y + 2$ is none of these.

5. $y - 2 = 5(x + 1)$ is in point-slope form.

7.
$$2x = 5y + 10$$
$$2x - 5y = 10 \qquad \text{Subtracting } 5y \text{ from both sides}$$

9.
$$y = 2x + 7$$
$$-2x + y = 7 \qquad \text{Subtracting } 2x \text{ from both sides}$$
$$2x - y = -7 \qquad \text{Multiplying } -1 \text{ to both sides}$$

11.
$$y - 2 = 3(x + 7)$$
$$y - 2 = 3x + 21 \qquad \text{Using the distributive law}$$
$$y = 3x + 23 \qquad \text{Adding 2}$$
$$-3x + y = 23 \qquad \text{Subtracting } 3x$$
$$3x - y = -23 \qquad \text{Multiplying by } -1$$

13.
$$2x - 7y = 8$$
$$-7y = -2x + 8 \qquad \text{Subtracting } 2x$$
$$-\frac{1}{7}(-7y) = -\frac{1}{7}(-2x + 8) \qquad \text{Multiplying } -\frac{1}{7}$$
$$y = \frac{2}{7}x - \frac{8}{7} \qquad \text{Using distributive law}$$

15.
$$8x = y + 3$$
$$8x - 3 = y \qquad \text{Subtracting 3}$$
$$y = 8x - 3 \qquad \text{rewriting}$$

17.
$$9y = 5 - 8x$$
$$\frac{1}{9}(9y) = \frac{1}{9}(-8x + 5)$$
$$y = -\frac{8}{9}x + \frac{5}{9}$$

19.
$$2 - 3y = 5y + 6$$
$$-4 - 3y = 5y \qquad \text{Subtracting 6}$$
$$-4 = 8y \qquad \text{Adding } 3y$$
$$-\frac{4}{8} = y \qquad \text{Multiplying } \frac{1}{8}$$
$$-\frac{1}{2} = y \qquad \text{Simplifying}$$
$$y = -\frac{1}{2}$$

Chapter 3 Review

1. True, see page 151 of the text.

3. False, slope-intercept form is $y = mx + b$.

5. True, see page 183 of the text.

7. True, see page 185 of the text.

9. True, see page 203 in the text.

11. **Familiarize.** From the pie chart we see that 23.8% of the searches using Yahoo. We let $x =$ the number of searches using Yahoo, in billions in July 2006.

Translate. We reword the problem.

What is 23.8% of 5.6
$$x = 23.8\% \cdot 5.6$$

Carry out.
$$x = 0.238 \cdot 5 \cdot 6 \approx 1.3$$

Check. We can repeat the calculations.

State. About 1.3 billion searches were done using Yahoo.

13.-15. We plot the points $(5, -1)$, $(2, 3)$ and $(-4, 0)$.

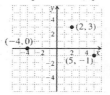

17. Since the first coordinate is positive and the second point is negative, the point $(15.3, -13.8)$ is in quadrant IV.

19. Point A is 5 units left and 1 unit down. The coordinates of A $(-5, -1)$.

21. Point C is 3 units right and 0 units up or down. The coordinates of C are $(3, 0)$

23. a) We substitute 3 for x and 1 for y.
$$\begin{array}{c|c} y = 2x + 7 \\ \hline 1 & 2(3) + 7 \\ & 6 + 7 \\ \hline \overset{?}{1 = 13} & \text{FALSE} \end{array}$$

No, the pair $(3, 1)$ is not a solution.
b) We substitute -3 for x and 1 for y.
$$\begin{array}{c|c} y = 2x + 7 \\ \hline 1 & 2(-3) + 7 \\ & -6 + 7 \\ \hline \overset{?}{1 = 1} & \text{TRUE} \end{array}$$

Yes, the pair $(-3, 1)$ is a solution.

25. $y = x - 5$

The y-intercept is $(0, -5)$. We find two other points.

When $x = 5$, $y = 5 - 5 = 0$.

When $x = 3$, $y = 3 - 5 = -2$.

x	y
0	-5
5	0
3	-2

We plot these points, draw the line and label the graph $y = x - 5$.

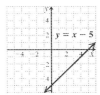

27. $y = -x + 4$

The y-intercept is $(0, 4)$. We find two other points.

When $x = 4$, $y = -4 + 4 = 0$.

When $x = 2$, $y = -2 + 4 = 2$.

x	y
0	4
4	0
2	2

We plot these points, draw the line and label the graph $y = -x + 4$.

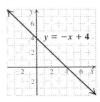

29. $4x + 5 = 3$

$x = -\frac{1}{2}$

Any order pair $\left(-\frac{1}{2}, y\right)$ is a solution. The variable x must be $-\frac{1}{2}$, but the y variable can be any number. A few are listed below.

When $x = 5$, $y = 5 - 5 = 0$.

When $x = 3$, $y = 3 - 5 = -2$.

x	y
$-\frac{1}{2}$	0
$-\frac{1}{2}$	2
$-\frac{1}{2}$	-2

We plot these points, draw the line and label the graph $4x + 5 = 3$.

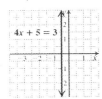

31. We graph $v = -\frac{1}{4}t + 9$ by selecting values of t and calculating the values for v.

If $t = 0$, $v = -\frac{1}{4}(0) + 9 = 9$.

If $t = 4$, $v = -\frac{1}{4}(4) + 9 = 8$.

If $t = 8$, $v = -\frac{1}{4}(8) + 9 = 7$.

t	v
0	9
4	8
8	7

We plot these points, draw the line and label the graph.

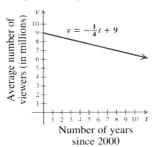

Since $2008 - 2000 = 8$. Locate the point on the line above 8 and find the corresponding value on the vertical axis. The value is 7, so we estimate about 7 million daily viewers in 2008.

33. The points (60 mi, 5 gal) and (120 mi, 10 gal) are on the graph. This tells $120 - 60$ mi and $10 - 5 = 5$gal. The rate is

$$\frac{60 \text{ mi}}{5 \text{ gal}} = 12 \text{ mpg}$$

35. We can use any two points on the line, such as $(-1, -2)$ and $(2, 5)$.

$$m = \frac{\text{change in } y}{\text{change in } x}$$

$$= \frac{5 - (-2)}{2 - (-1)} = \frac{7}{3}$$

37. $(-2, 5)$ and $(3, -1)$

$$m = \frac{-1 - 5}{3 - (-2)} = \frac{-6}{5}$$

39. $(-3, 0)$ and $(-3, 5)$

$$m = \frac{5 - 0}{-3 - (-3)} = \frac{5}{0}, \text{ undefined}$$

41. The grade is expressed as a percent

$$m = \frac{1}{12} \approx 0.08\overline{3} \approx 8.\overline{3}\%$$

43. Rewrite the equation in slope-intercept form.

$3x + 5y = 45$:

$$5y = -3x + 45$$

$$y = -\frac{3}{5}x + 9$$

The slope is $-\frac{3}{5}$ and the $y-$intercept is $(0, 9)$.

45. $y - y_1 = m(x - x_1)$

We substitute $-\frac{1}{3}$ for m, -2 for x_1, and 9 for y_1.

$$y - 9 = -\frac{1}{3}(x - (-2))$$

47. $y - y_1 = m(x - x_1)$

We substitute 5 for m, 3 for x_1, and -10 for y_1.

$$y - (-10) = 5(x - 3)$$
$$y + 10 = 5x - 15$$
$$y = 5x - 25$$

49. $2x + y = 4$

The y-intercept is $(0, 4)$. We find two other points.

When $x = 1$, $2(1) + y = 4$, $y = 2$.

When $x = -2$, $2(-2) + y = 4$, $y = 8$.

x	y
0	4
1	2
-2	8

Plot these points, draw the line, and label the graph $2x + y = 4$.

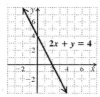

51. $x = -2$

Any ordered pair $(-2, y)$ is a solution. The variable x must be -2, but the y variable can be any number. A few are listed below. Plot these points and draw the graph.

x	y
-2	0
-2	1
-2	2

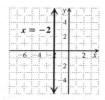

53. *Writing Exercise.* Two perpendicular lines share the same y-intercept if their point of intersection is on the y-axis.

55. $y = mx + 3$, we substitute $(-2, 5)$ and solve.

$$5 = m(-2) + 3$$
$$5 = -2m + 3$$
$$2 = -2m$$
$$-1 = m$$

57. Plot the three given points and observe that the coordinates of the fourth vertex is $(-2, -3)$. The length of the rectangel is $7 - (-2) = 9$ units, and the width is $2 - (-3) = 5$ units.
$A = lw = 9(5) = 45$ sq units
$P = 2l + 2w = 2(9) + 2(5) = 28$ units

Chapter 3 Test

1. First determine the number of student volunteers:
$25\% \times 1200 = 0.25 \times 1200$
$$= 300$$

From the chart, 31.6% of the students will volunteer in education or youth services, so
$31.6\% \times 300 = 0.316 \times 300$
$$= 94.8$$

Therefore, about 95 students will volunteer in education or youth services.

3. The point having coordinates $(-2, -10)$ is located in quadrant III.

5. Point A has coordinates $(3, 4)$.

7. Point C has coordinates $(-5, 2)$.

9. $2x - 4y = -8$

We rewrite the equation in slope-intercept form.
$$2x - 4y = -8$$
$$-4y = -2x - 8$$
$$4y = 2x + 8$$
$$y = \tfrac{1}{2}x + 2$$

Slope: $\tfrac{1}{2}$; y-intercept: $(0, 2)$. First we plot the y-intercept $(0, 2)$. We can start at the y-intercept and use the slope $\tfrac{1}{2}$ to find another point. We move up 1 unit and right 2 units to get a new point $(2, 3)$. Thinking of the slope as $\tfrac{-1}{-2}$ we can start at $(0, 2)$ and move down 1 unit and left 2 units to get another point $(-2, 1)$. To finish, we draw and label

the line.

11. $y = \tfrac{3}{4}x$

We rewrite this equation in slope-intercept form.

$y = \tfrac{3}{4}x + 0$

Slope: $\tfrac{3}{4}$; y-intercept: $(0, 0)$

First we plot the y-intercept $(0, 0)$. We can start at the y-intercept and using the slope as $\tfrac{3}{4}$ we find another point. We move up 3 units and right 4 units to get a new point $(4, 3)$. Thinking of the slope as $\tfrac{-3}{-4}$ we can start at $(0, 0)$ and move down 3 units and left 4 units to get another point $(-4, -3)$. To finish, we draw and label the line.

13. $x = -1$ This is a vertical line with x-intercept $(-1, 0)$.

15. $(-5, 6)$ and $(-1, -3)$

$$m = \frac{-3 - 6}{-1 - (-5)} = \frac{-3 - 6}{-1 + 5} = \frac{-9}{4}$$

17.
$$\text{rate} = \frac{\text{change in distance}}{\text{change in time}}$$

$$= \frac{6 \text{ km} - 3 \text{ km}}{2\text{:}24 \text{ P.M.} - 2\text{:}15 \text{ P.M.}}$$

$$= \frac{3 \text{ km}}{9 \text{min}} = \frac{1}{3} \text{ km/min}$$

19. $5x - y = 30$

To find the x-intercept, we let $y = 0$ and solve for x.
$$5x - 0 \cdot 0 = 30$$
$$5x - 0 = 30$$
$$5x = 30$$
$$x = 6$$

The x-intercept is $(6, 0)$.
To find the y-intercept, we let $x = 0$ and solve for y.
$$5 \cdot 0 - y = 30$$
$$-y = 30$$
$$y = -30$$

The y-intercept is $(0, -30)$.

21. Slope: $-\frac{1}{3}$; y-intercept: $(0, -11)$.

The slope-intercept equation is $y = -\frac{1}{3}x - 11$.

23. $y = \frac{1}{4}x - 2$
Slope: $\frac{1}{4}$; y-intercept: $(0, -2)$.
First we plot the y-intercept and use the slope $\frac{1}{4}$ to find another point. We move up 1 unit and right 4 units to get a new point $(4, -1)$. To finish, we draw and label the line.

25. We first find the slope of the line having equation $2x - 5y = 6$ by writing it in slope- intercept form.
$$2x - 5y = 6$$
$$-5y = -2x + 6$$
$$5y = 2x - 6$$
$$y = \frac{2}{5}x - \frac{6}{5}$$

This line has slope $\frac{2}{5}$ and any line parallel to this line would also have slope $\frac{2}{5}$.

Second, we must find the y-intercept of the line $3x + y = 9$ by writing it in slope- intercept form.
$$3x + y = 9$$
$$y = -3x + 9$$

The y-intercept of this line is $(0, 9)$.

We use the point-slope form of the line to determine the equation of a line having slope $\frac{2}{5}$ and containing the point $(0, 9)$.
$$y - 9 = \frac{2}{5}(x - 0)$$
$$y - 9 = \frac{2}{5}x$$
$$y = \frac{2}{5}x + 9$$

27. $(-2, 14)$ and $(17, -5)$

First determine the slope of the line containing these two points. $m = \frac{-5 - 14}{17 - (-2)} = \frac{-19}{19} = -1$

Using the coordinates of either point and the point-slope formula, we can determine the equation of the line containing the two points. We choose the point $(-2, 14)$.
$$y - 14 = -1(x - (-2))$$
$$y - 14 = -1(x + 2)$$
$$y - 14 = -x - 2$$
$$y = -x + 12$$

Any points whose coordinates satisfy this equation lie on the same line. Answers will vary, but $(0, 12)$, $(-3, 15)$, and $(5, 7)$ all satisfy this equation and are therefore points on the same line.

Chapter 4

Polynomials

Exercise Set 4.1

---

1. By the rule for raising a product to a power on page 226, choice (e) is correct.

3. By the power rule on page 225, choice (b) is correct.

5. By the definition of 0 as an exponent on page 224, choice (g) is correct.

7. By the rule for raising a quotient to a power on page 223, choice (c) is correct.

9. The base is $2x$. The exponent is 5.

11. The base is x. The exponent is 3.

13. The base is $\dfrac{4}{y}$. The exponent is 7.

15. $d^3 \cdot d^{10} = d^{3+10} = d^{13}$

17. $a^6 \cdot a = a^6 \cdot a^1 = a^{6+1} = a^7$

19. $6^5 \cdot 6^{10} = 6^{5+10} = 6^{15}$

21. $(3y)^4(3y)^8 = (3y)^{4+8} = (3y)^{12}$

23. $(8n)(8n)^9 = (8n)^1(8n)^9 = (8n)^{1+9} = (8n)^{10}$

25. $(a^2 b^7)(a^3 b^2) = a^2 b^7 a^3 b^2$ Using an associative law

$\qquad = a^2 a^3 b^7 b^2$ Using a commutative law

$\qquad = a^5 b^9$ Adding exponents

27. $(x+3)^5(x+3)^8 = (x+3)^{5+8} = (x+3)^{13}$

29. $r^3 \cdot r^7 \cdot r^0 = r^{3+7+0} = r^{10}$

31. $(mn^5)(m^3 n^4) = mn^5 m^3 n^4$

$\qquad\qquad\qquad = m^1 m^3 n^5 n^4$

$\qquad\qquad\qquad = m^{1+3} n^{5+4}$

$\qquad\qquad\qquad = m^4 n^9$

33. $\dfrac{7^5}{7^2} = 7^{5-2} = 7^3$ Subtracting exponents

35. $\dfrac{t^8}{t} = t^{8-1} = t^7$ Subtracting exponents

37. $\dfrac{(5a)^7}{(5a)^6} = (5a)^{7-6} = (5a)^1 = 5a$

39. $\dfrac{(x+y)^8}{(x+y)^8}$

Observe that we have an expression divided by itself. Thus, the result is 1.

We could also do this exercise as follows:

$\dfrac{(x+y)^8}{(x+y)^8} = (x+y)^{8-8} = (x+y)^0 = 1$

41. $\dfrac{(r+s)^{12}}{(r+s)^4} = (r+s)^{12-4} = (r+s)^8$

43. $\dfrac{12d^9}{15d^2} = \dfrac{12}{15} d^{9-2} = \dfrac{4}{5} d^7$

45. $\dfrac{8a^9 b^7}{2a^2 b} = \dfrac{8}{2} \cdot \dfrac{a^9}{a^2} \cdot \dfrac{b^7}{b^1} = 4a^{9-2} b^{7-1} = 4a^7 b^6$

46. $3r^8 s^6$

47. $\dfrac{x^{12} y^9}{x^0 y^2} = x^{12-0} y^{9-2} = x^{12} y^7$

49. When $t = 15$, $t^0 = 15^0 = 1$. (Any nonzero number raised to the 0 power is 1.)

51. When $x = -22$, $5x^0 = 5(-22)^0 = 5 \cdot 1 = 5$.

53. $7^0 + 4^0 = 1 + 1 = 2$

55. $(-3)^1 - (-3)^0 = -3 - 1 = -4$

57. $(x^3)^{11} = x^{3 \cdot 11} = x^{33}$ Multiplying exponents

59. $(5^8)^4 = 5^{8 \cdot 4} = 5^{32}$ Multiplying exponents

61. $(t^{20})^4 = t^{20 \cdot 4} = t^{80}$

63. $(10x)^2 = 10^2 x^2 = 100x^2$

65. $(-2a)^3 = (-2)^3 a^3 = -8a^3$

67. $(-5n^7)^2 = (-5)^2 (n^7)^2 = 25n^{7 \cdot 2} = 25n^{14}$

69. $(a^2 b)^7 = (a^2)^7 (b^7) = a^{14} b^7$

71. $(r^5 t)^3 (r^2 t^8) = (r^5)^3 (t)^3 r^2 t^8 = r^{15} t^3 r^2 t^8 = r^{17} t^{11}$

73. $(2x^5)^3 (3x^4) = 2^3 (x^5)^3 (3x^4) = 8x^{15} \cdot 3x^4 = 24x^{19}$

75. $\left(\dfrac{x}{5}\right)^3 = \dfrac{x^3}{5^3} = \dfrac{x^3}{125}$

77. $\left(\dfrac{7}{6n}\right)^2 = \dfrac{7^2}{(6n)^2} = \dfrac{49}{6^2 n^2} = \dfrac{49}{36n^2}$

79. $\left(\dfrac{a^3}{b^8}\right)^6 = \dfrac{(a^3)^6}{(b^8)^6} = \dfrac{a^{18}}{b^{48}}$

81. $\left(\dfrac{x^2 y}{z^3}\right)^4 = \dfrac{(x^2 y)^4}{(z^3)^4} = \dfrac{(x^2)^4 (y^4)}{z^{12}} = \dfrac{x^8 y^4}{z^{12}}$

83. $\left(\dfrac{a^3}{-2b^5}\right)^4 = \dfrac{(a^3)^4}{(-2b^5)^4} = \dfrac{a^{12}}{(-2)^4 (b^5)^4} = \dfrac{a^{12}}{16b^{20}}$

85. $\left(\dfrac{5x^7 y}{-2z^4}\right)^3 = \dfrac{(5x^7 y)^3}{(-2z^4)^3} = \dfrac{5^3 (x^7)^3 y^3}{-2^3 (z^4)^3} = \dfrac{125x^{21} y^3}{-8z^{12}}$

87. $\left(\dfrac{4x^3 y^5}{3z^7}\right)^0$

Observe that for $x \neq 0$, $y \neq 0$, and $z \neq 0$, we have a nonzero number raised to the 0 power. Thus, the result is 1.

89. *Writing Exercise.* -5^2 is the opposite of the square of 5, or the opposite of 25, so it is -25; $(-5)^2$ is the square of the opposite of 5, or $-5(-5)$, so it is 25.

91. $9x + 2y - x - 2y = 9x - x + 2y - 2y = 8x$

93. $-3x + (-2) - 5 - (-x)$
$= -3x + x - 2 - 5$
$= -2x - 7$

95. $4 + x^3 = 4 + 10^3 = 4 + 1000 = 1004$

97. *Writing Exercise.* Any number raised to an even power is nonnegative. Any nonnegative number raised to an odd power is nonnegative. Any negative number raised to an odd power is negative. Thus, a must be a negative number, and n must be an odd number.

99. *Writing Exercise.* Let $s =$ the length of a side of the smaller square. Then $3s =$ the length of a side of the larger square. The area of the smaller square is s^2, and the area of the larger square is $(3s)^2$, or $9s^2$, so the area of the larger square is 9 times the area of the smaller square.

101. Choose any number except 0. For example, let $x = 1$.
$3x^2 = 3 \cdot 1^2 = 3 \cdot 1 = 3$, but
$(3x)^2 = (3 \cdot 1)^2 = 3^2 = 9$.

103. Choose any number except 0 or 1. For example, let $t = -1$.
Then $\dfrac{t^6}{t^2} = \dfrac{(-1)^6}{(-1)^2} = \dfrac{1}{1} = 1$, but $t^3 = (-1)^3 = -1$.

105. $y^{4x} \cdot y^{2x} = y^{4x+2x} = y^{6x}$

107. $\dfrac{x^{5t}(x^t)^2}{(x^{3t})^2} = \dfrac{x^{5t}x^{2t}}{x^{6t}} = \dfrac{x^{7t}}{x^{6t}} = x^t$

109. $\dfrac{t^{26}}{t^x} = t^x$
$t^{26-x} = t^x$
$26 - x = x$ Equating exponents
$26 = 2x$
$13 = x$
The solution is 13.

111. Since the bases are the same, the expression with the larger exponent is larger. Thus, $4^2 < 4^3$.

113. $4^3 = 64$, $3^4 = 81$, so $4^3 < 3^4$.

115. $25^8 = (5^2)^8 = 5^{16}$
$125^5 = (5^3)^5 = 5^{15}$
$5^{16} > 5^{15}$, or $25^8 > 125^5$.

117. $2^{22} = 2^{10} \cdot 2^{10} \cdot 2^2 \approx 10^3 \cdot 10^3 \cdot 4 \approx 1000 \cdot 1000 \cdot 4 \approx 4,000,000$
Using a calculator, we find that $2^{22} = 4,194,304$. The difference between the exact value and the approximation is $4,194,304 - 4,000,000$, or $194,304$.

119. $2^{31} = 2^{10} \cdot 2^{10} \cdot 2^{10} \cdot 2 \approx 10^3 \cdot 10^3 \cdot 10^3 \cdot 2$
$\approx 1000 \cdot 1000 \cdot 1000 \cdot 2 = 2,000,000,000$
Using a calculator, we find that $2^{31} = 2,147,483,648$. The difference between the exact value and the approximation is $2,147,483,648 - 2,000,000,000 = 147,483,648$.

121. $1.5 \text{ MB} = 1.5 \times 1000 \text{ KB}$
$= 1.5 \times 1000 \times 1 \times 2^{10}$ bytes
$= 1,536,000$ bytes
$\approx 1,500,000$ bytes

Exercise Set 4.2

---

1. The only expression with 4 terms is (b).

3. Expression (h) has three terms and they are written in descending order.

5. Expression (g) has two terms, and the degree of the leading term is 7.

7. Expression (a) has two terms, but it is not a binomial because $\dfrac{2}{x^2}$ is not a monomial.

9. $8x^3 - 11x^2 + 6x + 1$
The terms are $8x^3$, $-11x^2$, $6x$, and 1.

11. $-t^6 - 3t^3 + 9t - 4$
The terms are $-t^6$, $-3t^3$, $9t$, and -4.

13. $8x^4 + 2x$

Term	Coefficient	Degree
$8x^4$	8	4
$2x$	2	1

15. $9t^2 - 3t + 4$

Term	Coefficient	Degree
$9t^2$	9	2
$-3t$	-3	1
4	4	0

17. $6a^5 + 9a + a^3$

Term	Coefficient	Degree
$6a^5$	6	5
$9a$	9	1
a^3	1	3

19. $x^4 - x^3 + 4x - 3$

Term	Coefficient	Degree
x^4	1	4
$-x^3$	-1	3
$4x$	4	1
-3	-3	0

21. $5t + t^3 + 8t^4$
a)

Term	$5t$	t^3	$8t^4$
Degree	1	3	4

b) The term of highest degree is $8t^4$. This is the leading term. Then the leading coefficient is 8.

c) Since the term of highest degree is $8t^4$, the degree of the polynomial is 4.

23. $3a^2 - 7 + 2a^4$

a)

Term	$3a^2$	-7	$2a^4$
Degree	2	0	4

b) The term of highest degree is $2a^4$. This is the leading term. Then the leading coefficient is 2.

c) Since the term of highest degree is $2a^4$, the degree of the polynomial is 4.

25. $8 + 6x^2 - 3x - x^5$

a)

Term	8	$6x^2$	$-3x$	$-x^5$
Degree	0	2	1	5

b) The term of highest degree is $-x^5$. This is the leading term. Then the leading coefficient is -1 since $-x^5 = -1 \cdot x^5$.

c) Since the term of highest degree is $-x^5$, the degree of the polynomial is 5.

27. $7x^2 + 8x^5 - 4x^3 + 6 - \dfrac{1}{2}x^4$

Term	Coefficient	Degree of Term	Degree of Polynomial
$8x^5$	8	5	
$-\dfrac{1}{2}x^4$	$-\dfrac{1}{2}$	4	
$-4x^3$	-4	3	5
$7x^2$	7	2	
6	6	0	

29. Three monomials are added, so $x^2 - 23x + 17$ is a trinomial.

31. The polynomial $x^3 - 7x + 2x^2 - 4$ is a polynomial with no special name because it is composed of four monomials.

33. Two monomials are added, so $y + 8$ is a binomial.

35. The polynomial 17 is a monomial because it is the product of a constant and a variable raised to a whole number power. (In this case the variable is raised to the power 0.)

37. $5n^2 + n + 6n^2 = (5 + 6)n^2 + n = 11n^2 + n$

39. $3a^4 - 2a + 2a + a^4 = (3 + 1)a^4 + (-2 + 2)a$
$= 4a^4 + 0a = 4a^4$

41. $7x^3 - 11x + 5x + x^2 = 7x^3 + x^2 + (-11 + 5)x$
$= 7x^3 + x^2 - 6x$

43. $4b^3 + 5b + 7b^3 + b^2 - 6b$
$= (4 + 7)b^3 + b^2 + (5 - 6)b$
$= 11b^3 + b^2 - b$

45. $10x^2 + 2x^3 - 3x^3 - 4x^2 - 6x^2 - x^4$
$= -x^4 + (2 - 3)x^3 + (10 - 4 - 6)x^2 = -x^4 - x^3$

47. $\dfrac{1}{5}x^4 + 7 - 2x^2 + 3 - \dfrac{2}{15}x^4 + 2x^2$
$= \left(\dfrac{1}{5} - \dfrac{2}{15}\right)x^4 + (-2 + 2)x^2 + (7 + 3)$
$= \left(\dfrac{3}{15} - \dfrac{2}{15}\right)x^4 + 0x^2 + 10 = \dfrac{1}{15}x^4 + 10$

49. $8.3a^2 + 3.7a - 8 - 9.4a^2 + 1.6a + 0.5$
$= (8.3 - 9.4)a^2 + (3.7 + 1.6)a - 8 + 0.5$
$= -1.1a^2 + 5.3a - 7.5$

51. For $x = 3$: $-4x + 9 = -4 \cdot 3 + 9$
$= -12 + 9$
$= -3$
For $x = -3$: $-4x + 9 = -4(-3) + 9$
$= 12 + 9$
$= 21$

53. For $x = 3$: $2x^2 - 3x + 7 = 2 \cdot 3^2 - 3 \cdot 3 + 7$
$= 2 \cdot 9 - 3 \cdot 3 + 7$
$= 18 - 9 + 7$
$= 16$
For $x = -3$: $2x^2 - 3x + 7 = 2(-3)^2 - 3(-3) + 7$
$= 2 \cdot 9 - 3(-3) + 7$
$= 18 + 9 + 7$
$= 34$

55. For $x = 3$:
$-3x^3 + 7x^2 - 4x - 8 = -3 \cdot 3^3 + 7 \cdot 3^2 - 4 \cdot 3 - 8$
$= -3 \cdot 27 + 7 \cdot 9 - 12 - 8$
$= -81 + 63 - 12 - 8$
$= -38$
For $x = -3$:
$-3x^3 + 7x^2 - 4x - 8 = -3(-3)^3 + (-3)^2 - 4(-3) - 8$
$= -3(-27) + 7 \cdot 9 + 12 - 8$
$= 148$

57. For $x = 3$: $2x^4 - \dfrac{1}{9}x^3 = 2 \cdot 3^4 - \dfrac{1}{9} \cdot 3^3$
$= 2 \cdot 81 - \dfrac{1}{9} \cdot 27 = 162 - 3 = 159$

For $x = -3$: $2x^4 - \dfrac{1}{9}x^3 = 2(-3)^4 - \dfrac{1}{9}(-3)^3$
$= 2 \cdot 81 - \dfrac{1}{9}(-27)$
$= 162 + 3$
$= 165$

59. For $x = 3$: $-x - x^2 - x^3 = -3 - 3^2 - 3^3 = -3 - 9 - 27 = -39$
For $x = -3$: $-x - x^2 - x^3 = -(-3) - (-3)^2 - (-3)^3$
$= 3 - 9 + 27 = 21$

61. Since 2006 is 2 years after 2004, we evaluate the polynomial for $t = 2$.
$$0.4t + 1.13 = 0.4(2) + 1.13$$
$$= 0.8 + 1.13$$
$$= 1.93$$

The amount spent on shoes for college in 2006 is about $1.93 billion.

63. $11.12t^2 = 11.12(10)^2 = 11.12(100) = 1112$

A skydiver has fallen approximately 1112 ft 10 seconds after jumping from a plane.

65. $2\pi r = 2(3.14)(10)$ Substituting 3.14 for π and 10 for r
$$= 62.8$$

The circumference is 62.8 cm.

67. $\pi r^2 = 3.14(7)^2$ Substituting 3.14 for π and 7 for r
$$= 3.14(49)$$
$$= 153.86$$

The area is 153.86 m^2.

69. Since 2006 is 3 years after 2003, we first locate 3 on the horizontal axis. From there we move vertically to the graph and then horizontally to the K-axis. This locates an K-value of about 75. Thus the number of kidney-paired donations in 2006 is about 75 donations.

71. Locate 10 on the horizontal axis. From there move vertically to the graph and then horizontally to the M-axis. This locates an M-value of about 9. Thus, about 9 words were memorized in 10 minutes.

73. Locate 8 on the horizontal axis. From there move vertically to the graph and then horizontally to the M-axis. This locates an M-value of about 6. Thus, the value of $-0.001t^3 + 0.1t^2$ for $t = 8$ is approximately 6.

75. Locate 4 on the horizontal axis. From there move vertically to the graph and then horizontally to the B-axis. This locates an BMI-value of about 16.

Locate 14 on the horizontal axis. From there move vertically to the graph and then horizontally to the B-axis. This locates an BMI-value of about 19.

77. *Writing Exercise.* A term is a number, a variable, or a product of numbers and variables which may be raised to powers whereas a monomial is a number, a variable, or a product of numbers and variables raised to *whole number* powers. For example, the term $5x^{-2}y^4$ is not a monomial.

79. $2x + 5 - (x + 8) = 2x + 5 - x - 8 = x - 3$

81. $4a + 3 - (-2a + 6) = 4a + 3 + 2a - 6 = 6a - 3$

83. $4t^4 + 8t - (5t^4 - 9t) = 4t^4 + 8t - 5t^4 + 9t = -t^4 + 17t$

85. *Writing Exercise.* Yes; the evaluation will yield a sum of products of integers which must be an integer.

87. Answers may vary. Choose an ax^5-term where a is an even integer. Then choose three other terms with different degrees, each less than degree 5, and coefficients $a+2$, $a+4$, and $a + 6$, respectively, when the polynomial is written in descending order. One such polynomial is $2x^5 + 4x^4 + 6x^3 + 8$.

89. Find the total revenue from the sale of 30 monitors:
$$250x - 0.5x^2 = 250(30) - 0.5(30)^2$$
$$= 250(30) - 0.5(900)$$
$$= 7500 - 450$$
$$= \$7050$$

Find the total cost of producing 30 monitors:
$$4000 + 0.6x^2 = 4000 + 0.6(30)^2$$
$$= 4000 + 0.6(900)$$
$$= 4000 + 540$$
$$= \$4540$$

Subtract the cost from the revenue to find the profit:
$7050 - \$4540 = \$2510$

91.
$$(3x^2)^3 + 4x^2 \cdot 4x^4 - x^4(2x)^2 + [(2x)^2]^3 - 100x^2(x^2)^2$$
$$= 27x^6 + 4x^2 \cdot 4x^4 - x^4 \cdot 4x^2 + (2x)^6 - 100x^2 \cdot x^4$$
$$= 27x^6 + 16x^6 - 4x^6 + 64x^6 - 100x^6$$
$$= 3x^6$$

93. First locate 16 on the vertical axis. Then move horizontally to the graph. We meet the curve at 2 places. At each place move down vertically to the horizontal axis and read the corresponding x-value. We see that the ages for a 16 BMI are 3 and 8.

95. We first find q, the quiz average, and t, the test average.
$$q = \frac{60 + 85 + 72 + 91}{4} = \frac{308}{4} = 77$$
$$t = \frac{89 + 93 + 90}{3} = \frac{272}{3} \approx 90.7$$
Now we substitute in the polynomial.
$$A = 0.3q + 0.4t + 0.2f + 0.1h$$
$$= 0.3(77) + 0.4(90.7) + 0.2(84) + 0.1(88)$$
$$= 23.1 + 36.28 + 16.8 + 8.8$$
$$= 84.98$$
$$\approx 85.0$$

97. When $t = 3$, $-t^2 + 10t - 18 = -3^2 + 10 \cdot 3 - 18 = -9 + 30 - 18 = 3$.
When $t = 4$, $-t^2 + 10t - 18 = -4^2 + 10 \cdot 4 - 18 = -16 + 40 - 18 = 6$.
When $t = 5$, $-t^2 + 10t - 18 = -5^2 + 10 \cdot 5 - 18 = -25 + 50 - 18 = 7$.
When $t = 6$, $-t^2 + 10t - 18 = -6^2 + 10 \cdot 6 - 18 = -36 + 60 - 18 = 6$.
When $t = 7$, $-t^2 + 10t - 18 = -7^2 + 10 \cdot 7 - 18 = -49 + 70 - 18 = 3$.

We complete the table. Then we plot the points and connect them with a smooth curve.

t	$-t^2 + 10t - 18$
3	3
4	6
5	7
6	6
7	3

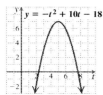

Exercise Set 4.3

1. Since the right-hand side has collected like terms, the correct expression is x^2 to make
$$(3x^2 + 2) + (6x^2 + 7) = (3 + 6)x^2 + (2 + 7).$$

3. Since the right-hand side is the result of using subtraction (the distributive law), the correct operation is $-$ to make
$$(9x^3 - x^2) - (3x^2 + x^2) = 9x^3 - x^2 - 3x^2 - x^2.$$

5. $(3x + 2) + (x + 7) = (3 + 1)x + (2 + 7) = 4x + 9$

7. $(2t + 7) + (-8t + 1) = (2 - 8)t + (7 + 1) = -6t + 8$

9. $(x^2 + 6x + 3) + (-4x^2 - 5) = (1 - 4)x^2 + 6x + (3 - 5)$
$$= -3x^2 + 6x - 2$$

11. $(7t^2 - 3t - 6) + (2t^2 + 4t + 9)$
$$= (7 + 2)t^2 + (-3 + 4)t + (-6 + 9) = 9t^2 + t + 3$$

13. $(4m^3 - 7m^2 + m - 5) + (4m^3 + 7m^2 - 4m - 2)$
$$= (4 + 4)m^3 + (-7 + 7)m^2 + (1 - 4)m + (-5 - 2)$$
$$= 8m^3 - 3m - 7$$

15. $(3 + 6a + 7a^2 + a^3) + (4 + 7a - 8a^2 + 6a^3)$
$$= (1 + 6)a^3 + (7 - 8)a^2 + (6 + 7)a + (3 + 4)$$
$$= 7a^3 - a^2 + 13a + 7$$

17. $(3x^6 + 2x^4 - x^3 + 5x) + (-x^6 + 3x^3 - 4x^2 + 7x^4)$
$$= (3 - 1)x^6 + (2 + 7)x^4 + (-1 + 3)x^3 - 4x^2 + 5x$$
$$= 2x^6 + 9x^4 + 2x^3 - 4x^2 + 5x$$

19. $\left(\frac{3}{5}x^4 + \frac{1}{2}x^3 - \frac{2}{3}x + 3\right) + \left(\frac{2}{5}x^4 - \frac{1}{4}x^3 - \frac{3}{4}x^2 - \frac{1}{6}x\right)$
$$= \left(\frac{3}{5} + \frac{2}{5}\right)x^4 + \left(\frac{1}{2} - \frac{1}{4}\right)x^3 - \frac{3}{4}x^2 + \left(-\frac{2}{3} - \frac{1}{6}\right)x + 3$$
$$= x^4 + \left(\frac{2}{4} - \frac{1}{4}\right)x^3 - \frac{3}{4}x^2 + \left(\frac{-4}{6} - \frac{1}{6}\right)x + 3$$
$$= x^4 + \frac{1}{4}x^3 - \frac{3}{4}x^2 - \frac{5}{6}x + 3$$

21. $(5.3t^2 - 6.4t - 9.1) + (4.2t^3 - 1.8t^2 + 7.3)$
$$= 4.2t^3 + (5.3 - 1.8)t^2 - 6.4t + (-9.1 + 7.3)$$
$$= 4.2t^3 + 3.5t^2 - 6.4t - 1.8$$

23.
$$\begin{array}{r} -4x^3 + 8x^2 + 3x - 2 \\ -4x^2 + 3x + 2 \\ \hline -4x^3 + 4x^2 + 6x + 0 \end{array}$$
$$-4x^3 + 4x^2 + 6x$$

25.
$$\begin{array}{r} 0.05x^4 + 0.12x^3 - 0.5x^2 \\ -0.02x^3 + 0.02x^2 + 2x \\ 1.5x^4 \qquad + 0.01x^2 \qquad + 0.15 \\ 0.25x^3 \qquad\qquad + 0.85 \\ -0.25x^4 \qquad + 10x^2 \qquad - 0.04 \\ \hline 1.3x^4 + 0.35x^3 + 9.53x^2 + 2x + 0.96 \end{array}$$

27. Two forms of the opposite of $-3t^3 + 4t - 7$ are
i) $-(-3t^3 + 4t - 7)$ and
ii) $3t^3 - 4t + 7$. (Changing the sign of every term.)

29. Two forms for the opposite of $x^4 - 8x^3 + 6x$ are
i) $-(x^4 - 8x^3 + 6x)$ and
ii) $-x^4 + 8x^3 - 6x$. (Changing the sign of every term)

31. We change the sign of every term inside parentheses.
$$-(9x - 10) = -9x + 10$$

33. We change the sign of every term inside parentheses.
$$-(3a^4 - 5a^2 + 1.2) = -3a^4 + 5a^2 - 1.2$$

35. We change the sign of every term inside parentheses.
$$-\left(-4x^4 + 6x^2 + \frac{3}{4}x - 8\right) = 4x^4 - 6x^2 - \frac{3}{4}x + 8$$

37. $(3x + 1) - (5x + 8)$
$$= 3x + 1 - 5x - 8 \quad \text{Changing the sign of every term inside parentheses}$$
$$= -2x - 7$$

39. $(-9t + 12) - (t^2 + 3t - 1)$
$$= -9t + 12 - t^2 - 3t + 1$$
$$= -t^2 - 12t + 13$$

41. $(4a^2 + a - 7) - (3 - 8a^3 - 4a^2) = 4a^2 + a - 7 - 3 + 8a^3 + 4a^2$
$$= 8a^3 + 8a^2 + a - 10$$

43. $(1.2x^3 + 4.5x^2 - 3.8x) - (-3.4x^3 - 4.7x^2 + 23)$
$$= 1.2x^3 + 4.5x^2 - 3.8x + 3.4x^3 + 4.7x^2 - 23$$
$$= 4.6x^3 + 9.2x^2 - 3.8x - 23$$

45. $(7x^3 - 2x^2 + 6) - (6 - 2x^2 + 7x^3)$

Observe that we are subtracting the polynomial $7x^3 - 2x^2 + 6$ from itself. The result is 0.

47. $(3 + 5a + 3a^2 - a^3) - (2 + 4a - 9a^2 + 2a^3)$
$$= 3 + 5a + 3a^2 - a^3 - 2 - 4a + 9a^2 - 2a^3$$
$$= 1 + a + 12a^2 - 3a^3$$

49. $\left(\frac{5}{8}x^3 - \frac{1}{4}x - \frac{1}{3}\right) - \left(-\frac{1}{2}x^3 + \frac{1}{4}x - \frac{1}{3}\right)$
$$= \frac{5}{8}x^3 - \frac{1}{4}x - \frac{1}{3} + \frac{1}{2}x^3 - \frac{1}{4}x + \frac{1}{3}$$
$$= \frac{9}{8}x^3 - \frac{2}{4}x$$
$$= \frac{9}{8}x^3 - \frac{1}{2}x$$

51. $(0.07t^3 - 0.03t^2 + 0.01t) - (0.02t^3 + 0.04t^2 - 1)$
$= 0.07t^3 - 0.03t^2 + 0.01t - 0.02t^3 - 0.04t^2 + 1$
$= 0.05t^3 - 0.07t^2 + 0.01t + 1$

53.
$$
\begin{array}{r}
x^3 + 3x^2 + 1 \\
-(x^3 + x^2 - 5) \\
\hline
\end{array}
$$
$$
\begin{array}{r}
x^3 + 3x^2 + 1 \\
-x^3 - x^2 + 5 \\
\hline
2x^2 + 6
\end{array}
$$

55.
$$
\begin{array}{r}
4x^4 - 2x^3 \\
-(7x^4 + 6x^3 + 7x^2) \\
\hline
\end{array}
$$
$$
\begin{array}{r}
4x^4 - 2x^3 \\
-7x^4 - 6x^3 - 7x^2 \\
\hline
-3x^4 - 8x^3 - 7x^2
\end{array}
$$

57. a)

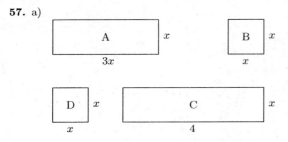

Familiarize. The area of a rectangle is the product of the length and the width.

Translate. The sum of the areas is found as follows:

$$\text{Area of } A + \text{Area of } B + \text{Area of } C + \text{Area of } D$$
$$= 3x \cdot x + x \cdot x + 4 \cdot x + x \cdot x$$

Carry out. We collect like terms.
$$3x^2 + x^2 + 4x + x^2 = 5x^2 + 4x$$

Check. We can go over our calculations. We can also assign some value to x, say 2, and carry out the computation of the area in two ways.

Sum of areas: $3 \cdot 2 \cdot 2 + 2 \cdot 2 + 4 \cdot 2 + 2 \cdot 2$
$$= 12 + 4 + 8 + 4 = 28$$

Substituting in the polynomial:
$$5(2)^2 + 4 \cdot 2 = 20 + 8 = 28$$

Since the results are the same, our solution is probably correct.

State. A polynomial for the sum of the areas is $5x^2 + 4x$.

b)   For $x = 5$: $5x^2 + 4x = 5 \cdot 5^2 + 4 \cdot 5$
$$= 5 \cdot 25 + 4 \cdot 5 = 125 + 20 = 145$$

When $x = 5$, the sum of the areas is 145 square units.

For $x = 7$: $5x^2 + 4x = 5 \cdot 7^2 + 4 \cdot 7$
$$= 5 \cdot 49 + 4 \cdot 7 = 245 + 28 = 273$$

When $x = 7$, the sum of the areas is 273 square units.

59. The perimeter is the sum of the lengths of the sides.
$$4y + 4 + 7 + 2y + 7 + 6 + (3y + 2) + 7y$$
$$= (4 + 2 + 3 + 7)y + (4 + 7 + 7 + 6 + 2)$$
$$= 16y + 26$$

61.

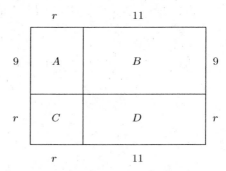

The length and width of the figure can be expressed as $r + 11$ and $r + 9$, respectively. The area of this figure (a rectangle) is the product of the length and width. An algebraic expression for the area is $(r + 11) \cdot (r + 9)$.

The algebraic expressions $9r + 99 + r^2 + 11r$ and $(r + 11) \cdot (r + 9)$ represent the same area.

The area of the figure can be found by adding the areas of the four rectangles A, B, C, and D. The area of a rectangle is the product of the length and the width.

$$\text{Area of } A + \text{Area of } B + \text{Area of } C + \text{Area of } D$$
$$= 9 \cdot r + 11 \cdot 9 + r \cdot r + 11 \cdot r$$
$$= 9r + 99 + r^2 + 11r$$

An algebraic expression for the area of the figure is $9r + 99 + r^2 + 11r$.

63.

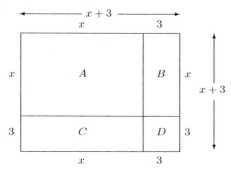

The length and width of the figure can each be expressed as $x + 3$. The area can be expressed as $(x + 3) \cdot (x + 3)$, or $(x + 3)^2$. Another way to express the area is to find an expression for the sum of the areas of the four rectangles A, B, C, and D. The area of each rectangle is the product of its length and width.

$$
\begin{array}{ccccccc}
& \text{Area} & + & \text{Area} & + & \text{Area} & + & \text{Area} \\
& \text{of } A & & \text{of } B & & \text{of } C & & \text{of } D \\
= & x \cdot x & + & 3 \cdot x & + & 3 \cdot x & + & 3 \cdot 3 \\
= & x^2 & + & 3x & + & 3x & + & 9
\end{array}
$$

The algebraic expressions $(x + 3)^2$ and $x^2 + 3x + 3x + 9$ represent the same area.

$$(x + 3)^2 = x^2 + 3x + 3x + 9$$

65. Recall that the area of a rectangle is length times width.

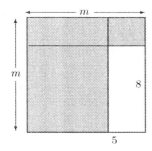

$$
\begin{array}{ccc}
\text{Area of entire} & - & \text{Area not} & = & \text{Shaded} \\
\text{square} & & \text{shaded} & & \text{area} \\
m \cdot m & - & 8 \cdot 5 & = \text{Shaded area} \\
& m^2 - 40 & = \text{Shaded area}
\end{array}
$$

67. Recall that the area of a circle is the product of π and the square of the radius, r^2.

$$A = \pi r^2.$$

The area of a square is the length of on side, s, squared.

$$A = s^2.$$

$$
\begin{array}{ccc}
\text{Area} & - & \text{Area} & = & \text{Shaded} \\
\text{of circle} & & \text{of square} & & \text{area} \\
\pi r^2 & - & 7 \cdot 7 & = & \text{Shaded area} \\
\pi r^2 & - & 49 & = & \text{Shaded area}
\end{array}
$$

69. Familiarize. Recall that the area of a rectangle is the product of the length and the width and that, consequently, the area of a square with side s is s^2. The remaining floor area is the area of the entire floor less the area of the bath enclosure, in square feet.

Translate.

$$
\begin{array}{ccc}
\text{Area of} & - & \text{Area of bath} & = & \text{Remaining} \\
\text{entire floor} & & \text{enclosure} & & \text{floor area} \\
x^2 & - & 2 \cdot 6 & = & \text{Remaining floor area}
\end{array}
$$

Carry out. We simplify the expression.

$$x^2 - 2 \cdot 6 = x^2 - 12$$

Check. We go over the calculations. The answer checks.

State. A polynomial for the remaining floor area is $(x^2 - 12)$ ft^2.

71. Familiarize. Recall that the area of a square with side z is z^2. Recall that the area of a circle with radius 6, or half the diameter is πr^2 or $\pi \cdot 6^2$ The remaining area of the garden is the area of the garden less the area of the patio, in square feet.

Translate.

$$
\begin{array}{ccc}
\text{Area of} & - & \text{Area of} & \text{is} & \text{Remaining} \\
\text{garden} & & \text{patio} & & \text{garden area} \\
z^2 & - & \pi \cdot 6^2 & = & \text{Remaining garden area}
\end{array}
$$

Carry out. We simplify the expression.

$$\left(z^2 - 36\pi\right) \text{ft}^2.$$

Check. We go over the calculations. The answer checks.

State. A polynomial for the remaining area of the garden is $(z^2 - 36\pi)$ ft^2.

73. Familiarize. Recall that the area of a square with side s is s^2 and the area of a circle with radius r is πr^2. The radius of the circle is half the diameter, or $\dfrac{d}{2}$ m. The area of the mat outside the circle is the area of the entire mat less the area of the circle, in square meters.

Translate.

$$
\begin{array}{ccc}
\text{Area} & - & \text{Area of} & \text{is} & \text{Area outside} \\
\text{of mat} & & \text{circle} & & \text{the circle} \\
12^2 & - & \pi \cdot \left(\dfrac{d}{2}\right)^2 & = & \text{Area outside the circle}
\end{array}
$$

Carry out. We simplify the expression.

$$12^2 - \pi \cdot \left(\frac{d}{2}\right)^2 = 144 - \pi \cdot \frac{d^2}{4} = 144 - \frac{d^2}{4}\pi$$

Check. We go over the calculations. The answer checks.

State. A polynomial for the area of the mat outside the wrestling circle is $\left(144 - \dfrac{d^2}{4}\pi\right)$ m^2.

75. *Writing Exercise.* We use the parentheses in $\left(x^2 - 64\pi\right)$ ft^2 to indicate that the units, square feet, apply to the entire quantity in the expression $x^2 - 64\pi$.

77. $2\left(x^2 - x + 3\right) = 2x^2 - 2x + 6$

79. $x^2 \cdot x^6 = x^{2+6} = x^8$

81. $2n \cdot n^2 = 2n^{1+2} = 2n^3$.

83. *Writing Exercise*. The polynomials are opposites.

85. $(6t^2 - 7t) + (3t^2 - 4t + 5) - (9t - 6)$
$= 6t^2 - 7t + 3t^2 - 4t + 5 - 9t + 6$
$= 9t^2 - 20t + 11$

87. $4\left(x^2 - x + 3\right) - 2\left(2x^2 + x - 1\right)$
$= 4x^2 - 4x + 12 - 4x^2 - 2x + 2$
$= (4 - 4)x^2 + (-4 - 2)x + (12 + 2)$
$= -6x + 14$

89. $(345.099x^3 - 6.178x) - (94.508x^3 - 8.99x)$
$= 345.099x^3 - 6.178x - 94.508x^3 + 8.99x$
$= 250.591x^3 + 2.812x$

91. Familiarize. The surface area is $2lw + 2lh + 2wh$, where $l =$ length, $w =$ width, and $h =$ height of the rectangular solid. Here we have $l = 3$, $w = w$, and $h = 7$.

Translate. We substitute in the formula above.
$$2 \cdot 3 \cdot w + 2 \cdot 3 \cdot 7 + 2 \cdot w \cdot 7$$

Carry out. We simplify the expression.
$2 \cdot 3 \cdot w + 2 \cdot 3 \cdot 7 + 2 \cdot w \cdot 7$
$= 6w + 42 + 14w$
$= 20w + 42$

Check. We can go over the calculations. We can also assign some value to w, say 6, and carry out the computation in two ways.

Using the formula: $2 \cdot 3 \cdot 6 + 2 \cdot 3 \cdot 7 + 2 \cdot 6 \cdot 7 =$
$36 + 42 + 84 = 162$

Substituting in the polynomial: $20 \cdot 6 + 42 =$
$120 + 42 = 162$

Since the results are the same, our solution is probably correct.

State. A polynomial for the surface area is $20w + 42$.

93. Familiarize. The surface area is $2lw + 2lh + 2wh$, where $l =$ length, $w =$ width, and $h =$ height of the rectangular solid. Here we have $l = x$, $w = x$, and $h = 5$.

Translate. We substitute in the formula above.
$$2 \cdot x \cdot x + 2 \cdot x \cdot 5 + 2 \cdot x \cdot 5$$

Carry out. We simplify the expression.
$2 \cdot x \cdot x + 2 \cdot x \cdot 5 + 2 \cdot x \cdot 5$
$= 2x^2 + 10x + 10x$
$= 2x^2 + 20x$

Check. We can go over the calculations. We can also assign some value to x, say 3, and carry out the computation in two ways.

Using the formula: $2 \cdot 3 \cdot 3 + 2 \cdot 3 \cdot 5 + 2 \cdot 3 \cdot 5 =$
$18 + 30 + 30 = 78$

Substituting in the polynomial: $2 \cdot 3^2 + 20 \cdot 3 =$
$2 \cdot 9 + 60 = 18 + 60 = 78$

Since the results are the same, our solution is probably correct.

State. A polynomial for the surface area is $2x^2 + 20x$.

95. Length of top edges: $x + 6 + x + 6$, or $2x + 12$
Length of bottom edges: $x + 6 + x + 6$, or $2x + 12$
Length of vertical edges: $4 \cdot x$, or $4x$
Total length of edges: $2x + 12 + 2x + 12 + 4x = 8x + 24$

97. *Writing Exercise*. Yes; $4(-x)^7 - 6(-x)^3 + 2(-x) =$
$-4x^7 + 6x^3 - 2x = -(4x^7 - 6x^3 + 2x)$.

Exercise Set 4.4

1. $3x^2 \cdot 2x^4 = (3 \cdot 2)(x^2 \cdot x^4) = 6x^6$
Choice (c) is correct.

3. $4x^3 \cdot 2x^5 = (4 \cdot 2)(x^3 \cdot x^5) = 8x^8$
Choice (d) is correct.

5. $4x^6 + 2x^6 = (4 + 2)x^6 = 6x^6$
Choice (c) is correct.

7. $(3x^5)7 = (3 \cdot 7)x^5 = 21x^5$

9. $(-x^3)(x^4) = (-1 \cdot x^3)(x^4) = -1(x^3 \cdot x^4) = -1 \cdot x^7 = -x^7$

11. $(-x^6)(-x^2) = (-1 \cdot x^6)(-1 \cdot x^2) = (-1)(-1)(x^6 \cdot x^2) = x^8$

13. $4t^2(9t^2) = (4 \cdot 9)(t^2 \cdot t^2) = 36t^4$

15. $(0.3x^3)(-0.4x^6) = 0.3(-0.4)(x^3 \cdot x^6) = -0.12x^9$

17. $\left(-\dfrac{1}{4}x^4\right)\left(\dfrac{1}{5}x^8\right) = \left(-\dfrac{1}{4} \cdot \dfrac{1}{5}\right)(x^4 \cdot x^8) = -\dfrac{1}{20}x^{12}$

19. $(-5n^3)(-1) = (-5)(-1)n^3 = 5n^3$

21. $11x^5(-4x^5) = (-11 \cdot 4)(x^5 \cdot x^5) = -44x^{10}$

23. $(-4y^5)(6y^2)(-3y^3) = -4(6)(-3)(y^5 \cdot y^2 \cdot y^3) = 72y^{10}$

25. $5x(4x + 1) = 5x(4x) + 5x(1) = 20x^2 + 5x$

27. $(a - 9)3a = a \cdot 3a - 9 \cdot 3a = 3a^2 - 27a$

29. $x^2(x^3 + 1) = x^2(x^3) + x^2(1)$
$= x^5 + x^2$

31. $-3n(2n^2 - 8n + 1)$
$= (-3n)(2n^2) + (-3n)(-8n) + (-3n)(1)$
$= -6n^3 + 24n^2 - 3n$

33. $-5t^2(3t + 6) = -5t^2(3t) - 5t^2(6) = -15t^3 - 30t^2$

35. $\dfrac{2}{3}a^4\left(6a^5 - 12a^3 - \dfrac{5}{8}\right)$
$= \dfrac{2}{3}a^4(6a^5) - \dfrac{2}{3}a^4(12a^3) - \dfrac{2}{3}a^4\left(\dfrac{5}{8}\right)$
$= \dfrac{12}{3}a^9 - \dfrac{24}{3}a^7 - \dfrac{10}{24}a^4$
$= 4a^9 - 8a^7 - \dfrac{5}{12}a^4$

37. $(x + 3)(x + 4) = (x + 3)x + (x + 3)4$
$= x \cdot x + 3 \cdot x + x \cdot 4 + 3 \cdot 4$
$= x^2 + 3x + 4x + 12$
$= x^2 + 7x + 12$

39. $(t + 7)(t - 3) = (t + 7)t + (t + 7)(-3)$
$$= t \cdot t + 7 \cdot t + t(-3) + 7(-3)$$
$$= t^2 + 7t - 3t - 21$$
$$= t^2 + 4t - 21$$

41. $(a - 0.6)(a - 0.7) = (a - 0.6)a + (a - 6)(-0.7)$
$$= a \cdot a - 0.6 \cdot a + a(-0.7) + (-0.6)(-0.7)$$
$$= a^2 - 0.6a - 0.7a + 0.42$$
$$= a^2 - 1.3a + 0.42$$

43. $(x + 3)(x - 3) = (x + 3)x + (x + 3)(-3)$
$$= x \cdot x + 3 \cdot x + x(-3) + 3(-3)$$
$$= x^2 + 3x - 3x - 9$$
$$= x^2 - 9$$

45. $(4 - x)(7 - 2x) = (4 - x)7 + (4 - x)(-2x)$
$$= 4 \cdot 7 - x \cdot 7 + 4(-2x) - x(-2x)$$
$$= 28 - 7x - 8x + 2x^2$$
$$= 28 - 15x + 2x^2$$

47. $\left(t + \dfrac{3}{2}\right)\left(t + \dfrac{4}{3}\right) = \left(t + \dfrac{3}{2}\right)t + \left(t + \dfrac{3}{2}\right)\left(\dfrac{4}{3}\right)$
$$= t \cdot t + \frac{3}{2} \cdot t + t \cdot \frac{4}{3} + \frac{3}{2} \cdot \frac{4}{3}$$
$$= t^2 + \frac{3}{2}t + \frac{4}{3}t + 2$$
$$= t^2 + \frac{9}{6}t + \frac{8}{6}t + 2$$
$$= t^2 + \frac{17}{6}t + 2$$

49. $\left(\dfrac{1}{4}a + 2\right)\left(\dfrac{3}{4}a - 1\right)$
$$= \left(\frac{1}{4}a + 2\right)\left(\frac{3}{4}a\right) + \left(\frac{1}{4}a + 2\right)(-1)$$
$$= \frac{1}{4}a\left(\frac{3}{4}a\right) + 2 \cdot \frac{3}{4}a + \frac{1}{4}a(-1) + 2(-1)$$
$$= \frac{3}{16}a^2 + \frac{3}{2}a - \frac{1}{4}a - 2$$
$$= \frac{3}{16}a^2 + \frac{6}{4}a - \frac{1}{4}a - 2$$
$$= \frac{3}{16}a^2 + \frac{5}{4}a - 2$$

51. Illustrate $x(x + 5)$ as the area of a rectangle with width x and length $x + 5$.

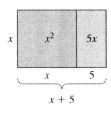

53. Illustrate $(x + 1)(x + 2)$ as the area of a rectangle with width $x + 1$ and length $x + 2$.

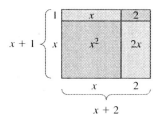

55. Illustrate $(x + 5)(x + 3)$ as the area of a rectangle with length $x + 5$ and width $x + 3$.

57. $(x^2 - x + 3)(x + 1)$
$$= (x^2 - x + 3)x + (x^2 - x + 3)1$$
$$= x^3 - x^2 + 3x + x^2 - x + 3$$
$$= x^3 + 2x + 3$$

A partial check can be made by selecting a convenient replacement for x, say 1, and comparing the values of the original expression and the result.

$(1^2 - 1 + 3)(1 + 1)$ $1^3 + 6 \cdot 1 + 3$
$= (1 - 1 + 3)(1 + 1)$ $= 1 + 6 + 3$
$= 3 \cdot 2$ $= 6$
$= 6$

Since the value of both expressions is 6, the multiplication is very likely correct.

59. $(2a + 5)(a^2 - 3a + 2)$
$$= (2a + 5)a^2 - (2a + 5)(3a) + (2a + 5)2$$
$$= 2a \cdot a^2 + 5 \cdot a^2 - 2a \cdot 3a - 5 \cdot 3a + 2a \cdot 2 + 5 \cdot 2$$
$$= 2a^3 + 5a^2 - 6a^2 - 15a + 4a + 10$$
$$= 2a^3 - a^2 - 11a + 10$$

A partial check can be made as in Exercise 57.

61. $(y^2 - 7)(3y^4 + y + 2)$
$$= (y^2 - 7)(3y^4) + (y^2 - 7)y + (y^2 - 7)(2)$$
$$= y^2 \cdot 3y^4 - 7 \cdot 3y^4 + y^2 \cdot y - 7 \cdot y + y^2 \cdot 2 - 7 \cdot 2$$
$$= 3y^6 - 21y^4 + y^3 - 7y + 2y^2 - 14$$
$$= 3y^6 - 21y^4 + y^3 + 2y^2 - 7y - 14$$

A partial check can be made as in Exercise 57.

63. $(3x + 2)(7x + 4x + 1) = (3x + 2)(11x + 1)$
$$= (3x + 2)(11x) + (3x + 2)(1)$$
$$= 3x \cdot 11x + 2 \cdot 11x + 3x \cdot 1 + 2 \cdot 1$$
$$= 33x^2 + 22x + 3x + 2$$
$$= 33x^2 + 25x + 2$$

65.

$$\begin{array}{l}
x^2 + 5x - 1 \quad \text{Line up like terms}\\
\underline{x^2 - x + 3} \quad \text{in columns}\\
3x^2 + 15x - 3 \quad \text{Multiplying by 3}\\
-x^3 - 5x^2 + x \quad \text{Multiplying by } -x\\
\underline{x^4 + 5x^3 - x^2 \qquad\quad \text{Multiplying by } x^2}\\
x^4 + 4x^3 - 3x^2 + 16x - 3
\end{array}$$

A partial check can be made as in Exercise 57.

67.

$$\begin{array}{l}
5t^2 - t + \tfrac{1}{2}\\
\underline{2t^2 + t - 4}\\
-20t^2 + 4t - 2 \quad \text{Multiplying by } -4\\
5t^3 - t^2 + \tfrac{1}{2}t \quad \text{Multiplying by } t\\
\underline{10t^4 - 2t^3 + t^2 \qquad \text{Multiplying by } 2t^2}\\
10t^4 + 3t^3 - 20t^2 + \tfrac{9}{2}t - 2
\end{array}$$

A partial check can be made as in Exercise 57.

69. We will multiply horizontally while still aligning like terms.

$(x + 1)(x^3 + 7x^2 + 5x + 4)$

$= x^4 + 7x^3 + 5x^2 + 4x$ $\quad$ Multiplying by x
$\quad\ + x^3 + 7x^2 + 5x + 4$ $\quad$ Multiplying by 1
$= x^4 + 8x^3 + 12x^2 + 9x + 4$

A partial check can be made as in Exercise 57.

71. *Writing Exercise.* No; the distributive law is the basis for polynomial multiplication.

73. $(9 - 3)(9 + 3) + 3^2 - 9^2 = (6)(12) + 3^2 - 9^2$
$= (6)(12) + 9 - 81 = 72 + 9 - 81 = 0$

75. $5 + \dfrac{7 + 4 + 2 \cdot 5}{7}$

$= 5 + \dfrac{7 + 4 + 10}{7}$

$= 5 + \dfrac{21}{7}$

$= 5 + 3$

$= 8$

77. $(4 + 3 \cdot 5 + 5) \div 3 \cdot 4$

$= (4 + 15 + 5) \div 3 \cdot 4$

$= 24 \div 3 \cdot 4$

$= 8 \cdot 4$

$= 32$

79. *Writing Exercise.* $(A+B)(C+D)$ will be a trinomial when there is exactly one pair of like terms among AC, AD, BC, and BD.

81. The shaded area is the area of the large rectangle, $6y(14y - 5)$ less the area of the unshaded rectangle, $3y(3y + 5)$. We have:

$6y(14y - 5) - 3y(3y + 5)$

$= 84y^2 - 30y - 9y^2 - 15y$

$= 75y^2 - 45y$

83. Let $n =$ the missing number.

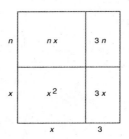

The area of the figure is $x^2 + 3x + nx + 3n$. This is equivalent to $x^2 + 8x + 15$, so we have $3x + nx = 8x$ and $3n = 15$. Solving either equation for n, we find that the missing number is 5.

85.

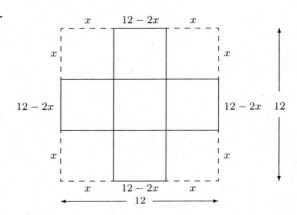

The dimensions, in inches, of the box are $12 - 2x$ by $12 - 2x$ by x. The volume is the product of the dimensions (volume = length × width × height):

Volume $= (12 - 2x)(12 - 2x)x$
$= (144 - 48x + 4x^2)x$
$= (144x - 48x^2 + 4x^3)$ in^3, or
$(4x^3 - 48x^2 + 144x)$ in^3

The outside surface area is the sum of the area of the bottom and the areas of the four sides. The dimensions, in inches, of the bottom are $12 - 2x$ by $12 - 2x$, and the dimensions, in inches, of each side are x by $12 - 2x$.

Surface area = Area of bottom + 4 · Area of each side
$= (12 - 2x)(12 - 2x) + 4 \cdot x(12 - 2x)$
$= 144 - 24x - 24x + 4x^2 + 48x - 8x^2$
$= 144 - 48x + 4x^2 + 48x - 8x^2$
$= (144 - 4x^2)$ in^2, or $(-4x^2 + 144)$ in^2

87.

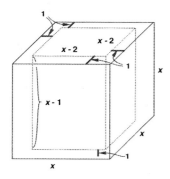

The interior dimensions of the open box are $x - 2$ cm by $x - 2$ cm by $x - 1$ cm.

$$\begin{aligned}
\text{Interior volume} &= (x - 2)(x - 2)(x - 1) \\
&= (x^2 - 4x + 4)(x - 1) \\
&= (x^3 - 5x^2 + 8x - 4) \text{ cm}^3
\end{aligned}$$

89. Let $x =$ the width of the garden. Then $2x =$ the length of the garden.

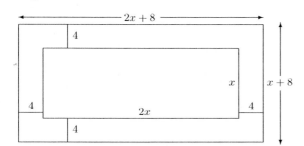

Area of garden and sidewalk together	is	Area of garden alone	+	256 ft^2
↓	↓	↓	↓	↓
$(2x + 8)(x + 8)$	$=$	$2x \cdot x$	$+$	256

$$2x^2 + 24x + 64 = 2x^2 + 256$$
$$24x = 192$$
$$x = 8$$

The dimensions are 8 ft by 16 ft.

91. $(x - 2)(x - 7) - (x - 7)(x - 2)$

First observe that, by the commutative law of multiplication, $(x - 2)(x - 7)$ and $(x - 7)(x - 2)$ are equivalent expressions. Then when we subtract $(x - 7)(x - 2)$ from $(x - 2)(x - 7)$, the result is 0.

93. $(x + 2)(x + 4)(x - 5)$
$$\begin{aligned}
&= (x^2 + 2x + 4x + 8)(x - 5) \\
&= (x^2 + 6x + 8)(x - 5) \\
&= x^3 + 6x^2 + 8x - 5x^2 - 30x - 40 \\
&= x^3 + x^2 - 22x - 40
\end{aligned}$$

95. $(x - a)(x - b) \cdots (x - x)(x - y)(x - z)$
$$\begin{aligned}
&= (x - a)(x - b) \cdots 0 \cdot (x - y)(x - z) \\
&= 0
\end{aligned}$$

1. It is true that FOIL is simply a memory device for finding the product of two binomials.

3. This statement is false. See Example 2(d).

5. $(x^2 + 2)(x + 3)$
$$\begin{array}{cccc} \text{F} & \text{O} & \text{I} & \text{L} \end{array}$$
$$= x^2 \cdot x + x^2 \cdot 3 + 2 \cdot x + 2 \cdot 3$$
$$= x^3 + 2x + 3x^2 + 6, \text{ or } x^3 + 3x^2 + 2x + 6$$

7. $(t^4 - 2)(t + 7)$
$$\begin{array}{cccc} \text{F} & \text{O} & \text{I} & \text{L} \end{array}$$
$$= t^4 \cdot t + t^4 \cdot 7 - 2 \cdot t - 2 \cdot 7$$
$$= t^5 + 7t^4 - 2t - 14$$

9. $(y + 2)(y - 3)$
$$\begin{array}{cccc} \text{F} & \text{O} & \text{I} & \text{L} \end{array}$$
$$= y \cdot y + y \cdot (-3) + 2 \cdot y + 2 \cdot (-3)$$
$$= y^2 - 3y + 2y - 6$$
$$= y^2 - y - 6$$

11. $(3x + 2)(3x + 5)$
$$\begin{array}{cccc} \text{F} & \text{O} & \text{I} & \text{L} \end{array}$$
$$= 3x \cdot 3x + 3x \cdot 5 + 2 \cdot 3x + 2 \cdot 5$$
$$= 9x^2 + 15x + 6x + 10$$
$$= 9x^2 + 21x + 10$$

13. $(5x - 3)(x + 4)$
$$\begin{array}{cccc} \text{F} & \text{O} & \text{I} & \text{L} \end{array}$$
$$= 5x \cdot x + 5x \cdot 4 - 3 \cdot x - 3 \cdot 4$$
$$= 5x^2 + 20x - 3x - 12$$
$$= 5x^2 + 17x - 12$$

15. $(3 - 2t)(5 - t)$
$$\begin{array}{cccc} \text{F} & \text{O} & \text{I} & \text{L} \end{array}$$
$$= 3 \cdot 5 + 3 \cdot t - 2t \cdot 5 + (-2t)(-t)$$
$$= 15 - 3t - 10t + 2t^2$$
$$= 15 - 13t + 2t^2$$

17. $(x^2 + 3)(x^2 - 7)$
$$\begin{array}{cccc} \text{F} & \text{O} & \text{I} & \text{L} \end{array}$$
$$= x^2 \cdot x^2 - x^2 \cdot 7 + 3 \cdot x^2 - 3 \cdot 7$$
$$= x^4 - 7x^2 + 3x^2 - 21$$
$$= x^4 - 4x^2 - 21$$

19. $\left(p - \dfrac{1}{4}\right)\left(p + \dfrac{1}{4}\right)$
$$\begin{array}{cccc} \text{F} & \text{O} & \text{I} & \text{L} \end{array}$$
$$= p \cdot p + p \cdot \frac{1}{4} + \left(-\frac{1}{4}\right) \cdot p + \left(-\frac{1}{4}\right) \cdot \frac{1}{4}$$
$$= p^2 + \frac{1}{4}p - \frac{1}{4}p - \frac{1}{16}$$
$$= p^2 - \frac{1}{16}$$

21. $(x - 0.3)(x - 0.3)$

$\qquad$ F $\qquad$ O $\qquad$ I $\qquad$ L

$= x \cdot x - x \cdot 0.3 + -0.3 \cdot x + (-0.3)(-0.3)$

$= x^2 - 0.3x - 0.3x + 0.09$

$= x^2 - 0.6x + 0.09$

23. $(-3n + 2)(n + 7)$

$\qquad$ F $\qquad$ O $\qquad$ I $\qquad$ L

$= -3n \cdot n - 3n \cdot 7 + 2 \cdot n + 2 \cdot 7$

$= -3n^2 - 21n + 2n + 14$

$= -3n^2 - 19n + 14$

25. $(x + 10)(x + 10)$

$\qquad$ F $\qquad$ O $\qquad$ I $\qquad$ L

$= x^2 + 10x + 10x + 100$

$= x^2 + 10x + 10x + 100$

$= x^2 + 20x + 100$

27. $(1 - 3t)(1 + 5t^2)$

$\qquad$ F $\qquad$ O $\qquad$ I $\qquad$ L

$= 1 + 5t^2 - 3t - 15t^3$

$= 1 - 3t - 5t^2 - 15t^3$

29. $(x^2 + 3)(x^3 - 1)$

$\qquad$ F $\qquad$ O $\qquad$ I $\qquad$ L

$= x^5 - x^2 + 3x^3 - 3$, or $x^5 + 3x^3 - x^2 - 3$

31. $(3x^2 - 2)(x^4 - 2)$

$\qquad$ F $\qquad$ O $\qquad$ I $\qquad$ L

$= 3x^6 - 6x^2 - 2x^4 + 4$, or $3x^6 - 2x^4 - 6x^2 + 4$

33. $(2t^3 + 5)(2t^3 + 5)$

$\qquad$ F $\qquad$ O $\qquad$ I $\qquad$ L

$= 4t^6 + 10t^3 + 10t^3 + 25$

$= 4t^6 + 20t^3 + 25$

35. $(8x^3 + 5)(x^2 + 2)$

$\qquad$ F $\qquad$ O $\qquad$ I $\qquad$ L

$= 8x^5 + 16x^3 + 5x^2 + 10$

37. $(10x^2 + 3)(10x^2 - 3)$

$\qquad$ F $\qquad$ O $\qquad$ I $\qquad$ L

$= 100x^4 - 30x^2 + 30x^2 - 9$

$= 100x^4 - 9$

39. $(x + 8)(x - 8)$ Product of sum and difference of the same two terms

$= x^2 - 8^2$

$= x^2 - 64$

41. $(2x + 1)(2x - 1)$ Product of sum and difference of the same two terms

$= (2x)^2 - 1^2$

$= 4x^2 - 1$

43. $(5m^2 + 4)(5m^2 - 4)$ Product of sum and difference of the same two terms

$= (5m)^2 - 4^2$

$= 25m^4 - 16$

45. $(9a^3 + 1)(9a^3 - 1)$

$= (9a^3)^2 - 1^2$

$= 81a^6 - 1$

47. $(x^4 + 0.1)(x^4 - .01)$

$= (x^4)^2 - 0.1^2$

$= x^8 - 0.01$

49. $\left(t - \dfrac{3}{4}\right)\left(t + \dfrac{3}{4}\right)$

$= t^2 - \left(\dfrac{3}{4}\right)^2$

$= t^2 - \dfrac{9}{16}$

51. $(x + 3)^2$

$= x^2 + 2 \cdot x \cdot 3 + 3^2$ $\qquad$ Square of a binomial

$= x^2 + 6x + 9$

53. $(7x^3 - 1)^2$ $\qquad$ Square of a binomial

$= (7x^3)^2 - 2 \cdot 7x^3 \cdot 1 + (-1)^2$

$= 49x^6 - 14x^3 + 1$

55. $\left(a - \dfrac{2}{5}\right)^2$ $\qquad$ Square of a binomial

$= a^2 - 2 \cdot a \cdot \dfrac{2}{5} + \left(\dfrac{2}{5}\right)^2$

$= a^2 - \dfrac{4}{5}a + \dfrac{4}{25}$

57. $(t^4 + 3)^2$ $\qquad$ Square of a binomial

$= (t^4)^2 + 2 \cdot t^4 \cdot 3 + 3^2$

$= t^8 + 6t^4 + 9$

59. $(2 - 3x^4)^2 = 2^2 - 2 \cdot 2 \cdot 3x^4 + (3x^4)^2$

$= 4 - 12x^4 + 9x^8$

61. $(5 + 6t^2)^2 = 5^2 + 2 \cdot 5 \cdot 6t^2 + (6t^2)^2$

$= 25 + 60t^2 + 36t^4$

63. $(7x - 0.3)^2 = (7x)^2 - 2(7x)(0.3) + (0.3)^2$

$= 49x^2 - 4.2x + 0.09$

65. $7n^3(2n^2 - 1)$

$= 7n^3 \cdot 2n^2 - 7n^3 \cdot 1$ Multiplying each term of

$= 14n^5 - 7n^3$ $\qquad$ the binomial by the monomial

67. $(a - 3)(a^2 + 2a - 4)$

$= a^3 + 2a^2 - 4a$ $\qquad$ Multiplying horizontally
$\quad - 3a^2 - 6a + 12$ and aligning like terms

$= a^3 - a^2 - 10a + 12$

69. $(7 - 3x^4)(7 - 3x^4)$

$= 7^2 - 2 \cdot 7 \cdot 3x^4 + (-3x^4)^2$ $\qquad$ Squaring a binomial

$= 49 - 42x^4 + 9x^8$

71. $5x(x^2 + 6x - 2)$

$= 5x \cdot x^2 + 5x \cdot 6x + 5x(-2)$ $\qquad$ Multiplying each
$\qquad$ term of the trinomial
$\qquad$ by the monomial

$= 5x^3 + 30x^2 - 10x$

73. $(q^5 + 1)(q^5 - 1)$
$= (q^5)^2 - 1^2$
$= q^{10} - 1$

75. $3t^2(5t^3 - t^2 + t)$
$= 3t^2 \cdot 5t^3 + 3t^2(-t^2) + 3t^2 \cdot t$ Multiplying each
                                               term of the trinomial
                                               by the monomial
$= 15t^5 - 3t^4 + 3t^3$

77. $(6x^4 - 3x)^2$ Squaring a binomial
$= (6x^4)^2 - 2 \cdot 6x^4 \cdot 3x + (-3x^2)$
$= 36x^8 - 36x^5 + 9x^2$

79. $(9a + 0.4)(2a^3 + 0.5)$ Product of two
                                           binomials; use FOIL
$= 9a \cdot 2a^3 + 9a \cdot 0.5 + 0.4 \cdot 2a^3 + 0.4 \cdot 0.5$
$= 18a^4 + 4.5a + 0.8a^3 + 0.2,$ or
    $18a^4 + 0.8a^3 + 4.5a + 0.2$

81. $\left(\dfrac{1}{5} - 6x^4\right)\left(\dfrac{1}{5} + 6x^4\right)$
$= \left(\dfrac{1}{5}\right)^2 - (6x^4)^2$
$= \dfrac{1}{25} - 36x^8$

83. $(a+1)(a^2 - a + 1)$

  $= a^3 - a^2 + a$ Multiplying horizontally
         $a^2 - a + 1$ and aligning like terms
  $= a^3$ $+ 1$

85.

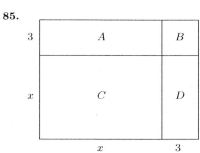

We can find the shaded area in two ways.

Method 1: The figure is a square with side $x + 3$, so the area is $(x + 3)^2 = x^2 + 6x + 9$.

Method 2: We add the areas of A, B, C, and D.

$3 \cdot x + 3 \cdot 3 + x \cdot x + x \cdot 3 = 3x + 9 + x^2 + 3x$
$= x^2 + 6x + 9.$

Either way we find that the total shaded area is $x^2 + 6x + 9$.

87.

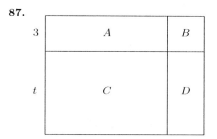

We can find the shaded area in two ways.

Method 1: The figure is a rectangle with dimensions $t + 3$ by $t + 4$, so the area is

$(t + 3)(t + 4) = t^2 + 4t + 3t + 12 = t^2 + 7t + 12.$

Method 2: We add the areas of A, B, C, and D.

$3 \cdot t + 3 \cdot 4 + t \cdot t + t \cdot 4 = 3t + 12 + t^2 + 4t = t^2 + 7t + 12.$

Either way, we find that the area is $t^2 + 7t + 12.$

89.

We can find the shaded area in two ways.

Method 1: The figure is a square with side $a + 5$, so the area is $(a + 5)^2 = a^2 + 10a + 25.$

Method 2: We add the areas of A, B, C, and D.

$5 \cdot a + 5 \cdot 5 + a \cdot a + 5 \cdot a = 5a + 25 + a^2 + 5a = a^2 + 10a + 25.$

Either way, we find that the total shaded area is $a^2 + 10a + 25.$

91.

We can find the shaded area in two ways.

Method 1: The figure is a rectangle with dimensions $x + 7$ by $x + 3$, so the area is $(x + 7)(x + 3)$
$= x^2 + 3x + 7x + 21 = x^2 + 10x + 21.$

Method 2: We add the areas of A, B, C, and D.

$x \cdot x + x \cdot 7 + 3 \cdot x + 3 \cdot 7 = x^2 + 10x + 21.$

Either way, we find that the total shaded area is $x^2 + 10x + 21.$

93.

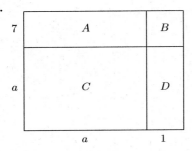

We can find the shaded area in two ways.

Method 1: The figure is a rectangle with dimensions $a + 1$ by $a + 7$, so the area is

$(a + 1)(a + 7) = a^2 + 7a + a + 7 = a^2 + 8a + 7$

Method 2: We add the areas of A, B, C, and D.

$a \cdot a + a \cdot 1 + 7 \cdot a + 7 \cdot 1 = a^2 + a + 7a + 7 = a^2 + 8a + 7.$

Either way, we find that the total shaded area is $a^2 + 8a + 7$.

95.

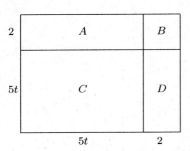

We can find the shaded area in two ways.

Method 1: The figure is a square with side $5t + 2$, so the area is $(5t + 2)^2 = 25t^2 + 20t + 4$.

Method 2: We add the areas of A, B, C, and D.

$5t \cdot 5t + 5t \cdot 2 + 2 \cdot 5t + 2 \cdot 2 = 25t^2 + 10t + 10t + 4 = 25t^2 + 20t + 4$.

Either way, we find that the total shaded area is $25t^2 + 20t + 4$.

97. We draw a square with side $x + 5$.

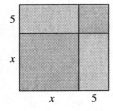

99. We draw a square with side $t + 9$.

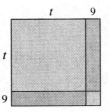

101. We draw a square with side $3 + x$.

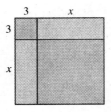

103. *Writing Exercise.* It's a good idea to study the other special products, because they allow for faster computations than the FOIL method.

105. *Familiarize.* Let $w =$ the energy, in kilowatt-hours per month, used by the washing machine. Then $21w =$ the amount of energy used by the refrigerator, and $11w =$ the amount of energy used by the freezer.

Translate.

Washing Machine	and	refrigerator	and	freezer	is	Total energy
↓	↓	↓	↓	↓	↓	↓
w	+	$21w$	+	$11w$	=	189

Solve. We solve the equation.

$$w + 21t + 11w = 297$$
$$33w = 297$$
$$w = 9$$

Then $21w = 21 \cdot 9 = 189$

and $11w = 11 \cdot 9 = 99$

Check. The energy used by the refrigerator, 189 kWh, is 21 times the energy used by the washing machine. The energy used by the freezer, 99 kWh is 11 times the energy used by the washing machine. Also, $9 + 189 + 99 = 297$, the total energy used.

State. The washing machine used 9 kWh, the refrigerator used 189 kWh, and the freezer used 99 kWh.

107. $5xy = 8$

$y = \dfrac{8}{5x}$ Dividing both sides by $5x$

109. $ax - by = c$

$ax = by + c$ Adding by to both sides

$x = \dfrac{by + c}{a}$ Dividing both sides by a

111. *Writing Exercise.* The computation $(20-1)(20+1) = 400-1 = 399$ is easily performed mentally and is equivalent to the computation $19 \cdot 21$.

113. $(4x^2+9)(2x+3)(2x-3)$
$= (4x^2+9)(4x^2-9)$
$= 16x^4 - 81$

115. $(3t-2)^2(3t+2)^2$
$= [(3t-2)(3t+2)]^2$
$= (9t^2-4)^2$
$= 81t^4 - 72t^2 + 16$

117. $(t^3-1)^4(t^3+1)^4$
$= [(t^3-1)(t^3+1)]^4$
$= (t^6-1)^4$
$= [(t^6-1)^2]^2$
$= (t^{12} - 2t^6 + 1)^2$
$= (t^{12} - 2t^6 + 1)(t^{12} - 2t^6 + 1)$
$= t^{24} - 2t^{18} + t^{12} - 2t^{18} + 4t^{12} - 2t^6 +$
$\qquad t^{12} - 2t^6 + 1$
$= t^{24} - 4t^{18} + 6t^{12} - 4t^6 + 1$

119. $18 \times 22 = (20-2)(20+2) = 20^2 - 2^2$
$= 400 - 4 = 396$

121. $(x+2)(x-5) = (x+1)(x-3)$
$x^2 - 5x + 2x - 10 = x^2 - 3x + x - 3$
$x^2 - 3x - 10 = x^2 - 2x - 3$
$\quad -3x - 10 = -2x - 3 \qquad$ Adding $-x^2$
$\quad -3x + 2x = 10 - 3 \qquad$ Adding $2x$ and 10
$\qquad\qquad -x = 7$
$\qquad\qquad\quad x = -7$

The solution is -7.

123.

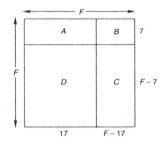

The area of the entire figure is F^2. The area of the un-shaded region, C, is $(F-7)(F-17)$. Then one expression for the area of the shaded region is
$F^2 - (F-7)(F-17)$.

To find a second expression we add the areas of regions A, B, and D. We have:
$17 \cdot 7 + 7(F-17) + 17(F-7)$
$= 119 + 7F - 119 + 17F - 119$
$= 24F - 119$

It is possible to find other equivalent expressions also.

125. The dimensions of the shaded area, regions A and D together, are $y+1$ by $y-1$ so the area is $(y+1)(y-1)$.

To find another expression we add the areas of regions A and D. The dimensions of region A are y by $y-1$, and the dimensions of region D are $y-1$ by 1, so the sum of the areas is $y(y-1) + (y-1)(1)$, or $y(y-1) + y - 1$.

It is possible to find other equivalent expressions also.

127.

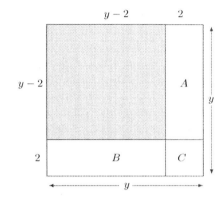

The shaded area is $(y-2)^2$. We find it as follows:

$$\begin{array}{l}
\frac{\text{Shaded}}{\text{area}} = \frac{\text{Area of}}{\text{square}} - \frac{\text{Area}}{\text{of } A} - \frac{\text{Area}}{\text{of } B} - \frac{\text{Area}}{\text{of } C} \\[2mm]
(y-2)^2 = y^2 - 2(y-2) - 2(y-2) - 2 \cdot 2 \\[2mm]
(y-2)^2 = y^2 - 2y + 4 - 2y + 4 - 4 \\[2mm]
(y-2)^2 = y^2 - 4y + 4
\end{array}$$

129. ▨

Chapter 4 Connecting the Concepts

1. $(3x^2 - 2x + 6) + (5x - 3) \qquad$ Addition
$= 3x^2 - 2x + 5x + 6 - 3$
$= 3x^2 + 3x + 3$

3. $6x^3(8x^2 - 7) \qquad$ Multiplication
$= 48x^5 - 42x^3$

5. $(9x^3 - 7x + 3) - (5x^2 - 10) \qquad$ Subtraction
$= 9x^3 - 7x + 3 - 5x^2 + 10$
$= 9x^3 - 5x^2 - 7x + 13$

7. $(9x+1)(9x-1) \qquad$ Multiplication
$= (9x)^2 - 1^2$
$= 81x^2 - 1 \qquad (A+B)(A-B) = A^2 - B^2$

9. $(4x^2 - x - 7) - (10x^2 - 3x + 5)$
$= 4x^2 - x - 7 - 10x^2 + 3x - 5$
$= -6x^2 + 2x - 12$

11. $8x^5(5x^4 - 6x^3 + 2) = 40x^9 - 48x^8 + 16x^5$

13. $(2m - 1)^2 = (2m)^2 - 2(2m)(1) + 1^2$
$= 4m^2 - 4m + 1$

15. $(5x^3 - 6x^2 - 2x) + (6x^2 + 2x + 3)$
$= 5x^3 - 6x^2 - 2x + 6x^2 + 2x + 3$
$= 5x^3 + 3$

17. $(4y^3 + 7)^2 = (4y^3)^2 + 2 \cdot 4y^3 \cdot 7 + 7^2$
$= 16y^6 + 56y^3 + 49$

19. $(4t^2 - 5)(4t^2 + 5)$
$= (4t^2)^2 - 5^2 \qquad (A+B)(A-B) = A^2 - B^2$
$= 16t^4 - 25$

Exercise Set 4.6

1. $(3x + 5y)^2$ is the square of a binomial, choice (a).

3. $(5a + 6b)(-6b + 5a)$, or $(5a + 6b)(5a - 6b)$ is the product of the sum and difference of the same two terms, choice (b).

5. $(r - 3s)(5r + 3s)$ is neither the square of a binomial nor the product of the sum and difference of the same two terms, so choice (c) is appropriate.

7. $(4x - 9y)(4x - 9y)$, or $(4x - 9y)^2$ is the square of a binomial, choice (a).

9. We replace x by 5 and y by -2.
$x^2 - 2y^2 + 3xy = 5^2 - 2(-2)^2 + 3 \cdot 5(-2)$
$= 25 - 8 - 30.$
$= -13.$

11. We replace x by 2, y by -3, and z by -4.
$xy^2z - z = 2(-3)^2(-4) - (-4) = -72 + 4 = -68$

13. Evaluate the polynomial for $h = 160$ and $A = 20$.
$0.041h - 0.018A - 2.69$
$= 0.041(160) - 0.018(20) - 2.69$
$= 6.56 - 0.36 - 2.69$
$= 3.51$

The woman's lung capacity is 3.51 liters.

15. Evaluate the polynomial for $w = 125$, $h = 64$, and $a = 27$.
$917 + 6w + 6h - 6a$
$= 917 + 6(125) + 6(64) - 6(27)$
$= 917 + 750 + 384 - 162$
$= 1889$

The daily caloric needs are 1889 calories.

17. Evaluate the polynomial for $h = 7$, $r = 1\frac{1}{2} = \frac{3}{2}$, and $\pi \approx 3.14$.
$2\pi rh + \pi r^2 \approx 2(3.14)\left(\frac{3}{2}\right)(7) + 3.14\left(\frac{3}{2}\right)^2$
$\approx 65.94 + 7.065$
≈ 73.005

The surface area is about 73.005 in².

19. Evaluate the polynomial for $h = 50$, $v = 18$, and $t = 2$.
$h + vt - 4.9t^2$
$= 50 + 18 \cdot 2 - 4.9(2)^2$
$= 50 + 36 - 19.6$
$= 66.4$

The ball will be 66.4 m above the ground 2 seconds after it is thrown.

21. $3x^2y + 5xy + 2y^2 - 11$

Term	Coefficient	Degree	
$3x^2y$	3	3	
$-5xy$	-5	2	
$2y^2$	2	2	
-11	-11	0	(Think: $-11 = -11x^0$)

The degree of the polynomial is the degree of the term of highest degree. The term of highest degree is $3x^2y$. Its degree is 3, so the degree of the polynomial is 3.

23. $7 - abc + a^2b + 9ab^2$

Term	Coefficient	Degree
7	7	0
$-abc$	-1	3
a^2b	1	3
$9ab^2$	9	3

The terms of highest degree are $-abc$, a^2b and $9ab^2$. Each has degree 3. The degree of the polynomial is 3.

25. $3r + s - r - 7s = (3-1)r + (1-7)s = 2r - 6s$

27. $5xy^2 - 2x^2y + x + 3x^2$

There are <u>no</u> like terms, so none of the terms can be combined.

29. $6u^2v - 9uv^2 + 3vu^2 - 2v^2u + 11u^2$
$= (6+3)u^2v + (-9-2)uv^2 + 11u^2$
$= 9u^2v - 11uv^2 + 11u^2$

31. $5a^2c - 2ab^2 + a^2b - 3ab^2 + a^2c - 2ab^2$
$= (5+1)a^2c + (-2-3-2)ab^2 + a^2b$
$= 6a^2c - 7ab^2 + a^2b$

33. $(6x^2 - 2xy + y^2) + (5x^2 - 8xy - 2y^2)$
$= (6+5)x^2 + (-2-8)xy + (1-2)y^2$
$= 11x^2 - 10xy - y^2$

35. $(3a^4 - 5ab + 6ab^2) - (9a^4 + 3ab - ab^2)$
$= 3a^4 - 5ab + 6ab^2 - 9a^4 - 3ab + ab^2$
   Adding the opposite
$= (3-9)a^4 + (-5-3)ab + (6+1)ab^2$
$= -6a^4 - 8ab + 7ab^2$

37. $(5r^2 - 4rt + t^2) + (-6r^2 - 5rt - t^2) + (-5r^2 + 4rt - t^2)$

Observe that the polynomials $5r^2 - 4rt + t^2$ and $-5r^2 + 4rt - t^2$ are opposites. Thus, their sum is 0 and the sum in the exercise is the remaining polynomial, $-6r^2 - 5rt - t^2$.

39. $(x^3 - y^3) - (-2x^3 + x^2y - xy^2 + 2y^3)$
$= x^3 - y^3 + 2x^3 - x^2y + xy^2 - 2y^3$
$= 3x^3 - 3y^3 - x^2y + xy^2$, or
$\qquad 3x^3 - x^2y + xy^2 - 3y^3$

41. $(2y^4x^3 - 3y^3x) + (5y^4x^3 - y^3x) - (9y^4x^3 - y^3x)$
$= (2 + 5 - 9)y^4x^3 + (-3 - 1 + 1)y^3x$
$= -2y^4x^3 - 3y^3x$

43. $(4x + 5y) + (-5x + 6y) - (7x + 3y)$
$= 4x + 5y - 5x + 6y - 7x - 3y$
$= (4 - 5 - 7)x + (5 + 6 - 3)y$
$= -8x + 8y$

45. $(4c - d)(3c + 2d)$
$\overset{F}{}\overset{O}{}\overset{I}{}\overset{L}{}$
$= 12c^2 + 8cd - 3cd - 2d^2$
$= 12c^2 + 5cd - 2d^2$

47. $(xy - 1)(xy + 5) = x^2y^2 + 5xy - xy - 5$
$\quad\quad\quad\quad\quad\quad\overset{F}{}\overset{O}{}\overset{I}{}\overset{L}{}$
$= x^2y^2 + 4xy - 5$

49. $(2a - b)(2a + b)$ $[(A + B)(A - B) = A^2 - B^2]$
$= 4a^2 - b^2$

51. $(5rt - 2)(4rt - 3) = \overset{F}{20r^2t^2} - \overset{O}{15rt} - \overset{I}{8rt} + \overset{L}{6}$
$= 20r^2t^2 - 23rt + 6$

53. $(m^3n + 8)(m^3n - 6)$
$\quad\quad\overset{F}{}\quad\overset{O}{}\quad\overset{I}{}\quad\overset{L}{}$
$= m^6n^2 - 6m^3n + 8m^3n - 48$
$= m^6n^2 + 2m^3n - 48$

55. $(6x - 2y)(5x - 3y)$
$\quad\quad\overset{F}{}\quad\overset{O}{}\quad\overset{I}{}\quad\overset{L}{}$
$= 30x^2 - 18xy - 10xy + 6y^2$
$= 30x^2 - 28xy + 6y^2$

57. $(pq + 0.1)(-pq + 0.1)$
$= (0.1 + pq)(0.1 - pq)$ $[(A + B)(A - B) = A^2 - B^2]$
$= 0.01 - p^2q^2$

59. $(x + h)^2$
$= x^2 + 2xh + h^2$ $\quad [(A + B)^2 = A^2 + 2AB + B^2]$

61. $(4a - 5b)^2$
$= 16a^2 - 40ab + 25b^2$ $\quad [(A - B)^2 = A^2 - 2AB + B^2]$

63. $(ab + cd^2)(ab - cd^2) = (ab)^2 - (cd^2)^2$
$= a^2b^2 - c^2d^4$

65. $(2xy + x^2y + 3)(xy + y^2)$
$= (2xy + x^2y + 3)(xy) + (2xy + x^2y + 3)(y^2)$
$= 2x^2y^2 + x^3y^2 + 3xy + 2xy^3 + x^2y^3 + 3y^2$

67. $(a + b - c)(a + b + c)$
$= [(a + b) - c][(a + b) + c]$
$= (a + b)^2 - c^2$
$= a^2 + 2ab + b^2 - c^2$

69. $[a + b + c][a - (b + c)]$
$= [a + (b + c)][a - (b + c)]$
$= a^2 - (b + c)^2$
$= a^2 - (b^2 + 2bc + c^2)$
$= a^2 - b^2 - 2bc - c^2$

71. The figure is a square with side $x + y$. Thus the area is $(x + y)^2 = x^2 + 2xy + y^2$.

73. The figure is a triangle with base $ab + 2$ and height $ab - 2$. Its area is $\frac{1}{2}(ab + 2)(ab - 2) = \frac{1}{2}(a^2b^2 - 4) = \frac{1}{2}a^2b^2 - 2$.

75. The figure is a rectangle with dimensions $a + b + c$ by $a + d + c$. Its area is
$(a + b + c)(a + d + c)$
$= [(a + c) + b][(a + c) + d]$
$= (a + c)^2 + (a + c)d + b(a + c) + bd$
$= a^2 + 2ac + c^2 + ad + cd + ab + bc + bd$

77. The figure is a parallelogram with base $m - n$ and height $m + n$. Its area is $(m - n)(m + n) = m^2 - n^2$.

79. We draw a rectangle with dimensions $r + s$ by $u + v$.

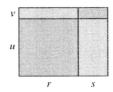

81. We draw a rectangle with dimensions $a + b + c$ by $a + d + f$.

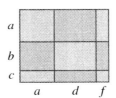

83. *Writing Exercise.* Yes; consider $a + b + c + d$. This is a polynomial in 4 variables but it has degree 1.

85.
$$\begin{array}{r} x^2 - 3x - 7 \\ -(+ 5x - 3) \\ \hline x^2 - 8x - 4 \end{array}$$

87.
$$\begin{array}{r} 3x^2 + x + 5 \\ -(3x^2 + 3x) \\ \hline -2x + 5 \end{array}$$

89.
$$\begin{array}{r} 5x^3 - 2x^2 + 1 \\ -(5x^3 - 15x^2) \\ \hline 13x^2 + 1 \end{array}$$

91. *Writing Exercise.* The leading term of a polynomial is the term of highest degree. When a polynomial has several variables it is possible that more than one term has the highest degree as in Exercise 23.

93. It is helpful to add additional labels to the figure.

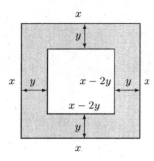

The area of the large square is $x \cdot x$, or x^2. The area of the small square is $(x - 2y)(x - 2y)$, or $(x - 2y)^2$.

$$\begin{array}{c}\text{Area of}\\\text{shaded}\\\text{region}\end{array} = \begin{array}{c}\text{Area of large}\\\text{square}\end{array} - \begin{array}{c}\text{Area of small}\\\text{square}\end{array}$$

$$\begin{array}{ccc}\text{Area of}\\\text{shaded} &= & x^2 & - & (x-2y)^2\\\text{region}\end{array}$$

$$= x^2 - (x^2 - 4xy + 4y^2)$$
$$= x^2 - x^2 + 4xy - 4y^2$$
$$= 4xy - 4y^2$$

95. The unshaded region is a circle with radius $a - b$. Then the shaded area is the area of a circle with radius a less the area of a circle with radius $a - b$. Thus, we have:

$$\text{Shaded area} = \pi a^2 - \pi(a - b)^2$$
$$= \pi a^2 - \pi(a^2 - 2ab + b^2)$$
$$= \pi a^2 - \pi a^2 + 2\pi ab - \pi b^2$$
$$= 2\pi ab - \pi b^2$$

97. The figure can be thought of as a cube with side x, a rectangular solid with dimensions x by x by y, a rectangular solid with dimensions x by y by y, and a rectangular solid with dimensions y by y by $2y$. Thus the volume is

$$x^3 + x \cdot x \cdot y + x \cdot y \cdot y + y \cdot y \cdot 2y, \text{ or}$$
$$x^3 + x^2y + xy^2 + 2y^3.$$

99. The surface area of the solid consists of the surface area of a rectangular solid with dimensions x by x by h less the areas of 2 circles with radius r plus the lateral surface area of a right circular cylinder with radius r and height h. Thus, we have

$$2x^2 + 2xh + 2xh - 2\pi r^2 + 2\pi rh, \text{ or}$$
$$2x^2 + 4xh - 2\pi r^2 + 2\pi rh.$$

101. *Writing Exercise.* The height of the observatory is 40 ft and its radius is 30/2, or 15 ft, so the surface area is $2\pi rh + \pi r^2 \approx 2(3.14)(15)(40) + (3.14)(15)^2 \approx 4474.5$ ft^2. Since 4474.5 ft^2/250 ft^2 = 17.898, 18 gallons of paint should be purchased.

103. For the formula $= 2 * A4 + 3 * B4$, we substitute 5 for $A4$ and 10 for $B4$.

$$= 2 * A4 + 3 * B4 = 2 \cdot 5 + 3 \cdot 10$$
$$= 10 + 30$$
$$= 40$$

The value of $D4$ is 40.

105. Replace t with 2 and multiply.

$$P(1 + r)^2$$
$$= P(1 + 2r + r^2)$$
$$= P + 2Pr + Pr^2$$

107. Substitute \$10,400 for P, 8.5%, or 0.085 for r, and 5 for t.

$$P(1 + r)^t$$
$$= \$10,400(1 + 0.085)^5$$
$$\approx \$15,638.03$$

Exercise Set 4.7

1. $\dfrac{40x^6 - 25x^3}{5} = \dfrac{40x^6}{5} - \dfrac{25x^3}{5}$

$$= \frac{40}{5}x^6 - \frac{25}{5}x^3 \quad \text{Dividing coefficients}$$

$$= 8x^6 - 5x^3$$

To check, we multiply the quotient by 5:

$$(8x^6 - 5x^3)5 = 40x^6 - 25x^3.$$

The answer checks.

3. $\dfrac{u - 2u^2 + u^7}{u}$

$$= \frac{u}{u} - \frac{2u^2}{u} + \frac{u^7}{u}$$
$$= 1 - 2u + u^6$$

Check: We multiply.

$$u(1 - 2u + u^6) = u - 2u^2 + u^7$$

5. $(18t^3 - 24t^2 + 6t) \div (3t)$

$$= \frac{18t^3 - 24t^2 + 6t}{3t}$$
$$= \frac{18t^3}{3t} - \frac{24t^2}{3t} + \frac{6t}{3t}$$
$$= 6t^2 - 8t + 2$$

Check: We multiply.

$$3t(6t^2 - 8t + 2) = 18t^3 - 24t^2 + 6t$$

7. $(42x^5 - 36x^3 + 9x^2) \div (6x^2)$

$$= \frac{42x^5 - 36x^3 + 9x^2}{6x^2}$$
$$= \frac{42x^5}{6x^2} - \frac{36x^3}{6x^2} + \frac{9x^2}{6x^2}$$
$$= 7x^3 - 6x + \frac{3}{2}$$

Check: We multiply.

$$6x^2\left(7x^3 - 6x + \frac{3}{2}\right) = 42x^5 - 36x^3 + 9x^2$$

9. $(32t^5 + 16t^4 - 8t^3) \div (-8t^3)$

$$= \frac{32t^5 + 16t^4 - 8t^3}{-8t^3}$$
$$= \frac{32t^5}{-8t^3} + \frac{16t^4}{-8t^3} - \frac{8t^3}{-8t^3}$$
$$= -4t^2 - 2t + 1$$

Check: We multiply.

$$-8t^3(-4t^2 - 2t + 1) = 32t^5 + 16t^4 - 8t^3$$

11. $\dfrac{8x^2 - 10x + 1}{2x}$

$$= \frac{8x^2}{2x} - \frac{10x}{2x} + \frac{1}{2x}$$

$$= 4x - 5 + \frac{1}{2x}$$

Check: We multiply.

$$2x\left(4x - 5 + \frac{1}{2x}\right) = 8x^2 - 10x + 1$$

13. $\dfrac{5x^3y + 10x^5y^2 + 15x^2y}{5x^2y}$

$$= \frac{5x^3y}{5x^2y} + \frac{10x^5y^2}{5x^2y} + \frac{15x^2y}{5x^2y}$$

$$= x + 2x^3y + 3$$

Check: We multiply.

$$5x^2y(x + 2x^3y + 3) = 5x^3y + 10x^5y^2 + 15x^2y$$

15. $\dfrac{9r^2s^2 + 3r^2s - 6rs^2}{-3rs}$

$$= \frac{9r^2s^2}{-3rs} + \frac{3r^2s}{-3rs} - \frac{6rs^2}{-3rs}$$

$$= -3rs - r + 2s$$

Check: We multiply.

$$-3rs(-3rs - r + 2s) = 9r^2s^2 + 3r^2s - 6rs^2$$

17.
$$
\begin{array}{r}
x - 6 \\
x - 2 \overline{\smash{\big)}\ x^2 + 8x + 12} \\
\underline{x^2 - 2x} \\
-6x + 12 \quad \leftarrow (x^2 + 8x) - (x^2 - 2x) = -6x \\
\underline{-6x + 12} \\
0 \quad \leftarrow (-6x + 12) - (-6x + 12) = 0
\end{array}
$$

The answer is $x - 6$.

19.
$$
\begin{array}{r}
t - 5 \\
t - 5 \overline{\smash{\big)}\ t^2 - 10t - 20} \\
\underline{t^2 - 5t} \\
-5t - 20 \quad \leftarrow (t^2 - 10t) - (t^2 - 5t) = \\
-5t \\
\underline{-5t + 25} \\
-45 \quad \leftarrow (-5t - 20) - (-5t + 25) = \\
-45
\end{array}
$$

The answer is $t - 5 + \dfrac{-45}{t - 5}$.

21.
$$
\begin{array}{r}
2x - 1 \\
x + 6 \overline{\smash{\big)}\ 2x^2 + 11x - 5} \\
\underline{2x^2 + 12x} \\
-x - 5 \quad \leftarrow (2x^2 + 11x) - (2x^2 + 12x) = -x \\
\underline{-x - 6} \\
1 \quad \leftarrow (-x - 5) - (-x - 6) = 1
\end{array}
$$

The answer is $2x - 1 + \dfrac{1}{x + 6}$.

23.
$$
\begin{array}{r}
t^2 - 3t + 9 \\
t + 3 \overline{\smash{\big)}\ t^3 + 0t^2 + 0t + 27} \quad \leftarrow\text{Writing in the missing} \\
\text{terms} \\
\underline{t^3 + 3t^2} \\
-3t^2 + 0t \quad \leftarrow t^3 - (t^3 + 3t^2) = -3t^2 \\
\underline{-3t^2 - 9t} \\
9t + 27 \quad \leftarrow -3t^2 - (-3t^2 - 9t) = 9t \\
\underline{9t + 27} \\
0 \quad \leftarrow (9t + 27) = -(9t + 27) = 0
\end{array}
$$

The answer is $t^2 - 3t + 9$.

25.
$$
\begin{array}{r}
a + 5 \\
a - 5 \overline{\smash{\big)}\ a^2 + 0a - 21} \quad \leftarrow\text{Writing in the missing term} \\
\underline{a^2 - 5a} \\
5a - 21 \quad \leftarrow a^2 - (a^2 - 5a) = 5a \\
\underline{5a - 25} \\
4 \quad \leftarrow (5a^2 - 21) - (5a - 25) = 4
\end{array}
$$

The answer is $a + 5 + \dfrac{4}{a - 5}$.

27.
$$
\begin{array}{r}
x - 3 \\
5x - 1 \overline{\smash{\big)}\ 5x^2 - 16x + 0} \quad \leftarrow\text{Writing in the missing term} \\
\underline{5x^2 - x} \\
-15x + 0 \quad \leftarrow (5x^2 - 16x) - (5x^2 - x) = \\
-15x \\
\underline{-15x + 3} \\
-3 \quad \leftarrow (-15x + 0) - (-15x + 3) = -3
\end{array}
$$

The answer is $x - 3 - \dfrac{3}{5x - 1}$.

29.
$$
\begin{array}{r}
3a + 1 \\
2a + 5 \overline{\smash{\big)}\ 6a^2 + 17a + 8} \\
\underline{6a^2 + 15a} \\
2a + 8 \quad \leftarrow (6a^2 + 17a) - (6a^2 + 15a) = 2a \\
\underline{2a + 5} \\
3 \quad \leftarrow (2a + 8) - (2a + 5) = 3
\end{array}
$$

The answer is $3a + 1 + \dfrac{3}{2a + 5}$.

31.
$$
\begin{array}{r}
t^2 - 3t + 1 \\
2t - 3 \overline{\smash{\big)}\ 2t^3 - 9t^2 + 11t - 3} \\
\underline{2t^3 - 3t^2} \\
-6t^2 + 11t \quad \leftarrow (2t^3 - 9t^2) - (2t^3 - 3t^2) = \\
-6t^2 \\
\underline{-6t^2 + 9t} \\
2t - 3 \quad \leftarrow (-6t^2 + 11t) - \\
(-6t^2 + 9t) = 2t \\
\underline{2t - 3} \\
0 \quad \leftarrow (2t - 3) - (2t - 3) = 0
\end{array}
$$

The answer is $t^2 - 3t + 1$.

33.
$$
\begin{array}{r}
x^2 + 1 \\
x - 1 \overline{\smash{\big)}\ x^3 - x^2 + x - 1} \\
\underline{x^3 - x^2} \\
x - 1 \quad \leftarrow (x^3 - x^2) - (x^3 - x^2) = 0 \\
\underline{x - 1} \\
0 \quad \leftarrow (x - 1) - (x - 1) = 0
\end{array}
$$

The answer is $x^2 + 1$.

35.

$$
\begin{array}{r}
t^2 \qquad -1 \\
t^2+5\overline{\smash{\big)}\,t^4+0t^3+4t^2+3t-6} \leftarrow \text{Writing in the}\\
\underline{t^4 \qquad +5t^2} \qquad\quad \text{missing term}\\
-t^2+3t-6 \;\leftarrow\; (t^4-4t^2)\\
\underline{-t^2 \qquad -5} \quad -(t^4+5t^2)=-t^2\\
3t-1 \;\leftarrow\; (-t^2+3t-6)\\
-(-t^2-5)=3t-1
\end{array}
$$

The answer is $t^2-1+\dfrac{3t-1}{t^2+5}$.

37.

$$
\begin{array}{r}
3x^2 \qquad -3 \\
2x^2+1\overline{\smash{\big)}\,6x^4+0x^3-3x^2+x-4} \leftarrow \text{Writing in the}\\
\underline{6x^4 \qquad +3x^2} \qquad\quad \text{missing term}\\
-6x^2+x-4 \;\leftarrow\; (6x^4-3x^2)-\\
\underline{-6x^2 \qquad -3} \quad (6x^4+3x^2)=-6x^2\\
x-1 \;\leftarrow\; (-6x^2+x-4)\\
-(-6x^2-3)=x-1
\end{array}
$$

The answer is $3x^2-3+\dfrac{x-1}{2x^2+1}$.

39. *Writing Exercise.* The distributive law is used in each step when a term of the quotient is multiplied by the divisor and when the subtraction is performed. The distributive law is also used when the quotient is checked.

41. $y^8 \cdot y^5 = y^{8+5} = y^{13}$

43. $\dfrac{y^8}{y^5} = y^{8-5} = y^3$

45. $(a^2b)\left(a^3b^4\right) = a^{2+3}b^{1+4} = a^5b^5$

47. $\left(\dfrac{3}{8n^3}\right)^2 = \dfrac{3^2}{8^2n^{3\cdot2}} = \dfrac{9}{64n^6}$.

49. *Writing Exercise.* Find the product of $x-5$ and a binomial of the form $ax+b$. Then add 3 to this result.

51.
$$
\begin{aligned}
&\left(10x^{9k}-32x^{6k}+28x^{3k}\right) \div \left(2x^{3k}\right)\\
&= \frac{10x^{9k}-32x^{6k}+28x^{3k}}{2x^{3k}}\\
&= \frac{10x^{9k}}{2x^{3k}} - \frac{32x^{6k}}{2x^{3k}} + \frac{28x^{3k}}{2x^{3k}}\\
&= 5x^{9k-3k} - 16x^{6k-3k} + 14x^{3k-3k}\\
&= 5x^{6k} - 16x^{3k} + 14
\end{aligned}
$$

53.

$$
\begin{array}{r}
3t^{2h}+2t^h \quad -5 \\
2t^h+3\overline{\smash{\big)}\,6t^{3h}+13t^{2h}-4t^h-15}\\
\underline{6t^{3h}+9t^{2h}}\\
4t^{2h}-4t^h\\
\underline{4t^{2h}+6t^h}\\
-10t^h-15\\
\underline{-10t^h-15}\\
0
\end{array}
$$

The answer is $3t^{2h}+2t^h-5$.

55.

$$
\begin{array}{r}
a \quad +3 \\
5a^2-7a-2\overline{\smash{\big)}\,5a^3+8a^2-23a-1}\\
\underline{5a^3-7a^2-2a}\\
15a^2-21a-1\\
\underline{15a^2-21a-6}\\
5
\end{array}
$$

The answer is $a+3+\dfrac{5}{5a^2-7a-2}$.

57.
$$
\begin{aligned}
&(4x^5-14x^3-x^2+3)+\\
&\quad(2x^5+3x^4+x^3-3x^2+5x)\\
&= 6x^5+3x^4-13x^3-4x^2+5x+3
\end{aligned}
$$

$$
\begin{array}{r}
2x^2+ \quad x \quad -3 \\
3x^3-2x-1\overline{\smash{\big)}\,6x^5+3x^4-13x^3-4x^2+5x+3}\\
\underline{6x^5 \qquad -4x^3-2x^2}\\
3x^4-9x^3-2x^2+5x\\
\underline{3x^4 \qquad -2x^2-x}\\
-9x^3 \qquad +6x+3\\
\underline{-9x^3 \qquad +6x+3}\\
0
\end{array}
$$

The answer is $2x^2+x-3$.

59.

$$
\begin{array}{r}
x \quad -3 \\
x-1\overline{\smash{\big)}\,x^2-4x+c}\\
\underline{x^2-x}\\
-3x+c\\
\underline{-3x+3}\\
c-3
\end{array}
$$

We set the remainder equal to 0.

$$c-3 = 0$$
$$c = 3$$

Thus, c must be 3.

61.

$$
\begin{array}{r}
c^2x+ \quad (2c+c^2) \\
x-1\overline{\smash{\big)}\,c^2x^2+2cx+1}\\
\underline{c^2x^2-c^2x}\\
(2c+c^2)x+1\\
\underline{(2c+c^2)x-(2c+c^2)}\\
1+(2c+c^2)
\end{array}
$$

We set the remainder equal to 0.

$$c^2+2c+1 = 0$$
$$(c+1)^2 = 0$$
$$c+1=0 \quad or \quad c+1=0$$
$$c=-1 \quad or \qquad c=-1$$

Thus, c must be -1.

Exercise Set 4.8

1. $\left(\dfrac{x^3}{y^2}\right)^{-2} = \left(\dfrac{y^2}{x^3}\right)^2 = \dfrac{\left(y^2\right)^2}{\left(x^3\right)^2} = \dfrac{y^4}{x^6} \Rightarrow$ (c)

3. $\left(\dfrac{y^{-2}}{x^{-3}}\right)^{-3} = \dfrac{\left(y^{-2}\right)^{-3}}{\left(x^{-3}\right)^{-3}} = \dfrac{y^6}{x^9} \Rightarrow$ (a)

5. $2^{-3} = \dfrac{1}{2^3} = \dfrac{1}{8}$

7. $(-2)^{-6} = \dfrac{1}{(-2)^6} = \dfrac{1}{64}$

9. $t^{-9} = \dfrac{1}{t^9}$

11. $xy^{-2} = x \cdot \dfrac{1}{y^2} = \dfrac{x}{y^2}$

13. $r^{-5}t = \dfrac{1}{r^5} \cdot t = \dfrac{t}{r^5}$

15. $\frac{1}{a^{-8}} = a^8$

17. $7^{-1} = \frac{1}{7^1} = \frac{1}{7}$

19. $\left(\frac{3}{5}\right)^{-2} = \left(\frac{5}{3}\right)^2 = \frac{5^2}{3^2} = \frac{25}{9}$

21. $\left(\frac{x}{2}\right)^{-5} = \left(\frac{2}{x}\right)^5 = \frac{2^5}{x^5} = \frac{32}{x^5}$

23. $\left(\frac{s}{t}\right)^{-7} = \left(\frac{t}{s}\right)^7 = \frac{t^7}{s^7}$

25. $\frac{1}{9^2} = 9^{-2}$

27. $\frac{1}{y^3} = y^{-3}$

29. $\frac{1}{5} = \frac{1}{5^1} = 5^{-1}$

31. $\frac{1}{t} = \frac{1}{t^1} = t^{-1}$

33. $2^{-5} \cdot 2^8 = 2^{-5+8} = 2^3$, or 8

35. $x^{-3} \cdot x^{-9} = x^{-12} = \frac{1}{x^{12}}$

37. $t^{-3} \cdot t = t^{-3} \cdot t^1 = t^{-3+1} = t^{-2} = \frac{1}{t^2}$

39. $\left(n^{-5}\right)^3 = n^{-5\cdot3} = n^{-15} = \frac{1}{n^{15}}$

41. $\left(t^{-3}\right)^{-6} = t^{-3(-6)} = t^{18}$

43. $\left(t^4\right)^{-3} = t^{4(-3)} = t^{-12} = \frac{1}{t^{12}}$

45. $(mn)^{-7} = \frac{1}{(mn)^7} = \frac{1}{m^7 n^7}$

47. $(3x^{-4})^2 = 3^2(x^{-4})^2 = 9x^{-8} = \frac{9}{x^8}$

49. $(5r^{-4}t^3)^2 = 5^2(r^{-4})^2(t^3)^2 = 25r^{-8}t^6 = \frac{25t^6}{r^8}$

51. $\frac{t^{12}}{t^{-2}} = t^{12-(-2)} = t^{14}$

53. $\frac{y^{-7}}{y^{-3}} = y^{-7-(-3)} = y^{-4} = \frac{1}{y^4}$

55. $\frac{15y^{-7}}{3y^{-10}} = 5y^{-7-(-10)} = 5y^3$

57. $\frac{2x^6}{x} = 2\frac{x^6}{x^1} = 2x^{6-1} = 2x^5$

59. $-\frac{15a^{-7}}{10b^{-9}} = -\frac{3b^9}{2a^7}$

61. $\frac{t^{-7}}{t^{-7}}$

Note that we have an expression divided by itself. Thus, the result is 1. We could also find this result as follows:
$$\frac{t^{-7}}{t^{-7}} = t^{-7-(-7)} = t^0 = 1.$$

63. $\frac{8x^{-3}}{y^{-7}z^{-1}} = \frac{8y^7 z}{x^3}$

65. $\frac{3t^4}{s^{-2}u^{-4}} = 3s^2 t^4 u^4$

67. $(x^4 y^5)^{-3} = (x^4)^{-3}(y^5)^{-3} = x^{-12}y^{-15} = \frac{1}{x^{12}y^{15}}$

69. $\left(3m^{-5}n^{-3}\right)^{-2} = 3^{-2}m^{-5(-2)}n^{-3(-2)} = 3^{-2}m^{10}n^6$
$= \frac{m^{10}n^6}{9}$

71. $(a^{-5}b^7c^{-2})(a^{-3}b^{-2}c^6) = a^{-5+(-3)}b^{7+(-2)}c^{-2+6}$
$= a^{-8}b^5c^4 = \frac{b^5 c^4}{a^8}$

73. $\left(\frac{a^4}{3}\right)^{-2} = \left(\frac{3}{a^4}\right)^2 = \frac{3^2}{(a^4)^2} = \frac{9}{a^8}$

75. $\left(\frac{m^{-1}}{n^{-4}}\right)^3 = \frac{(m^{-1})^3}{(n^{-4})^3} = \frac{m^{-3}}{n^{-12}} = \frac{n^{12}}{m^3}$

77. $\left(\frac{2a^2}{3b^4}\right)^{-3} = \left(\frac{3b^4}{2a^2}\right)^3 = \frac{(3b^4)^3}{(2a^2)^3} = \frac{3^3(b^4)^3}{2^3(a^2)^3} = \frac{27b^{12}}{8a^6}$

79. $\left(\frac{5x^{-2}}{3y^{-2}z}\right)^0$

Any nonzero expression raised to the 0 power is equal to 1. Thus, the answer is 1.

81. $4.92 \times 10^3 = 4.920 \times 10^3$ The decimal point moves right three places
$= 4920$

83. 8.92×10^{-3}

Since the exponent is negative, the decimal point will move to the left.

.008.92   The decimal point moves left 3 places.

$8.92 \times 10^{-3} = 0.00892$

85. 9.04×10^8

Since the exponent is positive, the decimal point will move to the right.

9.04000000.   8 places

$9.04 \times 10^8 = 904{,}000{,}000$

87. $3.497 \times 10^{-6} = 0000003.497 \times 10^{-6}$ The decimal point moves left six places
$= 0.000003497$

89. 4.209×10^7

Since the exponent is positive, the decimal point will move to the right.

4.2090000.   7 places

$4.209 \times 10^7 = 42{,}090{,}000$

91. $36,000,000 = 3.6 \times 10^m$ We move the decimal point 7 places to the right. Thus m is 7.

$= 3.6 \times 10^7$

93. $0.00583 = 5.83 \times 10^m$

To write 5.83 as 0.00583 we move the decimal point 3 places to the left. Thus, m is -3 and

$0.00583 = 5.83 \times 10^{-3}$.

95. $78,000,000,000 = 7.8 \times 10^m$

To write 7.8 as 78,000,000,000 we move the decimal point 10 places to the right. Thus, m is 10 and

$78,000,000,000 = 7.8 \times 10^{10}$.

97. $0.000000527 = 5.27 \times 10^m$

To write 5.27 as 0.000000527 we move the decimal point 7 places to the left. Thus, m is -7 and

$0.000000527 = 5.27 \times 10^{-7}$.

99. $0.000001032 = 1.032 \times 10^m$ We move the decimal point 6 places to the left. Thus $m = -6$.

$= 1.032 \times 10^{-6}$

101. $1,094,000,000,000,000 = 1.094 \times 10^m$

To write 1.094 as 1,094,000,000,000,000 we move the decimal point 15 places to the right. Thus, m is 15 and

$1,094,000,000,000,000 = 1.094 \times 10^{15}$.

103. $(3 \times 10^5)(2 \times 10^8) = (3 \cdot 2) \times (10^5 \cdot 10^8)$

$= 6 \times 10^{5+8}$

$= 6 \times 10^{13}$

105. $(3.8 \times 10^9)(6.5 \times 10^{-2}) = (3.8 \cdot 6.5) \times (10^9 \cdot 10^{-2})$

$= 24.7 \times 10^7$

The answer is not yet in scientific notation since 24.7 is not a number between 1 and 10. We convert to scientific notation.

$24.7 \times 10^7 = (2.47 \times 10) \times 10^7 = 2.47 \times 10^8$

107. $(8.7 \times 10^{-12})(4.5 \times 10^{-5})$

$= (8.7 \cdot 4.5) \times (10^{-12} \cdot 10^{-5})$

$= 39.15 \times 10^{-17}$

The answer is not yet in scientific notation since 39.15 is not a number between 1 and 10. We convert to scientific notation.

$39.15 \times 10^{-17} = (3.915 \times 10) \times 10^{-17} = 3.915 \times 10^{-16}$

109. $\dfrac{8.5 \times 10^8}{3.4 \times 10^{-5}} = \dfrac{8.5}{3.4} \times \dfrac{10^8}{10^{-5}}$

$= 2.5 \times 10^{8-(-5)}$

$= 2.5 \times 10^{13}$

111. $(4.0 \times 10^3) \div (8.0 \times 10^8) = \dfrac{4.0}{8.0} \times \dfrac{10^3}{10^8}$

$= 0.5 \times 10^{3-8}$

$= 0.5 \times 10^{-5}$

$= 5.0 \times 10^{-6}$

113. $\dfrac{7.5 \times 10^{-9}}{2.5 \times 10^{12}} = \dfrac{7.5}{2.5} \times \dfrac{10^{-9}}{10^{12}}$

$= 3.0 \times 10^{-9-12}$

$= 3.0 \times 10^{-21}$

115. *Writing Exercise.* $3^{-29} = \dfrac{1}{3^{29}}$ and $2^{-29} = \dfrac{1}{2^{29}}$. Since $3^{29} > 2^{29}$, we have $\dfrac{1}{3^{29}} < \dfrac{1}{2^{29}}$.

117. $3x - 4y = 12$

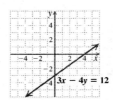

119. $3y - 2 = 7$

$y = 3$

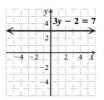

121. $m = \dfrac{5 - 2}{-7 - 3} = \dfrac{-3}{10}$

123. $y = -5x - 10$

125. *Writing Exercise.* x^{-n} represents a negative integer when x is negative, $\dfrac{1}{x}$ is an integer, and n is an odd number.

127. $\dfrac{1}{1.25 \times 10^{-6}} = \dfrac{1}{1.25} \times \dfrac{1}{10^{-6}} = 0.8 \times 10^6$

$= (8 \times 10^{-1}) \times 10^6 = 8 \times 10^5$

129. $8^{-3} \times 32 \div 16^2 = (2^3)^{-3} \cdot 2^5 \div (2^4)^2$

$= 2^{-9} \cdot 2^5 \div 2^8 = 2^{-4} \div 2^8 = 2^{-12}$

131. $\dfrac{125^{-4}(25^2)^4}{125} = \dfrac{(5^3)^{-4}((5^2)^2)^4}{5^3}$

$= \dfrac{5^{-12}(5^4)^4}{5^3} = \dfrac{5^{-12} \cdot 5^{16}}{5^3} = \dfrac{5^4}{5^3} = 5^1 = 5$

133. $\left[(5^{-3})^2\right]^{-1} = 5^{(-3)(2)(-1)} = 5^6$

135. $3^{-1} + 4^{-1} = \dfrac{1}{3} + \dfrac{1}{4} = \dfrac{4}{12} + \dfrac{3}{12} = \dfrac{7}{12}$

137. $\dfrac{27^{-2}(81^2)^3}{9^8} = \dfrac{(3^3)^{-2}((3^4)^2)^3}{(3^2)^8}$

$= \dfrac{3^{-6} \cdot 3^{24}}{3^{16}} = \dfrac{3^{18}}{3^{16}} = 3^2 = 9$

139. $\dfrac{5.8 \times 10^{17}}{(4.0 \times 10^{-13})(2.3 \times 10^4)}$

$= \dfrac{5.8}{(4.0 \cdot 2.3)} \times \dfrac{10^{17}}{(10^{-13} \cdot 10^4)}$

$\approx 0.6304347826 \times 10^{17-(-13)-4}$

$\approx (6.304347826 \times 10^{-1}) \times 10^{26}$

$\approx 6.304347826 \times 10^{25}$

141. $\dfrac{(2.5 \times 10^{-8})(6.1 \times 10^{-11})}{1.28 \times 10^{-3}}$

$= \dfrac{(2.5 \cdot 6.1)}{1.28} \times \dfrac{(10^{-8} \cdot 10^{-11})}{10^{-3}}$

$= 11.9140625 \times 10^{-8+(-11)-(-3)}$

$= 11.9140625 \times 10^{-16}$

$= (1.19140625 \times 10) \times 10^{-16}$

$= 1.19140625 \times 10^{-15}$

143. *Familiarize*. Let $n =$ the number of car miles.

Translate. We reword the problem.

$$\underbrace{\text{Miles for one tree}}_{\downarrow} \quad \underbrace{\text{times}}_{\downarrow} \quad \underbrace{\text{the number of trees}}_{\downarrow} \quad \underbrace{\text{is}}_{\downarrow} \quad \underbrace{n}_{\downarrow}$$
$$\;\; 500 \qquad\quad \times \qquad\quad 600{,}000 \qquad = \;\; n$$

Carry out. We solve the equation.

$500 \times 600{,}000 = n$

$300{,}000{,}000 = n$

$n = 3 \times 10^8$

Check. Review the computation, The answer is reasonable.

State. Trees can clean 3×10^8 miles of car traffic in a year.

145. *Familiarize*. Let $c =$ the total cost of the condominiums.

Translate. We reword the problem.

$$\underbrace{\text{What}}_{\downarrow} \; \underbrace{\text{is}}_{\downarrow} \; \underbrace{\substack{\text{price} \\ \$2100 \text{ per ft}^2}}_{\downarrow} \; \underbrace{\text{times}}_{\downarrow} \; \underbrace{\substack{\text{space} \\ 110{,}000 \text{ ft}^2}}_{\downarrow}$$
$$\;\; c \quad = \qquad (2100) \qquad\; \times \qquad (110{,}000)$$

Carry out.

$c = 2100 \times 1.1 \times 10^5$

$c = 231{,}000{,}000$

Check. We review calculations.

State. $c = \$2.31 \times 10^8$ is the total cost.

147. *Familiarize*. Let $m =$ the number of minutes.

Convert. 3.3 hours $= 3.3 \times 60 = 198$ minutes

Translate.

$$\underbrace{\text{What}}_{\downarrow} \; \underbrace{\text{is}}_{\downarrow} \; \underbrace{\substack{\text{number of} \\ \text{patients}}}_{\downarrow} \; \underbrace{\text{times}}_{\downarrow} \; \underbrace{\substack{\text{average time} \\ \text{in minutes}}}_{\downarrow}$$
$$\;\; m \quad = \quad 115{,}000{,}000 \quad\; \times \qquad 198$$

$m = 2.277 \times 10^{10}$

Check. We review calculations.

State. In 2005 patients spent 2.277×10^{10} minutes in emergency rooms.

Chapter 4 Connecting the Concepts

1. $x^4 x^{10} = x^{4+10} = x^{14}$

3. $\dfrac{x^{-4}}{x^{10}} = x^{-4-10} = x^{-14} = \dfrac{1}{x^{14}}$

5. $\left(x^{-4}\right)^{-10} = x^{-4(-10)} = x^{40}$

7. $\dfrac{1}{c^{-8}} = c^8$

9. $\left(2x^3 y\right)^4 = 2^4 \left(x^3\right)^4 y^4 = 16x^{12} y^4$

11. $\left(3xy^{-1} z^5\right)^0 = 1$

13. $\left(\dfrac{a^3}{b^4}\right)^5 = \dfrac{a^{3\cdot 5}}{b^{4\cdot 5}} = \dfrac{a^{15}}{b^{20}}$

15. $\dfrac{30x^4 y^3}{12xy^7} = \dfrac{5x^{4-1} y^{3-7}}{2} = \dfrac{5}{2} x^3 y^{-4} = \dfrac{5x^3}{2y^4}$

17. $\dfrac{7p^{-5}}{xt^{-6}} = \dfrac{7t^6}{xp^5}$

19. $\left(2p^2 q^4\right)\left(3pq^5\right)^2 = 2p^2 q^4 \cdot 3^2 p^2 q^{5\cdot 2}$
$= 2p^2 q^4 \cdot 9p^2 q^{10}$
$= 18p^4 q^{14}$

Chapter 4 Review

1. True; see page 239 in the text.

3. True; see page 256 in the text.

5. False; the degree of the polynomial is the degree of the leading term.

7. True; see page 267 in the text.

9. $n^3 \cdot n^8 \cdot n = n^{3+8+1} = n^{12}$

11. $t^6 \cdot t^0 = t^{6+0} = t^6$

13. $\dfrac{(a+b)^4}{(a+b)^4} = 1$

15. $\left(-2xy^2\right)^3 = (-2)^3 x^{1\cdot 3} y^{2\cdot 3} = -8x^3 y^6$

17. $\left(a^2 b\right)(ab)^5 = a^2 b \cdot a^5 b^5 = a^{2+5} b^{1+5} = a^7 b^6$

19. $8x^2 - x + \dfrac{2}{3}$

The terms are $8x^2$, $-x$, and $\dfrac{2}{3}$.

21. $9x^2 - x + 7$
The coefficients are 9, -1, and 7.

23. $4t^2 + 6 + 15t^5$

a)

Term	$4t^2$	6	$15t^5$
Degree	2	0	5

b) the term of highest degree is $15t^5$. This is the leading term. Then the leading coefficient is 15 since $15t^5 = 15 \cdot t^5$.

c) Since the term of highest degree is $15t^5$, the degree of the polynomial is 5.

25. Three monomials are added, so $4x^3 - 5x + 3$ is a trinomial.

27. There is just one monomial, so $7y^2$ is a monomial.

29. $\dfrac{3}{4}x^3 + 4x^2 - x^3 + 7 = \left(\dfrac{3}{4} - 1\right)x^3 + 4x^2 + 7$
$= -\dfrac{1}{4}x^3 + 4x^2 + 7$

31. $-a + \dfrac{1}{3} + 20a^5 - 1 - 6a^5 - 2a^2$
$= (20 - 6)a^5 - 2a^2 - a + \left(\dfrac{1}{3} - 1\right)$
$= 14a^5 - 2a^2 - a - \dfrac{2}{3}$

33. For $x = -2$: $\quad x^2 - 3x + 6$
$= (-2)^2 - 3(-2) + 6$
$= 4 + 6 + 6$
$= 16$

35. $\left(5a^5 - 2a^3 - 9a^2\right) + \left(2a^5 + a^3\right) + \left(-a^5 - 3a^2\right)$
$= 5a^5 - 2a^3 - 9a^2 + 2a^5 + a^3 - a^5 - 3a^2$
$= (5 + 2 - 1)a^5 + (-2 + 1)a^3 + (-9 - 3)a^2$
$= 6a^5 - a^3 - 12a^2$

37. $\left(3x^5 - 4x^4 + 2x^2 + 3\right) - \left(2x^5 - 4x^4 + 3x^3 + 4x^2 - 5\right)$
$= 3x^5 - 4x^4 + 2x^2 + 3 - 2x^5 + 4x^4 - 3x^3 - 4x^2 + 5$
$= (3 - 2)x^5 + (-4 + 4)x^4 - 3x^3 + (2 - 4)x^2 + (3 + 5)$
$= x^5 - 3x^3 - 2x^2 + 8$

39.
$$2x^5 \qquad - x^3 \qquad + x + 3$$
$$\underline{-(3x^5 - x^4 + 4x^3 + 2x^2 - x + 3)}$$

$$2x^5 \qquad - x^3 \qquad + x + 3$$
$$\underline{-3x^5 + x^4 - 4x^3 - 2x^2 + x - 3}$$
$$-x^5 + x^4 - 5x^3 - 2x^2 + 2x$$

41. $5x^2\left(-6x^3\right) = 5(-6)x^{2+3} = -30x^5$

43. $(a - 7)(a + 4)$
$\quad$ F $\quad$ O $\quad$ I $\quad$ L
$= a^2 + 4a - 7a - 28$
$= a^2 - 3a - 28$

45. $\left(4x^2 - 5x + 1\right)(3x - 2)$
$= \left(4x^2 - 5x + 1\right)3x + \left(4x^2 - 5x + 1\right)(-2)$
$= 12x^3 - 15x^2 + 3x - 8x^2 + 10x - 2$
$= 12x^3 + (-15 - 8)x^2 + (3 + 10)x - 2$
$= 12x^3 - 23x^2 + 13x - 2$

47. $3t^2(5t^3 - 2t^2 + 4t) = 15t^5 - 6t^4 + 12t^3$

49. $(x - 0.8)(x - 0.5)$
$\quad$ F $\quad$ O $\quad$ I $\quad$ L
$= x^2 - 0.5x - 0.8x + 0.4$
$= x^2 - 1.3x + 0.4$

51. $(4y^3 - 5)^2 = (4y^3)^2 - 2 \cdot 4y^3 \cdot 5 + 5^2 \qquad (A - B)^2$
$\hspace{8cm} = A - 2AB + B^2$
$\qquad\qquad = 16y^6 - 40y^3 + 25$

53. $\left(a - \dfrac{1}{2}\right)\left(a + \dfrac{2}{3}\right)$
$\quad$ F $\quad$ O $\quad$ I $\quad$ L
$= a^2 + \dfrac{2}{3}a - \dfrac{1}{2}a - \dfrac{1}{2} \cdot \dfrac{2}{3}$
$= a^2 + \dfrac{1}{6}a - \dfrac{1}{3}$

55. For $x = -1$ and $y = 2$:
$2 - 5xy + y^2 - 4xy^3 + x^6$
$= 2 - 5(-1)(2) + 2^2 - 4(-1)(2)^3 + (-1)^6$
$= 2 + 10 + 4 + 32 + 1$
$= 49$

57. $a^3b^8c^2 - c^{22} + a^5c^{10}$

Term	Coefficient	Degree
$a^3b^8c^2$	1	13
$-c^{22}$	-1	22
a^5c^{10}	1	15

The term of highest degree is $-c^{22}$. Its degree is 22, so the degree of the polynomial is 22.

59. $6m^3 + 3m^2n + 4mn^2 + m^2n - 5mn^2$
$= 6m^3 + (3 + 1)m^2n + (4 - 5)mn^2$
$= 6m^3 + 4m^2n - mn^2$

61. $\left(6x^3y^2 - 4x^2y - 6x\right) - \left(-5x^3y^2 + 4x^2y + 6x^2 - 6\right)$
$= 6x^3y^2 - 4x^2y - 6x + 5x^3y^2 - 4x^2y - 6x^2 + 6$
$= (6 + 5)x^3y^2 + (-4 - 4)x^2y - 6x^2 - 6x + 6$
$= 11x^3y^2 - 8x^2y - 6x^2 - 6x + 6$

63. $\left(5ab - cd^2\right)^2 = (5ab)^2 - 2 \cdot 5ab \cdot cd^2 + \left(cd^2\right)^2$
$= 25a^2b^2 - 10abcd^2 + c^2d^4$

65. $\left(3y^5 - y^2 + 12y\right) \div (3y)$
$= \dfrac{3y^5 - y^2 + 12y}{3y}$
$= \dfrac{3y^5}{3y} - \dfrac{y^2}{3y} + \dfrac{12y}{3y}$
$= y^4 - \dfrac{1}{3}y + 4$

67.

$$
\begin{array}{r}
t^3 \qquad + 2t - 3 \\
t+1 \overline{\smash{\big)}\, t^4+t^3+2t^2 - t - 3} \\
\underline{t^4+t^3} \qquad\qquad \\
2t^2 - t \qquad \\
\underline{2t^2+2t} \qquad \\
-3t - 3 \\
\underline{-3t - 3} \\
0
\end{array}
$$

The answer is $t^3 + 2t - 3$.

69. $\dfrac{1}{a^9} = a^{-9}$

71. $\dfrac{6a^{-5}b}{3a^8 b^{-8}} = 2a^{-5-8}b^{1+8} = 2a^{-13}b^9 = \dfrac{2b^9}{a^{13}}$

73. $\left(2x^{-3}y\right)^{-2} = (2)^{-2}\,x^{-3(-2)}y^{-2} = 4^{-1}x^6 y^{-2} = \dfrac{x^6}{4y^2}$

75. 4.7×10^8 Since the exponent is positive, the decimal point will move to the right

4.70000000.

$\llcorner$_____$\uparrow$ 8 places

$4.7 \times 10^8 = 470,000,000$

77. $\left(3.8 \times 10^4\right)\left(5.5 \times 10^{-1}\right) = 20.9 \times 10^3$
$= 2.09 \times 10^4$

79. *Writing Exercise.* In the expression $5x^3$, the exponent refers only to the x. In the expression $(5x)^3$, the entire expression $5x$ is the base.

81. a) For $\left(x^5 - 6x^2 + 3\right)\left(x^4 + 3x^3 + 7\right)$, the highest terms of each factor are x^5 and x^4. Their product is x^9, which is degree 9. Thus, the degree of the product is 9.
b) For $\left(x^7 - 4\right)^4$, the term of highest degree is x^7, which is degree 7. Taking that term to the fourth power results in a term of degree 28.

83. Let $c = $ the coefficient of x^3.
Let $2c = $ the coefficient of x^4.
Let $2c - 3 = $ the coefficient of x.
Let $2c - 3 - 7 = $ the constant $\cdot$ (the remaining term)
The coefficient of x^2 is 0.

Solve:
$$
\begin{aligned}
2c + c + 0 + 2c - 3 + 2c - 3 - 7 &= 15 \\
7c - 13 &= 15 \\
7c &= 28 \\
c &= 4
\end{aligned}
$$

Coefficient of x^3, $c = 4$
Coefficient of x^4, $2c = 2 \cdot 4 = 8$
Coefficient of x^2, 0
Coefficient of x, $2c - 3 = 2 \cdot 4 - 3 = 8 - 3 = 5$
Constant, $2c - 3 - 7 = 2 \cdot 4 - 10 = 8 - 10 = -2$
The polynomial is $8x^4 + 4x^3 + 5x - 2$.

85.
$$
\begin{aligned}
(x - 7)(x + 10) &= (x - 4)(x - 6) \\
x^2 + 10x - 7x - 70 &= x^2 - 6x - 4x + 24 \\
x^2 + 3x - 70 &= x^2 - 10x + 24 \\
3x - 70 &= -10x + 24 && \text{Subtracting } x^2 \\
3x &= -10x + 94 && \text{Adding 70} \\
13x &= 94 && \text{Adding } 10x \\
x &= \frac{94}{13}
\end{aligned}
$$

The solution is $\dfrac{94}{13}$.

Chapter 4 Test

1. $x^7 \cdot x \cdot x^5 = x^{7+1+5} = x^{13}$

3. $\dfrac{(3m)^4}{(3m)^4} = 1$

5. $(-3y^2)^3 = (-3)^3 \cdot (y^2)^3 = -27y^{2\cdot 3} = -27y^6$

7. $\dfrac{24a^7 b^4}{20a^2 b} = \dfrac{24}{20}a^{7-2}b^{4-1}$
$= \dfrac{6}{5}a^5 b^3$

9. Two monomials are added so $4x^2 y - 7y^3$ is a binomial.

11. $2t^3 - t + 7t^5 + 4$
The degrees of the terms are $3, 1, 5, 0$; the leading term is $7t^5$; the leading coefficient is 7; the degree of the polynomial is 5.

13. $4a^2 - 6 + a^2 = (4 + 1)a^2 - 6 = 5a^2 - 6$

15. $3 - x^2 + 8x + 5x^2 - 6x - 2x + 4x^3$
$= 4x^3 + (-1 + 5)x^2 + (8 - 6 - 2)x + 3$
$= 4x^3 + 4x^2 + 3$

17. $\left(x^4 + \dfrac{2}{3}x + 5\right) + \left(4x^4 + 5x^2 + \dfrac{1}{3}x\right)$
$= (1 + 4)x^4 + 5x^2 + \left(\dfrac{2}{3}x + \dfrac{1}{3}\right)x + 5$
$= 5x^4 + 5x^2 + x + 5$

19. $(t^3 - 0.3t^2 - 20) - (t^4 - 1.5t^3 + 0.3t^2 - 11)$
$= t^3 - 0.3t^2 - 20 - t^4 + 1.5t^3 - 0.3t^2 + 11$
$= -t^4 + (1 + 1.5)t^3 + (-0.3 - 0.3)t^2$
$\quad + (-20 + 11)$
$= -t^4 + 2.5t^3 - 0.6t^2 - 9$

21. $\left(x - \dfrac{1}{3}\right)^2 = x^2 - 2 \cdot x \cdot \dfrac{1}{3} + \dfrac{1}{3}^2$
$= x^2 - \dfrac{2}{3}x + \dfrac{1}{9}$

23. $(3b + 5)(2b - 1) = 6b^2 - 3b + 10b - 5$
$= 6b^2 - 7b - 5$

25. $(8 - y)(6 + 5y) = 48 + 40y - 6y - 5y^2$
$= 48 + 34y - 5y^2$

27. $(8a^3 + 3)^2 = (8a^3)^2 + 2(8a^3)(3) + 3^2$
$= 64a^6 + 48a^3 + 9$

29. $2x^3y - y^3 + xy^3 + 8 - 6x^3y - x^2y^2 + 11$

$\quad = (2-6)x^3y - x^2y^2 + xy^3 - y^3 + (8+11)$
$\quad = -4x^3y - x^2y^2 + xy^3 - y^3 + 19$

31. $(3x^5 - y)(3x^5 + y) = (3x^5)^2 - y^2$
$\qquad\qquad\qquad\quad = 9x^{10} - y^2$

33.

$$
\begin{array}{r}
2x^2 \quad -4x \quad -2 \\
3x+2\overline{)\;6x^3 \;-8x^2 -14x +13} \\
\underline{-6x^3 \;-4x^2} \\
-12x^2 -14x \\
\underline{12x^2 \;+8x} \\
-6x +13 \\
\underline{6x \;+4} \\
17
\end{array}
$$

The answer is $2x^2 - 4x - 2$, R 17 or $2x^2 - 4x - 2 + \dfrac{17}{3x+2}$.

35. $\dfrac{1}{5^6} = 5^{-6}$

37. $\dfrac{9x^3y^2}{3x^8y^{-3}} = 3x^{3-8}y^{2+3} = 3x^{-5}y^5 = \dfrac{3y^5}{x^5}$

39. $\left(\dfrac{ab}{c}\right)^{-3} = \left(\dfrac{c}{ab}\right)^3 = \dfrac{c^3}{a^3b^3}$

41. $5 \times 10^{-8} = 0.00000005$

43. $(2.4 \times 10^5)(5.4 \times 10^{16}) = (2.4 \times 5.4)(10^5 \times 10^{16})$
$\qquad\qquad\qquad\qquad\qquad = 12.96 \times 10^{5+16}$
$\qquad\qquad\qquad\qquad\qquad = 1.296 \times 10^1 \times 10^{21}$
$\qquad\qquad\qquad\qquad\qquad = 1.296 \times 10^{1+21}$

45. $\qquad x^2 + (x-7)(x+4) = 2(x-6)^2$
$\quad x^2 + x^2 + 4x - 7x - 28 = 2[x^2 - 2\cdot x \cdot 6 + 6^2]$
$\quad x^2 + x^2 + 4x - 7x - 28 = 2[x^2 - 12x + 36]$
$\qquad\qquad 2x^2 - 3x - 28 = 2x^2 - 24x + 72$
$\qquad\qquad\quad -3x - 28 = -24x + 72$
$\qquad\qquad\quad 21x - 28 = 72$
$\qquad\qquad\qquad 21x = 100$
$\qquad\qquad\qquad\quad x = \dfrac{100}{21}$

47. *Familiarize.* We need to multiply to determine the hours wasted. There are 60 sec in a minute and 60 min in an hr. $1 \text{ sec} = \dfrac{1}{60} \cdot \dfrac{1}{60} = \dfrac{1}{3600}$hr. So $4 \text{ sec} = \dfrac{4}{3600}$hr

Translate. We multiply the number of spam emails by the time wasted on each spam email. 12.4 billion $= 1.24 \times 10^{10}$.

Carry out

$s = 1.24 \times 10^{10} \cdot \dfrac{4}{3600} \approx 1.4 \times 10^7$hr

Check. We recalculate to check our solution. The answer checks.

State. About 1.4×10^7hr each day are wasted due to spam.

Chapter 5

Polynomials and Factoring

---

1. Since $7a \cdot 5ab = 35a^2b$, choice (h) is most appropriate.

3. $5x + 10 = 5(x + 2)$ and $4x + 8 = 4(x + 2)$, so $x + 2$ is a common factor of $5x + 10$ and $4x + 8$ and choice (b) is most appropriate.

5. $3x^2(3x^2 - 1) = 9x^4 - 3x^2$, so choice (c) is most appropriate.

7. $3a + 6a^2 = 3a(1 + 2a)$, so $1 + 2a$ is a factor of $3a + 6a^2$ and choice (d) is most appropriate.

9. Answers may vary. $14x^3 = (14x)(x^2) = (7x^2)(2x) = (-2)(-7x^3)$

11. Answers may vary. $-15a^4 = (-15)(a^4) = (-5a)(3a^3) = (-3a^2)(5a^2)$

13. Answers may vary. $25t^5 = (5t^2)(5t^3) = (25t)(t^4) = (-5t)(-5t^4)$

15. $8x + 24 = 8 \cdot x + 8 \cdot 3$
$ = 8(x + 3)$

17. $6x - 30 = 6 \cdot x - 6 \cdot 5$
$ = 6(x - 5)$

19. $2x^2 + 2x - 8 = 2 \cdot x^2 + 2 \cdot x - 2 \cdot 4$
$ = 2(x^2 + x - 4)$

21. $3t^2 + t = t \cdot 3t + t \cdot 1$
$ = t(3t + 1)$

23. $-5y^2 - 10y = -5y \cdot y - 5y \cdot 2$
$ = -5y(y + 2)$

25. $x^3 + 6x^2 = x^2 \cdot x + x^2 \cdot 6$
$ = x^2(x + 6)$

27. $16a^4 - 24a^2 = 8a^2 \cdot 2a^2 - 8a^2 \cdot 3$
$ = 8a^2(2a^2 - 3)$

29. $-6t^6 + 9t^4 - 4t^2 = -t^2 \cdot 6t^4 - t^2\left(-9t^2\right) - t^2 \cdot 4$
$ = -t^2\left(6t^4 - 9t^2 + 4\right)$

31. $6x^8 + 12x^6 - 24x^4 + 30x^2$
$= 6x^2 \cdot x^6 + 6x^2 \cdot 2x^4 - 6x^2 \cdot 4x^2 + 6x^2 \cdot 5$
$= 6x^2(x^6 + 2x^4 - 4x^2 + 5)$

33. $x^5y^5 + x^4y^3 + x^3y^3 - x^2y^2$
$= x^2y^2 \cdot x^3y^3 + x^2y^2 \cdot x^2y + x^2y^2 \cdot xy - x^2y^2 \cdot 1$
$= x^2y^2(x^3y^3 + x^2y + xy - 1)$

35. $-35a^3b^4 + 10a^2b^3 - 15a^3b^2$
$= -5a^2b^2 \cdot 7ab^2 - 5a^2b^2\left(-2b\right) - 5a^2b^2 \cdot 3a$
$= -5a^2b^2\left(7ab^2 - 2b + 3a\right)$

37. $\quad n\left(n - 6\right) + 3\left(n - 6\right)$
$= (n - 6)\left(n + 3\right)$ Factoring out the common binomial factor $n - 6$

39. $\quad x^2(x + 3) - 7(x + 3)$
$= (x + 3)(x^2 - 7)$ Factoring out the common binomial factor $x + 3$

41. $\quad y^2\left(2y - 9\right) + (2y - 9)$
$= y^2\left(2y - 9\right) + 1\left(2y - 9\right)$
$= (2y - 9)\left(y^2 + 1\right)$ Factoring out the common factor $2y - 9$

43. $\quad x^3 + 2x^2 + 5x + 10$
$= \left(x^3 + 2x^2\right) + (5x + 10)$
$= x^2\left(x + 2\right) + 5\left(x + 2\right)$ Factoring each binomial
$= (x + 2)\left(x^2 + 5\right)$ Factoring out the common factor $x + 2$

45. $\quad 5a^3 + 15a^2 + 2a + 6$
$= \left(5a^3 + 15a^2\right) + (2a + 6)$
$= 5a^2(a + 3) + 2(a + 3)$ Factoring each binomial
$= (a + 3)(5a^2 + 2)$ Factoring out the common factor $a + 3$

47. $\quad 9n^3 - 6n^2 + 3n - 2$
$= 3n^2\left(3n - 2\right) + 1\left(3n - 2\right)$
$= (3n - 2)\left(3n^2 + 1\right)$

49. $\quad 4t^3 - 20t^2 + 3t - 15$
$= 4t^2(t - 5) + 3(t - 5)$
$= (t - 5)(4t^2 + 3)$

51. $7x^3 + 5x^2 - 21x - 15 = x^2\left(7x + 5\right) - 3\left(7x + 5\right)$
$ = (7x + 5)\left(x^2 - 3\right)$

53. $6a^3 + 7a^2 + 6a + 7 = a^2\left(6a + 7\right) + 1\left(6a + 7\right)$
$ = (6a + 7)\left(a^2 + 1\right)$

55. $2x^3 + 12x^2 - 5x - 30 = 2x^2\left(x + 6\right) - 5\left(x + 6\right)$
$ = (x + 6)\left(2x^2 - 5\right)$

57. We try factoring by grouping.
$p^3 + p^2 - 3p + 10 = p^2(p + 1) - (3p - 10)$, or
$p^3 - 3p + p^2 + 10 = p(p^2 - 3) + p^2 + 10$

Because we cannot find a common binomial factor, this polynomial cannot be factored using factoring by grouping.

59. $y^3 + 8y^2 - 2y - 16 = y^2(y+8) - 2(y+8) = (y+8)(y^2 - 2)$

61. $2x^3 - 8x^2 - 9x + 36 = 2x^2(x-4) - 9(x-4)$
$$= (x-4)(2x^2 - 9)$$

63. *Writing Exercise.* Yes; the opposite of a factor is also a factor so both can be correct.

65. $(x+2)(x+7)$
$$\quad\quad\text{F}\quad\quad\text{O}\quad\quad\text{I}\quad\quad\text{L}$$
$$= x \cdot x + x \cdot 7 + 2 \cdot x + 2 \cdot 7$$
$$= x^2 + 7x + 2x + 14$$
$$= x^2 + 9x + 14$$

67. $(x+2)(x-7)$
$$\quad\quad\text{F}\quad\quad\text{O}\quad\quad\text{I}\quad\quad\text{L}$$
$$= x \cdot x + x \cdot (-7) + 2 \cdot x - 2 \cdot 7$$
$$= x^2 - 7x + 2x + 14$$
$$= x^2 - 5x - 14$$

69. $(a-1)(a-3)$
$$\quad\quad\text{F}\quad\quad\text{O}\quad\quad\text{I}\quad\quad\text{L}$$
$$= a \cdot a - a \cdot 3 - 1 \cdot a - 1\,(-3)$$
$$= a^2 - 3a - a + 3$$
$$= a^2 - 4a + 3$$

71. $(t-5)(t+10)$
$$\quad\quad\text{F}\quad\quad\text{O}\quad\quad\text{I}\quad\quad\text{L}$$
$$= t \cdot t + t \cdot 10 - 5 \cdot t - 5 \cdot 10$$
$$= t^2 + 10t - 5t - 50$$
$$= t^2 + 5t - 50$$

73. *Writing Exercise.* This is a good idea, because it is unlikely that Azrah will choose two replacement values that give the same value for non-equivalent expressions.

75. $4x^5 + 6x^2 + 6x^3 + 9 = 2x^2(2x^3 + 3) + 3(2x^3 + 3)$
$$= (2x^3 + 3)(2x^2 + 3)$$

77. $2x^4 + 2x^3 - 4x^2 - 4x = 2x(x^3 + x^2 - 2x - 2)$
$$= 2x\left(x^2(x+1) - 2(x+1)\right)$$
$$= 2x(x+1)(x^2 - 2)$$

79. $5x^5 - 5x^4 + x^3 - x^2 + 3x - 3$
$$= 5x^4(x-1) + x^2(x-1) + 3(x-1)$$
$$= (x-1)(5x^4 + x^2 + 3)$$

We could also do this exercise as follows:
$$5x^5 - 5x^4 + x^3 - x^2 + 3x - 3$$
$$= (5x^5 + x^3 + 3x) - (5x^4 + x^2 + 3)$$
$$= x(5x^4 + x^2 + 3) - 1(5x^4 + x^2 + 3)$$
$$= (5x^4 + x^2 + 3)(x-1)$$

81. Answers may vary. $8x^4y^3 - 24x^3y^3 + 16x^2y^4$

Exercise Set 5.2

1. If c is positive, then p and q have the same sign. If both are negative, then b is negative; if both are positive then c is positive. Thus we replace each blank with "positive."

3. If p is negative and q is negative, then b is negative because it is the sum of two negative numbers and c is positive because it is the product of two negative numbers.

5. Since c is negative, it is the product of a negative and a positive number. Then because c is the product of p and q and we know that p is negative, q must be positive.

7. $x^2 + 8x + 16$

Since the constant term and the coefficient of the middle term are both positive, we look for a factorization of 16 in which both factors are positive. Their sum must be 8.

Pairs of factors	Sums of factors
1, 16	17
2, 8	10
4, 4	8

The numbers we want are 4 and 4.
$$x^2 + 8x + 16 = (x+4)(x+4)$$

9. $x^2 + 11x + 10$

Since the constant term and the coefficient of the middle term are both positive, we look for a factorization of 10 in which both factors are positive. Their sum must be 11.

Pairs of factors	Sums of factors
1, 10	11
2, 5	10

The numbers we want are 1 and 10.
$$x^2 + 11x + 10 = (x+1)(x+10)$$

11. $x^2 + 10x + 21$

Since the constant term and the coefficient of the middle term are both positive, we look for a factorization of 21 in which both factors are positive. Their sum must be 10.

Pairs of factors	Sums of factors
1, 21	22
3, 7	10

The numbers we want are 3 and 7.
$$x^2 + 10x + 21 = (x+3)(x+7)$$

13. $t^2 - 9t + 14$

Since the constant term is positive and the coefficient of the middle term is negative, we look for a factorization of 14 in which both factors are negative. Their sum must be -9.

Pairs of factors	Sums of factors
-1, -14	-15
-2, -7	-9

The numbers we want are -2 and -7.
$$t^2 - 9t + 14 = (t-2)(t-7)$$

15. $b^2 - 5b + 4$

Since the constant term is positive and the coefficient of the middle term is negative, we look for a factorization of 4 in which both factors are negative. Their sum must be -5.

Pairs of factors	Sums of factors
$-1, -4$	-5
$-2, -2$	-4

The numbers we want are -1 and -4.

$b^2 - 5b + 4 = (b-1)(b-4)$.

17. $a^2 - 7a + 12$

Since the constant term is positive and the coefficient of the middle term is negative, we look for a factorization of 12 in which both factors are negative. Their sum must be -7.

Pairs of factors	Sums of factors
$-1, -12$	-13
$-2, -6$	-8
$-3, -4$	-7

The numbers we need are -3 and -4.

$a^2 - 7a + 12 = (a-3)(a-4)$.

19. $d^2 - 7d + 10$

Since the constant term is positive and the coefficient of the middle term is negative, we look for a factorization of 10 in which both factors are negative. Their sum must be -7.

Pairs of factors	Sums of factors
$-1, -10$	-11
$-2, -5$	-7

The numbers we want are -2 and -5.

$d^2 - 7d + 10 = (d-2)(d-5)$.

21. $x^2 - 2x - 15$

The constant term, -15, must be expressed as the product of a negative number and a positive number. Since the sum of those two numbers must be negative, the negative number must have the greater absolute value.

Pairs of factors	Sums of factors
$1, -15$	-14
$3, -5$	-2

The numbers we need are 3 and -5.

$x^2 - 2x - 15 = (x+3)(x-5)$.

23. $x^2 + 2x - 15$

The constant term, -15, must be expressed as the product of a negative number and a positive number. Since the sum of those two numbers must be positive, the positive number must have the greater absolute value.

Pairs of factors	Sums of factors
$-1, 15$	14
$-3, 5$	2

The numbers we need are -3 and 5.

$x^2 + 2x - 15 = (x-3)(x+5)$.

25. $2x^2 - 14x - 36 = 2(x^2 - 7x - 18)$

After factoring out the common factor, 2, we consider $x^2 - 7x - 18$. The constant term, -18, must be expressed as the product of a negative number and a positive number. Since the sum of those two numbers must be negative, the negative number must have the greater absolute value.

Pairs of factors	Sums of factors
$1, -18$	-17
$2, -9$	-7
$3, -6$	-3

The numbers we need are 2 and -9. The factorization of $x^2 - 7x - 18$ is $(x-9)(x+2)$. We must not forget the common factor, 2. Thus, $2x^2 - 14x - 36 = 2(x^2 - 7x - 18) = 2(x-9)(x+2)$.

27. $-x^3 + 6x^2 + 16x = -x(x^2 - 6x - 16)$

After factoring out the common factor, $-x$, we consider $x^2 - 6x - 16$. The constant term, -16, must be expressed as the product of a negative number and a positive number. Since the sum of those two numbers must be negative, the negative number must have the greater absolute value.

Pairs of factors	Sums of factors
$1, -16$	-15
$2, -8$	-6
$4, -4$	0

The numbers we need are 2 and -8. The factorization of $x^2 - 6x - 16$ is $(x+2)(x-8)$. We must not forget the common factor, $-x$. Thus, $-x^3 + 6x^2 + 16x = -x(x^2 - 6x - 16) = -x(x+2)(x-8)$.

29. $4y - 45 + y^2 = y^2 + 4y - 45$

The constant term, -45, must be expressed as the product of a negative number and a positive number. Since the sum of those two numbers must be positive, the positive number must have the greater absolute value.

Pairs of factors	Sums of factors
$-1, 45$	44
$-3, 15$	12
$-5, 9$	4

The numbers we need are -5 and 9.

$4y - 45 + y^2 = (y-5)(y+9)$

31. $x^2 - 72 + 6x = x^2 + 6x - 72$

The constant term, -72, must be expressed as the product of a negative number and a positive number. Since the sum of those two numbers must be positive, the positive number must have the greater absolute value.

Pairs of factors	Sums of factors
−1, 72	71
−2, 36	34
−3, 24	21
−4, 18	14
−6, 12	6

The numbers we need are −6 and 12.

$x^2 - 72 + 6x = (x - 6)(x + 12)$

33. $-5b^2 - 35b + 150 = -5(b^2 + 7b - 30)$

After factoring out the common factor, −5, we consider $b^2 + 7b - 30$. The constant term, −30, must be expressed as the product of a negative number and a positive number. Since the sum of those two numbers must be positive, the positive number must have the greater absolute value.

Pairs of factors	Sums of factors
−1, 30	29
−2, 15	13
−3, 10	7
−5, 6	1

The numbers we need are −3 and 10. The factorization of $b^2 + 7b - 30$ is $(b - 3)(b + 10)$. We must not forget the common factor. Thus, $-5b^2 - 35b + 150 = -5(b^2 + 7b - 30) = -5(b - 3)(b + 10)$.

35. $x^5 - x^4 - 2x^3 = x^3(x^2 - x - 2)$

After factoring out the common factor, x^3, we consider $x^2 - x - 2$. The constant term, −2, must be expressed as the product of a negative number and a positive number. Since the sum of those two numbers must be negative, the negative number must have the greater absolute value. The only possible factors that fill these requirements are 1 and −2. These are the numbers we need. The factorization of $x^2 - x - 2$ is $(x + 1)(x - 2)$. We must not forget the common factor, x^3. Thus, $x^5 - x^4 - 2x^3 = x^3(x^2 - x - 2) = x^3(x + 1)(x - 2)$.

37. $x^2 + 5x + 10$

Since the constant term and the coefficient of the middle term are both positive, we look for a factorization of 10 in which both factors are positive. Their sum must be 5. The only possible pairs of positive factors are 1 and 10, and 2 and 5 but neither sum is 5. Thus, this polynomial is not factorable into polynomials with integer coefficients. It is prime.

39. $32 + 12t + t^2 = t^2 + 12t + 32$

Since the constant term is positive and the coefficient of the middle term is positive, we look for a factorization of 32 in which both terms are positive. Their sum must be 12.

Pairs of factors	Sums of factors
1,32	33
2,16	18
4, 8	12

The numbers we want are 4 and 8.

$32 + 12t + t^2 = (t + 4)(t + 8)$.

41. $x^2 + 20x + 99$

We look for two factors, both positive, whose product is 99 and whose sum is 20.

They are 9 and 11: $9 \cdot 11 = 99$ and $9 + 11 = 20$.

$x^2 + 20 + 99 = (x + 9)(x + 11)$

43. $3x^3 - 63x^2 - 300x = 3x(x^2 - 21x - 100)$

After factoring out the common factor, $3x$, we consider $x^2 - 21x - 100$. We look for two factors, one positive, one negative, whose product is −100 and whose sum is −21.

They are −25 and 4: $-25(4) = -100$ and $-25 + 4 = -21$.

$x^2 - 21x - 100 = (x - 25)(x + 4)$, so $3x^3 - 63x^2 - 300x = 3x(x - 25)(x + 4)$.

45. $-2x^2 + 42x + 144 = -2(x^2 - 21x - 72)$

After factoring out the common factor, −2, we consider $x^2 - 21x - 72$. We look for two factors, one positive, and one negative, whose product is −72 and whose sum is −21. They are −24 and 3.

$x^2 - 21x - 72 = (x - 24)(x + 3)$, so $-2x^2 + 42x + 144 = -2(x^2 - 21x - 72) = -2(x - 24)(x + 3)$.

47. $y^2 - 20y + 96$

We look for two factors, both negative, whose product is 96 and whose sum is −20. They are −8 and −12.

$y^2 - 20y + 96 = (y - 8)(y - 12)$

49. $-a^6 - 9a^5 + 90a^4 = -a^4(a^2 + 9a - 90)$

After factoring out the common factor, $-a^4$, we consider $a^2 + 9a - 90$. We look for two factors, one positive and one negative, whose product is −90 and whose sum is 9. They are −6 and 15.

$a^2 + 9a - 90 = (a - 6)(a + 15)$, so $-a^6 - 9a^5 + 90a^4 = -a^4(a - 6)(a + 15)$.

51. $t^2 + \frac{2}{3}t + \frac{1}{9}$

We look for two factors, both positive, whose product is $\frac{1}{9}$ and whose sum is $\frac{2}{3}$. They are $\frac{1}{3}$ and $\frac{1}{3}$.

$t^2 + \frac{2}{3}t + \frac{1}{9} = \left(t + \frac{1}{3}\right)\left(t + \frac{1}{3}\right)$, or $\left(t + \frac{1}{3}\right)^2$

53. $11 + w^2 - 4w = w^2 - 4w + 11$

Since the constant term is positive and the coefficient of the middle term is negative, we look for a factorization of 11 in which both factors are negative. Their sum must be −4. The only possible pair of factors is −1 and −11, but their sum is not −4. Thus, this polynomial is not factorable into polynomials with integer coefficients. It is prime.

55. $p^2 + 7pq + 10q^2$

Think of −7q as a "coefficient" of p. Then we look for factors of $10q^2$ whose sum is −7q. They are −5q and −2q.

$p^2 + 7pq + 10q^2 = (p - 5q)(p - 2q)$.

57. $m^2 + 5mn + 5n^2 = m^2 + 5nm + 5n^2$

We look for factors of $5n^2$ whose sum is $5n$. The only reasonable possibilities are shown below.

Pairs of factors	Sums of factors
$5n, \quad n$	$6n$
$-5n, -n$	$-6n$

There are no factors whose sum is $5n$. Thus, the polynomial is not factorable into polynomials with integer coefficients. It is prime.

59. $s^2 - 4st - 12t^2 = s^2 - 4ts - 12t^2$

We look for factors of $-12t^2$ whose sum is $-4t$. They are $-6t$ and $2t$.

$s^2 - 4st - 12t^2 = (s - 6t)(s + 2t)$

61. $6a^{10} + 30a^9 - 84a^8 = 6a^8(a^2 + 5a - 14)$

After factoring out the common factor, $6a^8$, we consider $a^2 + 5a - 14$. We look for two factors, one positive and one negative, whose product is -14 and whose sum is 5. They are -2 and 7.

$a^2 + 5a - 14 = (a - 2)(a + 7)$, so $6a^{10} + 30a^9 - 84a^8$
$= 6a^8(a - 2)(a + 7)$.

63. *Writing Exercise.* Since both constants are negative, the middle term will be negative so $(x - 17)(x - 18)$ cannot be a factorization of $x^2 + 35x + 306$.

65. $(2x + 3)(3x + 4)$
$\qquad$ F $\qquad$ O $\qquad$ I $\qquad$ L
$= 2x \cdot 3x + 2x \cdot 4 + 3 \cdot 3x + 3 \cdot 4$
$= 6x^2 + 8x + 9x + 12$
$= 6x^2 + 17x + 12$

67. $(2x - 3)(3x + 4)$
$\qquad$ F $\qquad$ O $\qquad$ I $\qquad$ L
$= 2x \cdot 3x + 2x \cdot 4 - 3 \cdot 3x - 3 \cdot 4$
$= 6x^2 + 8x - 9x - 12$
$= 6x^2 - x - 12$

69. $(5x - 1)(x - 7)$
$\qquad$ F $\qquad$ O $\qquad$ I $\qquad$ L
$= 5x \cdot x - 5x \cdot 7 - 1 \cdot x - 1(-7)$
$= 5x^2 - 35x - x + 7$
$= 5x^2 - 36x + 7$

71. *Writing Exercise.* There is a finite number of pairs of numbers with the correct product, but there are infinitely many pairs with the correct sum.

73. $a^2 + ba - 50$

We look for all pairs of integer factors whose product is -50. The sum of each pair is represented by b.

Pairs of factors whose product is -50	Sums of factors
$-1, \ 50$	49
$1, -50$	-49
$-2, \ 25$	23
$2, -25$	-23
$-5, \ 10$	5
$5, -10$	-5

The polynomial $a^2 + ba - 50$ can be factored if b is 49, -49, 23, -23, 5, or -5.

75. $y^2 - 0.2y - 0.08$

We look for two factors, one positive and one negative, whose product is -0.08 and whose sum is -0.2. They are -0.4 and 0.2.

$y^2 - 0.2y - 0.08 = (y - 0.4)(y + 0.2)$

77. $-\dfrac{1}{3}a^3 + \dfrac{1}{3}a^2 + 2a = -\dfrac{1}{3}a(a^2 - a - 6)$

After factoring out the common factor, $-\dfrac{1}{3}a$, we consider $a^2 - a - 6$. We look for two factors, one positive and one negative, whose product is -6 and whose sum is -1. They are 2 and -3.

$a^2 - a - 6 = (a + 2)(a - 3)$, so
$-\dfrac{1}{3}a^3 + \dfrac{1}{3}a^2 + 2a = -\dfrac{1}{3}a(a + 2)(a - 3)$.

79. $x^{2m} + 11x^m + 28 = (x^m)^2 + 11x^m + 28$

We look for numbers p and q such that $x^{2m} + 11x^m + 28 = (x^m + p)(x^m + q)$. We find two factors, both positive, whose product is 28 and whose sum is 11. They are 4 and 7.

$x^{2m} + 11x^m + 28 = (x^m + 4)(x^m + 7)$

81. $(a + 1)x^2 + (a + 1)3x + (a + 1)2$
$= (a + 1)(x^2 + 3x + 2)$

After factoring out the common factor $a + 1$, we consider $x^2 + 3x + 2$. We look for two factors, whose product is 2 and whose sum is 3. They are 1 and 2.

$x^2 + 3x + 2 = (x + 1)(x + 2)$, so
$(a + 1)x^2 + (a + 1)3x + (a + 1)2$
$= (a + 1)(x + 1)(x + 2)$.

83. $6x^2 + 36x + 54 = 6(x^2 + 6x + 9) = 6(x + 3)(x + 3) = 6(x + 3)^2$

Since the surface area of a cube with sides is given by $6s^2$, we know that this cube has side $x + 3$. The volume of a cube with side s is given by s^3, so the volume of this cube is $(x + 3)^3$, or $x^3 + 9x^2 + 27x + 27$.

85. The shaded area consists of the area of a rectangle with sides x and $x + x$, or $2x$, and $\dfrac{3}{4}$ of the area of a circle with radius x. It can be expressed as follows:

$x \cdot 2x + \dfrac{3}{4}\pi x^2 = 2x^2 + \dfrac{3}{4}\pi x^2 = x^2\left(2 + \dfrac{3}{4}\pi\right)$, or
$\dfrac{1}{4}x^2(8 + 3\pi)$

87. The shaded area consists of the area of a square with side $x + x + x$, or $3x$, less the area of a semicircle with radius x. It can be expressed as follows:

$$3x \cdot 3x - \frac{1}{2}\pi x^2 = 9x^2 - \frac{1}{2}\pi x^2 = x^2\left(9 - \frac{1}{2}\pi\right)$$

89. $x^2 + 4x + 5x + 20 = x^2 + 9x + 20 = (x + 4)(x + 5)$

Exercise Set 5.3

1. Since $(6x-1)(2x+3) = 12x^2+16x-3$, choice (c) is correct.

3. Since $(7x+1)(2x-3) = 14x^2-19x-3$, choice (d) is correct.

5. $2x^2 + 7x - 4$

(1) There is no common factor (other than 1 or -1).

(2) Because $2x^2$ can be factored as $2x \cdot x$, we have this possibility:

$$(2x + \quad)(x + \quad)$$

(3) There are 3 pairs of factors of -4 and they can be listed two ways:

$$-4,1 \quad 4,-1 \quad 2,-2$$
$$\text{and} \quad 1,-4 \quad -1,4 \quad -2,2$$

(4) Look for Outer and Inner products resulting from steps (2) and (3) for which the sum is $7x$. We can immediately reject all possibilities in which a factor has a common factor, such as $(2x - 4)$ or $(2x + 2)$, because we determined at the outset that there is no common factor other than 1 and -1. We try some possibilities:

$$(2x + 1)(x - 4) = 2x^2 - 7x - 4$$
$$(2x - 1)(x + 4) = 2x^2 + 7x - 4$$

The factorization is $(2x - 1)(x + 4)$.

7. $3x^2 - 17x - 6$

(1) There is no common factor (other than 1 or -1).

(2) Because $3x^2$ can be factored as $3x \cdot x$, we have this possibility:

$$(3x + \quad)(x + \quad)$$

(3) There are 4 pairs of factors of -6 and they can be listed two ways:

$$-6,1 \quad 6,-1 \quad -3,2 \quad 3,-2$$
$$\text{and} \quad 1,-6 \quad -1,6 \quad 2,-3 \quad -2,3$$

(4) Look for Outer and Inner products resulting from steps (2) and (3) for which the sum is $-17x$. We can immediately reject all possibilities in which either factor has a common factor, such as $(3x - 6)$ or $(3x + 3)$, because at the outset we determined that there is no common factor other than 1 or -1. We try some possibilities:

$$(3x + 2)(x - 3) = 3x^2 - 7x - 6$$
$$(3x + 1)(x - 6) = 3x^2 - 17x - 6$$

The factorization is $(3x + 1)(x - 6)$.

9. $4t^2 + 12t + 5$

(1) There is no common factor (other than 1 or -1).

(2) Because $4t^2$ can be factored as $4t \cdot t$ or $2t \cdot 2t$, we have these possibilities:

$$(4t + \quad)(t + \quad) \text{ and } (2t + \quad)(2t + \quad)$$

(3) There are 2 pairs of factors of 5 and they can be listed two ways:

$$5,1 \quad -5,-1$$
$$\text{and} \quad 1,5 \quad -1,-5$$

(4) Look for Outer and Inner products resulting from steps (2) and (3) for which the sum is $12t$. We try some possibilities:

$$(4t + 1)(t + 5) = 4t^2 + 21t + 5$$
$$(2t + 1)(2t + 5) = 4t^2 + 12t + 5$$

The factorization is $(2t + 1)(2t + 5)$.

11. $15a^2 - 14a + 3$

(1) There is no common factor (other than 1 or -1).

(2) Because $15a^2$ can be factored as $15a \cdot a$ or $5a \cdot 3a$, we have these possibilities:

$$(15a + \quad)(a + \quad) \text{ and } (5a + \quad)(3a + \quad)$$

(3) There are 2 pairs of factors of 3 and they can be listed two ways:

$$3,1 \quad -3,-1$$
$$\text{and} \quad 1,3 \quad -1,-3$$

(4) Look for Outer and Inner products resulting from steps (2) and (3) for which the sum is $-14a$. We can immediately reject all possibilities in which either factor has a common factor, such as $(15a + 3)$ or $(3a - 3)$, because at the outset we determined that there is no common factor other than 1 or -1. Since the sign of the middle term is negative and the sign of the last term is positive, the factors of 3 must both be negative. We try some possibilities:

$$(15a - 1)(a - 3) = 15a^2 - 46a + 3$$
$$(5a - 3)(3a - 1) = 15a^2 - 14a + 3$$

The factorization is $(5a - 3)(3a - 1)$.

13. $6x^2 + 17x + 12$

(1) There is no common factor (other than 1 or -1).

(2) Because $6x^2$ can be factored as $6x \cdot x$ and $3x \cdot 2x$, we have these possibilities:

$$(6x + \quad)(x + \quad) \text{ and } (3x + \quad)(2x + \quad)$$

(3) Since all coefficients are positive, we need consider only positive pairs of factors of 12. There are 3 pairs and they can be listed two ways:

$$1,12 \quad 2,6 \quad 3,4$$
$$\text{and} \quad 12,1 \quad 6,2 \quad 4,3$$

(4) We can immediately reject all possibilities in which either factor has a common factor, such as $(6x + 12)$ or $(3x + 3)$, because at the outset we determined that there is no common factor other than 1 or -1. We try some possibilities:

$(6x + 1)(x + 12) = 6x^2 + 73x + 12$

$(3x + 4)(2x + 3) = 6x^2 + 17x + 12$

The factorization is $(3x + 4)(2x + 3)$.

15. $6x^2 - 10x - 4$

(1) We factor out the largest common factor, 2:

$2(3x^2 - 5x - 2)$.

Then we factor the trinomial $3x^2 - 5x - 2$.

(2) Because $3x^2$ can be factored as $3x \cdot x$, we have this possibility:

$(3x + \quad)(x + \quad)$

(3) There are 2 pairs of factors of -2 and they can be listed two ways:

$$-2, 1 \quad 2, -1$$
$$\text{and} \quad 1, -2 \quad -1, 2$$

(4) Look for Outer and Inner products resulting from steps (2) and (3) for which the sum is $-5x$. We try some possibilities:

$(3x - 2)(x + 1) = 3x^2 + x - 2$

$(3x + 2)(x - 1) = 3x^2 - x - 2$

$(3x + 1)(x - 2) = 3x^2 - 5x - 2$

The factorization of $3x^2 - 5x - 2$ is $(3x+1)(x-2)$. We must include the common factor in order to get a factorization of the original trinomial.

$6x^2 - 10x - 4 = 2(3x + 1)(x - 2)$

17. $7t^3 + 15t^2 + 2t$

(1) We factor out the common factor, t:

$t(7t^2 + 15t + 2)$.

Then we factor the trinomial $7t^2 + 15t + 2$.

(2) Because $7t^2$ can be factored as $7t \cdot t$, we have this possibility:

$(7t + \quad)(t + \quad)$

(3) Since all coefficients are positive, we need consider only positive factors of 2. There is only 1 such pair and it can be listed two ways:

$$2, 1 \quad 1, 2$$

(4) Look for Outer and Inner products resulting from steps (2) and (3) for which the sum is $15t$. We try some possibilities:

$(7t + 2)(t + 1) = 7t^2 + 9t + 2$

$(7t + 1)(t + 2) = 7t^2 + 15t + 2$

The factorization of $7t^2 + 15t + 2$ is $(7t+1)(t+2)$. We must include the common factor in order to get a factorization of the original trinomial.

$7t^3 + 15t^2 + 2t = t(7t + 1)(t + 2)$

19. $10 - 23x + 12x^2 = 12x^2 - 23x + 10$

(1) There is no common factor (other than 1 or -1).

(2) Because $12x^2$ can be factored as $12x \cdot x$, $6x \cdot 2x$, and $4x \cdot 3x$, we have these possibilities:

$(12x + \quad)(x + \quad)$, $(6x + \quad)(2x + \quad)$, and $(4x + \quad)(3x + \quad)$

(3) Since the sign of the middle term is negative but the sign of the last term is positive, we need consider only negative factors of 10.

$$-10, -1 \quad -5, -2$$
$$\text{and} \quad -1, -10 \quad -2, -5$$

(4) We can immediately reject all possibilities in which either factor has a common factor, such as $(2x - 10)$ or $(4x - 2)$, because we determined at the outset that there is no common factor other than 1 or -1. We try some possibilities:

$(12x - 5)(x - 2) = 12x^2 - 29x + 10$

$(4x - 1)(3x - 10) = 12x^2 - 43x + 10$

$(4x - 5)(3x - 2) = 12x^2 - 23x + 10$

The factorization is $(4x - 5)(3x - 2)$.

21. $-35x^2 - 34x - 8$

(1) We factor out -1 in order to have a trinomial with a positive leading coefficient.

$-35x^2 - 34x - 8 = -1(35x^2 + 34x + 8)$

Now we factor $35x^2 + 34x + 8$.

(2) Because $35x^2$ can be factored as $35x \cdot x$ or $7x \cdot 5x$, we have these possibilities:

$(35x + \quad)(x + \quad)$ and $(7x + \quad)(5x + \quad)$

(3) Since all coefficients are positive, we need consider only positive pairs of factors of 8. There are 2 such pairs and they can be listed two ways:

$$8, 1 \quad 4, 2$$
$$\text{and} \quad 1, 8 \quad 2, 4$$

(4) We try some possibilities:

$(35x + 8)(x + 1) = 35x^2 + 43x + 8$

$(7x + 8)(5x + 1) = 35x^2 + 47x + 8$

$(7x + 4)(5x + 2) = 35x^2 + 34x + 8$

The factorization of $35x^2 + 34x + 8$ is $(7x + 4)(5x + 2)$.

We must include the factor of -1 in order to get a factorization of the original trinomial.

$-35x^2 - 34x - 8 = -1(7x + 4)(5x + 2)$, or $-(7x + 4)(5x + 2)$.

23. $4 + 6t^2 - 13t = 6t^2 - 13t + 4$

(1) There is no common factor (other than 1 or -1).

(2) Because $6t^2$ can be factored as $6t \cdot t$ or $3t \cdot 2t$, we have these possibilities:

$(6t + \quad)(t + \quad)$ and $(3t + \quad)(2t + \quad)$

(3) Since the sign of the middle term is negative but the sign of the last term is positive, we need to consider only negative factors of 4. There is only 1 such pair and it can be listed two ways:

$-4, -1$ and $-1, -4$

(4) We can immediately reject all possibilities in which either factor has a common factor, such as $(6t-4)$ or $(2t-4)$, because we determined at the outset that there is no common factor other than 1 or -1. We try some possibilities:

$$(6t-1)(t-4) = 6t^2 - 25t + 4$$
$$(3t-4)(2t-1) = 6t^2 - 11t + 4$$

These are the only possibilities that do not contain a common factor. Since neither is the desired factorization, we must conclude that $4 + 6t^2 - 13t$ is prime.

25. $25x^2 + 40x + 16$

(1) There is no common factor (other than 1 or -1).

(2) Because $25x^2$ can be factored as $25x \cdot x$ or $5x \cdot 5x$, we have these possibilities:

$$(25x + \quad)(x + \quad) \text{ and } (5x + \quad)(5x + \quad)$$

(3) Since all coefficients are positive, we need consider only positive pairs of factors of 16. There are 3 such pairs and two of them can be listed two ways:

$$16, 1 \quad 8, 2 \quad 4, 4$$
$$\text{and} \quad 1, 16 \quad 2, 8$$

(4) We try some possibilities:

$$(25x + 16)(x + 1) = 25x^2 + 41x + 16$$
$$(5x + 8)(5x + 2) = 25x^2 + 50x + 16$$
$$(5x + 4)(5x + 4) = 25x^2 + 40x + 16$$

The factorization is $(5x + 4)(5x + 4)$, or $(5x + 4)^2$.

27. $20y^2 + 59y - 3$

(1) There is no common factor (other than 1 or -1).

(2) Because $20y^2$ can be factored as $20y \cdot y$, $10y \cdot 2y$, or $5y \cdot 4y$, we have these possibilities:

$$(20y + \quad)(y + \quad) \text{ and } (10y + \quad)(2y + \quad) \text{ and}$$
$$(5y + \quad)(4y + \quad)$$

(3) There are 2 such pairs of factors of -3, which can be listed two ways:

$$-3, 1 \quad 3, -1$$
$$\text{and} \quad 1, -3 \quad -1, 3$$

(4) Look for Outer and Inner products resulting from steps (2) and (3) for which the sum is 59y. We try some possibilities:

$$(20y - 3)(y + 1) = 20y^2 + 17y - 3$$
$$(10y - 3)(2y + 1) = 20y^2 + 4y - 3$$
$$(5y - 1)(4y + 3) = 20y^2 + 11y - 3$$
$$(20y - 1)(y + 3) = 20y^2 + 59y - 3$$

The factorization is $(20y - 1)(y + 3)$.

29. $14x^2 + 73x + 45$

(1) There is no common factor (other than 1 or -1).

(2) Because $14x^2$ can be factored as $14x \cdot x$, and $7x \cdot 2x$, we have two possibilities:

$$(14x + \quad)(x + \quad), \text{ and } (7x + \quad)(2x + \quad)$$

(3) Since all coefficients are positive, we need consider only positive pairs of factors of 45. There are 3 such pairs and they can be listed two ways.

$$45, 1 \quad 15, 3 \quad 9, 5$$
$$\text{and} \quad 1, 45 \quad 3, 15 \quad 5, 9$$

(4) Look for Outer and Inner products from steps (2) and (3) for which the sum is 73x. We try some possibilities:

$$(14x + 45)(x + 1) = 14x^2 + 59x + 45$$
$$(14x + 15)(x + 3) = 14x^2 + 57x + 45$$
$$(7x + 1)(2x + 45) = 14x^2 + 317x + 45$$
$$(7x + 5)(2x + 9) = 14x^2 + 73x + 45$$

The factorization is $(7x + 5)(2x + 9)$.

31. $-2x^2 + 15 + x = -2x^2 + x + 15$

(1) We factor out -1 in order to have a trinomial with a positive leading coefficient.

$$-2x^2 + x + 15 = -1(2x^2 - x - 15)$$

Now we factor $2x^2 - x - 15$.

(2) Because $2x^2$ can be factored as $2x \cdot x$ we have this possibility:

$$(2x + \quad)(x + \quad)$$

(3) There are 4 pairs of factors of -15 and they can be listed two ways:

$$-15, 1 \quad 15, -1 \quad -5, 3 \quad 5, -3$$
$$\text{and} \quad 1, -15 \quad -1, 15 \quad 3, -5 \quad -3, 5$$

(4) We try some possibilities:

$$(2x - 15)(x + 1) = 2x^2 - 13x - 15$$
$$(2x - 5)(x + 3) = 2x^2 + x - 15$$
$$(2x + 5)(x - 3) = 2x^2 - x - 15$$

The factorization of $2x^2 - x - 15$ is $(2x+5)(x-3)$. We must include the factor of -1 in order to get a factorization of the original trinomial.

$$-2x^2 + 15 + x = -1(2x + 5)(x - 3), \text{ or}$$
$$-(2x + 5)(x - 3)$$

33. $-6x^2 - 33x - 15$

(1) Factor out -3. This not only removes the largest common factor, 3. It also produces a trinomial with a positive leading coefficient.

$$-3(2x^2 + 11x + 5)$$

Then we factor the trinomial $2x^2 + 11x + 5$.

(2) Because $2x^2$ can be factored as $2x \cdot x$ we have this possibility:

$$(2x + \quad)(x + \quad)$$

(3) Since all coefficients are positive, we need consider only positive pairs of factors of 5. There is one such pair and it can be listed two ways:

$$5, 1 \quad \text{and} \quad 1, 5$$

(4) We try some possibilities:

$$(2x + 5)(x + 1) = 2x^2 + 7x + 5$$
$$(2x + 1)(x + 5) = 2x^2 + 11x + 5$$

The factorization of $2x^2+11x+5$ is $(2x+1)(x+5)$. We must include the common factor in order to get a factorization of the original trinomial.

$$-6x^2 - 33x - 15 = -3(2x + 1)(x + 5)$$

35. $10a^2 - 8a - 18$

(1) Factor out the common factor, 2:

$$2(5a^2 - 4a - 9)$$

Then we factor the trinomial $5a^2 - 4a - 9$.

(2) Because $5a^2$ can be factored as $5a \cdot a$, we have this possibility:

$$(5a + \quad)(a + \quad)$$

(3) There are 3 pairs of factors of -9, and they can be listed two ways.

$$-9, 1 \quad 9, -1 \quad 3, -3$$
$$\text{and} \quad 1, -9 \quad -1, 9 \quad -3, 3$$

(4) Look for Outer and Inner products resulting from steps (2) and (3) for which the sum is $-4a$. We try some possibilities:

$$(5a - 3)(a + 3) = 5a^2 + 12a - 9$$
$$(5a + 9)(a - 1) = 5a^2 + 4a - 9$$
$$(a + 1)(5a - 9) = 5a^2 - 4a - 9$$

The factorization of $5a^2 - 4a - 9$ is $(a+1)(5a-9)$. We must include the common factor in order to get a factorization of the original trinomial.

$$2(a + 1)(5a - 9)$$

37. $12x^2 + 68x - 24$

(1) Factor out the common factor, 4:

$$4(3x^2 + 17x - 6)$$

Then we factor the trinomial $3x^2 + 17x - 6$.

(2) Because $3x^2$ can be factored as $3x \cdot x$ we have this possibility:

$$(3x + \quad)(x + \quad)$$

(3) There are 4 pairs of factors of -6 and they can be listed two ways:

$$6, -1 \quad -6, 1 \quad 3, -2 \quad -3, 2$$
$$\text{and} \quad -1, 6 \quad 1, -6 \quad -2, 3 \quad 2, -3$$

(4) We can immediately reject all possibilities in which either factor has a common factor, such as $(3x + 6)$ or $(3x - 3)$, because we determined at the outset that there is no common factor other than 1 or -1. We try some possibilities:

$$(3x - 1)(x + 6) = 3x^2 + 17x - 6$$

The factorization of $3x^2 + 17x - 6$ is $(3x - 1)(x + 6)$. We must include the common factor in order to get a factorization of the original trinomial.

$$12x^2 + 68x - 24 = 4(3x - 1)(x + 6)$$

39. $4x + 1 + 3x^2 = 3x^2 + 4x + 1$

(1) There is no common factor (other than 1 or -1).

(2) Because $3x^2$ can be factored as $3x \cdot x$ we have this possibility:

$$(3x + \quad)(x + \quad)$$

(3) Since all coefficients are positive, we need consider only positive pairs of factors of 1. There is one such pair: 1,1.

(4) We try the possible factorization:

$$(3x + 1)(x + 1) = 3x^2 + 4x + 1$$

The factorization is $(3x + 1)(x + 1)$.

41. $x^2 + 3x - 2x - 6 = x(x + 3) - 2(x + 3)$
$$= (x + 3)(x - 2)$$

43. $8t^2 - 6t - 28t + 21 = 2t(4t - 3) - 7(4t - 3)$
$$= (4t - 3)(2t - 7)$$

45. $6x^2 + 4x + 15x + 10 = 2x(3x + 2) + 5(3x + 2)$
$$= (3x + 2)(2x + 5)$$

47. $2y^2 + 8y - y - 4 = 2y(y + 4) - 1(y + 4)$
$$= (y + 4)(2y - 1)$$

49. $6a^2 - 8a - 3a + 4 = 2a(3a - 4) - 1(3a - 4)$
$$= (3a - 4)(2a - 1)$$

51. $16t^2 + 23t + 7$

(1) First note that there is no common factor (other than 1 or -1).

(2) Multiply the leading coefficient, 16, and the constant, 7:

$$16 \cdot 7 = 112$$

(3) We look for factors of 112 that add to 23. Since all coefficients are positive, we need to consider only positive factors.

Pairs of factors	Sums of factors
1, 112	113
2, 56	58
4, 28	32
8, 14	22
16, 7	23

The numbers we need are 16 and 7.

(4) Rewrite the middle term:

$$23t = 16t + 7t$$

(5) Factor by grouping:

$$16t^2 + 23t + 7 = 16t^2 + 16t + 7t + 7$$
$$= 16t(t + 1) + 7(t + 1)$$
$$= (t + 1)(16t + 7)$$

53. $-9x^2 - 18x - 5$

(1) We factor out -1 in order to have a trinomial with a positive leading coefficient.

$$-9x^2 - 18x - 5 = -1(9x^2 + 18x + 5)$$

Now we factor $9x^2 + 18x + 5$.

(2) Multiply the leading coefficient, 9, and the constant, 5:

$$9 \cdot 5 = 45$$

(3) We look for factors of 45 that add to 18. Since all coefficients are positive, we need to consider only positive factors.

Pairs of factors	Sums of factors
1, 45	46
3, 15	18
5, 9	14

The numbers we need are 3 and 15.

(4) Rewrite the middle term:

$$18x = 3x + 15x$$

(5) Factor by grouping:

$$9x^2 + 18x + 5 = 9x^2 + 3x + 15x + 5$$
$$= 3x(3x + 1) + 5(3x + 1)$$
$$= (3x + 1)(3x + 5)$$

We must include the factor of -1 in order to get a factorization of the original trinomial.

$-9x^2 - 18x - 5 = -1(3x + 1)(3x + 5)$, or
$-(3x + 1)(3x + 5)$

55. $10x^2 + 30x - 70$

(1) Factor out the largest common factor, 10:

$$10x^2 + 30x - 70 = 10(x^2 + 3x - 7)$$

Since $x^2 + 3x - 7$ is prime, this trinomial cannot be factored further.

57. $18x^3 + 21x^2 - 9x$

(1) Factor out the largest common factor, $3x$:

$$18x^3 + 21x^2 - 9x = 3x(6x^2 + 7x - 3)$$

(2) To factor $6x^2 + 7x - 3$ by grouping we first multiply the leading coefficient, 6, and the constant, -3:

$$6(-3) = -18$$

(3) We look for factors of -18 that add to 7.

Pairs of factors	Sums of factors
$-1, 18$	17
$1, -18$	-17
$-2, 9$	7
$2, -9$	-7
$-3, 6$	3
$3, -6$	-3

The numbers we need are -2 and 9.

(4) Rewrite the middle term:

$$7x = -2x + 9x$$

(5) Factor by grouping:

$$6x^2 + 7x - 3 = 6x^2 - 2x + 9x - 3$$
$$= 2x(3x - 1) + 3(3x - 1)$$
$$= (3x - 1)(2x + 3)$$

The factorization of $6x^2 + 7x - 3$ is $(3x - 1)(2x + 3)$. We must include the common factor in order to get a factorization of the original trinomial:

$$18x^3 + 21x^2 - 9x = 3x(3x - 1)(2x + 3)$$

59. $89x + 64 + 25x^2 = 25x^2 + 89x + 64$

(1) First note that there is no common factor (other than 1 or -1).

(2) Multiply the leading coefficient, 25, and the constant, 64:

$$25 \cdot 64 = 1600$$

(3) We look for factors of 1600 that add to 89. Since all coefficients are positive, we need to consider only positive factors. The numbers we need are 25 and 64.

(4) Rewrite the middle term:

$$89x = 25x + 64x$$

(5) Factor by grouping:

$$25x^2 + 89x + 64 = 25x^2 + 25x + 64x + 64$$
$$= 25x(x + 1) + 64(x + 1)$$
$$= (x + 1)(25x + 64)$$

61. $168x^3 + 45x^2 + 3x$

(1) Factor out the largest common factor, $3x$:

$$168x^3 + 45x^2 + 3x = 3x(56x^2 + 15x + 1)$$

(2) To factor $56x^2 + 15x + 1$ we first multiply the leading coefficient, 56, and the constant, 1:

$$56 \cdot 1 = 56$$

(3) We look for factors of 56 that add to 15. Since all coefficients are positive, we need to consider only positive factors. The numbers we need are 7 and 8.

(4) Rewrite the middle term:

$$15x = 7x + 8x$$

(5) Factor by grouping:

$$56x^2 + 15x + 1 = 56x^2 + 7x + 8x + 1$$
$$= 7x(8x + 1) + 1(8x + 1)$$
$$= (8x + 1)(7x + 1)$$

The factorization of $56x^2 + 15x + 1$ is $(8x + 1)(7x + 1)$. We must include the common factor in order to get a factorization of the original trinomial:

$$168x^3 + 45x^2 + 3x = 3x(8x + 1)(7x + 1)$$

63. $-14t^4 + 19t^3 + 3t^2$

(1) Factor out $-t^2$. This not only removes the largest common factor, t^2. It also produces a trinomial with a positive leading coefficient.

$$-14t^4 + 19t^3 + 3t^2 = -t^2(14t^2 - 19t - 3)$$

(2) To factor $14t^2 - 19t - 3$ we first multiply the leading coefficient, 14, and the constant, -3:

$$14(-3) = -42$$

(3) We look for factors of -42 that add to -19. The numbers we need are -21 and 2.

(4) Rewrite the middle term:

$$-19t = -21t + 2t$$

(5) Factor by grouping:

$$14t^2 - 19t - 3 = 14t^2 - 21t + 2t - 3$$
$$= 7t(2t - 3) + 1(2t - 3)$$
$$= (2t - 3)(7t + 1)$$

The factorization of $14t^2 - 19t - 3$ is $(2t-3)(7t+1)$. We must include the common factor in order to get a factorization of the original trinomial:

$$-14t^4 + 19t^3 + 3t^2 = -t^2(2t-3)(7t+1)$$

65. $132y + 32y^2 - 54 = 32y^2 + 132y - 54$

(1) Factor out the largest common factor, 2:

$$32y^2 + 132y - 54 = 2(16y^2 + 66y - 27)$$

(2) To factor $16y^2 + 66y - 27$ we first multiply the leading coefficient, 16, and the constant, -27:

$$16(-27) = -432$$

(3) We look for factors of -432 that add to 66. The numbers we need are 72 and -6.

(4) Rewrite the middle term:

$$66y = 72y - 6y$$

(5) Factor by grouping:

$$16y^2 + 66y - 27 = 16y^2 + 72y - 6y - 27$$
$$= 8y(2y+9) - 3(2y+9)$$
$$= (2y+9)(8y-3)$$

The factorization of $16y^2 + 66y - 27$ is $(2y+9)(8y-3)$. We must include the common factor in order to get a factorization of the original trinomial:

$$132y + 32y^2 - 54 = 2(2y+9)(8y-3)$$

67. $2a^2 - 5ab + 2b^2$

(1) There is no common factor (other than 1 or -1).

(2) Multiply the leading coefficient, 2, and the constant, 2:

$$2 \cdot 2 = 4$$

(3) We look for factors of 4 that add to -5. The numbers we need are -1 and -4.

(4) Rewrite the middle term:

$$-5ab = -ab - 4ab$$

(5) Factor by grouping:

$$2a^2 - 5ab + 2b^2 = 2a^2 - ab - 4ab + 2b^2$$
$$= a(2a-b) - 2b(2a-b)$$
$$= (2a-b)(a-2b)$$

69. $8s^2 + 22st + 14t^2$

(1) Factor out the largest common factor, 2:

$$8s^2 + 22st + 14t^2 = 2(4s^2 + 11st + 7t^2)$$

(2) Multiply the leading coefficient, 4, and the constant, 7:

$$4 \cdot 7 = 28$$

(3) We look for factors of 28 that add to 11. The numbers we need are 4 and 7.

(4) Rewrite the middle term:

$$11st = 4st + 7st$$

(5) Factor by grouping:

$$4s^2 + 11st + 7t^2 = 4s^2 + 4st + 7st + 7t^2$$
$$= 4s(s+t) + 7t(s+t)$$
$$= (s+t)(4s+7t)$$

The factorization of $4s^2 + 11st + 7t^2$ is $(s+t)(4s+7t)$. We must include the common factor in order to get a factorization of the original trinomial:

$$8s^2 + 22st + 14t^2 = 2(s+t)(4s+7t)$$

71. $27x^2 - 72xy + 48y^2$

(1) Factor out the largest common factor, 3:

$$27x^2 - 72xy + 48y^2 = 3(9x^2 - 24xy + 16y^2)$$

(2) To factor $9x^2 - 24xy + 16y^2$, we first multiply the leading coefficient, 9, and the constant, 16:

$$9 \cdot 16 = 144$$

(3) We look for factors of 144 that add to -24. The numbers we need are -12 and -12.

(4) Rewrite the middle term:

$$-24xy = -12xy - 12xy$$

(5) Factor by grouping:

$$9x^2 - 24xy + 16y^2 = 9x^2 - 12xy - 12xy + 16y^2$$
$$= 3x(3x-4y) - 4y(3x-4y)$$
$$= (3x-4y)(3x-4y)$$

The factorization of $9x^2 - 24xy + 16y^2$ is $(3x-4y)(3x-4y)$. We must include the common factor in order to get a factorization of the original trinomial:

$$27x^2 - 72xy + 48y^2 = 3(3x-4y)(3x-4y) \text{ or, } 3(3x-4y)^2$$

73. $-24a^2 + 34ab - 12b^2$

(1) Factor out -2. This not only removes the largest common factor, 2. It also produces a trinomial with a positive leading coefficient.

$$-24a^2 + 34ab - 12b^2 = -2(12a^2 - 17ab + 6b^2)$$

(2) To factor $12a^2 - 17ab + 6b^2$, we first multiply the leading coefficient, 12, and the constant, 6:

$$12 \cdot 6 = 72$$

(3) We look for factors of 72 that add to -17. The numbers we need are -8 and -9.

(4) Rewrite the middle term:

$$-17ab = -8ab - 9ab$$

(5) Factor by grouping:

$$12a^2 - 17ab + 6b^2 = 12a^2 - 8ab - 9ab + 6b^2$$
$$= 4a(3a-2b) - 3b(3a-2b)$$
$$= (3a-2b)(4a-3b)$$

The factorization of $12a^2 - 17ab + 6b^2$ is $(3a-2b)(4a-3b)$. We must include the common factor in order to get a factorization of the original trinomial:

$$-24a^2 + 34ab - 12b^2 = -2(3a-2b)(4a-3b)$$

75. $19x^3 - 3x^2 + 14x^4 = 14x^4 + 19x^3 - 3x^2$

(1) Factor out the largest common factor, x^2:

$$x^2(14x^2 + 19x - 3)$$

(2) To factor $14x^2 + 19x - 3$ by grouping we first multiply the leading coefficient, 14, and the constant, -3:

$$14(-3) = -42$$

(3) We look for factors of -42 that add to 19. The numbers we need are 21 and -2.

(4) Rewrite the middle term:

$$19x = 21x - 2x$$

(5) Factor by grouping:

$$\begin{aligned}14x^2 + 19x - 3 &= 14x^2 + 21x - 2x - 3\\ &= 7x(2x + 3) - 1(2x + 3)\\ &= (2x + 3)(7x - 1)\end{aligned}$$

The factorization of $14x^2 + 19x - 3$ is $(2x+3)(7x-1)$. We must include the common factor in order to get a factorization of the original trinomial:

$$19x^3 - 3x^2 + 14x^4 = x^2(2x + 3)(7x - 1)$$

77. $18a^7 + 8a^6 + 9a^8 = 9a^8 + 18a^7 + 8a^6$

(1) Factor out the largest common factor, a^6:

$$9a^8 + 18a^7 + 8a^6 = a^6(9a^2 + 18a + 8)$$

(2) To factor $9a^2 + 18a + 8$ we first multiply the leading coefficient, 9, and the constant, 8:

$$9 \cdot 8 = 72$$

(3) Look for factors of 72 that add to 18. The numbers we need are 6 and 12.

(4) Rewrite the middle term:

$$18a = 6a + 12a$$

(5) Factor by grouping:

$$\begin{aligned}9a^2 + 18a + 8 &= 9a^2 + 6a + 12a + 8\\ &= 3a(3a + 2) + 4(3a + 2)\\ &= (3a + 2)(3a + 4)\end{aligned}$$

The factorization of $9a^2 + 18a + 8$ is $(3a + 2)(3a + 4)$. We must include the common factor in order to get a factorization of the original trinomial:

$$18a^7 + 8a^6 + 9a^8 = a^6(3a + 2)(3a + 4)$$

79. *Writing Exercise.* Kay has incorrectly changed the sign of the middle term when factoring out the largest common factor. Thus, the signs in both terms of the final factorization are wrong. The number of points that should be deducted will vary.

81. $$\begin{aligned}(x - 2)^2 &= (x)^2 + 2 \cdot x(-2) + (-2)^2\\ &\quad [(A - B)^2 = A^2 - 2AB + B^2]\\ &= x^2 - 4x + 4\end{aligned}$$

83. $$\begin{aligned}(x + 2)(x - 2) &= x^2 - 2^2\\ &\quad [(A+B)(A-B) = A^2 - B^2]\\ &= x^2 - 4\end{aligned}$$

85. $$\begin{aligned}(4a + 1)^2 &= (4a)^2 + 2(4a)(1) + 1^2\\ &\quad [(A + B)^2 = A^2 + 2AB + B^2]\\ &= 16a^2 + 8a + 1\end{aligned}$$

87. $$\begin{aligned}(3c - 10)^2 &= (3c)^2 - 2(3c)(10) + 10^2\\ &\quad [(A - B)^2 = A^2 - 2AB + B^2]\\ &= 9c^2 - 60c + 100\end{aligned}$$

89. $$\begin{aligned}(8n + 3)(8n - 3) &= (8n)^2 - 3^2\\ &\quad [(A+B)(A-B) = A^2 - B^2]\\ &= 64n^2 - 9\end{aligned}$$

91. For the trinomial $ax^2 + bx + c$ to be prime:

(1) Show that there is no common factor (other than 1 or -1)

(2) Multiply the leading coefficient, a, and the constant c.

$$a \cdot c = ac$$

(3) Show there are <u>no</u> factors of ac that add to b.

(4) The trinomial is prime.

93. $18x^2y^2 - 3xy - 10$

We will factor by grouping.

(1) There is no common factor (other than 1 or -1).

(2) Multiply the leading coefficient, 18, and the constant, -10:

$$18(-10) = -180$$

(3) We look for factors of -180 that add to -3. The numbers we want are -15 and 12.

(4) Rewrite the middle term:

$$-3xy = -15xy + 12xy$$

(5) Factor by grouping:

$$\begin{aligned}18x^2y^2 - 3xy - 10 &= 18x^2y^2 - 15xy + 12xy - 10\\ &= 3xy(6xy - 5) + 2(6xy - 5)\\ &= (6xy - 5)(3xy + 2)\end{aligned}$$

95. $9a^2b^3 + 25ab^2 + 16$

We cannot factor the leading term, $9a^2b^3$, in a way that will produce a middle term with variable factors ab^2, so this trinomial is prime.

97. $16t^{10} - 8t^5 + 1 = 16(t^5)^2 - 8t^5 + 1$

(1) There is no common factor (other than 1 or -1).

(2) Because $16t^{10}$ can be factored as $16t^5 \cdot t^5$ or $8t^5 \cdot 2t^5$ or $4t^5 \cdot 4t^5$, we have these possibilities:

$$(16t^5 +\ \ )(t^5 +\ \ ) \text{ and } (8t^5 +\ \ )(2t^5 +\ \ )$$

and $(4t^5 +\ \ )(4t^5 +\ \ )$

(3) Since the last term is positive and the middle term is negative we need consider only negative factors of 1. The only negative pair of factors is $-1, -1$.

(4) We try some possibilities:

$$\begin{aligned}(16t^5 - 1)(t^5 - 1) &= 16t^{10} - 17t^5 + 1\\ (8t^5 - 1)(2t^5 - 1) &= 16t^{10} - 10t^5 + 1\\ (4t^5 - 1)(4t^5 - 1) &= 16t^{10} - 8t^5 + 1\end{aligned}$$

The factorization is $(4t^5 - 1)(4t^5 - 1)$, or $(4t^5 - 1)^2$.

99. $-15x^{2m} + 26x^m - 8 = -15(x^m)^2 + 26x^m - 8$

(1) Factor out -1 in order to have a trinomial with a positive leading coefficient.

$$-15x^{2m} + 26x^m - 8 = -1(15x^{2m} - 26x^m + 8)$$

(2) Because $15x^{2m}$ can be factored as $15x^m \cdot x^m$, or $5x^m \cdot 3x^m$, we have these possibilities:

$$(15x^m +\ \ )(x^m +\ \ ) \text{ and } (5x^m +\ \ )(3x^m +\ \ )$$

(3) Since the last term is positive and the middle term is negative we need consider only negative factors of 8. There are 2 such pairs and they can be listed in two ways:

$$-8, -1 \quad -4, -2$$
$$\text{and} \quad -1, -8 \quad -2, -4$$

(4) We try some possibilities:

$$(15x^m - 8)(x^m - 1) = 15x^{2m} - 9x^m + 8$$
$$(5x^m - 8)(3x^m - 1) = 15x^{2m} - 29x^m + 8$$
$$(5x^m - 2)(3x^m - 4) = 15x^{2m} - 26x^m + 8$$

The factorization of $15x^{2m} - 26x^m + 8$ is $(5x^m - 2)(3x^m - 4)$. We must include the common factor to get a factorization of the original trinomial.

$$-15x^{2m} + 26x^m - 8 = -1(5x^m - 2)(3x^m - 4), \text{ or}$$
$$-(5x^m - 2)(3x^m - 4)$$

101. $3a^{6n} - 2a^{3n} - 1 = 3(a^{3n})^2 - 2a^{3n} - 1$

(1) There is no common factor (other than 1 or -1).

(2) Because $3a^{6n}$ can be factored as $3a^{3n} \cdot a^{3n}$, we have this possibility:

$$(3a^{3n} + \quad)(a^{3n} + \quad)$$

(3) The only one pair of factors of -1: $1, -1$.

(4) We try some possibilities:

$$(3a^{3n} - 1)(a^{3n} + 1) = 3a^{6n} + 2a^{3n} - 1$$
$$(3a^{3n} + 1)(a^{3n} - 1) = 3a^{6n} - 2a^{3n} - 1$$

The factorization of $3a^{6n} - 2a^{3n} - 1$ is $(3a^{3n} + 1)(a^{3n} - 1)$.

103. $7(t-3)^{2n} + 5(t-3)^n - 2 = 7[(t-3)^n]^2 + 5(t-3)^n - 2$

(1) There is no common factor (other than 1 or -1).

(2) Multiply the leading coefficient, 7, and the constant, -2:

$$7(-2) = -14$$

(3) Look for factors of -14 that add to 5. The numbers we want are 7 and -2.

(4) Rewrite the middle term:

$$5(t-3)^n = 7(t-3)^n - 2(t-3)^n$$

(5) Factor by grouping:

$$7(t-3)^{2n} + 5(t-3)^n - 2$$
$$= 7(t-3)^{2n} + 7(t-3)^n - 2(t-3)^n - 2$$
$$= 7(t-3)^n[(t-3)^n + 1] - 2[(t-3)^n + 1]$$
$$= [(t-3)^n + 1][7(t-3)^n - 2]$$

The factorization of $7(t-3)^{2n} + 5(t-3)^n - 2$ is

$$[(t-3)^n + 1][7(t-3)^n - 2]$$

Chapter 5 Connecting the Concepts

1. $\quad 6x^5 - 18x^2 \qquad$ Common factor: $6x^2$
$$= 6x^2 \cdot x^3 - 6x^2 \cdot 3$$
$$= 6x^2(x^3 - 3)$$

3. $\quad 2x^2 + 13x - 7 \qquad$ No common factor; factor with FOIL.
$$= (x + 7)(2x - 1)$$

5. $\quad 5x^2 + 40x - 100 \qquad$ Common factor: 5
$$= 5(x^2 + 8x - 20) \qquad$$ Factor with FOIL
$$= 5(x - 2)(x + 10)$$

7. $\quad 7x^2y - 21xy - 28y \qquad$ Common factor: $7y$
$$= 7y(x^2 - 3x - 4) \qquad$$ Factor with FOIL
$$= 7y(x - 4)(x + 1)$$

9. $\quad b^2 - 14b + 49 \qquad$ Perfect-square trinomial
$$= b^2 - 2 \cdot 7 \cdot b + 7^2$$
$$= (b - 7)^2$$

11. $\quad c^3 + c^2 - 4c - 4 \qquad$ No common factor; factor with grouping.
$$= c^2(c + 1) - 4(c + 1)$$
$$= (c + 1)(c^2 - 4)$$
$$= (c + 1)(c + 2)(c - 2)$$

13. $\quad t^2 + t - 10 \qquad$ No common factor
The trinomial is prime.

15. $\quad 15p^2 + 16pq + 4q^2 \qquad$ No common factor; factor with FOIL.
$$= (3p + 2q)(5p + 2q)$$

17. $\quad x^2 + 4x - 77 \qquad$ No common factor; factor with FOIL.
$$= (x + 11)(x - 7)$$

19. $\quad 5 + 3x - 2x^2 \qquad$ Common factor: -1; write in descending order
$$= -1(2x^2 - 3x - 5) \qquad$$ Factor with FOIL
$$= -1(2x - 5)(x + 1)$$

Exercise Set 5.4

1. $4x^2 + 49$ is not a trinomial. It is not a difference of squares because the terms do not have different signs. There is no common factor, so $4x^2 + 49$ is a prime polynomial.

3. $t^2 - 100 = t^2 - 10^2$, so $t^2 - 100$ is a difference of squares.

5. $9x^2 + 6x + 1 = (3x)^2 + 2 \cdot 3x \cdot 1 + 1^2$, so this is a perfect-square trinomial.

7. $2t^2 + 10t + 6$ does not contain a term that is a square so it is neither a perfect-square trinomial nor a difference of squares. (We could also say that it is not a difference of squares because it is not a binomial.) There is a common factor, 2, so this is not a prime polynomial. Thus it is none of the given possibilities.

9. $16t^2 - 25 = (4t)^2 - 5^2$, so $16t^2 - 25$ is a difference of squares.

11. $x^2 + 18x + 81$

(1) Two terms, x^2 and 81, are squares.

(2) Neither x^2 nor 81 is being subtracted.

(3) Twice the product of the square roots, $2 \cdot x \cdot 9$, is $18x$, the remaining term.

Thus, $x^2 + 18x + 81$ is a perfect-square trinomial.

13. $x^2 - 10x - 25$

(1) Two terms, x^2 and 25, are squares.

(2) There is a minus sign before 25, so $x^2 - 10x - 25$ is not a perfect-square trinomial.

15. $x^2 - 3x + 9$

(1) Two terms, x^2 and 9, are squares.

(2) There is no minus sign before x^2 or 9.

(3) Twice the product of the square roots, $2 \cdot x \cdot 3$, is $6x$. This is neither the remaining term nor its opposite, so $x^2 - 3x + 9$ is not a perfect-square trinomial.

17. $9x^2 + 25 - 30x$

(1) Two terms $9x^2$, and 25, are squares.

(2) Neither $9x^2$ nor 25 is being subtracted.

(3) Twice the product of the square roots, $2 \cdot 3x \cdot 5$, is $30x$, the opposite of the remaining term, $-30x$.

Thus, $9x^2 + 25 - 30x$ is a perfect-square trinomial.

19.
$$\begin{aligned}
& \quad\; x^2 + 16x + 64 \\
&= x^2 + 2 \cdot x \cdot 8 + 8^2 = (x+8)^2 \\
&\quad\; \uparrow \quad\; \uparrow \;\; \uparrow \;\; \uparrow \quad\; \uparrow \\
&= A^2 + 2 \quad A \quad B + B^2 = (A+B)^2
\end{aligned}$$

21.
$$\begin{aligned}
& \quad\; x^2 - 10x + 25 \\
&= x^2 - 2 \cdot x \cdot 5 + 5^2 = (x-5)^2 \\
&\quad\; \uparrow \quad\; \uparrow \;\; \uparrow \;\; \uparrow \quad\; \uparrow \\
&= A^2 - 2 \quad A \quad B + B^2 = (A-B)^2
\end{aligned}$$

23.
$$\begin{aligned}
5p^2 + 20p + 20 &= 5(p^2 + 4p + 4) \\
&= 5(p^2 + 2 \cdot p \cdot 2 + 2^2) \\
&= 5(p+2)^2
\end{aligned}$$

25.
$$\begin{aligned}
1 - 2t + t^2 &= 1^2 - 2 \cdot 1 \cdot t + t^2 \\
&= (1-t)^2
\end{aligned}$$

We could also factor as follows:
$$\begin{aligned}
1 - 2t + t^2 &= t^2 - 2t + 1 \\
&= t^2 - 2 \cdot t \cdot 1 + 1^2 \\
&= (t-1)^2
\end{aligned}$$

27.
$$\begin{aligned}
18x^2 + 12x + 2 &= 2(9x^2 + 6x + 1) \\
&= 2[(3x)^2 + 2 \cdot 3x \cdot 1 + 1^2] \\
&= 2(3x+1)^2
\end{aligned}$$

29.
$$\begin{aligned}
49 - 56y + 16y^2 &= 16y^2 - 56y + 49 \\
&= (4y)^2 - 2 \cdot 4y \cdot 7 + 7^2 \\
&= (4y-7)^2
\end{aligned}$$

We could also factor as follows:
$$\begin{aligned}
49 - 56y + 16y^2 &= 7^2 - 2 \cdot 7 \cdot 4y + (4y)^2 \\
&= (7-4y)^2
\end{aligned}$$

31.
$$\begin{aligned}
-x^5 + 18x^4 - 81x^3 &= -x^3(x^2 - 18x + 81) \\
&= -x^3(x^2 - 2 \cdot x \cdot 9 + 9^2) \\
&= -x^3(x-9)^2
\end{aligned}$$

33.
$$\begin{aligned}
2n^3 + 40n^2 + 200n &= 2n(n^2 + 20n + 100) \\
&= 2n(n^2 + 2 \cdot n \cdot 10 + 10^2) \\
&= 2n(n+10)^2
\end{aligned}$$

35.
$$\begin{aligned}
20x^2 + 100x + 125 &= 5(4x^2 + 20x + 25) \\
&= 5[(2x)^2 + 2 \cdot 2x \cdot 5 + 5^2] \\
&= 5(2x+5)^2
\end{aligned}$$

37. $49 - 42x + 9x^2 = 7^2 - 2 \cdot 7 \cdot 3x + (3x)^2 = (7-3x)^2$, or $(3x-7)^2$

39.
$$\begin{aligned}
16x^2 + 24x + 9 &= (4x)^2 + 2 \cdot 4x \cdot 3 + 3^2 \\
&= (4x+3)^2
\end{aligned}$$

41.
$$\begin{aligned}
2 + 20x + 50x^2 &= 2(1 + 10x + 25x^2) \\
&= 2[1^2 + 2 \cdot 1 \cdot 5x + (5x)^2] \\
&= 2(1+5x)^2, \text{ or } 2(5x+1)^2
\end{aligned}$$

43.
$$\begin{aligned}
9p^2 + 12pq + 4q^2 &= (3p)^2 + 2 \cdot 3p \cdot 2q + (2q)^2 \\
&= (3p+2q)^2
\end{aligned}$$

45. $a^2 - 12ab + 49b^2$

This is not a perfect square trinomial because $-2 \cdot a \cdot 7b = -14ab \neq -12ab$. Nor can it be factored using the methods of Sections 5.2 and 5.3. Thus, it is prime.

47.
$$\begin{aligned}
-64m^2 - 16mn - n^2 &= -1(64m^2 + 16mn + n^2) \\
&= -1[(8m)^2 + 2 \cdot 8m \cdot n + n^2] \\
&= -1(8m+n)^2, \text{ or } -(8m+n)^2
\end{aligned}$$

49.
$$\begin{aligned}
-32s^2 + 80st - 50t^2 &= -2(16s^2 - 40st + 25t^2) \\
&= -2[(4s)^2 - 2 \cdot 4s \cdot 5t + (5t)^2] \\
&= -2(4s-5t)^2
\end{aligned}$$

51. $x^2 - 100$

(1) The first expression is a square: x^2

The second expression is a square: $100 = 10^2$

(2) The terms have different signs.

Thus, $x^2 - 100$ is a difference of squares $x^2 - 10^2$.

53. $n^4 + 1$

(1) The first expression is a square: $n^4 = (n^2)^2$

The second expression is a square: $1 = 1^2$

(2) The terms do not have different signs.

Thus, $n^4 + 1$ is not a difference of squares.

55. $-1 + 64t^2$ or $64t^2 - 1$

(1) The first term is a square: $1 = 1^2$.

The second term is a square: $64t^2 = (8t)^2$.

(2) The terms have different signs.

Thus, $-1 + 64t^2$ is a difference of squares, $(8t)^2 - 1^2$.

57. $x^2 - 25 = x^2 - 5^2 = (x + 5)(x - 5)$

59. $p^2 - 9 = p^2 - 3^2 = (p + 3)(p - 3)$

61. $-49 + t^2 = t^2 - 49 = t^2 - 7^2 = (t + 7)(t - 7)$, or $(7 + t)(-7 + t)$

63. $6a^2 - 24 = 6(a^2 - 4) = 6(a^2 - 2^2) = 6(a + 2)(a - 2)$

65. $49x^2 - 14x + 1 = (7x)^2 - 2 \cdot 7x \cdot 1 + 1^2 = (7x - 1)^2$

67. $200 - 2t^2 = 2(100 - t^2) = 2(10^2 - t^2)$
$\qquad = 2(10 + t)(10 - t)$

69. $-80a^2 + 45 = -5(16a^2 - 9) = -5[(4a^2) - 3^2]$
$\qquad = -5(4a + 3)(4a - 3)$

71. $5t^2 - 80 = 5(t^2 - 16) = 5(t^2 - 4^2)$
$\qquad = 5(t + 4)(t - 4)$

73. $8x^2 - 162 = 2(4x^2 - 81) = 2[(2x)^2 - 9^2]$
$\qquad = 2(2x + 9)(2x - 9)$

75. $36x - 49x^3 = x(36 - 49x^2) = x[6^2 - (7x)^2]$
$\qquad = x(6 + 7x)(6 - 7x)$

77. $49a^4 - 20$

There is no common factor (other than 1 or -1). Since 20 is not a square, this is not a difference of squares. Thus, the polynomial is prime.

79. $\qquad t^4 - 1$
$\qquad = (t^2)^2 - 1^2$
$\qquad = (t^2 + 1)(t^2 - 1)$
$\qquad = (t^2 + 1)(t + 1)(t - 1)$ Factoring further;
$\qquad\qquad\qquad t^2 - 1$ is a difference of squares

81. $-3x^3 + 24x^2 - 48x = -3x(x^2 - 8x + 16)$
$\qquad\qquad\qquad\qquad = -3x(x^2 - 2 \cdot x \cdot 4 + 4^2)$
$\qquad\qquad\qquad\qquad = -3x(x - 4)^2$

83. $75t^3 - 27t = 3t(25t^2 - 9)$
$\qquad\qquad\qquad = 3[(5t)^2 - 3^2]$
$\qquad\qquad\qquad = 3(5t + 3)(5t - 3)$

85. $a^8 - 2a^7 + a^6 = a^6(a^2 - 2a + 1)$
$\qquad\qquad\qquad = a^6(a^2 - 2 \cdot a \cdot 1 + 1^2)$
$\qquad\qquad\qquad = a^6(a - 1)^2$

87. $10a^2 - 10b^2 = 10(a^2 - b^2)$
$\qquad\qquad\qquad = 10(a + b)(a - b)$

89. $16x^4 - y^4 = (4x^2)^2 - (y^2)^2$
$\qquad\qquad\qquad = (4x^2 + y^2)(4x^2 - y^2)$
$\qquad\qquad\qquad = (4x^2 + y^2)(2x + y)(2x - y)$

91. $18t^2 - 8s^2 = 2(9t^2 - 4s^2)$
$\qquad\qquad\qquad = 2[(3t)^2 - (2s)^2]$
$\qquad\qquad\qquad = 2(3t + 2s)(3t - 2s)$

93. *Writing Exercise.* Two terms must be squares. There must be no minus sign before either square. The remaining term must be twice the product of the square roots of the squares or must be the opposite of that product.

95. Two points on the line are $(-2, -5)$ and $(3, -6)$.
$$m = \frac{-6 - (-5)}{3 - (-2)} = \frac{-1}{5}$$

97. $2x - 5y = 10$

Find the y-intercept:
$$-5y = 10$$
$$y = -2$$

The y-intercept is $(0, -2)$.

Find the x-intercept:
$$2x = 10$$
$$x = 5$$

The x-intercept is $(5, 0)$.

Find find a third point, we replace y with 2 and solve for x.
$$2x - 5 \cdot 2 = 10$$
$$2x - 10 = 10$$
$$2x = 20$$
$$x = 10$$

The point is $(10, 2)$.

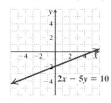

99. Graph: $y = \dfrac{2}{3}x - 1$

Because the equation is in the form $y = mx + b$, we know the y-intercept is $(0, -1)$. We find two other points on the line, substituting multiples of 3 for x to avoid fractions.

When $x = -3$, $y = \dfrac{2}{3}(-3) - 1 = -2 - 1 = -3$.

When $x = 6$, $y = \dfrac{2}{3} \cdot 6 - 1 = 4 - 1 = 3$.

x	y
0	-1
-3	-3
6	3

101. *Writing Exercise.*
$$(x + 3)(x - 3)$$
$$= (x - 3)(x + 3) \qquad \text{Using a commutative law}$$
$$= x^2 - 9$$

Since $x^2 - 9$ and $x^2 + 9$ are not equivalent the student's factorization of $x^2 + 9$ is incorrect. (Also it can be easily shown by multiplying that $(x + 3)(x - 3) \neq x^2 + 9$.) The student should recall that, if the greatest common factor has been removed, a sum of squares cannot be factored further.

103. $x^8 - 2^8 = (x^4 + 2^4)(x^4 - 2^4)$
$$= (x^4 + 2^4)(x^2 + 2^2)(x^2 - 2^2)$$
$$= (x^4 + 2^4)(x^2 + 2^2)(x + 2)(x - 2), \text{ or}$$
$$(x^4 + 16)(x^2 + 4)(x + 2)(x - 2)$$

105. $18x^3 - \dfrac{8}{25}x = 2x\left(9x^2 - \dfrac{4}{25}\right)$
$$= 2x\left(3x + \frac{2}{5}\right)\left(3x - \frac{2}{5}\right)$$

107. $(y - 5)^4 - z^8$
$$= [(y - 5)^2 + z^4][(y - 5)^2 - z^4]$$
$$= [(y - 5)^2 + z^4][y - 5 + z^2][y - 5 - z^2]$$
$$= (y^2 - 10y + 25 + z^4)(y - 5 + z^2)(y - 5 - z^2)$$

109. $-x^4 + 8x^2 + 9 = -1(x^4 - 8x^2 - 9)$
$$= -1(x^2 - 9)(x^2 + 1)$$
$$= -1(x + 3)(x - 3)(x^2 + 1)$$

111. $(y + 3)^2 + 2(y + 3) + 1$
$$= (y + 3)^2 + 2 \cdot (y + 3) \cdot 1 + 1^2$$
$$= [(y + 3) + 1]^2$$
$$= (y + 4)^2$$

113. $27p^3 - 45p^2 - 75p + 125$
$$= 9p^2(3p - 5) - 25(3p - 5)$$
$$= (3p - 5)(9p^2 - 25)$$
$$= (3p - 5)(3p + 5)(3p - 5), \text{ or}$$
$$(3p - 5)^2(3p + 5)$$

115. $81 - b^{4k} = 9^2 - (b^{2k})^2$
$$= (9 + b^{2k})(9 - b^{2k})$$
$$= (9 + b^{2k})[3^2 - (b^k)^2]$$
$$= (9 + b^{2k})(3 + b^k)(3 - b^k)$$

117. $x^2(x + 1)^2 - (x^2 + 1)^2$
$$= x^2(x^2 + 2x + 1) - (x^4 + 2x^2 + 1)$$
$$= x^4 + 2x^3 + x^2 - x^4 - 2x^2 - 1$$
$$= 2x^3 + x^2 - 2x^2 - 1$$
$$= 2x^3 - x^2 - 1$$

119. $y^2 + 6y + 9 - x^2 - 8x - 16$
$$= (y^2 + 6y + 9) - (x^2 + 8x + 16)$$
$$= (y + 3)^2 - (x + 4)^2$$
$$= [(y + 3) + (x + 4)][(y + 3) - (x + 4)]$$
$$= (y + 3 + x + 4)(y + 3 - x - 4)$$
$$= (y + x + 7)(y - x - 1)$$

121. For $c = a^2$, $2 \cdot a \cdot 3 = 24$. Then $a = 4$, so $c = 4^2 = 16$.

123. $(x + 1)^2 - x^2$
$$= [(x + 1) + x][(x + 1) - x]$$
$$= 2x + 1$$
$$= (x + 1) + x$$

Exercise Set 5.5

1. common factor

3. grouping

5. $5a^2 - 125$
$$= 5(a^2 - 25) \qquad \text{5 is a common factor.}$$
$$= 5(a + 5)(a - 5) \quad \text{Factoring the difference of squares}$$

7. $y^2 + 49 - 14y$
$$= y^2 - 14y + 49 \qquad \text{Perfect-square trinomial}$$
$$= (y - 7)^2$$

9. $3t^2 + 16t + 21$

There is no common factor (other than 1 or -1). This trinomial has three terms, but it is not a perfect-square trinomial. Multiply the leading coefficient and the constant, 3 and 21: $3 \cdot 21 = 63$. Try to factor 63 so that the sum of the factors is 16. The numbers we want are 7 and 9: $7 \cdot 9 = 63$ and $7 + 9 = 16$. Split the middle term and factor by grouping.
$$3t^2 + 16t + 21 = 3t^2 + 7t + 9t + 21$$
$$= t(3t + 7) + 3(3t + 7)$$
$$= (3t + 7)(t + 3)$$

11. $x^3 + 18x^2 + 81x$
$$= x(x^2 + 18x + 81) \qquad \text{x is a common factor.}$$
$$= x(x^2 + 2 \cdot x \cdot 9 + 9^2) \qquad \text{Perfect-square trinomial}$$
$$= x(x + 9)^2$$

13. $x^3 - 5x^2 - 25x + 125$
$$= x^2(x - 5) - 25(x - 5) \quad \text{Factoring by grouping}$$
$$= (x - 5)(x^2 - 25)$$
$$= (x - 5)(x + 5)(x - 5) \quad \text{Factoring the difference of squares}$$
$$= (x - 5)^2(x + 5)$$

15. $27t^3 - 3t$
$$= 3t(9t^2 - 1) \qquad \text{$3t$ is a common factor.}$$
$$= 3t[(3t)^2 - 1^2] \qquad \text{Difference of squares}$$
$$= 3t(3t + 1)(3t - 1)$$

17. $9x^3 + 12x^2 - 45x$
$$= 3x(3x^2 + 4x - 15) \quad \text{$3x$ is a common factor.}$$
$$= 3x(x + 3)(3x - 5) \quad \text{Factoring the trinomial}$$

19. $t^2 + 25$

The polynomial has no common factor and is not a difference of squares. It is prime.

21. $6y^2 + 18y - 240$

$= 6(y^2 + 3y - 40)$ 6 is a common factor.

$= 6(y + 8)(y - 5)$ Factoring the trinomial

23. $-2a^6 + 8a^5 - 8a^4$

$= -2a^4(a^2 - 4a + 4)$ Factoring out $-2a^4$

$= -2a^4(a - 2)^2$ Factoring the
                          perfect-square trinomial

25. $5x^5 - 80x$

$= 5x(x^4 - 16)$ $5x$ is a common factor.

$= 5x[(x^2)^2 - 4^2]$ Difference of squares

$= 5x(x^2 + 4)(x^2 - 4)$ Difference of squares

$= 5x(x^2 + 4)(x + 2)(x - 2)$

27. $t^4 - 9$ Difference of squares

$= (t^2 + 3)(t^2 - 3)$

29. $-x^6 + 2x^5 - 7x^4$

$= -x^4(x^2 - 2x + 7)$

The trinomial is prime, so this is the complete factorization.

31. $p^2 - q^2$ Difference of squares

$= (p + q)(p - q)$

33. $ax^2 + ay^2 = a(x^2 + y^2)$

35. $= 2\pi rh + 2\pi r^2$ $2\pi r$ is a common factor

$= 2\pi r(h + r)$

37. $(a + b)(5a) + (a + b)(3b)$

$= (a + b)(5a + 3b)$ $(a + b)$ is a common
                          factor.

39. $x^2 + x + xy + y$

$= x(x + 1) + y(x + 1)$ Factoring by grouping

$= (x + 1)(x + y)$

41. $a^2 - 2a - ay + 2y$

$= a(a - 2) - y(a - 2)$ Factoring by grouping

$= (a - 2)(a - y)$

43. $3x^2 + 13xy - 10y^2 = (3x - 2y)(x + 5y)$

45. $8m^3n - 32m^2n^2 + 24mn$

$= 8mn(m^2 - 4mn + 3)$ $8mn$ is a common factor

47. $4b^2 + a^2 - 4ab$

$= 4b^2 - 4ab + a^2$

$= (2b)^2 - 2 \cdot 2b \cdot a + a^2$ Perfect-square trinomial

$= (2b - a)^2$

This result can also be expressed as $(a - 2b)^2$.

49. $16x^2 + 24xy + 9y^2$

$= (4x)^2 + 2 \cdot 4x \cdot 3y + (3y)^2$ Perfect-square trinomial

$= (4x + 3y)^2$

51. $m^2 - 5m + 8$

We cannot find a pair of factors whose product is 8 and whose sum is -5, so $m^2 - 5m + 8$ is prime.

53. $a^4b^4 - 16$

$= (a^2b^2)^2 - 4^2$ Difference of squares

$= (a^2b^2 + 4)(a^2b^2 - 4)$ Difference of squares

$= (a^2b^2 + 4)(ab + 2)(ab - 2)$

55. $80cd^2 - 36c^2d + 4c^3$

$= 4c(20d^2 - 9cd + c^2)$ $4c$ is a common factor

$= 4c(4d - c)(5d - c)$ Factoring the trinomial

57. $3b^2 + 17ab - 6a^2 = (3b - a)(b + 6a)$

59. $-12 - x^2y^2 - 8xy$

$= -x^2y^2 - 8xy - 12$

$= -1(x^2y^2 + 8xy + 12)$

$= -1(xy + 2)(xy + 6),$ or $-(xy + 2)(xy + 6)$

61. $5p^2q^2 + 25pq - 30$

$= 5(p^2q^2 + 5pq - 6)$ 5 is a common factor·

$= 5(pq + 6)(pq - 1)$ Factoring the trinomial

63. $4ab^5 - 32b^4 + a^2b^6$

$= b^4(4ab - 32 + a^2b^2)$ b^4 is a common factor.

$= b^4(a^2b^2 + 4ab - 32)$

$= b^4(ab + 8)(ab - 4)$ Factoring the trinomial

65. $x^6 + x^5y - 2x^4y^2$

$= x^4(x^2 + xy - 2y^2)$ x^4 is a common factor.

$= x^4(x + 2y)(x - y)$ Factoring the trinomial

67. $36a^2 - 15a + \dfrac{25}{16}$

$= (6a)^2 - 2 \cdot 6a \cdot \dfrac{5}{4} + \left(\dfrac{5}{4}\right)^2$ Perfect-square trinomial

$= \left(6a - \dfrac{5}{4}\right)^2$

69. $\dfrac{1}{81}x^2 - \dfrac{8}{27}x + \dfrac{16}{9}$

$= \left(\dfrac{1}{9}x\right)^2 - 2 \cdot \dfrac{1}{9}x \cdot \dfrac{4}{3} + \left(\dfrac{4}{3}\right)^2$ Perfect-square trinomial

$= \left(\dfrac{1}{9}x - \dfrac{4}{3}\right)^2$

If we had factored out $\dfrac{1}{9}$ at the outset, the final result would have been $\dfrac{1}{9}\left(\dfrac{1}{3}x - 4\right)^2$.

71. $1 - 16x^{12}y^{12}$

$= (1 + 4x^6y^6)(1 - 4x^6y^6)$ Difference of squares

$= (1 + 4x^6y^6)(1 + 2x^3y^3)(1 - 2x^3y^3)$ Difference of squares

73. $4a^2b^2 + 12ab + 9$

$= (2ab)^2 + 2 \cdot 2ab \cdot 3 + 3^2$ Perfect-square trinomial

$= (2ab + 3)^2$

75. $z^4 + 6z^3 - 6z^2 - 36z$

$= z(z^3 + 6z^2 - 6z - 36)$ z is a common factor

$= z\left[z^2(z + 6) - 6(z + 6)\right]$ Factoring by grouping

$= z(z + 6)(z^2 - 6)$

77. *Writing Exercise.* Both are correct; $(x - 4)^2 = x^2 - 8x + 16 = 16 - 8x + x^2 = (4 - x)^2$

79. $8x - 9 = 0$

$8x = 9$ Adding 9 to both sides

$x = \dfrac{9}{8}$ Dividing both sides by 8

The solution is $\dfrac{9}{8}$.

81. $2x + 7 = 0$

$2x = -7$ Subtracting 7 from both sides

$x = -\dfrac{7}{2}$ Dividing both sides by 2

The solution is $-\dfrac{7}{2}$.

83. $3 - x = 0$

$3 = x$ Adding x to both sides

The solution is 3.

85. $2x - 5 = 8x + 1$

$-5 = 6x + 1$ Subtracting $2x$ from both sides

$-6 = 6x$ Subtracting 1 from both sides

$-1 = x$ Dividing both sides by 6

The solution is -1.

87. *Writing Exercise.* Find the product of a binomial and a trinomial. One example is found as follows:

$(x + 1)(2x^2 - 3x + 4)$

$= 2x^3 - 3x^2 + 4x + 2x^2 - 3x + 4$

$= 2x^3 - x^2 + x + 4$

89. $-(x^5 + 7x^3 - 18x)$

$= -x(x^4 + 7x^2 - 18)$

$= -x(x^2 + 9)(x^2 - 2)$

91. $-x^4 + 7x^2 + 18$

$= -1(x^4 - 7x^2 - 18)$

$= -1(x^2 + 2)(x^2 - 9)$

$= -1(x^2 + 2)(x + 3)(x - 3)$, or

$-(x^2 + 2)(x + 3)(x - 3)$

93. $y^2(y + 1) - 4y(y + 1) - 21(y + 1)$

$= (y + 1)(y^2 - 4y - 21)$

$= (y + 1)(y - 7)(y + 3)$

95. $(y + 4)^2 + 2x(y + 4) + x^2$

$= (y + 4)^2 + 2 \cdot (y + 4) \cdot x + x^2$ Perfect-square trinomial

$= \left[(y + 4) + x\right]^2$

$= (y + 4 + x)^2$

97. $2(a + 3)^4 - (a + 3)^3(b - 2) - (a + 3)^2(b - 2)^2$

$= (a + 3)^2[2(a + 3)^2 - (a + 3)(b - 2) - (b - 2)^2]$

$= (a + 3)^2[2(a + 3) + (b - 2)][(a + 3) - (b - 2)]$

$= (a + 3)^2(2a + 6 + b - 2)(a + 3 - b + 2)$

$= (a + 3)^2(2a + b + 4)(a - b + 5)$

99. $49x^4 + 14x^2 + 1 - 25x^6$

$= (7x^2 + 1)^2 - 25x^6$ Perfect-square trinomial

$= \left[(7x^2 + 1) - 5x^3\right]\left[(7x^2 + 1 + 5x^3)\right]$ Difference of squares

$= (7x^2 + 1 - 5x^3)(7x^2 + 1 + 5x^3)$

Exercise Set 5.6

1. Equations of the type $ax^2 + bx + c = 0$, with $a \neq 0$, are quadratic, so choice (c) is correct.

3. The principle of zero products states that $A \cdot B = 0$ if and only if $A = 0$ or $B = 0$, so choice (d) is correct.

5. $(x + 2)(x + 9) = 0$

We use the principle of zero products.

$x + 2 = 0$ *or* $x + 9 = 0$

$x = -2$ *or* $x = -9$

Check:

For -2:

$$\frac{(x + 2)(x + 9) = 0}{\begin{array}{c|c} (-2 + 2)(-2 + 9) & 0 \\ 0 \cdot 7 & \\ & 0 \overset{?}{=} 0 \quad \text{TRUE} \end{array}}$$

For -9:

$$\frac{(x + 2)(x + 9) = 0}{\begin{array}{c|c} (-9 + 2)(-9 + 9) & 0 \\ -7 \cdot 0 & \\ & 0 \overset{?}{=} 0 \quad \text{TRUE} \end{array}}$$

The solutions are -2 and -9.

7. $(x + 1)(x - 8) = 0$

$x + 1 = 0$ *or* $x - 8 = 0$

$x = -1$ *or* $x = 8$

Check:

For -1:

$$\frac{(x + 1)(x - 8) = 0}{\begin{array}{c|c} (-1 + 1)(-1 - 8) & 0 \\ 0 \cdot (-9) & \\ & 0 \overset{?}{=} 0 \quad \text{TRUE} \end{array}}$$

For 8:

$$\frac{(x+1)(x-8)=0}{(8+1)(8-8) \mid 0}$$
$$9 \cdot 0 \mid$$

$$0 \stackrel{?}{=} 0 \quad \text{TRUE}$$

The solutions are -1 and 8.

9. $(2t-3)(t+6)=0$

$2t-3=0 \quad or \quad t+6=0$

$2t=3 \quad or \quad t=-6$

$t=\dfrac{3}{2} \quad or \quad t=-6$

The solutions are $\dfrac{3}{2}$ and -6.

11. $4(7x-1)(10x-3)=0$

$(7x-1)(10x-3)=0 \qquad \text{Dividing both sides by 4}$

$7x-1=0 \quad or \quad 10x-3=0$

$7x=1 \quad or \quad 10x=3$

$x=\dfrac{1}{7} \quad or \quad x=\dfrac{3}{10}$

The solutions are $\dfrac{1}{7}$ and $\dfrac{3}{10}$.

13. $x(x-7)=0$

$x=0 \quad or \quad x-7=0$

$x=0 \quad or \quad x=7$

The solutions are 0 and 7.

15. $\left(\dfrac{2}{3}x-\dfrac{12}{11}\right)\left(\dfrac{7}{4}x-\dfrac{1}{12}\right)=0$

$\dfrac{2}{3}x-\dfrac{12}{11}=0 \qquad or \qquad \dfrac{7}{4}x-\dfrac{1}{12}=0$

$\dfrac{2}{3}x=\dfrac{12}{11} \qquad or \qquad \dfrac{7}{4}x=\dfrac{1}{12}$

$x=\dfrac{3}{2}\cdot\dfrac{12}{11} \qquad or \qquad x=\dfrac{4}{7}\cdot\dfrac{1}{12}$

$x=\dfrac{18}{11} \qquad or \qquad x=\dfrac{1}{21}$

The solutions are $\dfrac{18}{11}$ and $\dfrac{1}{21}$.

17. $6n(3n+8)=0$

$6n=0 \quad or \quad 3n+8=0$

$n=0 \quad or \quad 3n=-8$

$n=0 \quad or \quad n=-\dfrac{8}{3}$

The solutions are 0 and $-\dfrac{8}{3}$.

19. $(20-0.4x)(7-0.1x)=0$

$20-0.4x=0 \qquad or \quad 7-0.1x=0$

$-0.4x=-20 \quad or \qquad -0.1x=-7$

$x=50 \qquad or \qquad x=70$

The solutions are 50 and 70.

21. $x^2-7x+6=0$

$(x-6)(x-1)=0 \quad \text{Factoring}$

$x-6=0 \quad or \quad x-1=0 \quad \text{Using the principle}$
of zero products

$x=6 \quad or \qquad x=1$

The solutions are 6 and 1.

23. $x^2+4x-21=0$

$(x-3)(x+7)=0 \quad \text{Factoring}$

$x-3=0 \quad or \quad x+7=0 \qquad \text{Using the principle}$
of zero products

$x=3 \quad or \qquad x=-7$

The solutions are 3 and -7.

25. $n^2+11n+18=0$

$(n+9)(n+2)=0$

$n+9=0 \quad or \quad n+2=0$

$n=-9 \quad or \qquad n=-2$

The solutions are -9 and -2.

27. $x^2-10x=0$

$x(x-10)=0$

$x=0 \quad or \quad x-10=0$

$x=0 \quad or \qquad x=10$

The solutions are 0 and 10.

29. $6t+t^2=0$

$t(6+t)=0$

$t=0 \quad or \quad 6+t=0$

$t=0 \quad or \qquad t=-6$

The solutions are 0 and -6.

31. $x^2-36=0$

$(x+6)(x-6)=0$

$x+6=0 \quad or \quad x-6=0$

$x=-6 \quad or \qquad x=6$

The solutions are -6 and 6.

33. $4t^2=49$

$4t^2-49=0$

$(2t+7)(2t-7)=0$

$2t+7=0 \quad or \quad 2t-7=0$

$2t=-7 \quad or \qquad 2t=7$

$t=\dfrac{-7}{2} \quad or \qquad t=\dfrac{7}{2}$

The solutions are $\dfrac{-7}{2}$ and $\dfrac{7}{2}$.

35. $0=25+x^2+10x$

$0=x^2+10x+25 \quad \text{Writing in descending}$
order

$0=(x+5)(x+5)$

$x+5=0 \quad or \quad x+5=0$

$x=-5 \quad or \qquad x=-5$

The solution is -5.

37.
$$64 + x^2 = 16x$$
$$x^2 - 16x + 64 = 0$$
$$(x - 8)(x - 8) = 0$$
$$x - 8 = 0 \quad or \quad x - 8 = 0$$
$$x = 8 \quad or \quad x = 8$$
The solution is 8.

39.
$$4t^2 = 8t$$
$$4t^2 - 8t = 0$$
$$4t(t - 2) = 0$$
$$t = 0 \quad or \quad t - 2 = 0$$
$$t = 0 \quad or \quad t = 2$$
The solutions are 0 and 2.

41.
$$4y^2 = 7y + 15$$
$$4y^2 - 7y - 15 = 0$$
$$(4y + 5)(y - 3) = 0$$
$$4y + 5 = 0 \quad or \quad y - 3 = 0$$
$$4y = -5 \quad or \quad y = 3$$
$$y = \frac{-5}{4} \quad or \quad y = 3$$
The solutions are $\frac{-5}{4}$ and 3.

43.
$$(x - 7)(x + 1) = -16$$
$$x^2 - 6x - 7 = -16$$
$$x^2 - 6x + 9 = 0$$
$$(x - 3)(x - 3) = 0$$
$$x - 3 = 0 \quad or \quad x - 3 = 0$$
$$x = 3 \quad or \quad x = 3$$
The solution is 3.

45.
$$15z^2 + 7 = 20z + 7$$
$$15z^2 - 20z + 7 = 7$$
$$15z^2 - 20z = 0$$
$$5z(3z - 4) = 0$$
$$5z = 0 \quad or \quad 3z - 4 = 0$$
$$z = 0 \quad or \quad 3z = 4$$
$$z = 0 \quad or \quad z = \frac{4}{3}$$
The solutions are 0 and $\frac{4}{3}$.

47.
$$36m^2 - 9 = 40$$
$$36m^2 - 49 = 0$$
$$(6m + 7)(6m - 7) = 0$$
$$6m + 7 = 0 \quad or \quad 6m - 7 = 0$$
$$6m = -7 \quad or \quad 6m = 7$$
$$m = \frac{-7}{6} \quad or \quad m = \frac{7}{6}$$
The solutions are $\frac{-7}{6}$ or $\frac{7}{6}$.

49.
$$(x + 3)(3x + 5) = 7$$
$$3x^2 + 14x + 15 = 7$$
$$3x^2 + 14x + 8 = 0$$
$$(3x + 2)(x + 4) = 0$$
$$3x + 2 = 0 \quad or \quad x + 4 = 0$$
$$3x = -2 \quad or \quad x = -4$$
$$x = \frac{-2}{3} \quad or \quad x = -4$$
The solutions are $-\frac{2}{3}$ and -4.

51.
$$3x^2 - 2x = 9 - 8x$$
$$3x^2 + 6x - 9 = 0 \qquad \text{Adding } 8x \text{ and } -9$$
$$3(x^2 + 2x - 3) = 0$$
$$3(x + 3)(x - 1) = 0$$
$$x + 3 = 0 \quad or \quad x - 1 = 0$$
$$x = -3 \quad or \quad x = 1$$
The solutions are -3 and 1.

53.
$$(6a + 1)(a + 1) = 21$$
$$6a^2 + 7a + 1 = 21$$
$$6a^2 + 7a - 20 = 0$$
$$(3a - 4)(2a + 5) = 0$$
$$3a - 4 = 0 \quad or \quad 2a + 5 = 0$$
$$3a = 4 \quad or \quad 2a = -5$$
$$a = \frac{4}{3} \quad or \quad a = -\frac{5}{2}$$
The solutions are $\frac{4}{3}$ and $-\frac{5}{2}$.

55. The solutions of the equation are the first coordinates of the x-intercepts of the graph. From the graph we see that the x-intercepts are $(-1, 0)$ and $(4, 0)$, so the solutions of the equation are -1 and 4.

57. The solutions of the equation are the first coordinates of the x-intercepts of the graph. From the graph we see that the x-intercepts are $(-3, 0)$ and $(2, 0)$, so the solutions of the equation are -3 and 2.

59. We let $y = 0$ and solve for x.
$$0 = x^2 - x - 6$$
$$0 = (x - 3)(x + 2)$$
$$x - 3 = 0 \quad or \quad x + 2 = 0$$
$$x = 3 \quad or \quad x = -2$$
The x-intercepts are $(3, 0)$ and $(-2, 0)$.

61. We let $y = 0$ and solve for x.
$$0 = x^2 + 2x - 8$$
$$0 = (x + 4)(x - 2)$$
$$x + 4 = 0 \quad or \quad x - 2 = 0$$
$$x = -4 \quad or \quad x = 2$$
The x-intercepts are $(-4, 0)$ and $(2, 0)$.

63. We let $y = 0$ and solve for x.

$$0 = 2x^2 + 3x - 9$$
$$0 = (2x - 3)(x + 3)$$
$$2x - 3 = 0 \quad or \quad x + 3 = 0$$
$$2x = 3 \quad or \quad x = -3$$
$$x = \frac{3}{2} \quad or \quad x = -3$$

The x-intercepts are $\left(\frac{3}{2}, 0\right)$ and $(-3, 0)$.

65. *Writing Exercise.* The graph has no x-intercepts.

67. Let m and n represent the numbers. The sum of the two numbers is $m + n$. Thus the square of the sum of the two numbers is $(m + n)^2$.

69. Let x represent the first integer. Then $x + 1$ represents the second integer. Thus, the product is $x(x + 1)$.

71. *Familiarize.* We draw a picture. We let $x =$ the measure of the second angle. Then $4x =$ the measure of the first angle, and $x - 30 =$ the measure of the third angle.

Recall that the measures of the angles of any triangle add up to $180°$.

Translate.

$$\underbrace{\text{Measure of first angle}}_{4x} + \underbrace{\text{Measure of 2nd angle}}_{x} + \underbrace{\text{Measure of third angle}}_{(x - 30)} \text{ is } \underbrace{180°}_{180}$$

Carry out. we solve the equation

$$4x + x + (x - 30) = 180$$
$$6x - 30 = 180$$
$$6x = 210$$
$$x = 35$$

Possible answers for the angle measures are as follows:
First angle: $4x = 4(35°) = 140°$
Second angle: $x = 35 \deg$
Third angle: $x - 30 = 35 - 30 = 5 \deg$.

Check. Consider $140°$, $35°$ and $5°$. The first angle is four times the first, and the third is $30°$ less than the second. The sum of the angles is $180°$. These numbers check.

State. The measure of the first angle is $140°$, the measure of the second angle is $35°$, and the measure of the third angle is $5°$.

73. *Writing Exercise.* One solution of the equation is 0. Dividing both sides of the equation by x, leaving the solution $x = 3$, is equivalent to dividing by 0.

75. $(2x - 11)(3x^2 + 29x + 56) = 0$
$(2x - 11)(3x + 8)(x + 7) = 0$

$$2x - 11 = 0 \quad or \quad 3x + 8 = 0 \quad or \quad x + 7 = 0$$
$$2x = 11 \quad or \quad 3x = -8 \quad or \quad x = -7$$
$$x = \frac{11}{2} \quad or \quad x = -\frac{8}{3} \quad or \quad x = -7$$

The solutions are -7, $-\frac{8}{3}$, and $\frac{11}{2}$.

77. a)
$$x = -4 \quad or \quad x = 5$$
$$x + 4 = 0 \quad or \quad x - 5 = 0$$
$$(x + 4)(x - 5) = 0 \qquad \text{Principle of zero products}$$
$$x^2 - x - 20 = 0 \qquad \text{Multiplying}$$

b)
$$x = -1 \quad or \quad x = 7$$
$$x + 1 = 0 \quad or \quad x - 7 = 0$$
$$(x + 1)(x - 7) = 0$$
$$x^2 - 6x - 7 = 0$$

c)
$$x = \frac{1}{4} \quad or \quad x = 3$$
$$x - \frac{1}{4} = 0 \quad or \quad x - 3 = 0$$
$$\left(x - \frac{1}{4}\right)(x - 3) = 0$$
$$x^2 - \frac{13}{4}x + \frac{3}{4} = 0$$
$$4\left(x^2 - \frac{13}{4}x + \frac{3}{4}\right) = 4 \cdot 0 \qquad \text{Multiplying both sides by 4}$$
$$4x^2 - 13x + 3 = 0$$

d)
$$x = \frac{1}{2} \quad or \quad x = \frac{1}{3}$$
$$x - \frac{1}{2} = 0 \quad or \quad x - \frac{1}{3} = 0$$
$$\left(x - \frac{1}{2}\right)\left(x - \frac{1}{3}\right) = 0$$
$$x^2 - \frac{5}{6}x + \frac{1}{6} = 0$$
$$6x^2 - 5x + 1 = 0 \qquad \text{Multiplying by 6}$$

e)
$$x = \frac{2}{3} \quad or \quad x = \frac{3}{4}$$
$$x - \frac{2}{3} = 0 \quad or \quad x - \frac{3}{4} = 0$$
$$\left(x - \frac{2}{3}\right)\left(x - \frac{3}{4}\right) = 0$$
$$x^2 - \frac{17}{12}x + \frac{1}{2} = 0$$
$$12x^2 - 17x + 6 = 0 \qquad \text{Multiplying by 12}$$

f)
$$x = -1 \quad or \quad x = 2 \ or \quad x = 3$$
$$x + 1 = 0 \quad or \quad x - 2 = 0 \ or \quad x - 3 = 0$$
$$(x + 1)(x - 2)(x - 3) = 0$$
$$(x^2 - x - 2)(x - 3) = 0$$
$$x^3 - 4x^2 + x + 6 = 0$$

79.
$$a(9 + a) = 4(2a + 5)$$
$$9a + a^2 = 8a + 20$$
$$a^2 + a - 20 = 0 \qquad \text{Subtracting } 8a \text{ and } 20$$
$$(a + 5)(a - 4) = 0$$
$$a + 5 = 0 \quad or \quad a - 4 = 0$$
$$a = -5 \quad or \qquad a = 4$$

The solutions are -5 and 4.

81.
$$-x^2 + \frac{9}{25} = 0$$
$$x^2 - \frac{9}{25} = 0 \qquad \text{Multiplying by } -1$$
$$\left(x - \frac{3}{5}\right)\left(x + \frac{3}{5}\right) = 0$$
$$x - \frac{3}{5} = 0 \quad or \quad x + \frac{3}{5} = 0$$
$$x = \frac{3}{5} \quad or \qquad x = -\frac{3}{5}$$

The solutions are $\frac{3}{5}$ and $-\frac{3}{5}$.

83. $(t + 1)^2 = 9$

Observe that $t + 1$ is a number which yields 9 when it is squared. Thus, we have

$$t + 1 = -3 \quad or \quad t + 1 = 3$$
$$t = -4 \quad or \qquad t = 2$$

The solutions are -4 and 2.

We could also do this exercise as follows:

$$(t + 1)^2 = 9$$
$$t^2 + 2t + 1 = 9$$
$$t^2 + 2t - 8 = 0$$
$$(t + 4)(t - 2) = 0$$
$$t + 4 = 0 \quad or \quad t - 2 = 0$$
$$t = -4 \quad or \qquad t = 2$$

Again we see that the solutions are -4 and 2.

85. a) $2(x^2 + 10x - 2) = 2 \cdot 0$ Multiplying (a) by 2
$$2x^2 + 20x - 4 = 0$$

(a) and $2x^2 + 20x - 4 = 0$ are equivalent.

b) $(x - 6)(x + 3) = x^2 - 3x - 18$ Multiplying

(b) and $x^2 - 3x - 18 = 0$ are equivalent.

c) $5x^2 - 5 = 5(x^2 - 1) = 5(x + 1)(x - 1) =$
$(x + 1)5(x - 1) = (x + 1)(5x - 5)$

(c) and $(x + 1)(5x - 5) = 0$ are equivalent.

d) $2(2x - 5)(x + 4) = 2 \cdot 0$ Multiplying (d) by 2
$$2(x + 4)(2x - 5) = 0$$
$$(2x + 8)(2x - 5) = 0$$

(d) and $(2x + 8)(2x - 5) = 0$ are equivalent.

e) $4(x^2 + 2x + 9) = 4 \cdot 0$ Multiplying (e) by 4
$$4x^2 + 8x + 36 = 0$$

(e) and $4x^2 + 8x + 36 = 0$ are equivalent.

f) $3(3x^2 - 4x + 8) = 3 \cdot 0$ Multiplying (f) by 3
$$9x^2 - 12x + 24 = 0$$

(f) and $9x^2 - 12x + 24 = 0$ are equivalent.

87. *Writing Exercise.* Graph $y = -x^2 - x + 6$ and $y = 4$ on the same set of axes. The first coordinates of the points of intersection of the two graphs are the solutions of $-x^2 - x + 6 = 4$.

89. $2.33, 6.77$

91. $-4.59, -9.15$

93. $-3.76, 0$

Chapter 5 Connecting the Concepts

1. Expression

3. Equation

5. Expression

7. $\left(2x^3 - 5x + 1\right) + \left(x^2 - 3x - 1\right)$
$= 2x^3 + x^2 + (-5 - 3)x + (1 - 1)$
$= 2x^3 + x^2 - 8x$

9.
$$t^2 - 100 = 0$$
$$(t + 10)(t - 10) = 0$$
$$t + 10 = 0 \quad or \quad t - 10 = 0$$
$$t = -10 \quad or \qquad t = 10$$

The solutions are -10 and 10.

11. $n^2 - 10n + 9 = (n - 1)(n - 9)$ Factor with FOIL

13.
$$4t^2 + 20t + 25 = 0$$
$$(2t)^2 + 2 \cdot 2t \cdot 5 + 5^2 = 0$$
$$(2t + 5)(2t + 5) = 0$$
$$2t + 5 = 0 \quad or \quad 2t + 5 = 0$$
$$2t = -5 \quad or \qquad 2t = -5$$
$$t = \frac{-5}{2} \quad or \qquad t = \frac{-5}{2}$$

The solutions is $\frac{-5}{2}$.

15. $16x^2 - 81 = (4x + 9)(4x - 9)$ Difference of squares

17. $\left(a^2 - 2\right) - \left(5a^2 + a + 9\right)$
$= a^2 - 2 - 5a^2 - a - 9$
$= (1 - 5)a^2 - a + (-2 - 9)$
$= -4a^2 - a - 11$

19.
$$3x^2 + 5x + 2 = 0$$
$$(x + 1)(3x + 2) = 0$$
$$x + 1 = 0 \quad or \quad 3x + 2 = 0$$
$$x = -1 \quad or \qquad x = \frac{-2}{3}$$

The solutions are -1 and $-\frac{2}{3}$.

Exercise Set 5.7

1. Familiarize. Let x = the number.

Translate. We reword the problem.

The square of a number minus the number is 6.

$$x^2 \quad - \quad x \quad = 6$$

Carry out. We solve the equation.

$$x^2 - x = 6$$
$$x^2 - x - 6 = 0$$
$$(x - 3)(x + 2) = 0$$

$$x - 3 = 0 \quad or \quad x + 2 = 0$$
$$x = 3 \quad or \quad x = -2$$

Check. For 3: The square of 3 is 3^2, or 9, and $9 - 3 = 6$.
For -2: The square of -2, or 4 and
$4 - (-2) = 4 + 2 = 6$. Both numbers check.

State. The numbers are 3 and -2.

3. Familiarize. Let x = the length of the shorter leg, in m. Then $x + 2$ = the length of the longer leg.

Translate. We use the Pythagorean theorem.
$$a^2 + b^2 = c^2$$
$$x^2 + (x + 2)^2 = 10^2$$

Carry out. We solve the equation.
$$x^2 + (x + 2)^2 = 10^2$$
$$x^2 + x^2 + 4x + 4 = 100$$
$$2x^2 + 4x + 4 = 100$$
$$2x^2 + 4x - 96 = 0$$
$$2(x^2 + 2x - 48) = 0$$
$$2(x + 8)(x - 6) = 0$$

$$x + 8 = 0 \quad or \quad x - 6 = 0$$
$$x = -8 \quad or \quad x = 6$$

Check. The number -8 cannot be the length of a side because it is negative. When $x = 6$, then $x + 2 = 8$, and $6^2 + 8^2 = 36 + 64 = 100 = 10^2$, so the number 6 checks.

State. The lengths of the sides are 6 m, 8 m, and 10 m.

5. Familiarize. The parking spaces are consecutive integers. Let x = the smaller integer. Then $x + 1$ = the larger integer.

Translate. We reword the problem.

Smaller integer times larger integer is 132.

$$x \quad \cdot \quad (x + 1) \quad = 132$$

Carry out. We solve the equation.

$$x(x + 1) = 132$$
$$x^2 + x = 132$$
$$x^2 + x - 132 = 0$$
$$(x + 12)(x - 11) = 0$$
$$x + 12 = 0 \quad or \quad x - 11 = 0$$
$$x = -12 \quad or \quad x = 11$$

Check. The solutions of the equation are -12 and 11. Since a parking space number cannot be negative, we only need to check 11. When $x = 11$, then $x + 1 = 12$, and $11 \cdot 12 = 132$. This checks.

State. The parking space numbers are 11 and 12.

7. Familiarize. Let x = the smaller even integer. Then $x + 2$ = the larger even integer.

Translate. We reword the problem.

Smaller even integer times larger even integer is 168.

$$x \quad \cdot \quad (x + 2) \quad = 168$$

Carry out.
$$x(x + 2) = 168$$
$$x^2 + 2x = 168$$
$$x^2 + 2x - 168 = 0$$
$$(x + 14)(x - 12) = 0$$
$$x + 14 = 0 \quad or \quad x - 12 = 0$$
$$x = -14 \quad or \quad x = 12$$

Check. The solutions of the equation are -14 and 12. When x is -14, then $x + 2$ is -12 and $-14(-12) = 168$. The numbers -14 and -12 are consecutive even integers which are solutions of the problem. When x is 12, then $x + 2 = 14$ and $12 \cdot 14 = 168$. The numbers 12 and 14 are also consecutive even integers which are solutions of the problem.

State. We have two solutions, each of which consists of a pair of numbers: -14 and -12 or 12 and 14.

9. Familiarize. Let w = the width of the porch, in feet. Then $5w$ = the length. Recall that the area of a rectangle is Length $\cdot$ Width.

Translate.

The area of the rectangle is 180 ft^2.

$$5w \cdot w \quad = \quad 180$$

Carry out. We solve the equation.
$$5w \cdot w = 180$$
$$5w^2 = 180$$
$$5w^2 - 180 = 0$$
$$5(w^2 - 36) = 0$$
$$5(w + 6)(w - 6) = 0$$

$$w + 6 = 0 \quad or \quad w - 6 = 0$$
$$w = -6 \quad or \quad w = 6$$

Check. Since the width must be positive, -6 cannot be a solution. If the width is 6 ft, then the length is $5 \cdot 6$ ft, or 30 ft, and the area is 6 ft $\cdot$ 30 ft $= 180$ ft^2. Thus, 6 checks.

State. The porch is 30 ft long and 6 ft wide.

11. **Familiarize**. We make a drawing. Let $w =$ the width, in cm. Then $w + 2 =$ the length, in cm.

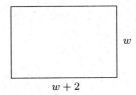

$$w + 2$$

Recall that the area of a rectangle is length times width.

Translate. We reword the problem.

Length times width is $\underbrace{24 \text{ cm}^2}$.

$\downarrow \quad \downarrow \quad \downarrow \quad \downarrow \quad \downarrow$

$(w + 2) \quad \cdot \quad w \quad = \quad 24$

Carry out. We solve the equation.

$$(w + 2)w = 24$$
$$w^2 + 2w = 24$$
$$w^2 + 2w - 24 = 0$$
$$(w + 6)(w - 4) = 0$$
$$w + 6 = 0 \quad or \quad w - 4 = 0$$
$$w = -6 \quad or \qquad w = 4$$

Check. Since the width must be positive, -6 cannot be a solution. If the width is 4 cm, then the length is $4 + 2$, or 6 cm, and the area is $6 \cdot 4$, or 24 cm^2. Thus, 4 checks.

State. The width is 4 cm, and the length is 6 cm.

13. **Familiarize**. Using the labels shown on the drawing in the text, we let $b =$ the base, in inches, and $b - 3 =$ the height, in inches. Recall that the formula for the area of a triangle is $\frac{1}{2} \cdot$ (base) $\cdot$ (height).

Translate.

$\frac{1}{2}$ times base times height is $\underbrace{54 \text{ in}^2}$.

$\downarrow \quad \downarrow \quad \downarrow \quad \downarrow \quad \downarrow \quad \downarrow \quad \downarrow$

$\frac{1}{2} \quad \cdot \quad (b) \quad \cdot \quad (b - 3) \quad = \quad 54$

Carry out.

$$\frac{1}{2}b(b - 3) = 54$$
$$b(b - 3) = 108 \qquad \text{Multiplying by 2}$$
$$b^2 - 3b = 108$$
$$b^2 - 3b - 108 = 0$$
$$(b - 12)(b + 9) = 0$$
$$b - 12 = 0 \quad or \quad b + 9 = 0$$
$$b = 12 \quad or \qquad b = -9$$

Check. Since the height must be positive, -9 cannot be a solution. If the base is 12 in, then the height is $12 - 3$, or 9 in, and the area is $\frac{1}{2} \cdot 12 \cdot 9$, or 54 in^2. Thus, 12 checks.

State. The base of the triangle is 12 in, and the height is 9 in.

15. **Familiarize**. Using the labels show on the drawing in the text, we let $x =$ the length of the foot of the sail, in ft, and $x + 5 =$ the height of the sail, in ft. Recall that the formula for the area of a triangle is $\frac{1}{2} \cdot$ (base) $\cdot$ (height).

Translate.

$\frac{1}{2}$ times base times height is $\underbrace{42 \text{ ft}^2}$.

$\downarrow \quad \downarrow \quad \downarrow \quad \downarrow \quad \downarrow \quad \downarrow \quad \downarrow$

$\frac{1}{2} \quad \cdot \quad x \quad \cdot \quad (x + 5) \quad = \quad 42$

Carry out.

$$\frac{1}{2}x(x + 5) = 42$$
$$x(x + 5) = 84 \quad \text{Multiplying by 2}$$
$$x^2 + 5x = 84$$
$$x^2 + 5x - 84 = 0$$
$$(x + 12)(x - 7) = 0$$
$$x + 12 = 0 \quad or \quad x - 7 = 0$$
$$x = -12 \quad or \qquad x = 7$$

Check. The solutions of the equation are -12 and 7. The length of the base of a triangle cannot be negative, so -12 cannot be a solution. Suppose the length of the foot of the sail is 7 ft. Then the height is $7 + 5$, or 12 ft, and the area is $\frac{1}{2} \cdot 7 \cdot 12$, or 42 ft^2. These numbers check.

State. The length of the foot of the sail is 7 ft, and the height is 12 ft.

17. **Familiarize and Translate**. We substitute 150 for A in the formula.

$$A = -50t^2 + 200t$$
$$150 = -50t^2 + 200t$$

Carry out. We solve the equation.

$$150 = -50t^2 + 200t$$
$$0 = -50t^2 + 200t - 150$$
$$0 = -50(t^2 - 4t + 3)$$
$$0 = -50(t - 1)(t - 3)$$
$$t - 1 = 0 \quad or \quad t - 3 = 0$$
$$t = 1 \quad or \qquad t = 3$$

Check. Since $-50 \cdot 1^2 + 200 \cdot 1 = -50 + 200 = 150$, the number 1 checks. Since $-50 \cdot 3^2 + 200 \cdot 3 = -450 + 600 = 150$, the number 3 checks also.

State. There will be about 150 micrograms of Albuterol in the bloodstream 1 minute and 3 minutes after an inhalation.

19. Familiarize. We will use the formula $N = 0.3t^2 + 0.6t$.

Translate. Substitute 36 for N.

$$36 = 0.3t^2 + 0.6t$$

Carry out.

$$
\begin{aligned}
36 &= 0.3t^2 + 0.6t \\
360 &= 3t^2 + 6t \qquad \text{Multiplying by 10} \\
0 &= 3t^2 + 6t - 360 \\
0 &= 3(t^2 + 2t - 120) \\
0 &= 3(t + 12)(t - 10)
\end{aligned}
$$

$$
\begin{aligned}
t + 12 &= 0 \quad \text{or} \quad t - 10 = 0 \\
t &= -12 \quad \text{or} \qquad t = 10
\end{aligned}
$$

Check. The solutions of the equation are -12 and 10. Since the number of years cannot be negative, -12 cannot be a solution. However, 10 checks since $0.3(10)^2 + 0.6(10) = 30 + 6 = 36$.

State. 10 years after 1998 is the year 2008.

21. Familiarize. We will use the formula $x^2 - x = N$.

Translate. Substitute 240 for N.

$$x^2 - x = 240$$

Carry out.

$$
\begin{aligned}
x^2 - x &= 240 \\
x^2 - x - 240 &= 0 \\
(x - 16)(x + 15) &= 0
\end{aligned}
$$

$$
\begin{aligned}
x - 16 &= 0 \quad \text{or} \quad x + 15 = 0 \\
x &= 16 \quad \text{or} \qquad x = -15
\end{aligned}
$$

Check. The solutions of the equation are 16 and -15. Since the number of teams cannot be negative, -15 cannot be a solution. But 16 checks since $16^2 - 16 = 256 - 16 = 240$.

State. There are 16 teams in the league.

22. Solve: $x^2 - x = 132$

$$x = -11 \quad \text{or} \quad x = 12$$

Since the number of teams must be positive, -11 cannot be a solution. The number 12 checks, so there are 12 teams in the league.

23. Familiarize. We will use the formula

$$H = \frac{1}{2}(n^2 - n).$$

Translate. Substitute 12 for n.

$$H = \frac{1}{2}(12^2 - 12)$$

Carry out. We do the computation on the right.

$$
\begin{aligned}
H &= \frac{1}{2}(12^2 - 12) \\
H &= \frac{1}{2}(144 - 12) \\
H &= \frac{1}{2}(132) \\
H &= 66
\end{aligned}
$$

Check. We can recheck the computation, or we can solve the equation $66 = \frac{1}{2}(n^2 - n)$. The answer checks.

State. 66 handshakes are possible.

25. Familiarize. We will use the formula $H = \frac{1}{2}(n^2 - n)$, since "high fives" can be substituted for handshakes.

Translate. Substitute 66 for H.

$$66 = \frac{1}{2}(n^2 - n)$$

Carry out.

$$
\begin{aligned}
66 &= \frac{1}{2}(n^2 - n) \\
132 &= n^2 - n \qquad \text{Multiplying by 2} \\
0 &= n^2 - n - 132 \\
0 &= (n - 12)(n + 11)
\end{aligned}
$$

$$
\begin{aligned}
n - 12 &= 0 \quad \text{or} \quad n + 11 = 0 \\
n &= 12 \quad \text{or} \qquad n = -11
\end{aligned}
$$

Check. The solutions of the equation are 12 and -11. Since the number of players cannot be negative, -11 cannot be a solution. However, 12 checks since $\frac{1}{2}(12^2 - 12) = \frac{1}{2}(144 - 12) = \frac{1}{2}(132) = 66$.

State. 12 players were on the team.

27. Familiarize. Let $h =$ the vertical height to which each brace reaches, in feet. We have a right triangle with hypotenuse 15 ft and legs 12 ft and h.

Translate. We use the Pythagorean theorem.

$$
\begin{aligned}
a^2 + b^2 &= c^2 \\
12^2 + h^2 &= 15^2
\end{aligned}
$$

Carry out. We solve the equation.

$$
\begin{aligned}
12^2 + h^2 &= 15^2 \\
144 + h^2 &= 225 \\
h^2 - 81 &= 0 \\
(h + 9)(h - 9) &= 0
\end{aligned}
$$

$$
\begin{aligned}
h + 9 &= 0 \quad \text{or} \quad h - 9 = 0 \\
h &= -9 \quad \text{or} \qquad h = 9
\end{aligned}
$$

Check. Since the vertical height must be positive, -9 cannot be a solution. If the height is 9 ft, then we have $12^2 + 9^2 = 144 + 81 = 225 = 15^2$. The number 9 checks.

State. Each brace reaches 9 ft vertically.

29. Familiarize. Let $w =$ the width of Main Street, in ft. We have a right triangle with hypotenuse 40 ft and legs of 24 ft and w.

Translate. We use the Pythagorean theorem.

$$a^2 + b^2 = c^2$$

Carry out.

$$
\begin{aligned}
24^2 + w^2 &= 40^2 \\
576 + w^2 &= 1600 \\
w^2 - 1024 &= 0 \\
(w - 32)(w + 32) &= 0
\end{aligned}
$$

$$
\begin{aligned}
w - 32 &= 0 \quad \text{or} \quad w + 32 = 0 \\
w &= 32 \quad \text{or} \qquad w = -32
\end{aligned}
$$

Check. Since the width must be positive, -32 cannot be a solution. If the width is 32 ft, then we have $24^2 + 32^2 = 576 + 1024 = 1600 = 40^2$. The number 32 checks.

State. The width of Main Street is 32 ft.

31. Familiarize. Let l = the length of the leg, in ft. Then $l + 200$ = the length of the hypotenuse in feet.

Translate. We use the Pythagorean theorem.
$$a^2 + b^2 = c^2$$
$$400^2 + l^2 = (l + 200)^2$$

Carry out.
$$400^2 + l^2 = l^2 + 400l + 40,000$$
$$160,000 + l^2 = l^2 + 400l + 40,000$$
$$120,000 = 400l$$
$$300 = l$$

Check. When $l = 300$, then $l + 200 = 500$, and $400^2 + 300^2 = 160,000 + 90,000 = 250,000 = 500^2$, so the number 300 checks.

State. The dimensions of the garden are 300 ft by 400 ft by 500 ft.

33. Familiarize. We label the drawing. Let x = the length of a side of the dining room, in ft. Then the dining room has dimensions x by x and the kitchen has dimensions x by 10. The entire rectangular space has dimension x by $x + 10$. Recall that we multiply these dimensions to find the area of the rectangle.

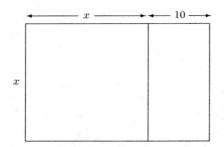

Translate.

The area of the rectangular space is 264 ft^2.

$$x(x + 10) = 264$$

Carry out. We solve the equation.
$$x(x + 10) = 264$$
$$x^2 + 10x = 264$$
$$x^2 + 10x - 264 = 0$$
$$(x + 22)(x - 12) = 0$$
$$x + 22 = 0 \quad or \quad x - 12 = 0$$
$$x = -22 \quad or \qquad x = 12$$

Check. Since the length of a side of the dining room must be positive, -22 cannot be a solution. If x is 12 ft, then

$x + 10$ is 22 ft, and the area of the space is $12 \cdot 22$, or 264 ft^2. The number 12 checks.

State. The dining room is 12 ft by 12 ft, and the kitchen is 12 ft by 10 ft.

35. Familiarize. We will use the formula $h = 48t - 16t^2$.

Translate. Substitute $\frac{1}{2}$ for t.
$$h = 48 \cdot \frac{1}{2} - 16\left(\frac{1}{2}\right)^2$$

Carry out. We do the computation on the right.
$$h = 48 \cdot \frac{1}{2} - 16\left(\frac{1}{2}\right)^2$$
$$h = 48 \cdot \frac{1}{2} - 16 \cdot \frac{1}{4}$$
$$h = 24 - 4$$
$$h = 20$$

Check. We can recheck the computation, or we can solve the equation $20 = 48t - 16t^2$. The answer checks.

State. The rocket is 20 ft high $\frac{1}{2}$ sec after it is launched.

37. Familiarize. We will use the formula $h = 48t - 16t^2$.

Translate. Substitute 32 for h.
$$32 = 48t - 16t^2$$

Carry out. We solve the equation.
$$32 = 48t - 16t^2$$
$$0 = -16t^2 + 48t - 32$$
$$0 = -16(t^2 - 3t + 2)$$
$$0 = -16(t - 1)(t - 2)$$
$$t - 1 = 0 \quad or \quad t - 2 = 0$$
$$t = 1 \quad or \qquad t = 2$$

Check. When $t = 1$, $h = 48 \cdot 1 - 16 \cdot 1^2 = 48 - 16 = 32$. When $t = 2$, $h = 48 \cdot 2 - 16 \cdot 2^2 = 96 - 64 = 32$. Both numbers check.

State. The rocket will be exactly 32 ft above the ground at 1 sec and at 2 sec after it is launched.

39. *Writing Exercise*. No; if we cannot factor the quadratic expression $ax^2 + bx + c$, $a \neq 0$, then we cannot solve the quadratic equation $ax^2 + bx + c = 0$.

41. $\dfrac{-3}{5} \cdot \dfrac{4}{7} = -\dfrac{3 \cdot 4}{5 \cdot 7} = \dfrac{-12}{35}$

43. $\dfrac{-5}{6} - \dfrac{1}{6} = \dfrac{-5}{6} + \dfrac{-1}{6} = -1$

45. $\dfrac{-3}{8} \cdot \left(\dfrac{-10}{15}\right) = \dfrac{\not{3} \cdot \not{2} \cdot \not{5}}{\not{2} \cdot 4 \cdot \not{3} \cdot \not{5}} = \dfrac{1}{4}$

47. $\dfrac{5}{24} + \dfrac{3}{28} = \dfrac{5}{24} \cdot \dfrac{7}{7} + \dfrac{3}{28} \cdot \dfrac{6}{6}$
$$= \dfrac{35}{168} + \dfrac{18}{168} = \dfrac{53}{168}$$

49. *Writing Exercise.* She could use the measuring sticks to draw a right angle as shown below. Then she could use the 3-ft and 4-ft sticks to extend one leg to 7 ft and the 4-ft and 5-ft sticks to extend the other leg to 9 ft.

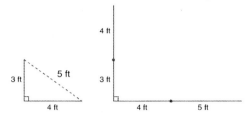

Next she could draw another right angle with either the 7-ft side or the 9-ft side as a side.

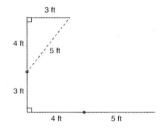

Then she could use the sticks to extend the other side to the appropriate length. Finally she would draw the remaining side of the rectangle.

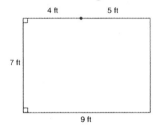

51. **Familiarize.** First we find the length of the other leg of the right triangle. Then we find the area of the triangle, and finally we multiply by the cost per square foot of the sailcloth. Let $x =$ the length of the other leg of the right triangle, in feet.

Translate. We use the Pythagorean theorem to find x.

$$a^2 + b^2 = c^2$$
$$x^2 + 24^2 = 26^2 \quad \text{Substituting}$$

Carry out.

$$x^2 + 24^2 = 26^2$$
$$x^2 + 576 = 676$$
$$x^2 - 100 = 0$$
$$(x + 10)(x - 10) = 0$$
$$x + 10 = 0 \quad or \quad x - 10 = 0$$
$$x = -10 \quad or \quad x = 10$$

Since the length of the leg must be positive, -10 cannot be a solution. We use the number 10. Find the area of the triangle:

$$\frac{1}{2}bh = \frac{1}{2} \cdot 10 \text{ ft} \cdot 24 \text{ ft} = 120 \text{ ft}^2$$

Finally, we multiply the area, 120 ft^2, by the price per square foot of the sailcloth, \$1.50:

$$120 \cdot (1.50) = 180$$

Check. Recheck the calculations. The answer checks.

State. A new main sail costs \$180.

53. **Familiarize.** We add labels to the drawing in the text.

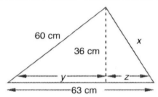

First we will use the Pythagorean theorem to find y. Then we will subtract to find z and, finally, we will use the Pythagorean theorem again to find x.

Translate. We use the Pythagorean theorem to find y.

$$a^2 + b^2 = c^2$$
$$y^2 + 36^2 = 60^2 \quad \text{Substituting}$$

Carry out.

$$y^2 + 36^2 = 60^2$$
$$y^2 + 1296 = 3600$$
$$y^2 - 2304 = 0$$
$$(y + 48)(y - 48) = 0$$
$$y + 48 = 0 \quad or \quad y - 48 = 0$$
$$y = -48 \quad or \quad y = 48$$

Since the length y cannot be negative, we use 48 cm. Then $z = 63 - 48 = 15$ cm.

Now we find x. We use the Pythagorean theorem again.

$$15^2 + 36^2 = x^2$$
$$225 + 1296 = x^2$$
$$1521 = x^2$$
$$0 = x^2 - 1521$$
$$0 = (x + 39)(x - 39)$$
$$x + 39 = 0 \quad or \quad x - 39 = 0$$
$$x = -39 \quad or \quad x = 39$$

Since the length x cannot be negative, we use 39 cm.

Check. We repeat all of the calculations. The answer checks.

State. The value of x is 39 cm.

55. **Familiarize.** Let $w =$ the width of the side turned up. Then $20 - 2w =$ the length, in inches of the base.

Recall that we multiply these dimensions to find the area of the rectangle.

Translate.

The area of the rectangular cross-section is 48 in^2.

$$w(20 - 2w) = 48$$

Carry out. We solve the equation.

$$w(20 - 2w) = 48$$
$$0 = 2w^2 - 20w + 48$$
$$0 = 2(w^2 - 10w + 24)$$
$$0 = 2(w - 6)(w - 4)$$

$$w - 6 = 0 \quad or \quad w - 4 = 0$$
$$w = 6 \quad or \quad w = 4$$

Check. If $w = 6$ in., $20 - 2(6) = 8$ in. and the area is

$$6 \text{ in.} \cdot 8 \text{ in.} = 48 \text{ in.}^2$$

If $w = 4$ in., $20 - 2(4) = 12$ in. and the area is

$$4 \text{ in.} \cdot 12 \text{ in.} = 48 \text{ in.}^2$$

State. The possible depths of the gutter are 4 in. or 6 in.

57. *Familiarize*. First we can use the Pythagorean theorem to find x, in ft. Then the height of the telephone pole is $x + 5$.

Translate. We use the Pythagorean theorem.

$$a^2 + b^2 = c^2$$
$$\left(\frac{1}{2}x + 1\right)^2 + x^2 = 34^2$$

Carry out. We solve the equation.

$$\left(\frac{1}{2}x + 1\right)^2 + x^2 = 34^2$$
$$\frac{1}{4}x^2 + x + 1 + x^2 = 1156$$
$$x^2 + 4x + 4 + 4x^2 = 4624 \quad \text{Multiplying by 4}$$
$$5x^2 + 4x + 4 = 4624$$
$$5x^2 + 4x - 4620 = 0$$
$$(5x + 154)(x - 30) = 0$$

$$5x + 154 = 0 \qquad or \quad x - 30 = 0$$
$$5x = -154 \quad or \qquad x = 30$$
$$x = -30.8 \quad or \qquad x = 30$$

Check. Since the length x must be positive, -30.8 cannot be a solution. If x is 30 ft, then $\frac{1}{2}x + 1$ is $\frac{1}{2} \cdot 30 + 1$, or 16 ft. Since $16^2 + 30^2 = 1156 = 34^2$, the number 30 checks. When x is 30 ft, then $x + 5$ is 35 ft.

State. The height of the telephone pole is 35 ft.

59. First substitute 18 for N in the given formula.

$$18 = -0.009t(t - 12)^3$$

Graph $y_1 = 18$ and $y_2 = -0.009x(x - 12)^3$ in the given window and use the TRACE feature to find the first coordinates of the points of intersection of the graphs. We find $x \approx 2$ hr and $x \approx 4.2$ hr.

61. Graph $y = -0.009x(x - 12)^3$ and use the TRACE feature to find the first coordinate of the highest point on the graph. We find $x = 3$ hr.

Chapter 5 Review

1. False. The largest common variable factor is the <u>smallest</u> power of the variable in the polynomial.

3. True. see p. 323

5. False. Some quadratic equations have two different solutions.

7. True. see p 346

9. $20x^3 = (4 \cdot 5)\left(x \cdot x^2\right) = (-2x)\left(-10x^2\right) = (20x)\left(x^2\right)$

11. $12x^4 - 18x^3 = 6x^3 \cdot 2x - 6x^3 \cdot 3 = 6x^3\left(2x - 3\right)$

13. $100t^2 - 1 = (10t + 1)(10t - 1)$

15. $x^2 + 14x + 49 = (x + 7)(x + 7) = (x + 7)^2$

17. $6x^3 + 9x^2 + 2x + 3$
$= 3x^2 \cdot 2x + 3x^2 \cdot 3 + 2x \cdot 1 + 3 \cdot 1$
$= 3x^2(2x + 3) + 1(2x + 3)$
$= (2x + 3)\left(3x^2 + 1\right)$

19. $25t^2 + 9 - 30t = 25t^2 - 30t + 9$
$= (5t - 3)(5t - 3) = (5t - 3)^2$

21. $81a^4 - 1 = \left(9a^2 + 1\right)\left(9a^2 - 1\right)$
$= \left(9a^2 + 1\right)(3a + 1)(3a - 1)$

23. $3x^2 - 27 = 3\left(x^2 - 9\right)$
$= 3(x + 3)(x - 3)$

25. $a^2b^4 - 64 = \left(ab^2 + 8\right)\left(ab^2 - 8\right)$

27. $75 + 12x^2 - 60x = 12x^2 - 60x + 75$
$= 3\left(4x^2 - 20x + 25\right)$
$= 3(2x - 5)(2x - 5) = 3(2x - 5)^2$

29. $-t^3 + t^2 + 42t = -t\left(t^2 - t - 42\right)$
$= -t(t - 7)(t + 6)$

31. $n^2 - 60 - 4n = n^2 - 4n - 60$
$= (n - 10)(n + 6)$

33. $4t^2 + 13t + 10 = (4t + 5)(t + 2)$

35. $7x^3 + 35x^2 + 28x = 7x\left(x^2 + 5x + 4\right)$
$= 7x(x + 1)(x + 4)$

37. $20x^2 - 20x + 5 = 5\left(4x^2 - 4x + 1\right)$
$= 5(2x - 1)(2x - 1)$
$= 5(2x - 1)^2$

39. $15 - 8x + x^2 = x^2 - 8x + 15$
$= (x - 3)(x - 5)$

41. $x^2y^2 + 6xy - 16 = (xy + 8)(xy - 2)$

43. $m^2 + 5m + mt + 5t = m(m + 5) + t(m + 5)$
$= (m + 5)(m + t)$

45. $6m^2 + 2mn + n^2 + 3mn = 2m(3m + n) + n(n + 3m)$
$= (3m + n)(2m + n)$

47. $(x - 9)(x + 11) = 0$

$x - 9 = 0 \quad or \quad x + 11 = 0$

$x = 9 \quad or \qquad x = -11$

49. $16x^2 = 9$

$16x^2 - 9 = 0$

$(4x + 3)(4x - 3) = 0$

$4x + 3 = 0 \quad or \quad 4x - 3 = 0$

$x = \dfrac{-3}{4} \quad or \qquad x = \dfrac{3}{4}$

51. $2x^2 - 7x = 30$

$2x^2 - 7x - 30 = 0$

$(2x + 5)(x - 6) = 0$

$2x + 5 = 0 \quad or \quad x - 6 = 0$

$x = \dfrac{-5}{2} \quad or \qquad x = 6$

53. $9t - 15t^2 = 0$

$3t(3 - 5t) = 0$

$3t = 0 \quad or \quad 3 - 5t = 0$

$t = 0 \quad or \qquad t = \dfrac{3}{5}$

55. *Familiarize* Let n = the number.

Translate.

The number squared is 12 more than the number

$$n^2 \qquad = 12 \quad + \qquad n$$

Carry out. We solve the equation.

$n^2 = 12 + n$

$n^2 - n - 12 = 0$

$(n - 4)(n + 3) = 0$

$n - 4 = 0 \quad or \quad n + 3 = 0$

$n = 4 \quad or \qquad n = -3$

Check.

$4^2 = 12 + 4$

$16 = 16 \qquad\qquad$ True

$(-3)^2 = 12 + (-3)$

$9 = 9 \qquad\qquad$ True

State. The solutions are 4 and -3.

57. We let $y = 0$ and solve for x

$0 = 2x^2 - 3x - 5$

$0 = (2x - 5)(x + 1)$

$2x - 5 = 0 \quad or \quad x + 1 = 0$

$x = \dfrac{5}{2} \quad or \qquad x = -1$

The x-intercepts are $\left(\dfrac{5}{2},\ 0\right)$ and $(-1, 0)$

59. *Familiarize*. Let d = the diagonal of the brace.

Translate. We use the Pythagorean theorem.

$a^2 + b^2 = c^2$

$8^2 + 6^2 = d^2$

Carry out. We solve the equation.

$8^2 + 6^2 = c^2$

$8^2 + 6^2 = d^2$

$100 = d^2$

$0 = d^2 - 100$

$0 = (d + 10)(d - 10)$

$= 0$

$d + 10 = 0 \quad or \quad d - 10 = 0$

$d = -10 \quad or \qquad d = 10$

Check. Since the diagonal must be positive, -10 cannot be a solution.

$8^2 + 6^2 = 10^2$

$100 = 100 \qquad$ True

State. The diagonal is 10 holes long.

61. *Writing Exercise.* The equations solved in this chapter have an x^2-term (are quadratic), whereas those solved previously have no x^2-term (are linear). The principle of zero products is used to solve quadratic equations and is not used to solve linear equations.

63. *Familiarize* Let n = the number.

Translate.

The cube of the number is twice the square of the number

$$n^3 \qquad = \qquad 2 \quad n^2$$

Carry out. We solve the equation.

$n^3 = 2n^2$

$n^3 - 2n^2 = 0$

$n^2(n - 2) = 0$

$n^2 = 0 \quad or \quad n - 2 = 0$

$n = 0 \quad or \qquad n = 2$

Check.

$n^3 = 2n^2$

$0^3 = 2(0)^2$

$0 = 0 \qquad\qquad$ True

$n^3 = 2n^2$

$2^3 = 2(2)^2$

$8 = 8 \qquad\qquad$ True

State. The number is 0 or 2.

65. *Familiarize*. Let s = the side of a square, in cm, $s + 5$ = side of the new square. Recall $A = x^2$.

Translate. The new square is $2\frac{1}{2}$ times the area of the original square

$(s + 5)^2 = 2\frac{1}{4}(s)^2$

Carry out. We solve the equation.

$$4s^2 + 40s + 100 = 9s^2$$
$$0 = 5s^2 - 40s - 100$$
$$0 = 5(s^2 - 8s - 20)$$
$$0 = 5(s - 10)(s + 2)$$

$$s - 10 = 0 \quad or \quad s + 2 = 0$$
$$s = 10 \quad or \quad s = -2$$

Check. Since the side of the square must be positive, -2 cannot be a solution.

$$(s + 5)^2 = 2\frac{1}{4}s^2$$
$$(10 + 5)^2 = \frac{9}{4}(10)^2$$
$$225 = 225 \qquad \text{True}$$

State. The original square has side 10 cm and area of 100 cm^2. The new square has side 15 cm and area of 225 cm^2.

67. $x^2 + 25 = 0$

No real solution, the sum of two squares cannot be factored.

Chapter 5 Test

1. Answers may vary.
$(3x^2)(4x^2), \ (-2x)(-6x^3), \ (12x^3)(x)$

3. $x^2 + 25 - 10x = x^2 - 10x + 25$
$$= (x - 5)(x - 5)$$
$$= (x - 5)^2$$

5. $x^3 + x^2 + 2x + 2 = x^2(x + 1) + 2(x + 1)$
$$= (x + 1)(x^2 + 2)$$

7. $a^3 + 3a^2 - 4a = a(a^2 + 3a - 4)$
$$= a(a + 4)(a - 1)$$

9. $4t^2 - 25 = (2t + 5)(2t - 5)$

11. $-6m^3 - 9m^2 - 3m = -3m(2m^2 + 3m + 1)$
$$= -3m(2m + 1)(m + 1)$$

13. $45r^2 + 60r + 20 = 5(9r^2 + 12r + 4)$
$$= 5(3r + 2)(3r + 2)$$
$$= 5(3r + 2)^2$$

15. $49t^2 + 36 + 84t = 49t^2 + 84t + 36$
$$= (7t + 6)(7t + 6)$$
$$= (7t + 6)^2$$

17. $x^2 + 3x + 6$ is prime.

19. $6t^3 + 9t^2 - 15t = 3t(2t^2 + 3t - 5)$
$$= 3t(2t + 5)(t - 1)$$

21. $x^2 - 6x + 5 = 0$
$$(x - 5)(x - 1) = 0$$
$$x - 5 = 0 \quad or \quad x - 1 = 0$$
$$x = 5 \quad or \quad x = 1$$

23.
$$4t - 10t^2 = 0$$
$$2t(2 - 5t) = 0$$
$$2t = 0 \quad or \quad 2 - 5t = 0$$
$$t = 0 \quad or \quad -5t = -2$$
$$t = 0 \quad or \quad t = \frac{2}{5}$$

The solutions are 0 and $\frac{2}{5}$.

25.
$$x(x - 1) = 20$$
$$x^2 - x = 20$$
$$x^2 - x - 20 = 0$$
$$(x - 5)(x + 4) = 0$$
$$x - 5 = 0 \quad or \quad x + 4 = 0$$
$$x = 5 \quad or \quad x = -4$$

The solutions are -4 and 5.

27. ***Familiarize***. We make a drawing, Let $w =$ the width, in m, Then $w + 6 =$ the length in m.

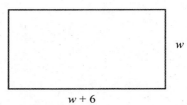

Recall the area of a rectangle is length times width.

Translate. We reword the problem

Length times width is 40 m^2
$(w + 6)$ · w = 40

Carry out. We solve the equation.
$$(w + 6)w = 40$$
$$w^2 + 6w = 40$$
$$w^2 + 6w - 40 = 0$$
$$(w + 10)(w - 4) = 0$$
$$w + 10 = 0 \quad or \quad w - 4 = 0$$
$$w = -10 \quad or \quad w = 4$$

Check. Since the width must be positive, -10 cannot be a solution. If the width is 4m, the length is $4 + 6$ or 10 m, and the area is $10 \cdot 4$, or 40 m^2. Thus 4 checks.
State. The width is 4 m and the length is 10 m.

29. ***Familiarize***. From the given drawing x is the distance in feet we are looking for.
Translate. We use the Pythagorean Theorem. $3^2 + 4^2 = x^2$
Carry out. We solve the equation:
$$3^2 + 4^2 = x^2$$
$$9 + 16 = x^2$$
$$25 = x^2$$
$$x = 5 \quad or \quad -5$$

Check. The number -5 is not a solution because distance cannot be negative. If $x = 5$, $3^2 + 4^2 = 9 + 16 = 5^2$, so the answer checks.

State. The distance is 5 ft.

31.
$$\begin{aligned}
(a+3)^2 - 2(a+3) - 35 &= [(a+3) - 7]\,[(a+3) + 5] \\
&= [a + 3 - 7][a + 3 + 5] \\
&= (a - 4)(a + 8)
\end{aligned}$$

Chapter 6

Rational Expressions and Equations

1. $x - 2 = 0$ when $x = 2$ and $x + 3 = 0$ when $x = -3$, so choice (e) is correct.

3. $a^2 - a - 12 = (a - 4)(a + 3)$; $a - 4 = 0$ when $a = 4$ and $a + 3 = 0$ when $a = -3$, so choice (d) is correct.

5. $2t - 1 = 0$ when $t = \dfrac{1}{2}$ and $3t + 4 = 0$ when $t = -\dfrac{4}{3}$, so choice (c) is correct.

7. $\dfrac{18}{-11x}$

We find the real number(s) that make the denominator 0. To do so we set the denominator equal to 0 and solve for x:
$$-11x = 0$$
$$x = 0$$

The expression is undefined for $x = 0$.

9. $\dfrac{y - 3}{y + 5}$

Set the denominator equal to 0 and solve for y:
$$y + 5 = 0$$
$$y = -5$$

The expression is undefined for $y = -5$.

11. $\dfrac{t - 5}{3t - 15}$

Set the denominator equal to 0 and solve for t:
$$3t - 15 = 0$$
$$3t = 15$$
$$t = 5$$

The expression is undefined for $t = 5$.

13. $\dfrac{x^2 - 25}{x^2 - 3x - 28}$

Set the denominator equal to 0 and solve for x:
$$x^2 - 3x - 28 = 0$$
$$(x - 7)(x + 4) = 0$$
$$x - 7 = 0 \quad or \quad x + 4 = 0$$
$$x = 7 \quad or \quad x = -4$$

The expression is undefined for $x = 7$ and $x = -4$.

15. $\dfrac{t^2 + t - 20}{2t^2 + 11t - 6}$

Set the denominator equal to 0 and solve for t:
$$2t^2 + 11t - 6 = 0$$
$$(2t - 1)(t + 6) = 0$$
$$2t - 1 = 0 \quad or \quad t + 6 = 0$$
$$2t = 1 \quad or \quad t = -6$$
$$t = \frac{1}{2} \quad or \quad t = -6$$

The expression is undefined for $t = \dfrac{1}{2}$ and $t = -6$.

17. $\dfrac{50a^2 b}{40ab^3}$

$= \dfrac{5a \cdot 10ab}{4b^2 \cdot 10ab}$ Factoring the numerator and denominator. Note the common factor of $10ab$.

$= \dfrac{5a}{4b^2} \cdot \dfrac{10ab}{10ab}$ Rewriting as a product of two rational expressions

$= \dfrac{5a}{4b^2} \cdot 1$ $\dfrac{10ab}{10ab} = 1$

$= \dfrac{5a}{4b^2}$ Removing the factor 1

19. $\dfrac{6t + 12}{6t - 18} = \dfrac{\cancel{6}(t + 2)}{\cancel{6}(t - 3)} = \dfrac{(t + 2)}{(t - 3)}$

21. $\dfrac{21t - 7}{24t - 8} = \dfrac{7(\cancel{3t - 1})}{8(\cancel{3t - 1})} = \dfrac{7}{8}$

23. $\dfrac{a^2 - 9}{a^2 + 4a + 3} = \dfrac{(a + 3)(a - 3)}{(a + 3)(a + 1)}$

$= \dfrac{a + 3}{a + 3} \cdot \dfrac{a - 3}{a + 1}$

$= 1 \cdot \dfrac{a - 3}{a + 1}$

$= \dfrac{a - 3}{a + 1}$

25. $\dfrac{-36x^8}{54x^5} = \dfrac{-2x^3 \cdot 18x^5}{3 \cdot 18x^5}$

$= \dfrac{-2x^3}{3} \cdot \dfrac{18x^5}{18x^5}$

$= \dfrac{-2x^3}{3}$

Check: Let $x = 1$.
$$\dfrac{-36x^8}{54x^5} = \dfrac{-36 \cdot 1^8}{54 \cdot 1^5} = \dfrac{-36}{54} = \dfrac{-2}{3}$$
$$\dfrac{-2x^3}{3} = \dfrac{-2 \cdot 1^3}{3} = \dfrac{-2}{3}$$

The answer is probably correct.

27. $\dfrac{-2y+6}{-8y} = \dfrac{-2(y-3)}{-2\cdot 4y}$

$\qquad\qquad = \dfrac{-2}{-2}\cdot\dfrac{y-3}{4y}$

$\qquad\qquad = 1\cdot\dfrac{y-3}{4y}$

$\qquad\qquad = \dfrac{y-3}{4y}$

Check: Let $x = 2$.

$\dfrac{-2y+6}{-8y} = \dfrac{-2\cdot 2+6}{-8\cdot 2} = \dfrac{2}{-16} = -\dfrac{1}{8}$

$\dfrac{y-3}{4y} = \dfrac{2-3}{4\cdot 2} = \dfrac{-1}{8} = -\dfrac{1}{8}$

The answer is probably correct.

29. $\dfrac{6a^2+3a}{7a^2+7a} = \dfrac{3\cancel{a}(2a+1)}{7\cancel{a}(a+1)}$

$\qquad\qquad = \dfrac{3(2a+1)}{7(a+1)}$

Check: Let $a = 1$.

$\dfrac{6a^2+3a}{7a^2+7a} = \dfrac{6\cdot 1^2+3\cdot 1}{7\cdot 1^2+7\cdot 1} = \dfrac{6+3}{7+7} = \dfrac{9}{14}$

$\dfrac{3(2a+1)}{7(a+1)} = \dfrac{3(2\cdot 1+1)}{7(1+1)} = \dfrac{3\cdot 3}{7\cdot 2} = \dfrac{9}{14}$

The answer is probably correct.

31. $\dfrac{t^2-16}{t^2-t-20} = \dfrac{(t-4)(t+4)}{(t-5)(t+4)}$

$\qquad\qquad = \dfrac{t-4}{t-5}\cdot\dfrac{t+4}{t+4}$

$\qquad\qquad = \dfrac{t-4}{t-5}\cdot 1$

$\qquad\qquad = \dfrac{t-4}{t-5}$

Check: Let $t = 1$.

$\dfrac{t^2-16}{t^2-t-20} = \dfrac{1^2-16}{1^2-1-20} = \dfrac{-15}{-20} = \dfrac{3}{4}$

$\dfrac{t-4}{t-5} = \dfrac{1-4}{1-5} = \dfrac{3}{4}$

The answer is probably correct.

33. $\dfrac{3a^2+9a-12}{6a^2-30a+24} = \dfrac{3(a^2+3a-4)}{6(a^2-5a+4)}$

$\qquad\qquad = \dfrac{3(a+4)(a-1)}{3\cdot 2(a-4)(a-1)}$

$\qquad\qquad = \dfrac{3(a-1)}{3(a-1)}\cdot\dfrac{a+4}{2(a-4)}$

$\qquad\qquad = 1\cdot\dfrac{a+4}{2(a-4)}$

$\qquad\qquad = \dfrac{a+4}{2(a-4)}$

Check: Let $a = 2$.

$\dfrac{3a^2+9a-12}{6a^2-30a+24} = \dfrac{3\cdot 2^2+9\cdot 2-12}{6\cdot 2^2-30\cdot 2+24} = \dfrac{18}{-12} = -\dfrac{3}{2}$

$\dfrac{a+4}{2(a-4)} = \dfrac{2+4}{2(2-4)} = \dfrac{6}{-4} = -\dfrac{3}{2}$

The answer is probably correct.

35. $\dfrac{x^2-8x+16}{x^2-16} = \dfrac{(x-4)(x-4)}{(x+4)(x-4)}$

$\qquad\qquad = \dfrac{x-4}{x+4}\cdot\dfrac{x-4}{x-4}$

$\qquad\qquad = \dfrac{x-4}{x+4}\cdot 1$

$\qquad\qquad = \dfrac{x-4}{x+4}$

Check: Let $x = 1$.

$\dfrac{x^2-8x+16}{x^2-16} = \dfrac{1^2-8\cdot 1+16}{1^2-16} = \dfrac{1-8+16}{1-16} = \dfrac{9}{-15} = -\dfrac{3}{5}$

$\dfrac{x-4}{x+4} = \dfrac{1-4}{1+4} = -\dfrac{3}{5}$

The answer is probably correct.

37. $\dfrac{t^2-1}{t+1} = \dfrac{(t+1)(t-1)}{t+1}$

$\qquad\qquad = \dfrac{t+1}{t+1}\cdot\dfrac{t-1}{1}$

$\qquad\qquad = 1\cdot\dfrac{t-1}{1}$

$\qquad\qquad = t-1$

Check: Let $t = 2$.

$\dfrac{t^2-1}{t+1} = \dfrac{2^2-1}{2+1} = \dfrac{3}{3} = 1$

$t-1 = 2-1 = 1$

The answer is probably correct.

39. $\dfrac{y^2+4}{y+2}$ cannot be simplified.

Neither the numerator nor the denominator can be factored.

41. $\dfrac{5x^2+20}{10x^2+40} = \dfrac{5(x^2+4)}{10(x^2+4)}$

$\qquad\qquad = \dfrac{1\cdot\cancel{5}\cdot\cancel{(x^2+4)}}{2\cdot\cancel{5}\cdot\cancel{(x^2+4)}}$

$\qquad\qquad = \dfrac{1}{2}$

Check: Let $x = 1$.

$\dfrac{5x^2+20}{10x^2+40} = \dfrac{5\cdot 1^2+20}{10\cdot 1^2+40} = \dfrac{25}{50} = \dfrac{1}{2}$

$\dfrac{1}{2} = \dfrac{1}{2}$

The answer is probably correct.

43. $\dfrac{y^2+6y}{2y^2+13y+6} = \dfrac{y(y+6)}{(2y+1)(y+6)}$

$\qquad\qquad = \dfrac{y}{2y+1}\cdot\dfrac{y+6}{y+6}$

$\qquad\qquad = \dfrac{y}{2y+1}\cdot 1$

$\qquad\qquad = \dfrac{y}{2y+1}$

Check: Let $y = 1$.

$\dfrac{y^2+6y}{2y^2+13y+6} = \dfrac{1^2+6\cdot 1}{2\cdot 1^2+13\cdot 1+6} = \dfrac{7}{21} = \dfrac{1}{3}$

$\dfrac{y}{2y+1} = \dfrac{1}{2\cdot 1+1} = \dfrac{1}{3}$

The answer is probably correct.

45. $\dfrac{4x^2 - 12x + 9}{10x^2 - 11x - 6} = \dfrac{(2x-3)(2x-3)}{(2x-3)(5x+2)}$

$\qquad = \dfrac{2x-3}{2x-3} \cdot \dfrac{2x-3}{5x+2}$

$\qquad = 1 \cdot \dfrac{2x-3}{5x+2}$

$\qquad = \dfrac{2x-3}{5x+2}$

Check: Let $t = 1$.

$\dfrac{4x^2 - 12x + 9}{10x^2 - 11x - 6} = \dfrac{4 \cdot 1^2 - 12 \cdot 1 + 9}{10 \cdot 1^2 - 11 \cdot 1 - 6} = \dfrac{1}{-7} = -\dfrac{1}{7}$

$\dfrac{2x-3}{5x+2} = \dfrac{2 \cdot 1 - 3}{5 \cdot 1 + 2} = \dfrac{-1}{7} = -\dfrac{1}{7}$

The answer is probably correct.

47. $\dfrac{10-x}{x-10} = \dfrac{-(x-10)}{x-10}$

$\qquad = \dfrac{-1}{1} \cdot \dfrac{x-10}{x-10} = -1 \cdot 1 = -1$

Check: Let $x = 1$.

$\dfrac{10-x}{x-10} = \dfrac{10-1}{1-10} = \dfrac{9}{-9} = -1$

The answer is probably correct.

49. $\dfrac{7t-14}{2-t} = \dfrac{7(t-2)}{-(t-2)}$

$\qquad = \dfrac{7}{-1} \cdot \dfrac{t-2}{t-2}$

$\qquad = \dfrac{7}{-1} \cdot 1$

$\qquad = -7$

Check: Let $t = 1$.

$\dfrac{7t-14}{2-t} = \dfrac{7 \cdot 1 - 14}{2-1} = \dfrac{-7}{1} = -7$

The answer is probably correct.

51. $\dfrac{a-b}{4b-4a} = \dfrac{a-b}{-4(a-b)}$

$\qquad = \dfrac{1}{-4} \cdot \dfrac{a-b}{a-b}$

$\qquad = -\dfrac{1}{4} \cdot 1$

$\qquad = -\dfrac{1}{4}$

Check: Let $a = 2$ and $b = 1$.

$\dfrac{a-b}{4b-4a} = \dfrac{2-1}{4 \cdot 1 - 4 \cdot 2} = \dfrac{1}{4-8} = -\dfrac{1}{-4}$

The answer is probably correct.

53. $\dfrac{3x^2 - 3y^2}{2y^2 - 2x^2} = \dfrac{3(x^2 - y^2)}{2(y^2 - x^2)}$

$\qquad = \dfrac{3(x^2 - y^2)}{2(-1)(x^2 - y^2)}$

$\qquad = \dfrac{3}{2(-1)} \cdot \dfrac{x^2 - y^2}{x^2 - y^2}$

$\qquad = \dfrac{3}{2(-1)} \cdot 1$

$\qquad = -\dfrac{3}{2}$

Check: Let $x = 1$ and $y = 2$.

$\dfrac{3x^2 - 3y^2}{2y^2 - 2x^2} = \dfrac{3 \cdot 1^2 - 3 \cdot 2^2}{2 \cdot 2^2 - 2 \cdot 1^2} = \dfrac{-9}{6} = -\dfrac{3}{2}$

The answer is probably correct.

55. $\dfrac{7s^2 - 28t^2}{28t^2 - 7s^2}$

Note that the numerator and denominator are opposites. Thus, we have an expression divided by its opposite, so the result is -1.

57. *Writing Exercise.* Simplifying removes a factor equal to 1, allowing us to rewrite an expression $a \cdot 1$ as a.

59. $-\dfrac{2}{15} \cdot \dfrac{10}{7} = -\dfrac{2 \cdot 10}{15 \cdot 7}$

$\qquad = -\dfrac{2 \cdot 2 \cdot \cancel{5}}{3 \cdot \cancel{5} \cdot 7}$

$\qquad = -\dfrac{4}{21}$

61. $\dfrac{5}{8} \div \left(-\dfrac{1}{6}\right) = \dfrac{5}{8} \cdot (-6)$

$\qquad = -\dfrac{5 \cdot 6}{8}$

$\qquad = -\dfrac{5 \cdot \cancel{2} \cdot 3}{\cancel{2} \cdot 4}$

$\qquad = -\dfrac{15}{4}$

63. $\dfrac{7}{9} - \dfrac{2}{3} \cdot \dfrac{6}{7} = \dfrac{7}{9} - \dfrac{4}{7} = \dfrac{7}{9} \cdot \dfrac{7}{7} - \dfrac{4}{7} \cdot \dfrac{9}{9}$

$\qquad = \dfrac{49}{63} - \dfrac{36}{63} = \dfrac{13}{63}$

65. *Writing Exercise.* Although a rational expression has been simplified incorrectly, it is possible that there are one or more values of the variable(s) for which the two expressions are the same. For example, $\dfrac{x^2 + x - 2}{x^2 + 3x + 2}$ could be simplified incorrectly as $\dfrac{x-1}{x+2}$, but evaluating the expressions for $x = 1$ gives 0 in each case. $\left(\text{The correct simplification is } \dfrac{x-1}{x+1}.\right)$

67. $\dfrac{16y^4 - x^4}{(x^2 + 4y^2)(x - 2y)}$

$\qquad = \dfrac{(4y^2 + x^2)(4y^2 - x^2)}{(x^2 + 4y^2)(x - 2y)}$

$\qquad = \dfrac{(4y^2 + x^2)(2y + x)(2y - x)}{(x^2 + 4y^2)(x - 2y)}$

$\qquad = \dfrac{(x^2 + 4y^2)(2y + x)(-1)(x - 2y)}{(x^2 + 4y^2)(x - 2y)}$

$\qquad = \dfrac{(x^2 + 4y^2)(x - 2y)}{(x^2 + 4y^2)(x - 2y)} \cdot \dfrac{(2y + x)(-1)}{1}$

$\qquad = -2y - x, \text{ or } -x - 2y, \text{ or } -(2y + x)$

69. $\dfrac{x^5 - 2x^3 + 4x^2 - 8}{x^7 + 2x^4 - 4x^3 - 8}$

$= \dfrac{x^3(x^2 - 2) + 4(x^2 - 2)}{x^4(x^3 + 2) - 4(x^3 + 2)}$

$= \dfrac{(x^2 - 2)(x^3 + 4)}{(x^3 + 2)(x^4 - 4)}$

$= \dfrac{(x^2 - 2)(x^3 + 4)}{(x^3 + 2)(x^2 + 2)(x^2 - 2)}$

$= \dfrac{\cancel{(x^2 - 2)}(x^3 + 4)}{(x^3 + 2)(x^2 + 2)\cancel{(x^2 - 2)}}$

$= \dfrac{x^3 + 4}{(x^3 + 2)(x^2 + 2)}$

71. $\dfrac{(t^4 - 1)(t^2 - 9)(t - 9)^2}{(t^4 - 81)(t^2 + 1)(t + 1)^2}$

$= \dfrac{(t^2 + 1)(t + 1)(t - 1)(t + 3)(t - 3)(t - 9)(t - 9)}{(t^2 + 9)(t + 3)(t - 3)(t^2 + 1)(t + 1)(t + 1)}$

$= \dfrac{\cancel{(t^2 + 1)}\cancel{(t + 1)}(t - 1)\cancel{(t + 3)}\cancel{(t - 3)}(t - 9)(t - 9)}{(t^2 + 9)\cancel{(t + 3)}\cancel{(t - 3)}\cancel{(t^2 + 1)}\cancel{(t + 1)}(t + 1)}$

$= \dfrac{(t - 1)(t - 9)(t - 9)}{(t^2 + 9)(t + 1)}$, or $\dfrac{(t - 1)(t - 9)^2}{(t^2 + 9)(t + 1)}$

73. $\dfrac{(x^2 - y^2)(x^2 - 2xy + y^2)}{(x + y)^2(x^2 - 4xy - 5y^2)}$

$= \dfrac{(x + y)(x - y)(x - y)(x - y)}{(x + y)(x + y)(x - 5y)(x + y)}$

$= \dfrac{\cancel{(x + y)}(x - y)(x - y)(x - y)}{\cancel{(x + y)}(x + y)(x - 5y)(x + y)}$

$= \dfrac{(x - y)^3}{(x + y)^2(x - 5y)}$

75. *Writing Exercise.*

$\dfrac{5(2x + 5) - 25}{10} = \dfrac{10x + 25 - 25}{10}$

$= \dfrac{10x}{10}$

$= x$

You get the same number you selected.

A person asked to select a number and then perform these operations would probably be surprised that the result is the original number.

Exercise Set 6.2

1. $\dfrac{3x}{8} \cdot \dfrac{x + 2}{5x - 1} = \dfrac{3x(x + 2)}{8(5x - 1)}$

3. $\dfrac{a - 4}{a + 6} \cdot \dfrac{a + 2}{a + 6} = \dfrac{(a - 4)(a + 2)}{(a + 6)(a + 6)}$, or $\dfrac{(a - 4)(a + 2)}{(a + 6)^2}$

5. $\dfrac{2x + 3}{4} \cdot \dfrac{x + 1}{x - 5} = \dfrac{(2x + 3)(x + 1)}{4(x - 5)}$

7. $\dfrac{n - 4}{n^2 + 4} \cdot \dfrac{n + 4}{n^2 - 4} = \dfrac{(n - 4)(n + 4)}{(n^2 + 4)(n^2 - 4)}$

9. $\dfrac{y + 6}{1 + y} \cdot \dfrac{y - 3}{y + 3} = \dfrac{(y + 6)(y - 3)}{(1 + y)(y + 3)}$

11. $\dfrac{8t^3}{5t} \cdot \dfrac{3}{4t}$

$= \dfrac{8t^3 \cdot 3}{5t \cdot 4t}$ Multiplying the numerators and the denominators

$= \dfrac{2 \cdot 4 \cdot t \cdot t \cdot t \cdot 3}{5 \cdot t \cdot 4 \cdot t}$ Factoring the numerator and the denominator

$= \dfrac{2 \cdot \cancel{4} \cdot t \cdot \cancel{t} \cdot \cancel{t} \cdot 3}{5 \cdot \cancel{t} \cdot \cancel{4} \cdot \cancel{t}}$ Removing a factor equal to 1

$= \dfrac{6t}{5}$ Simplifying

13. $\dfrac{3c}{d^2} \cdot \dfrac{8d}{6c^3}$

$= \dfrac{3c \cdot 8d}{d^2 \cdot 6c^3}$ Multiplying the numerators and the denominators

$= \dfrac{3 \cdot c \cdot 2 \cdot 4 \cdot d}{d \cdot d \cdot 3 \cdot 2 \cdot c \cdot c \cdot c}$ Factoring the numerator and the denominator

$= \dfrac{\cancel{3} \cdot \cancel{c} \cdot \cancel{2} \cdot 4 \cdot \cancel{d}}{\cancel{d} \cdot d \cdot \cancel{3} \cdot \cancel{2} \cdot \cancel{c} \cdot c \cdot c}$

$= \dfrac{4}{dc^2}$

15. $\dfrac{x^2 - 3x - 10}{(x - 2)^2} \cdot (x - 2) = \dfrac{(x^2 - 3x - 10)(x - 2)}{(x - 2)^2}$

$= \dfrac{(x - 5)(x + 2)(x - 2)}{(x - 2)(x - 2)}$

$= \dfrac{(x - 5)(x + 2)\cancel{(x - 2)}}{(x - 2)\cancel{(x - 2)}}$

$= \dfrac{(x - 5)(x + 2)}{x - 2}$

17. $\dfrac{n^2 - 6n + 5}{n + 6} \cdot \dfrac{n - 6}{n^2 + 36} = \dfrac{(n^2 - 6n + 5)(n - 6)}{(n + 6)(n^2 + 36)}$

$= \dfrac{(n - 5)(n - 1)(n - 6)}{(n + 6)(n^2 + 36)}$

(No simplification is possible.)

19. $\dfrac{a^2 - 9}{a^2} \cdot \dfrac{7a}{a^2 + a - 12} = \dfrac{(a + 3)(a - 3) \cdot 7 \cdot a}{a \cdot a(a + 4)(a - 3)}$

$= \dfrac{(a + 3)\cancel{(a - 3)} \cdot 7 \cdot \cancel{a}}{\cancel{a} \cdot a(a + 4)\cancel{(a - 3)}}$

$= \dfrac{7(a + 3)}{a(a + 4)}$

21. $\dfrac{4v - 8}{5v} \cdot \dfrac{15v^2}{4v^2 - 16v + 16}$

$= \dfrac{(4v - 8)15v^2}{5v(4v^2 - 16v + 16)}$

$= \dfrac{\cancel{4}(v - 2) \cdot \cancel{5} \cdot 3 \cdot v \cdot \cancel{v}}{\cancel{5}\cancel{v} \cdot \cancel{4}\cancel{(v - 2)}(v - 2)}$

$= \dfrac{3v}{v - 2}$

23. $\dfrac{t^2 + 2t - 3}{t^2 + 4t - 5} \cdot \dfrac{t^2 - 3t - 10}{t^2 + 5t + 6}$

$= \dfrac{(t^2 + 2t - 3)(t^2 - 3t - 10)}{(t^2 + 4t - 5)(t^2 + 5t + 6)}$

$= \dfrac{(t + 3)(t - 1)(t - 5)(t + 2)}{(t + 5)(t - 1)(t + 3)(t + 2)}$

$= \dfrac{\cancel{(t+3)}\cancel{(t-1)}(t - 5)\cancel{(t+2)}}{(t + 5)\cancel{(t-1)}\cancel{(t+3)}\cancel{(t+2)}}$

$= \dfrac{t - 5}{t + 5}$

25. $\dfrac{12y + 12}{5y + 25} \cdot \dfrac{3y^2 - 75}{8y^2 - 8}$

$= \dfrac{(12y + 12)(3y^2 - 75)}{(5y + 25)(8y^2 - 8)}$

$= \dfrac{3 \cdot \cancel{4}(y+1)\ 3(y+5)(y - 5)}{5(y+5)\ 2 \cdot \cancel{4}(y+1)(y - 1)}$

$= \dfrac{9(y - 5)}{10(y - 1)}$

27. $\dfrac{x^2 + 4x + 4}{(x - 1)^2} \cdot \dfrac{x^2 - 2x + 1}{(x + 2)^2} = \dfrac{(x + 2)^2(x - 1)^2}{(x - 1)^2(x + 2)^2} = 1$

29. $\dfrac{t^2 - 4t + 4}{t^2 - 7t + 6} \cdot \dfrac{2t^2 + 7t - 15}{t^2 - 10t + 25}$

$= \dfrac{(t^2 - 4t + 4)(2t^2 + 7t - 15)}{(2t^2 - 7t + 6)(t^2 - 10t + 25)}$

$= \dfrac{(t - 2)(t - 2)(2t - 3)(t + 5)}{(2t - 3)(t - 2)(t - 5)(t - 5)}$

$= \dfrac{\cancel{(t-2)}(t - 2)\cancel{(2t-3)}(t + 5)}{\cancel{(2t-3)}\cancel{(t-2)}(t - 5)(t - 5)}$

$= \dfrac{(t - 2)(t + 5)}{(t - 5)^2}$

31. $(10x^2 - x - 2) \cdot \dfrac{4x^2 - 8x + 3}{10x^2 - 11x - 6}$

$= \dfrac{(10x^2 - x - 2)(4x^2 - 8x + 3)}{(10x^2 - 11x - 6)}$

$= \dfrac{\cancel{(5x+2)}(2x - 1)(2x - 1)\cancel{(2x-3)}}{\cancel{(5x+2)}\cancel{(2x-3)}}$

$= (2x - 1)^2$

33. The reciprocal of $\dfrac{2x}{9}$ is $\dfrac{9}{2x}$ because $\dfrac{2x}{9} \cdot \dfrac{9}{2x} = 1$.

35. The reciprocal of $a^4 + 3a$ is $\dfrac{1}{a^4 + 3a}$ because

$\dfrac{a^4 + 3a}{1} \cdot \dfrac{1}{a^4 + 3a} = 1.$

37. The reciprocal of $\dfrac{x^2 + 2x - 5}{x^2 - 4x + 7}$ is $\dfrac{x^2 - 4x + 7}{x^2 + 2x - 5}$ because

$\dfrac{x^2 + 2x - 5}{x^2 - 4x + 7} \cdot \dfrac{x^2 - 4x + 7}{x^2 + 2x - 5} = 1.$

39. $\dfrac{5}{9} \div \dfrac{3}{4}$

$= \dfrac{5}{9} \cdot \dfrac{4}{3}$ Multiplying by the reciprocal of the divisor

$= \dfrac{5 \cdot 4}{9 \cdot 3}$

$= \dfrac{20}{27}$

No simplification is possible.

41. $\dfrac{x}{4} \div \dfrac{5}{x}$

$= \dfrac{x}{4} \cdot \dfrac{x}{5}$ Multiplying by the reciprocal of the divisor

$= \dfrac{x \cdot x}{4 \cdot 5}$

$= \dfrac{x^2}{20}$

43. $\dfrac{a^5}{b^4} \div \dfrac{a^2}{b} = \dfrac{a^5}{b^4} \cdot \dfrac{b}{a^2}$

$= \dfrac{a^5 \cdot b}{b^4 \cdot a^2}$

$= \dfrac{a^2 \cdot a^3 \cdot b}{b \cdot b^3 \cdot a^2}$

$= \dfrac{a^2 b}{a^2 b} \cdot \dfrac{a^3}{b^3}$

$= \dfrac{a^3}{b^3}$

45. $\dfrac{t - 3}{6} \div \dfrac{t + 1}{8} = \dfrac{t - 3}{6} \cdot \dfrac{8}{t + 1}$

$= \dfrac{(t - 3)(8)}{6 \cdot (t + 1)}$

$= \dfrac{(t - 3) \cdot 4 \cdot \cancel{2}}{\cancel{2} \cdot 3(t + 1)}$

$= \dfrac{4(t - 3)}{3(t + 1)}$

47. $\dfrac{4y - 8}{y + 2} \div \dfrac{y - 2}{y^2 - 4} = \dfrac{4y - 8}{y + 2} \cdot \dfrac{y^2 - 4}{y - 2}$

$= \dfrac{(4y - 8)(y^2 - 4)}{(y + 2)(y - 2)}$

$= \dfrac{4\cancel{(y-2)}\cancel{(y+2)}(y - 2)}{\cancel{(y+2)}\cancel{(y-2)}(1)}$

$= 4(y - 2)$

49. $\dfrac{a}{a - b} \div \dfrac{b}{b - a} = \dfrac{a}{a - b} \cdot \dfrac{b - a}{b}$

$= \dfrac{a(b - a)}{(a - b)(b)}$

$= \dfrac{a(-1)\cancel{(a-b)}}{\cancel{(a-b)}(b)}$

$= \dfrac{-a}{b} = -\dfrac{a}{b}$

51.
$$(n^2 + 5n + 6) \div \frac{n^2 - 4}{n + 3} = \frac{(n^2 + 5n + 6)}{1} \cdot \frac{(n + 3)}{n^2 - 4}$$
$$= \frac{(n^2 + 5n + 6)(n + 3)}{n^2 - 4}$$
$$= \frac{(n + 3)(n + 2)(n + 3)}{(n + 2)(n - 2)}$$
$$= \frac{(n + 3)^2}{n - 2}$$

53.
$$\frac{-3 + 3x}{16} \div \frac{x - 1}{5} = \frac{3x - 3}{16} \cdot \frac{5}{x - 1}$$
$$= \frac{(3x - 3) \cdot 5}{16(x - 1)}$$
$$= \frac{3(x - 1) \cdot 5}{16(x - 1)}$$
$$= \frac{3(x - 1) \cdot 5}{16(x - 1)}$$
$$= \frac{15}{16}$$

55.
$$\frac{x - 1}{x + 2} \div \frac{1 - x}{4 + x^2} = \frac{x - 1}{x + 2} \cdot \frac{4 + x^2}{1 - x}$$
$$= \frac{(x - 1)(4 + x^2)}{(x + 2)(1 - x)}$$
$$= \frac{(x - 1)(x^2 + 4)}{-1(x + 2)(x - 1)}$$
$$= -\frac{x^2 + 4}{x + 2} \text{ or } \frac{-x^2 - 4}{x + 2}$$

57.
$$\frac{a + 2}{a - 1} \div \frac{3a + 6}{a - 5} = \frac{a + 2}{a - 1} \cdot \frac{a - 5}{3a + 6}$$
$$= \frac{(a + 2)(a - 5)}{(a - 1)(3a + 6)}$$
$$= \frac{(a + 2)(a - 5)}{(a - 1) \cdot 3 \cdot (a + 2)}$$
$$= \frac{(a + 2)(a - 5)}{(a - 1) \cdot 3 \cdot (a + 2)}$$
$$= \frac{a - 5}{3(a - 1)}$$

59.
$$(2x - 1) \div \frac{2x^2 - 11x + 5}{4x^2 - 1}$$
$$= \frac{2x - 1}{1} \cdot \frac{4x^2 - 1}{2x^2 - 11x + 5}$$
$$= \frac{(2x - 1)(4x^2 - 1)}{1 \cdot (2x^2 - 11x + 5)}$$
$$= \frac{(2x - 1)(2x + 1)(2x - 1)}{1 \cdot (2x - 1)(x - 5)}$$
$$= \frac{(2x - 1)(2x + 1)(2x - 1)}{1 \cdot (2x - 1)(x - 5)}$$
$$= \frac{(2x - 1)(2x + 1)}{x - 5}$$

61.
$$\frac{w^2 - 14w + 49}{2w^2 - 3w - 14} \div \frac{3w^2 - 20w - 7}{w^2 - 6w - 16}$$
$$= \frac{w^2 - 14w + 49}{2w^2 - 3w - 14} \cdot \frac{w^2 - 6w - 16}{3w^2 - 20w - 7}$$
$$= \frac{(w^2 - 14w + 49)(w^2 - 6w - 16)}{(2w^2 - 3w - 14)(3w^2 - 20w - 7)}$$
$$= \frac{(w - 7)(w - 7)(w - 8)(w + 2)}{(2w - 7)(w + 2)(3w + 1)(w - 7)}$$
$$= \frac{(w - 7)(w - 8)}{(2w - 7)(3w + 1)}$$

63.
$$\frac{c^2 + 10c + 21}{c^2 - 2c - 15} \div (5c^2 + 32c - 21)$$
$$= \frac{c^2 + 10c + 21}{c^2 - 2c - 15} \cdot \frac{1}{5c^2 + 32c - 21}$$
$$= \frac{(c^2 + 10c + 21) \cdot 1}{(c^2 - 2c - 15)(5c^2 + 32c - 21)}$$
$$= \frac{(c + 7)(c + 3)}{(c - 5)(c + 3)(5c - 3)(c + 7)}$$
$$= \frac{(c + 7)(c + 3)}{(c + 7)(c + 3)} \cdot \frac{1}{(c - 5)(5c - 3)}$$
$$= \frac{1}{(c - 5)(5c - 3)}$$

65.
$$\frac{x - y}{x^2 + 2xy + y^2} \div \frac{x^2 - y^2}{x^2 - 5xy + 4y^2}$$
$$= \frac{x - y}{x^2 + 2xy + y^2} \cdot \frac{x^2 - 5xy + 4y^2}{x^2 - y^2}$$
$$= \frac{(x - y)(x - y)(x - 4y)}{(x + y)(x + y)(x + y)(x - y)}$$
$$= \frac{(x - y)(x - 4y)}{(x + y)^3}$$

67. *Writing Exercise.* Parentheses are required to ensure that numerators and denominators are multiplied correctly. That is, the product of $(x + 2)$ and $(3x - 1)$ and the product of $(5x - 7)$ and $(x + 4)$ in the denominator.

69.
$$\frac{3}{4} + \frac{5}{6} = \frac{3}{4} \cdot \frac{3}{3} + \frac{5}{6} \cdot \frac{2}{2}$$
$$= \frac{9}{12} + \frac{10}{12}$$
$$= \frac{19}{12}$$

71.
$$\frac{2}{9} - \frac{1}{6} = \frac{2}{9} \cdot \frac{2}{2} - \frac{1}{6} \cdot \frac{3}{3}$$
$$= \frac{4}{18} - \frac{3}{18}$$
$$= \frac{1}{18}$$

73. $2x^2 - x + 1 - (x^2 - x - 2) = 2x^2 - x + 1 - x^2 + x + 2 = x^2 + 3$

75. *Writing Exercise.* Yes; consider the product $\frac{a}{b} \cdot \frac{c}{d} = \frac{ac}{bd}$. The reciprocal of the product is $\frac{bd}{ac}$. This is equal to the product of the reciprocals of the two original factors: $\frac{b}{a} \cdot \frac{d}{c} = \frac{bd}{ac}$.

77. The reciprocal of $2\frac{1}{3}x$ is $\dfrac{1}{2\frac{1}{3}x} = \dfrac{1}{\frac{7x}{3}} = 1 \div \dfrac{7x}{3} = 1 \cdot \dfrac{3}{7x} = \dfrac{3}{7x}$

79. $(x - 2a) \div \dfrac{a^2x^2 - 4a^4}{a^2x + 2a^3} = \dfrac{x - 2a}{1} \cdot \dfrac{a^2x + 2a^3}{a^2x^2 - 4a^4}$

$\qquad = \dfrac{(x - 2a)(a^2x + 2a^3)}{(a^2x^2 - 4a^4)}$

$\qquad = \dfrac{(x - 2a)a^2(x + 2a)}{a^2(x - 2a)(x + 2a)} = 1$

81. $\dfrac{3x^2 - 2xy - y^2}{x^2 - y^2} \div (3x^2 + 4xy + y^2)^2$

$\qquad = \dfrac{3x^2 - 2xy - y^2}{x^2 - y^2} \cdot \dfrac{1}{(3x^2 + 4xy + y^2)^2}$

$\qquad = \dfrac{(3x + y)(x - y) \cdot 1}{(x + y)(x - y)(3x + y)(3x + y)(x + y)(x + y)}$

$\qquad = \dfrac{1}{(x + y)^3(3x + y)}$

83. $\dfrac{a^2 - 3b}{a^2 + 2b} \cdot \dfrac{a^2 - 2b}{a^2 + 3b} \cdot \dfrac{a^2 + 2b}{a^2 - 3b}$

Note that $\dfrac{a^2 - 3b}{a^2 + 2b} \cdot \dfrac{a^2 + 2b}{a^2 - 3b}$ is the product of reciprocals and thus is equal to 1. Then the product in the original exercise is the remaining factor, $\dfrac{a^2 - 2b}{a^2 + 3b}$.

85. $\dfrac{z^2 - 8z + 16}{z^2 + 8z + 16} \div \dfrac{(z-4)^5}{(z+4)^5} \div \dfrac{3z+12}{z^2-16}$

$\qquad = \dfrac{(z - 4)^2}{(z + 4)^2} \cdot \dfrac{(z + 4)^5}{(z - 4)^5} \cdot \dfrac{(z + 4)(z - 4)}{3(z + 4)}$

$\qquad = \dfrac{(z - 4)^2(z + 4)^2(z + 4)^3(z + 4)(z - 4)}{(z + 4)^2(z - 4)^2(z - 4)(z - 4)^2(3)(z + 4)}$

$\qquad = \dfrac{(z + 4)^3}{3(z - 4)^2}$

87. $\dfrac{3x + 3y + 3}{9x} \div \dfrac{x^2 + 2xy + y^2 - 1}{x^4 + x^2}$

$\qquad = \dfrac{3x + 3y + 3}{9x} \cdot \dfrac{x^4 + x^2}{x^2 + 2xy + y^2 - 1}$

$\qquad = \dfrac{3(x + y + 1)(x^2)(x^2 + 1)}{9x[(x + y) + 1][(x + y) - 1]}$

$\qquad = \dfrac{3(x + y + 1)(x)(x)(x^2 + 1)}{3 \cdot 3 \cdot x(x + y + 1)(x + y - 1)}$

$\qquad = \dfrac{3x(x + y + 1)}{3x(x + y + 1)} \cdot \dfrac{x(x^2 + 1)}{3(x + y - 1)}$

$\qquad = \dfrac{x(x^2 + 1)}{3(x + y - 1)}$

89. $\dfrac{a^4 - 81b^4}{a^2c - 6abc + 9b^2c} \cdot \dfrac{a + 3b}{a^2 + 9b^2} \div \dfrac{a^2 + 6ab + 9b^2}{(a - 3b)^2}$

$\qquad = \dfrac{(a^2 + 9b^2)(a + 3b)(a - 3b)}{c(a - 3b)^2} \cdot \dfrac{a + 3b}{a^2 + 9b^2} \cdot \dfrac{(a - 3b)^2}{(a + 3b)^2}$

$\qquad = \dfrac{a - 3b}{c}$

91. Enter $y_1 = \dfrac{x - 1}{x^2 + 2x + 1} \div \dfrac{x^2 - 1}{x^2 - 5x + 4}$ and $y_2 = \dfrac{x^2 - 5x + 4}{(x + 1)^3}$, display the values of y_1 and y_2 in a table, and compare the values. (See the Technology Connection on page 373 in the text.)

Exercise Set 6.3

1. To add two rational expressions when the denominators are the same, add <u>numerators</u> and keep the common <u>denominator</u>. (See page 377 in the text.)

3. The least common multiple of two denominators is usually referred to as the <u>least common denominator</u> and is abbreviated <u>LCD</u>. (See page 378 in the text.)

5. $\dfrac{3}{t} + \dfrac{5}{t} = \dfrac{8}{t}$ Adding numerators

7. $\dfrac{x}{12} + \dfrac{2x + 5}{12} = \dfrac{3x + 5}{12}$ Adding numerators

9. $\dfrac{4}{a + 3} + \dfrac{5}{a + 3} = \dfrac{9}{a + 3}$

11. $\dfrac{11}{4x - 7} - \dfrac{3}{4x - 7} = \dfrac{8}{4x - 7}$ Subtracting numerators

13. $\dfrac{3y + 8}{2y} - \dfrac{y + 1}{2y}$

$\qquad = \dfrac{3y + 8 - (y + 1)}{2y}$

$\qquad = \dfrac{3y + 8 - y - 1}{2y}$ Removing parentheses

$\qquad = \dfrac{2y + 7}{2y}$

15. $\dfrac{5x + 7}{x + 3} + \dfrac{x + 11}{x + 3}$

$\qquad = \dfrac{6x + 18}{x + 3}$ Adding numerators

$\qquad = \dfrac{6(x + 3)}{x + 3}$ Factoring

$\qquad = \dfrac{6(x + 3)}{x + 3}$ Removing a factor equal to 1

$\qquad = 6$

17. $\dfrac{5x + 7}{x + 3} - \dfrac{x + 11}{x + 3} = \dfrac{5x + 7 - (x + 11)}{x + 3}$

$\qquad = \dfrac{5x + 7 - x - 11}{x + 3}$

$\qquad = \dfrac{4x - 4}{x + 3}$

$\qquad = \dfrac{4(x - 1)}{x + 3}$

19. $\dfrac{a^2}{a - 4} + \dfrac{a - 20}{a - 4} = \dfrac{a^2 + a - 20}{a - 4}$

$\qquad = \dfrac{(a + 5)(a - 4)}{a - 4}$

$\qquad = \dfrac{(a + 5)(a - 4)}{a - 4}$

$\qquad = a + 5$

21. $\dfrac{y^2}{y+2} - \dfrac{5y+14}{y+2} = \dfrac{y^2 - (5y+14)}{y+2}$

$\qquad\qquad\qquad = \dfrac{y^2 - 5y - 14}{y+2}$

$\qquad\qquad\qquad = \dfrac{(y-7)(y+2)}{y+2}$

$\qquad\qquad\qquad = \dfrac{(y-7)\cancel{(y+2)}}{\cancel{y+2}}$

$\qquad\qquad\qquad = y - 7$

23. $\dfrac{t^2 - 5t}{t-1} + \dfrac{5t - t^2}{t-1}$

Note that the numerators are opposites, so their sum is 0.

Then we have $\dfrac{0}{t-1}$, or 0.

25. $\dfrac{x-6}{x^2+5x+6} + \dfrac{9}{x^2+5x+6} = \dfrac{x+3}{x^2+5x+6}$

$\qquad\qquad\qquad\qquad = \dfrac{x+3}{(x+3)(x+2)}$

$\qquad\qquad\qquad\qquad = \dfrac{\cancel{x+3}}{\cancel{(x+3)}(x+2)}$

$\qquad\qquad\qquad\qquad = \dfrac{1}{x+2}$

27. $\dfrac{t^2 - 5t}{t^2+6t+9} + \dfrac{4t - 12}{t^2+6t+9} = \dfrac{t^2 - t - 12}{t^2+6t+9}$

$\qquad\qquad\qquad\qquad = \dfrac{(t-4)(t+3)}{(t+3)^2}$

$\qquad\qquad\qquad\qquad = \dfrac{(t-4)\cancel{(t+3)}}{(t+3)\cancel{(t+3)}}$

$\qquad\qquad\qquad\qquad = \dfrac{t-4}{t+3}$

29. $\dfrac{2y^2 + 3y}{y^2 - 7y + 12} - \dfrac{y^2 + 4y + 6}{y^2 - 7y + 12}$

$\quad = \dfrac{2y^2 + 3y - (y^2 + 4y + 6)}{y^2 - 7y + 12}$

$\quad = \dfrac{2y^2 + 3y - y^2 - 4y - 6}{y^2 - 7y + 12}$

$\quad = \dfrac{y^2 - y - 6}{y^2 - 7y + 12}$

$\quad = \dfrac{(y-3)(y+2)}{(y-3)(y-4)}$

$\quad = \dfrac{\cancel{(y-3)}(y+2)}{\cancel{(y-3)}(y-4)}$

$\quad = \dfrac{y+2}{y-4}$

31. $\dfrac{3-2x}{x^2-6x+8} + \dfrac{7-3x}{x^2-6x+8}$

$\quad = \dfrac{10-5x}{x^2-6x+8}$

$\quad = \dfrac{5(2-x)}{(x-4)(x-2)}$

$\quad = \dfrac{5(-1)(x-2)}{(x-4)(x-2)}$

$\quad = \dfrac{5(-1)\cancel{(x-2)}}{(x-4)\cancel{(x-2)}}$

$\quad = \dfrac{-5}{x-4}, \text{ or } -\dfrac{5}{x-4}, \text{ or } \dfrac{5}{4-x}$

33. $\dfrac{x-9}{x^2+3x-4} - \dfrac{2x-5}{x^2+3x-4}$

$\quad = \dfrac{x-9-(2x-5)}{x^2+3x-4}$

$\quad = \dfrac{x-9-2x+5}{x^2+3x-4}$

$\quad = \dfrac{-x-4}{x^2+3x-4}$

$\quad = \dfrac{-(x+4)}{(x+4)(x-1)}$

$\quad = \dfrac{-1\cancel{(x+4)}}{\cancel{(x+4)}(x-1)}$

$\quad = \dfrac{-1}{x-1}, \text{ or } -\dfrac{1}{x-1}, \text{ or } \dfrac{1}{1-x}$

35. $15 = 3 \cdot 5$

$36 = 2 \cdot 2 \cdot 3 \cdot 3$

$\text{LCM} = 2 \cdot 2 \cdot 3 \cdot 3 \cdot 5 = 180$

37. $8 = 2 \cdot 2 \cdot 2$

$9 = 3 \cdot 3$

$\text{LCM} = 2 \cdot 2 \cdot 2 \cdot 3 \cdot 3, \text{ or } 72$

39. $6 = 2 \cdot 3$

$12 = 2 \cdot 2 \cdot 3$

$15 = 3 \cdot 5$

$\text{LCM} = 2 \cdot 2 \cdot 3 \cdot 5 = 60$

41. $18t^2 = 2 \cdot 3 \cdot 3 \cdot t \cdot t$

$6t^5 = 2 \cdot 3 \cdot t \cdot t \cdot t \cdot t \cdot t$

$\text{LCM} = 2 \cdot 3 \cdot 3 \cdot t \cdot t \cdot t \cdot t \cdot t = 18t^5$

43. $15a^4b^7 = 3 \cdot 5 \cdot a \cdot a \cdot a \cdot a \cdot b \cdot b \cdot b \cdot b \cdot b \cdot b \cdot b$

$10a^2b^8 = 2 \cdot 5 \cdot a \cdot a \cdot b \cdot b \cdot b \cdot b \cdot b \cdot b \cdot b \cdot b$

$\text{LCM} = 2 \cdot 3 \cdot 5 \cdot a \cdot a \cdot a \cdot a \cdot b \cdot b \cdot b \cdot b \cdot b \cdot b \cdot b \cdot b,$

$\qquad\qquad \text{or } 30a^4b^8$

45. $2(y-3) = 2 \cdot (y-3)$

$6(y-3) = 2 \cdot 3 \cdot (y-3)$

$\text{LCM} = 2 \cdot 3 \cdot (y-3), \text{ or } 6(y-3)$

47. $x^2 - 2x - 15 = (x-5)(x+3)$

$x^2 - 9 = (x-3)(x+3)$

$\text{LCM} = (x-5)(x-3)(x+3)$

49. $t^3 + 4t^2 + 4t = t(t^2 + 4t + 4) = t(t + 2)(t + 2)$

$t^2 - 4t = t(t - 4)$

$\text{LCM} = t(t + 2)(t + 2)(t - 4) = t(t + 2)^2(t - 4)$

51. $6xz^2 = 2 \cdot 3 \cdot x \cdot z \cdot z$

$8x^2y = 2 \cdot 2 \cdot 2 \cdot x \cdot x \cdot y$

$15y^3z = 3 \cdot 5 \cdot y \cdot y \cdot y \cdot z$

$\text{LCM} = 2 \cdot 2 \cdot 2 \cdot 3 \cdot 5 \cdot x \cdot x \cdot y \cdot y \cdot y \cdot z \cdot z = 120x^2y^3z^2$

53. $a + 1 = a + 1$

$(a - 1)^2 = (a - 1)(a - 1)$

$a^2 - 1 = (a + 1)(a - 1)$

$\text{LCM} = (a + 1)(a - 1)(a - 1) = (a + 1)(a - 1)^2$

55. $2n^2 + n - 1 = (2n - 1)(n + 1)$

$2n^2 + 3n - 2 = (2n - 1)(n + 2)$

$\text{LCM} = (2n - 1)(n + 1)(n + 2)$

56. $m^2 - 2m - 3 = (m - 3)(m + 1)$

$2m^2 + 3m + 1 = (2m + 1)(m + 1)$

$\text{LCM} = (2m + 1)(m + 1)(m - 3)$

57. $6x^3 - 24x^2 + 18x = 6x(x^2 - 4x + 3)$

$= 2 \cdot 3 \cdot x(x - 1)(x - 3)$

$4x^5 - 24x^4 + 20x^3 = 4x^3(x^2 - 6x + 5)$

$= 2 \cdot 2 \cdot x \cdot x \cdot x(x - 1)(x - 5)$

$\text{LCM} = 2 \cdot 2 \cdot 3 \cdot x \cdot x \cdot x(x - 1)(x - 3)(x - 5)$

$= 12x^3(x - 1)(x - 3)(x - 5)$

59. $6t^4 = 2 \cdot 3 \cdot t \cdot t \cdot t \cdot t$

$18t^2 = 2 \cdot 3 \cdot 3 \cdot t \cdot t$

The LCD is $2 \cdot 3 \cdot 3 \cdot t \cdot t \cdot t \cdot t$, or $18t^4$.

$\dfrac{5}{6t^4} \cdot \dfrac{3}{3} = \dfrac{15}{18t^4}$ and

$\dfrac{s}{18t^2} \cdot \dfrac{t^2}{t^2} = \dfrac{st^2}{18t^4}$

61. $3x^4y^2 = 3 \cdot x \cdot x \cdot x \cdot x \cdot y \cdot y$

$9xy^3 = 3 \cdot 3 \cdot x \cdot y \cdot y \cdot y$

The LCD is $3 \cdot 3 \cdot x \cdot x \cdot x \cdot x \cdot y \cdot y \cdot y$, or $9x^4y^3$.

$\dfrac{7}{3x^4y^2} \cdot \dfrac{3y}{3y} = \dfrac{21y}{9x^4y^3}$ and

$\dfrac{4}{9xy^3} \cdot \dfrac{x^3}{x^3} = \dfrac{4x^3}{9x^4y^3}$

63. The LCD is $(x + 2)(x - 2)(x + 3)$. (See Exercise 47.)

$\dfrac{2x}{x^2 - 4} = \dfrac{2x}{(x + 2)(x - 2)} \cdot \dfrac{x + 3}{x + 3}$

$= \dfrac{2x(x + 3)}{(x + 2)(x - 2)(x + 3)}$

$\dfrac{4x}{x^2 + 5x + 6} = \dfrac{4x}{(x + 3)(x + 2)} \cdot \dfrac{x - 2}{x - 2}$

$= \dfrac{4x(x - 2)}{(x + 3)(x + 2)(x - 2)}$

65. *Writing Exercise.* If the numbers have a common factor, their product contains that factor more than the greatest number of times it occurs in any one factorization. In this case, their product is not their least common multiple.

67. $-\dfrac{5}{8} = \dfrac{-5}{8} = \dfrac{5}{-8}$

69. $-(x - y) = -x + y$, or $y - x$

71. $-1(2x - 7) = -2x + 7$ or $7 - 2x$

73. *Writing Exercise.* The polynomials contain no common factors other than constants.

75. $\dfrac{6x - 1}{x - 1} + \dfrac{3(2x + 5)}{x - 1} + \dfrac{3(2x - 3)}{x - 1}$

$= \dfrac{6x - 1 + 6x + 15 + 6x - 9}{x - 1}$

$= \dfrac{18x + 5}{x - 1}$

77. $\dfrac{x^2}{3x^2 - 5x - 2} - \dfrac{2x}{3x + 1} \cdot \dfrac{1}{x - 2}$

$= \dfrac{x^2}{(3x + 1)(x - 2)} - \dfrac{2x}{(3x + 1)(x - 2)}$

$= \dfrac{x^2 - 2x}{(3x + 1)(x - 2)}$

$= \dfrac{x(x - 2)}{(3x + 1)(x - 2)}$

$= \dfrac{x}{3x + 1}$

79. The smallest number of strands that can be used is the LCM of 10 and 3.

$10 = 2 \cdot 5$

$3 = 3$

$\text{LCM} = 2 \cdot 5 \cdot 3 = 30$

The smallest number of strands that can be used is 30.

81. If the number of strands must also be a multiple of 4, we find the smallest multiple of 30 that is also a multiple of 4.

$1 \cdot 30 = 30$, not a multiple of 4

$2 \cdot 30 = 60 = 15 \cdot 4$, a multiple of 4

The smallest number of strands that can be used is 60.

83. $4x^2 - 25 = (2x + 5)(2x - 5)$

$6x^2 - 7x - 20 = (3x + 4)(2x - 5)$

$(9x^2 + 24x + 16)^2 = [(3x + 4)(3x + 4)]^2$

$= (3x + 4)(3x + 4)(3x + 4)(3x + 4)$

$\text{LCM} = (2x + 5)(2x - 5)(3x + 4)^4$

85. The first copier prints 20 pages per minute, which is $\dfrac{20}{60}$ or $\dfrac{1}{3}$ copy per second. The second copier prints 18 pages per minutes, which is $\dfrac{18}{60}$ or $\dfrac{3}{10}$ copy per second. The time it takes until the machines begin copying a page at exactly the same time again is the LCM of their copying rates.

$3 = 3$

$10 = 2 \cdot 5$

$\text{LCM} = 2 \cdot 3 \cdot 5 = 30$

It takes 30 seconds

87. The number of minutes after 5:00 A.M. when the shuttles will first leave at the same time again is the LCM of their departure intervals, 25 minutes and 35 minutes.

$25 = 5 \cdot 5$

$35 = 5 \cdot 7$

$\text{LCM} = 5 \cdot 5 \cdot 7, \text{ or } 175$

Thus, the shuttles will leave at the same time 175 minutes after 5:00 A.M., or at 7:55 A.M.

89. *Writing Exercise.* Evaluate both expressions for some value of the variable for which both are defined. If the results are the same, we can conclude that the answer is probably correct.

Exercise Set 6.4

1. To add or subtract when denominators are different, first find the <u>LCD</u>.

3. Add or subtract the <u>numerators</u>, as indicated. Write the sum or difference over the <u>LCD</u>.

5. $\dfrac{3}{x^2} + \dfrac{5}{x} = \dfrac{3}{x \cdot x} + \dfrac{5}{x}$ $\text{LCD} = x \cdot x, \text{ or } x^2$

$\qquad = \dfrac{3}{x \cdot x} + \dfrac{5}{x} \cdot \dfrac{x}{x}$

$\qquad = \dfrac{3 + 5x}{x^2}$

7. $\left.\begin{array}{l} 6r = 2 \cdot 3 \cdot r \\ 8r = 2 \cdot 2 \cdot 2 \cdot r \end{array}\right\} \text{LCD} = 2 \cdot 2 \cdot 2 \cdot 3 \cdot r, \text{ or } 24r$

$\dfrac{1}{6r} - \dfrac{3}{8r} = \dfrac{1}{6r} \cdot \dfrac{4}{4} - \dfrac{3}{8r} \cdot \dfrac{3}{3}$

$\qquad = \dfrac{4 - 9}{24r}$

$\qquad = \dfrac{-5}{24r}, \text{ or } -\dfrac{5}{24r}$

9. $\left.\begin{array}{l} uv^2 = u \cdot v \cdot v \\ u^3 v = u \cdot u \cdot u \cdot v \end{array}\right\} \text{LCD} = u \cdot u \cdot u \cdot v \cdot v, \text{ or } u^3 v^2$

$\dfrac{3}{uv^2} + \dfrac{4}{u^3 v} = \dfrac{3}{uv^2} \cdot \dfrac{u^2}{u^2} + \dfrac{4}{u^3 v} \cdot \dfrac{v}{v} = \dfrac{3u^2 + 4v}{u^3 v^2}$

11. $\left.\begin{array}{l} 3xy^2 = 3 \cdot x \cdot y \cdot y \\ x^2 y^3 = x \cdot x \cdot y \cdot y \cdot y \end{array}\right\} \begin{array}{l} \text{LCD} = 3 \cdot x \cdot x \cdot y \cdot y \cdot y, \text{ or} \\ 3x^2 y^3 \end{array}$

$\dfrac{-2}{3xy^2} - \dfrac{6}{x^2 y^3} = \dfrac{-2}{3xy^2} \cdot \dfrac{xy}{xy} - \dfrac{6}{x^2 y^3} \cdot \dfrac{3}{3} = \dfrac{-2xy - 18}{3x^2 y^3}$

13. $\left.\begin{array}{l} 8 = 2 \cdot 2 \cdot 2 \\ 6 = 2 \cdot 3 \end{array}\right\} \text{LCD} = 2 \cdot 2 \cdot 2 \cdot 3, \text{ or } 24$

$\dfrac{x + 3}{8} + \dfrac{x - 2}{6} = \dfrac{x + 3}{8} \cdot \dfrac{3}{3} + \dfrac{x - 2}{6} \cdot \dfrac{4}{4}$

$\qquad = \dfrac{3(x + 3) + 4(x - 2)}{24}$

$\qquad = \dfrac{3x + 9 + 4x - 8}{24}$

$\qquad = \dfrac{7x + 1}{24}$

15. $\left.\begin{array}{l} 6 = 2 \cdot 3 \\ 3 = 3 \end{array}\right\} \text{LCD} = 2 \cdot 3, \text{ or } 6$

$\dfrac{x - 2}{6} - \dfrac{x + 1}{3} = \dfrac{x - 2}{6} - \dfrac{x + 1}{3} \cdot \dfrac{2}{2}$

$\qquad = \dfrac{x - 2}{6} - \dfrac{2x + 2}{6}$

$\qquad = \dfrac{x - 2 - (2x + 2)}{6}$

$\qquad = \dfrac{x - 2 - 2x - 2}{6}$

$\qquad = \dfrac{-x - 4}{6}, \text{ or } \dfrac{-(x + 4)}{6}$

17. $\left.\begin{array}{l} 15a = 3 \cdot 5 \cdot a \\ 3a^2 = 3 \cdot a \cdot a \end{array}\right\} \text{LCD} = 5 \cdot 3a \cdot a, \text{ or } 15a^2$

$\dfrac{a + 3}{15a} + \dfrac{2a - 1}{3a^2} = \dfrac{a + 3}{15a} \cdot \dfrac{a}{a} + \dfrac{2a - 1}{3a^2} \cdot \dfrac{5}{5}$

$\qquad = \dfrac{a^2 + 3a + 10a - 5}{15a^2}$

$\qquad = \dfrac{a^2 + 13a - 5}{15a^2}$

19. $\left.\begin{array}{l} 3z = 3 \cdot z \\ 4z = 2 \cdot 2 \cdot z \end{array}\right\} \text{LCD} = 2 \cdot 2 \cdot 3 \cdot z, \text{ or } 12z$

$\dfrac{4z - 9}{3z} - \dfrac{3z - 8}{4z} = \dfrac{4z - 9}{3z} \cdot \dfrac{4}{4} - \dfrac{3z - 8}{4z} \cdot \dfrac{3}{3}$

$\qquad = \dfrac{16z - 36}{12z} - \dfrac{9z - 24}{12z}$

$\qquad = \dfrac{16z - 36 - (9z - 24)}{12z}$

$\qquad = \dfrac{16z - 36 - 9z + 24}{12z}$

$\qquad = \dfrac{7z - 12}{12z}$

21. $\left.\begin{array}{l} cd^2 = c \cdot d \cdot d \\ c^2 d = c \cdot c \cdot d \end{array}\right\} \text{LCD} = c \cdot c \cdot d \cdot d, \text{ or } c^2 d^2$

$\dfrac{3c + d}{cd^2} + \dfrac{c - d}{c^2 d} = \dfrac{3c + d}{cd^2} \cdot \dfrac{c}{c} + \dfrac{c - d}{c^2 d} \cdot \dfrac{d}{d}$

$\qquad = \dfrac{c(3c + d) + d(c - d)}{c^2 d^2}$

$\qquad = \dfrac{3c^2 + cd + cd - d^2}{c^2 d^2}$

$\qquad = \dfrac{3c^2 + 2cd - d^2}{c^2 d^2}$

23. $3xt^2 = 3 \cdot x \cdot t \cdot t$
$\left.\right\}$ LCD $= 3 \cdot x \cdot x \cdot t \cdot t$, or $3x^2t^2$
$x^2t = x \cdot x \cdot t$

$$\frac{4x+2t}{3xt^2} - \frac{5x-3t}{x^2t}$$

$$= \frac{4x+2t}{3xt^2} \cdot \frac{x}{x} - \frac{5x-3t}{x^2t} \cdot \frac{3t}{3t}$$

$$= \frac{4x^2+2tx}{3x^2t^2} - \frac{15xt-9t^2}{3x^2t^2}$$

$$= \frac{4x^2+2tx-(15xt-9t^2)}{3x^2t^2}$$

$$= \frac{4x^2+2tx-15xt+9t^2}{3x^2t^2}$$

$$= \frac{4x^2-13xt+9t^2}{3x^2t^2}$$

(Although $4x^2 - 13xt + 9t^2$ can be factored, doing so will not enable us to simplify the result further.)

25. The denominators cannot be factored, so the LCD is their product, $(x-2)(x+2)$.

$$\frac{3}{x-2} + \frac{3}{x+2} = \frac{3}{x-2} \cdot \frac{x+2}{x+2} + \frac{3}{x+2} \cdot \frac{x-2}{x-2}$$

$$= \frac{3(x+2)+3(x-2)}{(x-2)(x+2)}$$

$$= \frac{3x+6+3x-6}{(x-2)(x+2)}$$

$$= \frac{6x}{(x-2)(x+2)}$$

27. $\dfrac{t}{t+3} - \dfrac{1}{t-1}$ $\qquad$ LCD $= (t+3)(t-1)$

$$= \frac{t}{t+3} \cdot \frac{t-1}{t-1} - \frac{1}{t-1} \cdot \frac{t+3}{t+3}$$

$$= \frac{t^2-t}{(t+3)(t-1)} - \frac{t+3}{(t+3)(t-1)}$$

$$= \frac{t^2-t-(t+3)}{(t+3)(t-1)}$$

$$= \frac{t^2-t-t-3}{(t+3)(t-1)}$$

$$= \frac{t^2-2t-3}{(t+3)(t-1)}$$

(Although $t^2 - 2t - 3$ can be factored, doing so will not enable us to simplify the result further.)

29. $3x = 3 \cdot x$
$\left.\right\}$ LCD $= 3x(x+1)$
$x + 1 = x + 1$

$$\frac{3}{x+1} + \frac{2}{3x} = \frac{3}{x+1} \cdot \frac{3x}{3x} + \frac{2}{3x} \cdot \frac{x+1}{x+1}$$

$$= \frac{9x+2(x+1)}{3x(x+1)}$$

$$= \frac{9x+2x+2}{3x(x+1)}$$

$$= \frac{11x+2}{3x(x+1)}$$

31. $\dfrac{3}{2t^2-2t} - \dfrac{5}{2t-2}$

$$= \frac{3}{2t(t-1)} - \frac{5}{2(t-1)}$$

$$= \frac{3}{2t(t-1)} - \frac{5}{2(t-1)} \cdot \frac{t}{t}$$

$$= \frac{3-5t}{2t(t-1)}$$

33. $\dfrac{3a}{a^2-9} + \dfrac{a}{a+3}$

$$= \frac{3a}{(a+3)(a+3)} + \frac{a}{a+3}$$

$$\qquad\qquad \text{LCD} = (a+3)(a-3)$$

$$= \frac{3a}{(a+3)(a+3)} + \frac{a}{a+3} \cdot \frac{a-3}{a-3}$$

$$= \frac{3a+a(a-3)}{(a+3)(a-3)}$$

$$= \frac{3a+a^2-3a}{(a+3)(a-3)}$$

$$= \frac{a^2}{(a+3)(a-3)}$$

35. $\dfrac{6}{z+4} - \dfrac{2}{3z+12} = \dfrac{6}{z+4} - \dfrac{2}{3(z+4)}$

$$\qquad\qquad \text{LCD} = 3(z+4)$$

$$= \frac{6}{z+4} \cdot \frac{3}{3} - \frac{2}{3(z+4)}$$

$$= \frac{18}{3(z+4)} - \frac{2}{3(z+4)}$$

$$= \frac{16}{3(z+4)}$$

37. $\dfrac{5}{q-1} + \dfrac{2}{(q-1)^2}$

$$= \frac{5}{q-1} \cdot \frac{q-1}{q-1} + \frac{2}{(q-1)^2}$$

$$= \frac{5(q-1)+2}{(q-1)^2}$$

$$= \frac{5q-5+2}{(q-1)^2}$$

$$= \frac{5q-3}{(q-1)^2}$$

39. $\dfrac{3}{x+2} - \dfrac{8}{x^2-4}$

$$= \frac{3}{x+2} - \frac{8}{(x+2)(x-2)} \quad \text{LCD} = (x+2)(x-2)$$

$$= \frac{3}{x+2} \cdot \frac{x-2}{x-2} - \frac{8}{(x+2)(x-2)}$$

$$= \frac{3(x-2)-8}{(x+2)(x-2)}$$

$$= \frac{3x-6-8}{(x+2)(x-2)}$$

$$= \frac{3x-14}{(x+2)(x-2)}$$

41. $\dfrac{3a}{4a-20} + \dfrac{9a}{6a-30}$

$= \dfrac{3a}{2\cdot 2(a-5)} + \dfrac{9a}{2\cdot 3(a-5)}$

$\qquad\qquad\text{LCD} = 2\cdot 2\cdot 3(a-5)$

$= \dfrac{3a}{2\cdot 2(a-5)}\cdot\dfrac{3}{3} + \dfrac{9a}{2\cdot 3(a-5)}\cdot\dfrac{2}{2}$

$= \dfrac{9a+18a}{2\cdot 2\cdot 3(a-5)}$

$= \dfrac{27a}{2\cdot 2\cdot 3(a-5)}$

$= \dfrac{\cancel{3}\cdot 9\cdot a}{2\cdot 2\cdot \cancel{3}(a-5)}$

$= \dfrac{9a}{4(a-5)}$

43. $\dfrac{x}{x-5} + \dfrac{x}{5-x} = \dfrac{x}{x-5} + \dfrac{x}{5-x}\cdot\dfrac{-1}{-1}$

$\qquad\qquad\qquad = \dfrac{x}{x-5} + \dfrac{-x}{x-5}$

$\qquad\qquad\qquad = 0$

45. $\dfrac{6}{a^2+a-2} + \dfrac{4}{a^2-4a+3}$

$= \dfrac{6}{(a+2)(a-1)} + \dfrac{4}{(a-3)(a-1)}$

$\qquad\qquad\text{LCD} = (a+2)(a-1)(a-3)$

$= \dfrac{6}{(a+2)(a-1)}\cdot\dfrac{a-3}{a-3} + \dfrac{4}{(a-3)(a-1)}\cdot\dfrac{a+2}{a+2}$

$= \dfrac{6(a-3)+4(a+2)}{(a+2)(a-1)(a-3)}$

$= \dfrac{6a-18+4a+8}{(a+2)(a-1)(a-3)}$

$= \dfrac{10a-10}{(a+2)(a-1)(a-3)}$

$= \dfrac{10\cancel{(a-1)}}{(a+2)\cancel{(a-1)}(a-3)}$

$= \dfrac{10}{(a+2)(a-3)}$

47. $\dfrac{x}{x^2+9x+20} - \dfrac{4}{x^2+7x+12}$

$= \dfrac{x}{(x+4)(x+5)} - \dfrac{4}{(x+3)(x+4)}$

$\qquad\qquad\text{LCD} = (x+3)(x+4)(x+5)$

$= \dfrac{x}{(x+4)(x+5)}\cdot\dfrac{x+3}{x+3} - \dfrac{4}{(x+3)(x+4)}\cdot\dfrac{x+5}{x+5}$

$= \dfrac{x(x+3)-4(x+5)}{(x+3)(x+4)(x+5)}$

$= \dfrac{x^2+3x-4x-20}{(x+3)(x+4)(x+5)}$

$= \dfrac{x^2-x-20}{(x+3)(x+4)(x+5)}$

$= \dfrac{\cancel{(x+4)}(x-5)}{(x+3)\cancel{(x+4)}(x+5)}$

$= \dfrac{x-5}{(x+3)(x+5)}$

49. $\dfrac{3z}{z^2-4z+4} + \dfrac{10}{z^2+z-6}$

$= \dfrac{3z}{(z-2)^2} + \dfrac{10}{(z-2)(z+3)},$

$\qquad\qquad\text{LCD} = (z-2)^2(z+3)$

$= \dfrac{3z}{(z-2)^2}\cdot\dfrac{z+3}{z+3} + \dfrac{10}{(z-2)(z+3)}\cdot\dfrac{z-2}{z-2}$

$= \dfrac{3z(z+3)+10(z-2)}{(z-2)^2(z+3)}$

$= \dfrac{3z^2+9z+10z-20}{(z-2)^2(z+3)}$

$= \dfrac{3z^2+19z-20}{(z-2)^2(z+3)}$

51. $\dfrac{-7}{x^2+25x+24} - \dfrac{0}{x^2+11x+10}$

Note that $\dfrac{0}{x^2+11x+10} = 0$, so the difference is

$\dfrac{-7}{x^2+25x+24}.$

53. $\dfrac{5x}{4} - \dfrac{x-2}{-4} = \dfrac{5x}{4} - \dfrac{x-2}{-4}\cdot\dfrac{-1}{-1}$

$\qquad\qquad\qquad = \dfrac{5x}{4} - \dfrac{2-x}{4}$

$\qquad\qquad\qquad = \dfrac{5x-(2-x)}{4}$

$\qquad\qquad\qquad = \dfrac{5x-2+x}{4}$

$\qquad\qquad\qquad = \dfrac{6x-2}{4}$

$\qquad\qquad\qquad = \dfrac{2(3x-1)}{2\cdot 2}$

$\qquad\qquad\qquad = \dfrac{\cancel{2}(3x-1)}{\cancel{2}\cdot 2}$

$\qquad\qquad\qquad = \dfrac{3x-1}{2}$

55. $\dfrac{y^2}{y-3} + \dfrac{9}{3-y} = \dfrac{y^2}{y-3} + \dfrac{9}{3-y} \cdot \dfrac{-1}{-1}$

$$= \dfrac{y^2}{y-3} + \dfrac{-9}{-3+y}$$

$$= \dfrac{y^2-9}{y-3}$$

$$= \dfrac{(y+3)(\cancel{y-3})}{\cancel{y-3}}$$

$$= y+3$$

57. $\dfrac{c-5}{c^2-64} + \dfrac{c-5}{64-c^2} = \dfrac{c-5}{c^2-64} + \dfrac{c-5}{64-c^2} \cdot \dfrac{-1}{-1}$

$$= \dfrac{c-5}{c^2-64} + \dfrac{5-c}{c^2-64}$$

$$= \dfrac{c-5+5-c}{c^2-64}$$

$$= \dfrac{0}{c^2-64}$$

$$= 0$$

59. $\dfrac{4-p}{25-p^2} + \dfrac{p+1}{p-5}$

$$= \dfrac{4-p}{(5+p)(5-p)} + \dfrac{p+1}{p-5}$$

$$= \dfrac{4-p}{(5+p)(5-p)} \cdot \dfrac{-1}{-1} + \dfrac{p+1}{p-5}$$

$$= \dfrac{p-4}{(p+5)(p-5)} + \dfrac{p+1}{p-5} \quad \text{LCD} = (p+5)(p-5)$$

$$= \dfrac{p-4}{(p+5)(p-5)} + \dfrac{p+1}{p-5} \cdot \dfrac{p+5}{p+5}$$

$$= \dfrac{p-4+p^2+6p+5}{(p+5)(p-5)}$$

$$= \dfrac{p^2+7p+1}{(p+5)(p-5)}$$

61. $\dfrac{x}{x-4} - \dfrac{3}{16-x^2}$

$$= \dfrac{x}{x-4} - \dfrac{3}{(4+x)(4-x)}$$

$$= \dfrac{x}{x-4} \cdot \dfrac{-1}{-1} - \dfrac{3}{(4+x)(4-x)}$$

$$= \dfrac{-x}{4-x} - \dfrac{3}{(4+x)(4-x)} \quad \text{LCD} = (4-x)(4+x)$$

$$= \dfrac{-x}{4-x} \cdot \dfrac{4+x}{4+x} - \dfrac{3}{(4+x)(4-x)}$$

$$= \dfrac{-x(4+x)-3}{(4-x)(4+x)}$$

$$= \dfrac{-4x-x^2-3}{(4-x)(4+x)}$$

$$= \dfrac{-x^2-4x-3}{(4-x)(4+x)}, \text{ or } \dfrac{x^2+4x+3}{(x+4)(x-4)}$$

(Although x^2+4x+3 can be factored, doing so will not enable us to simplify the result further.)

63. $\dfrac{a}{a^2-1} + \dfrac{2a}{a-a^2} = \dfrac{a}{a^2-1} + \dfrac{2\cdot\cancel{a}}{\cancel{a}(1-a)}$

$$= \dfrac{a}{(a+1)(a-1)} + \dfrac{2}{1-a}$$

$$= \dfrac{a}{(a+1)(a-1)} + \dfrac{2}{1-a} \cdot \dfrac{-1}{-1}$$

$$= \dfrac{a}{(a+1)(a-1)} + \dfrac{-2}{a-1}$$

$$\text{LCD} = (a+1)(a-1)$$

$$= \dfrac{a}{(a+1)(a-1)} + \dfrac{-2}{a-1} \cdot \dfrac{a+1}{a+1}$$

$$= \dfrac{a-2a-2}{(a+1)(a-1)}$$

$$= \dfrac{-a-2}{(a+1)(a-1)}, \text{ or }$$

$$= \dfrac{a+2}{(1+a)(1-a)}$$

65. $\dfrac{4x}{x^2-y^2} - \dfrac{6}{y-x}$

$$= \dfrac{4x}{(x+y)(x-y)} - \dfrac{6}{y-x}$$

$$= \dfrac{4x}{(x+y)(x-y)} - \dfrac{6}{y-x} \cdot \dfrac{-1}{-1}$$

$$= \dfrac{4x}{(x+y)(x-y)} - \dfrac{-6}{x-y} \quad \text{LCD} = (x+y)(x-y)$$

$$= \dfrac{4x}{(x+y)(x-y)} - \dfrac{-6}{x-y} \cdot \dfrac{x+y}{x+y}$$

$$= \dfrac{4x-(-6)(x+y)}{(x+y)(x-y)}$$

$$= \dfrac{4x+6x+6y}{(x+y)(x-y)}$$

$$= \dfrac{10x+6y}{(x+y)(x-y)}$$

(Although $10x+6y$ can be factored, doing so will not enable us to simplify the result further.)

67. $\dfrac{x-3}{2-x} - \dfrac{x+3}{x+2} + \dfrac{x+6}{4-x^2}$

$$= \dfrac{x-3}{2-x} - \dfrac{x+3}{x+2} + \dfrac{x+6}{(2+x)(2-x)}$$

$$\text{LCD} = (2+x)(2-x)$$

$$= \dfrac{x-3}{2-x} \cdot \dfrac{2+x}{2+x} - \dfrac{x+3}{x+2} \cdot \dfrac{2-x}{2-x} + \dfrac{x+6}{(2+x)(2-x)}$$

$$= \dfrac{(x-3)(2+x)-(x+3)(2-x)+(x+6)}{(2+x)(2-x)}$$

$$= \dfrac{x^2-x-6-(-x^2-x+6)+x+6}{(2+x)(2-x)}$$

$$= \dfrac{x^2-x-6+x^2+x-6+x+6}{(2+x)(2-x)}$$

$$= \dfrac{2x^2+x-6}{(2+x)(2-x)}$$

$$= \dfrac{(2x-3)(\cancel{x+2})}{(\cancel{2+x})(2-x)}$$

$$= \dfrac{2x-3}{2-x}$$

69. $\dfrac{2x+5}{x+1} + \dfrac{x+7}{x+5} - \dfrac{5x+17}{(x+1)(x+5)}$

$$\text{LCD} = (x+1)(x+5)$$

$$= \frac{(2x+5)(x+5) + (x+7)(x+1) - (5x+17)}{(x+1)(x+5)}$$

$$= \frac{2x^2 + 15x + 25 + x^2 + 8x + 7 - 5x - 17}{(x+1)(x+5)}$$

$$= \frac{3x^2 + 18x + 15}{(x+1)(x+5)}$$

$$= \frac{3\cancel{(x+1)}\cancel{(x+5)}}{\cancel{(x+1)}\cancel{(x+5)}}$$

$$= 3$$

71. $\dfrac{1}{x+y} + \dfrac{1}{x-y} - \dfrac{2x}{x^2 - y^2}$

$$\text{LCD} = (x+y)(x-y)$$

$$= \frac{1}{x+y} \cdot \frac{x-y}{x-y} + \frac{1}{x-y} \cdot \frac{x+y}{x+y} - \frac{2x}{(x+y)(x-y)}$$

$$= \frac{(x-y) + (x+y) - 2x}{(x+y)(x-y)}$$

$$= 0$$

73. *Writing Exercise.* Using the least common denominator usually reduces the complexity of computations and requires less simplification of the sum or difference.

75. $\dfrac{-3}{8} \div \dfrac{11}{4} = \dfrac{-3}{8} \cdot \dfrac{4}{11} = -\dfrac{3 \cdot \cancel{4}}{2 \cdot \cancel{4} \cdot 11} = \dfrac{-3}{22}$

77. $\dfrac{\frac{3}{4}}{\frac{5}{6}} = \dfrac{3}{4} \cdot \dfrac{6}{5} = \dfrac{3 \cdot 3 \cdot \cancel{2}}{\cancel{2} \cdot 2 \cdot 5} = \dfrac{9}{10}$

79. $\dfrac{2x+6}{x-1} \div \dfrac{3x+9}{x-1} = \dfrac{2x+6}{x-1} \cdot \dfrac{x-1}{3x+9} = \dfrac{2\cancel{(x+3)}\cancel{(x-1)}}{\cancel{(x-1)}3\cancel{(x+3)}} = \dfrac{2}{3}$

81. *Writing Exercise.* Their sum is zero. Another explanation is that $-\left(\dfrac{1}{3-x}\right) = \dfrac{1}{-(3-x)} = \dfrac{1}{x-3}$.

83. $P = 2\left(\dfrac{3}{x+4}\right) + 2\left(\dfrac{2}{x-5}\right)$

$$= \frac{6}{x+4} + \frac{4}{x-5} \qquad \text{LCD} = (x+4)(x-5)$$

$$= \frac{6}{x+4} \cdot \frac{x-5}{x-5} + \frac{4}{x-5} \cdot \frac{x+4}{x+4}$$

$$= \frac{6x - 30 + 4x + 16}{(x+4)(x-5)}$$

$$= \frac{10x - 14}{(x+4)(x-5)}, \text{ or } \frac{10x-14}{x^2 - x - 20}$$

$$A = \left(\frac{3}{x+4}\right)\left(\frac{2}{x-5}\right) = \frac{6}{(x+4)(x-5)}$$

85. $\dfrac{x^2}{3x^2 - 5x - 2} - \dfrac{2x}{3x+1} \cdot \dfrac{1}{x-2}$

$$= \frac{x^2}{(3x+1)(x-2)} - \frac{2x}{(3x+1)(x-2)}$$

$$= \frac{x^2 - 2x}{(3x+1)(x-2)}$$

$$= \frac{x(x-2)}{(3x+1)(x-2)}$$

$$= \frac{x}{3x+1} \cdot \frac{x-2}{x-2}$$

$$= \frac{x}{3x+1}$$

87. We recognize that this is the product of the sum and difference of two terms: $(A+B)(A-B) = A^2 - B^2$.

$$\left(\frac{x}{x+7} - \frac{3}{x+2}\right)\left(\frac{x}{x+7} + \frac{3}{x+2}\right)$$

$$= \frac{x^2}{(x+7)^2} - \frac{9}{(x+2)^2} \quad \text{LCD} = (x+7)^2(x+2)^2$$

$$= \frac{x^2}{(x+7)^2} \cdot \frac{(x+2)^2}{(x+2)^2} - \frac{9}{(x+2)^2} \cdot \frac{(x+7)^2}{(x+7)^2}$$

$$= \frac{x^2(x+2)^2 - 9(x+7)^2}{(x+7)^2(x+2)^2}$$

$$= \frac{x^2(x^2 + 4x + 4) - 9(x^2 + 14x + 49)}{(x+7)^2(x+2)^2}$$

$$= \frac{x^4 + 4x^3 + 4x^2 - 9x^2 - 126x - 441}{(x+7)^2(x+2)^2}$$

$$= \frac{x^4 + 4x^3 - 5x^2 - 126x - 441}{(x+7)^2(x+2)^2}$$

89. $\left(\dfrac{a}{a-b} + \dfrac{b}{a+b}\right)\left(\dfrac{1}{3a+b} + \dfrac{2a+6b}{9a^2 - b^2}\right)$

$$= \frac{a}{(a-b)(3a+b)} + \frac{a(2a+6b)}{(a-b)(9a^2 - b^2)} +$$

$$\frac{b}{(a+b)(3a+b)} + \frac{b(2a+6b)}{(a+b)(9a^2 - b^2)}$$

$$= \frac{a}{(a-b)(3a+b)} + \frac{2a^2 + 6ab}{(a-b)(3a+b)(3a-b)} +$$

$$\frac{b}{(a+b)(3a+b)} + \frac{2ab + 6b^2}{(a+b)(3a+b)(3a-b)}$$

$$\text{LCD} = (a-b)(a+b)(3a+b)(3a-b)$$

$$= [a(a+b)(3a-b) + (2a^2 + 6ab)(a+b) +$$

$$b(a-b)(3a-b) + (2ab+6b^2)(a-b)]/$$

$$[(a-b)(a+b)(3a+b)(3a-b)]$$

$$= (3a^3 + 2a^2 b - ab^2 + 2a^3 + 8a^2 b + 6ab^2 + b^3 -$$

$$4ab^2 + 3a^2 b + 4ab^2 - 6b^3 + 2a^2 b)/$$

$$[(a-b)(a+b)(3a+b)(3a-b)]$$

$$= \frac{5a^3 + 15a^2 b + 5ab^2 - 5b^3}{(a-b)(a+b)(3a+b)(3a-b)}$$

$$= \frac{5\cancel{(a+b)}(a^2 + 2ab - b^2)}{(a-b)\cancel{(a+b)}(3a+b)(3a-b)}$$

$$= \frac{5(a^2 + 2ab - b^2)}{(a-b)(3a+b)(3a-b)}$$

91. Answers may vary. $\dfrac{a}{a-b} + \dfrac{3b}{b-a}$

93. *Writing Exercise.* Both y_1 and y_2 are undefined when $x = 5$.

Chapter 6 Connecting the Concepts

1.
$$= \frac{3}{5x} + \frac{2}{x^2} \qquad \text{Addition}$$
$$= \frac{3}{5x} \cdot \frac{x}{x} + \frac{2}{x^2} \cdot \frac{5}{5} \qquad \text{LCD} = 5x^2$$
$$= \frac{3x + 10}{5x^2}$$

3.
$$= \frac{3}{5x} \div \frac{2}{x^2} \qquad \text{Division}$$
$$= \frac{3}{5x} \cdot \frac{x^2}{2}$$
$$= \frac{3x}{10} \cdot \frac{x}{x}$$
$$= \frac{3x}{10}$$

5.
$$= \frac{2x - 6}{5x + 10} \cdot \frac{x + 2}{6x - 12} \qquad \text{Multiplication}$$
$$= \frac{2(x-3)(x+2)}{5(x+2) \cdot 3(x-2)}$$
$$= \frac{(x-3)}{15(x-2)}$$

7.
$$= \frac{2}{x-5} \div \frac{6}{x-5} \qquad \text{Division}$$
$$= \frac{2}{x-5} \cdot \frac{x-5}{6}$$
$$= \frac{1}{3} \cdot \frac{2(x-5)}{2(x-5)}$$
$$= \frac{1}{3}$$

9.
$$= \frac{2}{x+3} + \frac{3}{x+4} \qquad \text{Addition}$$
$$= \frac{2}{x+3} \cdot \frac{x+4}{x+4} + \frac{3}{x+4} \cdot \frac{x+3}{x+3} \qquad \text{LCD} = (x+3)(x+4)$$
$$= \frac{2(x+4) + 3(x+3)}{(x+3)(x+4)}$$
$$= \frac{2x+8+3x+9}{(x+3)(x+4)}$$
$$= \frac{5x+17}{(x+3)(x+4)}$$

11.
$$= \frac{3}{x-4} - \frac{2}{4-x} \qquad \text{Subtraction}$$
$$= \frac{3}{x-4} + \frac{2}{x-4}$$
$$= \frac{5}{x-4}$$

13.
$$= \frac{a}{6a-9b} - \frac{b}{4a-6b} \qquad \text{Subtraction}$$
$$= \frac{a}{3(2a-3b)} \cdot \frac{2}{2} - \frac{b}{2(2a-3b)} \cdot \frac{3}{3} \qquad \text{LCD} = 6(2a-3b)$$
$$= \frac{2a-3b}{6(2a-3b)}$$
$$= \frac{1}{6}$$

15.
$$= \frac{x+1}{x^2-7x+10} + \frac{3}{x^2-x-2} \qquad \text{Addition}$$
$$= \frac{x+1}{(x-5)(x-2)} \cdot \frac{x+1}{x+1}$$
$$\quad + \frac{3}{(x-2)(x+1)} \cdot \frac{x-5}{x-5} \qquad \text{LCD} = (x+1)(x-2)(x-5)$$
$$= \frac{x^2+2x+1+3x-15}{(x+1)(x-2)(x-5)}$$
$$= \frac{x^2+5x-14}{(x+1)(x-2)(x-5)}$$
$$= \frac{(x+7)(x-2)}{(x+1)(x-2)(x-5)}$$
$$= \frac{x+7}{(x+1)(x-5)}$$

17.
$$= \frac{t+2}{10} + \frac{2t+1}{15} \qquad \text{Addition}$$
$$= \frac{3(t+2)}{3(10)} + \frac{2(2t+1)}{2(15)} \qquad \text{LCD} = 30$$
$$= \frac{3t+6+4t+2}{30} = \frac{7t+8}{30}$$

19.
$$= \frac{a^2-2a+1}{a^2-4} \div (a^2-3a+2) \qquad \text{Division}$$
$$= \frac{(a-1)(a-1)}{(a+2)(a-2)} \cdot \frac{1}{(a-2)(a-1)}$$
$$= \frac{a-1}{(a+2)(a-2)^2}$$

Exercise Set 6.5

1. The LCD is the LCM of x^2, x, 2, and $4x$. It is $4x^2$.

$$\frac{\dfrac{5}{x^2} + \dfrac{1}{x}}{\dfrac{7}{2} - \dfrac{3}{4x}} \cdot \frac{4x^2}{4x^2} = \frac{\dfrac{5}{x^2} \cdot 4x^2 + \dfrac{1}{x} \cdot 4x^2}{\dfrac{7}{2} \cdot 4x^2 - \dfrac{3}{4x} \cdot 4x^2}$$

Choice (d) is correct.

3. We subtract to get a single rational expression in the numerator and add to get a single rational expression in the denominator.

$$\frac{\dfrac{4}{5x} - \dfrac{1}{10}}{\dfrac{8}{x^2} + \dfrac{7}{2}} = \frac{\dfrac{4}{5x} \cdot \dfrac{2}{2} - \dfrac{1}{10} \cdot \dfrac{x}{x}}{\dfrac{8}{x^2} \cdot \dfrac{2}{2} + \dfrac{7}{2} \cdot \dfrac{x^2}{x^2}}$$

$$= \frac{\dfrac{8}{10x} - \dfrac{x}{10x}}{\dfrac{16}{2x^2} + \dfrac{7x^2}{2x^2}}$$

$$= \frac{\dfrac{8-x}{10x}}{\dfrac{16+7x^2}{2x^2}}$$

Choice (b) is correct.

5. $\dfrac{1+\dfrac{1}{4}}{2+\dfrac{3}{4}}$ LCD is 4

$=\dfrac{1+\dfrac{1}{4}}{2+\dfrac{3}{4}}\cdot\dfrac{4}{4}$ Multiplying by $\dfrac{4}{4}$

$=\dfrac{\left(1+\dfrac{1}{4}\right)4}{\left(2+\dfrac{3}{4}\right)4}$ Multiplying numerator and denominator by 4

$=\dfrac{1\cdot4+\dfrac{1}{4}\cdot4}{2\cdot4+\dfrac{3}{4}\cdot4}$

$=\dfrac{4+1}{8+3}$

$=\dfrac{5}{11}$

7. $\dfrac{\dfrac{1}{2}+\dfrac{1}{3}}{\dfrac{1}{4}-\dfrac{1}{6}}$

$=\dfrac{\dfrac{1}{2}\cdot\dfrac{3}{3}+\dfrac{1}{3}\cdot\dfrac{2}{2}}{\dfrac{1}{4}\cdot\dfrac{3}{3}-\dfrac{1}{6}\cdot\dfrac{2}{2}}$ Getting a common denominator in numerator and in denominator

$=\dfrac{\dfrac{3}{6}+\dfrac{2}{6}}{\dfrac{3}{12}-\dfrac{2}{12}}$

$=\dfrac{\dfrac{5}{6}}{\dfrac{1}{12}}$ Adding in the numerator; subtracting in the denominator

$=\dfrac{5}{6}\cdot\dfrac{12}{1}$ Multiplying by the reciprocal of the divisor

$=\dfrac{5\cdot6\cdot2}{6}$

$=\dfrac{5\cdot\cancel{6}\cdot2}{\cancel{6}}$

$=10$

9. $\dfrac{\dfrac{x}{4}+x}{\dfrac{4}{x}+x}$ LCD is $4x$

$=\dfrac{\dfrac{x}{4}+x}{\dfrac{4}{x}+x}\cdot\dfrac{4x}{4x}$

$=\dfrac{\left(\dfrac{x}{4}+x\right)(4x)}{\left(\dfrac{4}{x}+x\right)(4x)}$

$=\dfrac{x^2+4x^2}{16+4x^2}$

$\dfrac{5x^2}{16+4x^2}$

11. $\dfrac{\dfrac{10}{t}}{\dfrac{2}{t^2}-\dfrac{5}{t}}$

$=\dfrac{\dfrac{10}{t}}{\dfrac{2}{t^2}-\dfrac{5}{t}}\cdot\dfrac{t^2}{t^2}$

$=\dfrac{\dfrac{10}{t}\cdot t^2}{\left(\dfrac{2}{t^2}-\dfrac{5}{t}\right)t^2}$

$=\dfrac{10t}{\dfrac{2}{t^2}\cdot t^2-\dfrac{5}{t}\cdot t^2}$

$=\dfrac{10t}{2-5t}$, or $\dfrac{-10t}{5t-2}$

13. $\dfrac{\dfrac{2a-5}{3a}}{\dfrac{a-7}{6a}}$

$=\dfrac{2a-5}{3a}\cdot\dfrac{6a}{a-7}$ Multiplying by the reciprocal of the divisor

$=\dfrac{(2a-5)\cdot2\cdot3a}{3a\cdot(a-7)}$

$=\dfrac{(2a-5)\cdot2\cdot\cancel{3a}}{\cancel{3a}\cdot(a-7)}$

$=\dfrac{2(2a-5)}{a-7}$

$=\dfrac{4a-10}{a-7}$

15. $\dfrac{\dfrac{x}{6}-\dfrac{3}{x}}{\dfrac{1}{3}+\dfrac{1}{x}}$ LCD is $6x$

$=\dfrac{\dfrac{x}{6}-\dfrac{3}{x}}{\dfrac{1}{3}+\dfrac{1}{x}}\cdot\dfrac{6x}{6x}$

$=\dfrac{\dfrac{x}{6}\cdot6x-\dfrac{3}{x}\cdot6x}{\dfrac{1}{3}\cdot6x+\dfrac{1}{x}\cdot6x}$

$=\dfrac{x^2-18}{2x+6}$

17. $\dfrac{\dfrac{1}{s}-\dfrac{1}{5}}{\dfrac{s-5}{s}}$ LCD is $5s$

$=\dfrac{\dfrac{1}{s}-\dfrac{1}{5}}{\dfrac{s-5}{s}}\cdot\dfrac{5s}{5s}$

$=\dfrac{\dfrac{1}{s}\cdot5s-\dfrac{1}{5}\cdot5s}{\left(\dfrac{s-5}{s}\right)(5s)}$

$=\dfrac{5-s}{(s-5)(5)}$

$=\dfrac{-\cancel{(s-5)}}{\cancel{(s-5)}(5)}$

$=-\dfrac{1}{5}$

19. $\dfrac{\dfrac{1}{t^2}+1}{\dfrac{1}{t}-1}$ LCD is t^2

$= \dfrac{\dfrac{1}{t^2}+1}{\dfrac{1}{t}-1} \cdot \dfrac{t^2}{t^2}$

$= \dfrac{\dfrac{1}{t^2}\cdot t^2 + 1\cdot t^2}{\dfrac{1}{t}\cdot t^2 - 1\cdot t^2}$

$= \dfrac{1+t^2}{t-t^2}$

(Although the denominator can be factored, doing so will not enable us to simplify further.)

21. $\dfrac{\dfrac{x^2}{x^2-y^2}}{\dfrac{x}{x+y}}$

$= \dfrac{x^2}{x^2-y^2} \cdot \dfrac{x+y}{x}$ Multiplying by the reciprocal of the divisor

$= \dfrac{x^2(x+y)}{(x^2-y^2)(x)}$

$= \dfrac{x\cdot x\cdot(x+y)}{(x+y)(x-y)(x)}$

$= \dfrac{\cancel{x}\cdot x\cdot\cancel{(x+y)}}{\cancel{(x+y)}(x-y)(\cancel{x})}$

$= \dfrac{x}{x-y}$

23. $\dfrac{\dfrac{7}{c^2}+\dfrac{4}{c}}{\dfrac{6}{c}-\dfrac{3}{c^3}}$ LCD is c^3

$= \dfrac{\dfrac{7}{c^2}+\dfrac{4}{c}}{\dfrac{6}{c}-\dfrac{3}{c^3}} \cdot \dfrac{c^3}{c^3}$

$= \dfrac{\dfrac{7}{c^2}\cdot c^3 + \dfrac{4}{c}\cdot c^3}{\dfrac{6}{c}\cdot c^3 - \dfrac{3}{c^3}\cdot c^3}$

$= \dfrac{7c+4c^2}{6c^2-3}$

(Although the numerator and the denominator can be factored, doing so will not enable us to simplify further.)

25. $\dfrac{\dfrac{2}{7a^4}-\dfrac{1}{14a}}{\dfrac{3}{5a^2}+\dfrac{2}{15a}} = \dfrac{\dfrac{2}{7a^4}\cdot\dfrac{2}{2}-\dfrac{1}{14a}\cdot\dfrac{a^3}{a^3}}{\dfrac{3}{5a^2}\cdot\dfrac{3}{3}+\dfrac{2}{15a}\cdot\dfrac{a}{a}}$

$= \dfrac{\dfrac{4-a^3}{14a^4}}{\dfrac{9+2a}{15a^2}}$

$= \dfrac{4-a^3}{14a^4} \cdot \dfrac{15a^2}{9+2a}$

$= \dfrac{15\cancel{a^2}(4-a^3)}{14a^2\cancel{a^2}(9+2a)}$

$= \dfrac{15(4-a^3)}{14a^2(9+2a)},\ \text{or}\ \dfrac{60-15a^3}{126a^2+28a^3}$

27. $\dfrac{\dfrac{x}{5y^3}+\dfrac{3}{10y}}{\dfrac{3}{10y}+\dfrac{x}{5y^3}}$

Observe that, by the commutative law of addition, the numerator and denominator are equivalent, so the result is 1.

29. $\dfrac{\dfrac{3}{ab^4}+\dfrac{4}{a^3b}}{\dfrac{5}{a^3b}-\dfrac{3}{ab}} = \dfrac{\dfrac{3}{ab^4}\cdot\dfrac{a^2}{a^2}+\dfrac{4}{a^3b}\cdot\dfrac{b^3}{b^3}}{\dfrac{5}{a^3b}-\dfrac{3}{ab}\cdot\dfrac{a^2}{a^2}}$

$= \dfrac{\dfrac{3a^2+4b^3}{a^3b^4}}{\dfrac{5-3a^2}{a^3b}}$

$= \dfrac{3a^2+4b^3}{a^3b^4} \cdot \dfrac{a^3b}{5-3a^2}$

$= \dfrac{\cancel{a^3b}(3a^2+4b^3)}{\cancel{a^3b}\cdot b^3(5-3a^2)}$

$= \dfrac{3a^2+4b^3}{b^3(5-3a^2)},\ \text{or}\ \dfrac{3a^2+4b^3}{5b^3-3a^2b^3}$

31. $\dfrac{2-\dfrac{3}{x^2}}{4+\dfrac{9}{x^4}} = \dfrac{2-\dfrac{3}{x^2}}{4+\dfrac{9}{x^4}} \cdot \dfrac{x^4}{x^4}$

$= \dfrac{2\cdot x^4 - \dfrac{3}{x^2}\cdot x^4}{4\cdot x^4 + \dfrac{9}{x^4}\cdot x^4}$

$= \dfrac{2x^4-3x^2}{4x^4+9}$

33. $\dfrac{t - \dfrac{9}{t}}{t + \dfrac{4}{t}} = \dfrac{t \cdot \dfrac{t}{t} - \dfrac{9}{t}}{t \cdot \dfrac{t}{t} + \dfrac{4}{t}}$

$= \dfrac{\dfrac{t^2 - 9}{t}}{\dfrac{t^2 + 4}{t}}$

$= \dfrac{t^2 - 9}{t} \cdot \dfrac{t}{t^2 + 4}$

$= \dfrac{t(t^2 - 9)}{t(t^2 + 4)}$

$= \dfrac{t^2 - 9}{t^2 + 4}$

35. $\dfrac{3 + \dfrac{4}{ab^3}}{\dfrac{3 + a}{a^2 b}} = \dfrac{3 + \dfrac{4}{ab^3}}{\dfrac{3 + a}{a^2 b}} \cdot \dfrac{a^2 b^3}{a^2 b^3}$

$= \dfrac{3 \cdot a^2 b^3 + \dfrac{4}{ab^3} \cdot a^2 b^3}{\dfrac{3 + a}{a^2 b} \cdot a^2 b^3}$

$= \dfrac{3a^2 b^3 + 4a}{b^2(3 + a)}, \text{ or } \dfrac{3a^2 b^3 + 4a}{3b^2 + ab^2}$

37. $\dfrac{t + 5 + \dfrac{3}{t}}{t + 2 + \dfrac{1}{t}}$ LCD is t

$= \dfrac{t + 5 + \dfrac{3}{t}}{t + 2 + \dfrac{1}{t}} \cdot \dfrac{t}{t}$

$= \dfrac{t \cdot t + 5 \cdot t + \dfrac{3}{t} \cdot t}{t \cdot t + 2 \cdot t + \dfrac{1}{t} \cdot t}$

$= \dfrac{t^2 + 5t + 3}{t^2 + 2t + 1}$

$= \dfrac{t^2 + 5t + 3}{(t + 1)^2}$

39. $\dfrac{x - 2 - \dfrac{1}{x}}{x - 5 - \dfrac{4}{x}} = \dfrac{x - 2 - \dfrac{1}{x}}{x - 5 - \dfrac{4}{x}} \cdot \dfrac{x}{x}$

$= \dfrac{x \cdot x - 2 \cdot x - \dfrac{1}{x} \cdot x}{x \cdot x - 5 \cdot x - \dfrac{4}{x} \cdot x}$

$= \dfrac{x^2 - 2x - 1}{x^2 - 5x - 4}$

41. *Writing Exercise.* Yes; Method 2, multiplying by the LCD, does not require division of rational expressions.

43. $3x - 5 + 2(4x - 1) = 12x - 3$

$3x - 5 + 8x - 2 = 12x - 3$

$11x - 7 = 12x - 3$

$-7 = x - 3$

$-4 = x$

The solution is -4.

45. $\quad \dfrac{3}{4}x - \dfrac{5}{8} = \dfrac{3}{8}x + \dfrac{7}{4}$ LCD is 8

$8\left(\dfrac{3}{4}x - \dfrac{5}{8}\right) = 8\left(\dfrac{3}{8}x + \dfrac{7}{4}\right)$

$8 \cdot \dfrac{3}{4}x - 8 \cdot \dfrac{5}{8} = 8 \cdot \dfrac{3}{8}x + 8 \cdot \dfrac{7}{4}$

$6x - 5 = 3x + 14$

$3x - 5 = 14$

$3x = 19$

$x = \dfrac{19}{3}$

The solution is $\dfrac{19}{3}$.

47. $\quad x^2 - 7x + 12 = 0$

$(x - 3)(x - 4) = 0$

$x - 3 = 0, \text{ or } x - 4 = 0$

$x = 3 \text{ or } x = 4$

49. *Writing Exercise.* Although either method could be used, Method 2 requires fewer steps.

51. $\dfrac{\dfrac{x - 5}{x - 6}}{\dfrac{x - 7}{x - 8}}$

This expression is undefined for any value of x that makes a denominator 0. We see that $x - 6 = 0$ when $x = 6$, $x - 7 = 0$ when $x = 7$, and $x - 8 = 0$ when $x = 8$, so the expression is undefined for the x-values 6, 7, and 8.

53. $\dfrac{\dfrac{2x + 3}{5x + 4}}{\dfrac{3}{7} - \dfrac{x^2}{21}}$

This expression is undefined for any value of x that makes a denominator 0. First we find the value of x for which $5x + 4 = 0$.

$5x + 4 = 0$

$5x = -4$

$x = -\dfrac{4}{5}$

Then we find the value of x for which $\dfrac{3}{7} - \dfrac{x^2}{21} = 0$:

$$\frac{3}{7} - \frac{x^2}{21} = 0$$

$$21\left(\frac{3}{7} - \frac{x^2}{21}\right) = 21 \cdot 0$$

$$21 \cdot \frac{3}{7} - 21 \cdot \frac{x^2}{21} = 0$$

$$9 - x^2 = 0$$

$$9 = x^2$$

$$\pm 3 = x$$

The expression is undefined for the x-values $-\frac{4}{5}$, -3 and 3.

55. For the complex rational expression

$\dfrac{\dfrac{A}{B}}{\dfrac{C}{D}}$ the LCD is BD.

$$= \frac{\dfrac{A}{B} \cdot BD}{\dfrac{C}{D} \cdot BD}$$

$$= \frac{\dfrac{ABD}{B}}{\dfrac{CBD}{D}} = \frac{\dfrac{A\cancel{B}D}{\cancel{B}}}{\dfrac{BC\cancel{D}}{\cancel{D}}}$$

$$= \frac{AD}{BC}$$

$$= \frac{A}{B} \cdot \frac{D}{C}$$

57. $\dfrac{\dfrac{x}{x+5} + \dfrac{3}{x+2}}{\dfrac{2}{x+2} - \dfrac{x}{x+5}} = \dfrac{\dfrac{x}{x+5} + \dfrac{3}{x+2}}{\dfrac{2}{x+2} - \dfrac{x}{x+5}} \cdot \dfrac{(x+5)(x+2)}{(x+5)(x+2)}$

$$= \frac{x(x+2) + 3(x+5)}{2(x+5) - x(x+2)}$$

$$= \frac{x^2 + 2x + 3x + 15}{2x + 10 - x^2 - 2x}$$

$$= \frac{x^2 + 5x + 15}{-x^2 + 10}$$

59. $\left[\dfrac{\dfrac{x-1}{x-1} - 1}{\dfrac{x+1}{x-1} + 1}\right]^5$

Consider the numerator of the complex rational expression:

$$\frac{x-1}{x-1} - 1 = 1 - 1 = 0$$

Since the denominator, $\dfrac{x+1}{x-1} + 1$ is not equal to 0, the simplified form of the original expression is 0.

61. $\dfrac{\dfrac{z}{1 - \dfrac{z}{2+2z}} - 2z}{\dfrac{2z}{5z-2} - 3} = \dfrac{\dfrac{z}{\dfrac{2+2z-z}{2+2z}} - 2z}{\dfrac{2z-15z+6}{5z-2}}$

$$= \frac{\dfrac{z}{\dfrac{2+z}{2+2z}} - 2z}{\dfrac{-13z+6}{5z-2}}$$

$$= \frac{z \cdot \dfrac{2+2z}{2+z} - 2z}{\dfrac{-13z+6}{5z-2}}$$

$$= \frac{\dfrac{z(2+2z) - 2z(2+z)}{2+z}}{\dfrac{-13z+6}{5z-2}}$$

$$= \frac{\dfrac{2z + 2z^2 - 4z - 2z^2}{2+z}}{\dfrac{-13z+6}{5z-2}}$$

$$= \frac{\dfrac{-2z}{2+z}}{\dfrac{-13z+6}{5z-2}}$$

$$= \frac{-2z}{2+z} \cdot \frac{5z-2}{-13z+6}$$

$$= \frac{-2z(5z-2)}{(2+z)(-13z+6)}, \text{ or}$$

$$\frac{2z(5z-2)}{(2+z)(13z-6)}$$

63. *Writing Exercise.* When a variable appears only in the numerator(s) of the rational expression(s) that are in the numerator of the complex rational expression, there will be no restrictions on the variables.

Exercise Set 6.6

1. The statement is false. See Example 2(c).

3. The statement is true. See page 404 in the text.

5. Because no variable appears in a denominator, no restrictions exist.

$$\frac{3}{5} - \frac{2}{3} = \frac{x}{6}, \text{ LCD} = 30$$

$$30\left(\frac{3}{5} - \frac{2}{3}\right) = 30 \cdot \frac{x}{6}$$

$$30 \cdot \frac{3}{5} - 30 \cdot \frac{2}{3} = 30 \cdot \frac{x}{6}$$

$$18 - 20 = 5x$$

$$-2 = 5x$$

$$\frac{-2}{5} = x$$

Check:

$$\frac{3}{5} - \frac{2}{3} = \frac{x}{6}$$

$\frac{3}{5} - \frac{2}{3}$	$\frac{-2}{6}$
$\frac{3}{5} - \frac{2}{3}$	$\frac{5}{6}$
$\frac{18}{30} - \frac{20}{30}$	$\frac{-2}{5} \cdot \frac{1}{6}$
$-\frac{2}{30} \overset{?}{=} \frac{-2}{30}$	TRUE

This checks, so the solution is $-\frac{2}{5}$.

7. Note that x cannot be 0.

$$\frac{1}{3} + \frac{5}{6} = \frac{1}{x}, \text{ LCD} = 6x$$

$$6x\left(\frac{1}{3} + \frac{5}{6}\right) = 6x \cdot \frac{1}{x}$$

$$6x \cdot \frac{1}{3} + 6x \cdot \frac{5}{6} = 6x \cdot \frac{1}{x}$$

$$2x + 5x = 6$$

$$7x = 6$$

$$x = \frac{6}{7}$$

Check:

$$\frac{1}{3} + \frac{5}{6} = \frac{1}{x}$$

$\frac{1}{3} + \frac{5}{6}$	$\frac{1}{\frac{6}{7}}$
$\frac{2}{6} + \frac{5}{6}$	$1 \cdot \frac{7}{6}$
$\frac{7}{6} \overset{?}{=} \frac{7}{6}$	TRUE

This checks, so the solution is $\frac{6}{7}$.

9. Note that t cannot be 0.

$$\frac{1}{8} + \frac{1}{12} = \frac{1}{t}, \text{ LCD} = 48t$$

$$48t\left(\frac{1}{8} + \frac{1}{12}\right) = 48t \cdot \frac{1}{t}$$

$$48t \cdot \frac{1}{8} + 48t \cdot \frac{1}{12} = 48t \cdot \frac{1}{t}$$

$$6t + 4t = 48$$

$$10t = 48$$

$$t = \frac{24}{5}$$

Check:

$$\frac{1}{8} + \frac{1}{12} = \frac{1}{t}$$

$\frac{1}{8} + \frac{1}{12}$	$\frac{1}{\frac{24}{5}}$
$\frac{3}{24} + \frac{2}{24}$	$1 \cdot \frac{5}{24}$
$\frac{5}{24} \overset{?}{=} \frac{5}{24}$	TRUE

This checks, so the solution is $\frac{24}{5}$.

11. Note that y cannot be 0.

$$y + \frac{4}{y} = -5 \text{ LCD } = y$$

$$y\left(y + \frac{4}{y}\right) = -5 \cdot y$$

$$y \cdot y + y \cdot \frac{4}{y} = -5 \cdot y$$

$$y^2 + 4 = -5y$$

$$y^2 + 5y + 4 = 0$$

$$(y + 4)(y + 1) = 0$$

$$y + 4 = 0 \text{ or } y + 1 = 0$$

$$y = -4 \text{ or } y = -1$$

Check:

$$y + \frac{4}{y} = -5$$

$-4 + \frac{4}{-4}$	-5
$-4 - 1$	
$-5 \overset{?}{=} -5$	TRUE

$$y + \frac{4}{y} = -5$$

$-1 + \frac{4}{-1}$	-5
$-1 - 4$	
$-5 \overset{?}{=} -5$	TRUE

Both of these check, so the solutions are -4 and -1.

13. Note that x cannot be 0.

$$\frac{x}{6} - \frac{6}{x} = 0, \text{ LCD} = 6x$$

$$6x\left(\frac{x}{6} - \frac{6}{x}\right) = 6x \cdot 0$$

$$6x \cdot \frac{x}{6} - 6x \cdot \frac{6}{x} = 6x \cdot 0$$

$$x^2 - 36 = 0$$

$$(x + 6)(x - 6) = 0$$

$$x + 6 = 0 \quad or \quad x - 6 = 0$$

$$x = -6 \quad or \qquad x = 6$$

Check:

$$\frac{x}{6} - \frac{6}{x} = 0 \qquad\qquad \frac{x}{6} - \frac{6}{x} = 0$$

$$\frac{-6}{6} - \frac{6}{-6} \;\bigg|\; 0 \qquad\qquad \frac{6}{6} - \frac{6}{6} \;\bigg|\; 0$$

$$-1 + 1 \qquad\qquad\qquad\quad 1 - 1$$

$$0 \overset{?}{=} 0 \;\; \text{TRUE} \qquad\quad 0 \overset{?}{=} 0 \;\; \text{TRUE}$$

Both of these check, so the two solutions are -6 and 6.

15. Note that x cannot be 0.

$$\frac{2}{x} = \frac{5}{x} - \frac{1}{4}, \;\; \text{LCD} = 4x$$

$$4x \cdot \frac{2}{x} = 4x \left(\frac{5}{x} - \frac{1}{4} \right)$$

$$4x \cdot \frac{2}{x} = 4x \cdot \frac{5}{x} - 4x \cdot \frac{1}{4}$$

$$8 = 20 - x$$

$$-12 = -x$$

$$12 = x$$

Check:

$$\frac{2}{x} = \frac{5}{x} - \frac{1}{4}$$

$$\frac{2}{12} \;\bigg|\; \frac{5}{12} - \frac{1}{4}$$

$$\bigg|\; \frac{5}{12} - \frac{3}{12}$$

$$\bigg|\; \frac{2}{12}$$

$$\frac{2}{12} \overset{?}{=} \frac{2}{12} \qquad\qquad \text{TRUE}$$

This checks, so the solution is 12.

17. Note that t cannot be 0.

$$\frac{5}{3t} + \frac{3}{t} = 1, \;\; \text{LCD} = 3t$$

$$3t \left(\frac{5}{3t} + \frac{3}{t} \right) = 3t \cdot 1$$

$$3t \cdot \frac{5}{3t} + 3t \cdot \frac{3}{t} = 3t \cdot 1$$

$$5 + 9 = 3t$$

$$14 = 3t$$

$$\frac{14}{3} = t$$

Check:

$$\frac{5}{3t} + \frac{3}{t} = 1$$

$$\frac{5}{3 \cdot \frac{14}{3}} + \frac{3}{\frac{14}{3}} \;\bigg|\; 1$$

$$\frac{5}{14} + \frac{9}{14}$$

$$\frac{14}{14}$$

$$1 \overset{?}{=} 1 \;\; \text{TRUE}$$

This checks, so the solution is $\frac{14}{3}$.

19. To avoid the division by 0, we must have $n - 6 \neq 0$, or $n \neq 6$.

$$\frac{n+2}{n-6} = \frac{1}{2}, \;\; \text{LCD} = 2(n-6)$$

$$2(n-6) \cdot \frac{n+2}{n-6} = 2(n-6) \cdot \frac{1}{2}$$

$$2(n+2) = n - 6$$

$$2n + 4 = n - 6$$

$$n = -10$$

Check:

$$\frac{n+2}{n-6} = \frac{1}{2}$$

$$\frac{-10+2}{-10-6} \;\bigg|\; \frac{1}{2}$$

$$\frac{-8}{-16}$$

$$\frac{1}{2} \overset{?}{=} \frac{1}{2} \;\; \text{TRUE}$$

This checks, so the solution is -10.

21. Note that x cannot be 0.

$$x + \frac{12}{x} = -7, \;\; \text{LCD is } x$$

$$x \left(x + \frac{12}{x} \right) = x \cdot (-7)$$

$$x \cdot x + x \cdot \frac{12}{x} = -7x$$

$$x^2 + 12 = -7x$$

$$x^2 + 7x + 12 = 0$$

$$(x+3)(x+4) = 0$$

$$x + 3 = 0 \quad or \quad x + 4 = 0$$

$$x = -3 \quad or \quad x = -4$$

Both numbers check, so the solutions are -3 and -4.

23. To avoid division by 0, we must have $x - 4 \neq 0$ and $x + 1 \neq 0$, or $x \neq 4$ and $x \neq -1$.

$$\frac{3}{x-4} = \frac{5}{x+1}, \text{ LCD} = (x-4)(x+1)$$

$$(x-4)(x+1) \cdot \frac{3}{x-4} = (x-4)(x+1) \cdot \frac{5}{x+1}$$

$$3(x+1) = 5(x-4)$$

$$3x + 3 = 5x - 20$$

$$23 = 2x$$

$$\frac{23}{2} = x$$

This checks, so the solution is $\frac{23}{2}$.

25. Because no variable appears in a denominator, no restrictions exist.

$$\frac{a}{6} - \frac{a}{10} = \frac{1}{6}, \text{ LCD} = 30$$

$$30\left(\frac{a}{6} - \frac{a}{10}\right) = 30 \cdot \frac{1}{6}$$

$$30 \cdot \frac{a}{6} - 30 \cdot \frac{a}{10} = 30 \cdot \frac{1}{6}$$

$$5a - 3a = 5$$

$$2a = 5$$

$$a = \frac{5}{2}$$

This checks, so the solution is $\frac{5}{2}$.

27. Because no variable appears in a denominator, no restrictions exist.

$$\frac{x+1}{3} - 1 = \frac{x-1}{2}, \text{ LCD} = 6$$

$$6\left(\frac{x+1}{3} - 1\right) = 6 \cdot \frac{x-1}{2}$$

$$6 \cdot \frac{x+1}{3} - 6 \cdot 1 = 6 \cdot \frac{x-1}{2}$$

$$2(x+1) - 6 = 3(x-1)$$

$$2x + 2 - 6 = 3x - 3$$

$$2x - 4 = 3x - 3$$

$$-1 = x$$

This checks, so the solution is -1.

29. To avoid division by 0, we must have $y - 3 \neq 0$, or $y \neq 3$.

$$\frac{y+3}{y-3} = \frac{6}{y-3}, \text{ LCD} = y - 3$$

$$(y-3) \cdot \frac{y+3}{y-3} = (y-3) \cdot \frac{6}{y-3}$$

$$y + 3 = 6$$

$$y = 3$$

Because of the restriction $y \neq 3$, the number 3 must be rejected as a solution. The equation has no solution.

31. To avoid division by 0, we must have $x + 4 \neq 0$ and $x \neq 0$, or $x \neq -4$ and $x \neq 0$.

$$\frac{3}{x+4} = \frac{5}{x}, \text{ LCD} = x(x+4)$$

$$x(x+4) \cdot \frac{3}{x+4} = x(x+4) \cdot \frac{5}{x}$$

$$3x = 5(x+4)$$

$$3x = 5x + 20$$

$$-2x = 20$$

$$x = -10$$

This checks, so the solution is -10.

33. To avoid division by 0, we must have $n + 2 \neq 0$ and $n + 1 \neq 0$, or $n \neq -2$ and $n \neq -1$.

$$\frac{n+1}{n+2} = \frac{n-3}{n+1}, \text{ LCD} = (n+2)(n+1)$$

$$(n+2)(n+1) \cdot \frac{n+1}{n+2} = (n+2)(n+1) \cdot \frac{n-3}{n+1}$$

$$(n+1)(n+1) = (n+2)(n-3)$$

$$n^2 + 2n + 1 = n^2 - n - 6$$

$$3n = -7$$

$$n = \frac{-7}{3}$$

This checks, so the solution is $\frac{-7}{3}$.

35. To avoid division by 0, we must have $t - 2 \neq 0$, or $t \neq 2$.

$$\frac{5}{t-2} + \frac{3t}{t-2} = \frac{4}{t^2-4t+4}, \text{ LCD is } (t-2)^2$$

$$(t-2)^2\left(\frac{5}{t-2} + \frac{3t}{t-2}\right) = (t-2)^2 \cdot \frac{4}{(t-2)^2}$$

$$5(t-2) + 3t(t-2) = 4$$

$$5t - 10 + 3t^2 - 6t = 4$$

$$3t^2 - t - 10 = 4$$

$$3t^2 - t - 14 = 0$$

$$(3t-7)(t+2) = 0$$

$$3t - 7 = 0 \quad or \quad t + 2 = 0$$

$$3t = 7 \quad or \quad t = -2$$

$$t = \frac{7}{3} \quad or \quad t = -2$$

Both numbers check. The solutions are $\frac{7}{3}$ and -2.

37. To avoid division by 0, we must have $x+5 \neq 0$ and $x-5 \neq 0$, or $x \neq -5$ and $x \neq 5$.
$$\frac{x}{x+5} - \frac{5}{x-5} = \frac{14}{x^2-25}.$$
$$\text{LCD} = (x+5)(x-5)$$

$$(x+5)(x-5) \cdot \qquad (x+5)(x-5) \cdot$$
$$\left(\frac{x}{x+5} - \frac{5}{x-5}\right) = \frac{14}{(x+5)(x-5)}$$
$$x(x-5) - 5(x+5) = 14$$
$$x^2 - 5x - 5x - 25 = 14$$
$$x^2 - 10x - 39 = 0$$
$$(x-13)(x+3) = 0$$
$$x-13 = 0 \text{ or } x+3 = 0$$
$$x = 13 \text{ or } x = -3$$

Both numbers check. The solutions are $-3, 13$.

39. To avoid division by 0, we must have $t-3 \neq 0$ and $t+3 \neq 0$, or $t \neq 3$ and $t \neq -3$.
$$\frac{5}{t-3} - \frac{30}{t^2-9} = 1,$$
$$\text{LCD} = (t-3)(t+3)$$

$$(t-3)(t+3) \cdot$$
$$\left(\frac{5}{t-3} - \frac{30}{t^2-9}\right) = (t-3)(t+3) \cdot 1$$
$$5(t+3) - 30 = (t-3)(t+3)$$
$$5t + 15 - 30 = t^2 - 9$$
$$0 = t^2 - 5t + 6$$
$$0 = (t-3)(t-2)$$
$$t-3 = 0 \text{ or } t-2 = 0$$
$$t = 3 \text{ or } t = 2$$

Because of the restriction $t \neq 3$, we must reject the number 3 as a solution. The number 2 checks, so it is the solution.

41. To avoid division by 0, we must have $6-a \neq 0$ (or equivalently $a-6 \neq 0$) or $a \neq 6$.
$$\frac{7}{6-a} = \frac{a+1}{a-6}$$
$$\frac{-1}{-1} \cdot \frac{7}{6-a} = \frac{a+1}{a-6}$$
$$\frac{-7}{a-6} = \frac{a+1}{a-6}, \quad \text{LCD} = a-6$$
$$(a-6) \cdot \frac{-7}{a-6} = (a-6) \cdot \frac{a+1}{a-6}$$
$$-7 = a+1$$
$$-8 = a$$

This checks.

43. $\dfrac{-2}{x+2} = \dfrac{x}{x+2}$

To avoid division by 0, we must have $x+2 \neq 0$, or $x \neq -2$. Now observe that the denominators are the same, so the numerators must be the same. Thus, we have $-2 = x$, but because of the restriction $x \neq -2$ this cannot be a solution. The equation has no solution.

45. Note that x cannot be 0.

$$\frac{12}{x} = \frac{x}{3}, \quad \text{LCD} = 3x$$
$$3x \cdot \frac{12}{x} = 3x \cdot \frac{x}{3}$$
$$36 = x^2$$
$$0 = x^2 - 36$$
$$0 = (x+6)(x-6)$$
$$x+6 = 0 \text{ or } x-6 = 0$$
$$x = -6 \text{ or } x = 6$$

This checks.

47. *Writing Exercise.* When solving rational equations, we multiply each side by the LCM of the denominators in order to clear fractions.

49. **Familiarize.** Let $x =$ the first odd integer. Then $x+2 =$ the next odd integer.

Translate.

The sum of two consecutive odd integers   is   276.
$$\underbrace{x + (x+2)}_{} \quad = \quad 276$$

Carry out. We solve the equation.
$$x + (x+2) = 276$$
$$2x + 2 = 276$$
$$2x = 274$$
$$x = 137$$

When $x = 137$, then $x+2 = 137 + 2 = 139$.

Check. The numbers 137 and 139 are consecutive odd integers and $137 + 139 = 276$. These numbers check.

State. The integers are 137 and 139.

51. **Familiarize.** Let $b =$ the base of the triangle, in cm. Then $b+3 =$ the height. Recall that the area of a triangle is given by $\frac{1}{2} \times$ base $\times$ height.

Translate.

The area of the triangle   is   54 cm^2.
$$\frac{1}{2} \cdot b \cdot (b+3) \quad = \quad 54$$

Carry out. We solve the equation.
$$\frac{1}{2}b(b+3) = 54$$
$$2 \cdot \frac{1}{2}b(b+3) = 2 \cdot 54$$
$$b(b+3) = 108$$
$$b^2 + 3b = 108$$
$$b^2 + 3b - 108 = 0$$
$$(b-9)(b+12) = 0$$
$$b-9 = 0 \quad or \quad b+12 = 0$$
$$b = 9 \quad or \qquad b = -12$$

Check. The length of the base cannot be negative so we need to check only 9. If the base is 9 cm, then the height is $9+3$, or 12 cm, and the area is $\frac{1}{2} \cdot 9 \cdot 12$, or 54 cm^2. The answer checks.

State. The base measures 9 cm, and the height measures 12 cm.

53. To find the rate, in centimeters per day, we divide the amount of growth by the number of days. From June 9 to June 24 is $24 - 9 = 15$ days.

$$\text{Rate, in cm per day} = \frac{0.9 \text{ cm}}{15 \text{ days}}$$
$$= 0.06 \text{ cm/day}$$
$$= 0.06 \text{ cm per day}$$

55. *Writing Exercise*. Begin with an equation. Then divide on both sides of the equation by an expression whose value is zero for at least one solution of the equation. See Exercises 43 and 44 for examples.

57. To avoid division by 0, we must have $x - 3 \neq 0$, or $x \neq 3$.

$$1 + \frac{x-1}{x-3} = \frac{2}{x-3} - x, \quad \text{LCD} = x - 3$$
$$(x-3)\left(1 + \frac{x-1}{x-3}\right) = (x-3)\left(\frac{2}{x-3} - x\right)$$
$$(x-3) \cdot 1 + (x-3) \cdot \frac{x-1}{x-3} = (x-3) \cdot \frac{2}{x-3} - (x-3)x$$
$$x - 3 + x - 1 = 2 - x^2 + 3x$$
$$2x - 4 = 2 - x^2 + 3x$$
$$x^2 - x - 6 = 0$$
$$(x-3)(x+2) = 0$$
$$x - 3 = 0 \quad or \quad x + 2 = 0$$
$$x = 3 \quad or \qquad x = -2$$

Because of the restriction $x \neq 3$, we must reject the number 3 as a solution. The number -2 checks, so it is the solution.

59. To avoid division by 0, we must have $x + 2 \neq 0$ and $x - 2 \neq 0$, or $x \neq -2$ and $x \neq 2$.

$$\frac{12 - 6x}{x^2 - 4} = \frac{3x}{x+2} - \frac{3 - 2x}{2 - x},$$
$$\frac{12 - 6x}{(x+2)(x-2)} = \frac{3x}{x+2} - \frac{3 - 2x}{2 - x} \cdot \frac{-1}{-1}$$
$$\frac{12 - 6x}{(x+2)(x-2)} = \frac{3x}{x+2} - \frac{2x - 3}{x - 2},$$
$$\text{LCD} = (x+2)(x-2)$$
$$(x+2)(x-2) \cdot \frac{12 - 6x}{(x+2)(x-2)} =$$
$$\qquad (x+2)(x-2)\left(\frac{3x}{x+2} - \frac{2x-3}{x-2}\right)$$
$$12 - 6x =$$
$$\qquad 3x(x-2) - (x+2)(2x-3)$$
$$12 - 6x =$$
$$\qquad 3x^2 - 6x - 2x^2 - x + 6$$
$$0 = x^2 - x - 6$$
$$0 = (x-3)(x+2)$$
$$x - 3 = 0 \quad or \quad x + 2 = 0$$
$$x = 3 \quad or \qquad x = -2$$

Because of the restriction $x \neq -2$, we must reject the number -2 as a solution. The number 3 checks, so it is the solution.

61. To avoid division by 0, we must have $a + 3 \neq 0$, or $a \neq -3$.

$$7 - \frac{a-2}{a+3} = \frac{a^2 - 4}{a+3} + 5, \quad \text{LCD} = a + 3$$
$$(a+3)\left(7 - \frac{a-2}{a+3}\right) = (a+3)\left(\frac{a^2-4}{a+3} + 5\right)$$
$$7(a+3) - (a-2) = a^2 - 4 + 5(a+3)$$
$$7a + 21 - a + 2 = a^2 - 4 + 5a + 15$$
$$6a + 23 = a^2 + 5a + 11$$
$$0 = a^2 - a - 12$$
$$0 = (a-4)(a+3)$$
$$a - 4 = 0 \quad or \quad a + 3 = 0$$
$$a = 4 \quad or \qquad a = -3$$

Because of the restriction $a \neq -3$, we must reject the number -3 as a solution. The number 4 checks, so it is the solution.

63. To avoid division by 0, we must have $x - 1 \neq 0$, or $x \neq 1$.

$$\frac{1}{x-1} + x - 5 = \frac{5x - 4}{x-1} - 6, \quad \text{LCD} = x - 1$$
$$(x-1)\left(\frac{1}{x-1} + x - 5\right) = (x-1)\left(\frac{5x-4}{x-1} - 6\right)$$
$$1 + x(x-1) - 5(x-1) = 5x - 4 - 6(x-1)$$
$$1 + x^2 - x - 5x + 5 = 5x - 4 - 6x + 6$$
$$x^2 - 6x + 6 = -x + 2$$
$$x^2 - 5x + 4 = 0$$
$$(x-1)(x-4) = 0$$
$$x - 1 = 0 \quad or \quad x - 4 = 0$$
$$x = 1 \quad or \qquad x = 4$$

Because of the restriction $x \neq 1$, we must reject the number 1 as a solution. The number 4 checks, so it is the solution.

65. Note that x cannot be 0.

$$\frac{\frac{1}{x} + 1}{x} = \frac{\frac{1}{x}}{2}$$
$$\left(\frac{1}{x} + 1\right) \cdot \frac{1}{x} = \frac{1}{x} \cdot \frac{1}{2}$$
$$\frac{1}{x^2} + \frac{1}{x} = \frac{1}{2x}, \quad \text{LCD} = 2x^2$$
$$2 + 2x = x$$
$$2 = -x$$
$$-2 = x$$

This checks.

67. ▓

Chapter 6 Connecting the Concepts

1. Expression; $\dfrac{4x^2 - 8x}{4x^2 + 4x} = \dfrac{4x(x-2)}{4x(x+1)} = \dfrac{x-2}{x+1}$

3. Equation; $\dfrac{3}{y} - \dfrac{1}{4} = \dfrac{1}{y}$ Note: $y \neq 0$

$$4y\left(\dfrac{3}{y} - \dfrac{1}{4}\right) = 4y\left(\dfrac{1}{y}\right) \qquad \text{LCD} = 4y$$

$$12 - y = 4$$
$$-y = -8$$
$$y = 8$$

The solution is 8.

5. Equation; $\dfrac{5}{x+3} = \dfrac{3}{x+2}$ Note $x \neq -2, -3$

$$(x+2)(x+3)\cdot\dfrac{5}{x+3} = (x+2)(x+3)\cdot\dfrac{3}{x+2}$$
$$\text{LCD} = (x+2)(x+3)$$

$$5(x+2) = 3(x+3)$$
$$5x + 10 = 3x + 9$$
$$5x = 3x - 1$$
$$2x = -1$$
$$x = \dfrac{-1}{2}$$

7. Expression; $\dfrac{2a}{a+1} - \dfrac{4a}{1-a^2}$

$$= \dfrac{2a}{a+1} - \dfrac{4a}{a^2-1}$$
$$= \dfrac{2a(a-1)+4a}{(a+1)(a-1)} \qquad \text{LCD} = (a+1)(a-1)$$
$$= \dfrac{2a^2 - 2a + 4}{(a+1)(a-1)}$$
$$= \dfrac{2a^2 - 2a}{(a+1)(a-1)}$$
$$= \dfrac{2a(a-1)}{(a+1)(a-1)}$$
$$= \dfrac{2a}{a+1}$$

9. Expression; $\dfrac{18x^2}{25} \div \dfrac{12x}{5} = \dfrac{18x^2}{25}\cdot\dfrac{5}{12x}$

$$= \dfrac{3x}{10}\cdot\dfrac{5\cdot 6x}{5\cdot 6x}$$
$$= \dfrac{3x}{10}$$

Exercise Set 6.7

1. Familiarize. The job takes Kelby 10 hr working alone and Rosa 8hr working alone. Then in 1 hr Kelby does $\dfrac{1}{10}$ of the job and Rosa does $\dfrac{1}{8}$ of the job. Working together they can do $\dfrac{1}{10} + \dfrac{1}{8}$, or $\dfrac{9}{40}$ of the job in 1 hr. In 4 hr, Kelby does $4\cdot\dfrac{1}{10}$ of the job and Rosa does $4\cdot\dfrac{1}{8}$ of the job. Working together they can do $4\cdot\dfrac{1}{10} + 4\cdot\dfrac{1}{8}$, or $\dfrac{9}{10}$, of the job in 4 hr. In 6 hr, Kelby does $6\cdot\dfrac{1}{10}$ of the job and Rosa does $6\cdot\dfrac{1}{8}$ of the job. Working together they can do $6\cdot\dfrac{1}{10} + 6\cdot\dfrac{1}{8}$, or $1\dfrac{7}{20}$ of the job which is more of the job than needs to be done. The answer is somewhere between 4 hr and 6 hr.

Translate. If they work together t hours, then Kelby does $t\cdot\dfrac{1}{10}$ of the job and Rosa does $t\cdot\dfrac{1}{8}$ of the job. We want a number t such that

$$\left(\dfrac{1}{10} + \dfrac{1}{8}\right)t = 1, \text{ or } \dfrac{9}{40}\cdot t = 1.$$

Carry out. We solve the equation.

$$\dfrac{9}{40}\cdot t = 1$$
$$\dfrac{40}{9}\cdot\dfrac{9}{40}\cdot t = 1\cdot\dfrac{40}{9}$$
$$t = \dfrac{40}{9}, \text{ or } 4\dfrac{4}{9}$$

Check. We can repeat the computations. We also expected the result to be between 4 hr 6 hr, as it is.

State. Working together, it takes Kelby and Rosa $\dfrac{40}{9}$ hr or $4\dfrac{4}{9}$ hr to do the job.

3. Familiarize. The job takes Oliver 75 hours working alone and Pat 100 hours working alone. Then in 1 hour Oliver does $\dfrac{1}{75}$ of the job and Pat does $\dfrac{1}{100}$ of the job. Working together, they can do $\dfrac{1}{75} + \dfrac{1}{100}$, or $\dfrac{7}{300}$ of the job in 1 hour. In 40 hours, Oliver does $40\cdot\dfrac{1}{75}$ of the job and Pat does $40\cdot\dfrac{1}{100}$ of the job. Working together they can do $40\cdot\dfrac{1}{75} + 40\cdot\dfrac{1}{100}$, or $\dfrac{14}{15}$ of the job in 40 hours. In 45 hours they can do $45\cdot\dfrac{1}{75} + 45\cdot\dfrac{1}{100}$, or $\dfrac{21}{20}$ or $1\dfrac{1}{20}$ of the job which is more of the job than needs to be done. The answer is somewhere between 40 hr and 45 hr.

Translate. If they work together t hours, then Oliver does $t\left(\dfrac{1}{75}\right)$ of the job and Pat does $t\left(\dfrac{1}{100}\right)$ of the job. We want some number t such that

$$\left(\dfrac{1}{75} + \dfrac{1}{100}\right)t = 1, \text{ or } \dfrac{7}{300}\cdot t = 1.$$

Carry out. We solve the equation.

$$\dfrac{7}{300}\cdot t = 1$$
$$\dfrac{300}{7}\cdot\dfrac{7}{300}\cdot t = \dfrac{300}{7}\cdot 1$$
$$t = \dfrac{300}{7}, \text{ or } 42\dfrac{6}{7}$$

Check. The check can be done by repeating the computations. We also expected the result would be between 40 hr and 45 hr.

State. Working together, it takes them $\dfrac{300}{7}$ hr, or $42\dfrac{6}{7}$ hr.

5. Familiarize. The job takes Barry 10 hours working alone and Deb 15 hours working alone. Then in 1 hour Barry does $\dfrac{1}{10}$ of the job and Deb does $\dfrac{1}{15}$ of the job. Working together they can do $\dfrac{1}{10} + \dfrac{1}{15}$, or $\dfrac{1}{6}$ of the job in 1 hour. In 4 hours, Barry does $4\left(\dfrac{1}{10}\right)$ of the job and Deb does $4\left(\dfrac{1}{15}\right)$

of the job. Working together they can do $4\left(\frac{1}{10}\right) + 4\left(\frac{1}{15}\right)$, or $\frac{2}{3}$ of the job in 4 hours. In 7 hours, Barry does $7\left(\frac{1}{10}\right)$ of the job and Deb does $7\left(\frac{1}{15}\right)$ of the job. Working together they can do $7\left(\frac{1}{10}\right) + 7\left(\frac{1}{15}\right)$, or $\frac{7}{6}$ of the job which is more of the job than needs to be done. The answer is somewhere between 4 hr and 7 hr.

Translate. If they work together t hours, then Barry does $t\left(\frac{1}{10}\right)$ of the job and Deb does $t\left(\frac{1}{15}\right)$ of the job. We want some number t such that

$$\left(\frac{1}{10} + \frac{1}{15}\right)t = 1, \text{ or } \frac{1}{6}t = 1.$$

Carry out. We solve the equation.

$$\frac{1}{6}t = 1$$
$$6 \cdot \frac{1}{6}t = 6 \cdot 1$$
$$t = 6$$

Check. We repeat computations. The answer checks. We also expected the result to be between 4 hr. and 7 hr.

State. Working together, it takes Barry and Deb 6 hrs.

7. **Familiarize**. The job takes Jorell 20 hours working alone and Ferdous 24 hours working alone. Then in 1 hour Jorell does $\frac{1}{20}$ of the job and Ferdous does $\frac{1}{24}$ of the job. Working together, they can do $\frac{1}{20} + \frac{1}{24}$, or $\frac{11}{120}$ of the job in 1 hour. In 8 hours, Jorell does $8\left(\frac{1}{20}\right)$ of the job and Ferdous does $8\left(\frac{1}{24}\right)$ of the job. Working together they can do $8\left(\frac{1}{20}\right) + 8\left(\frac{1}{24}\right)$, or $\frac{11}{15}$ of the job in 8 hours. In 12 hours Jorell does $12\left(\frac{1}{20}\right)$ of the job and Ferdous does $12\left(\frac{1}{24}\right)$ of the job. Working together they can do $12\left(\frac{1}{20}\right) + 12\left(\frac{1}{24}\right)$, or $\frac{11}{10}$ of the job which is more of the job then needs to be done. The answer is somewhere between 8 hours and 12 hours.

Translate. If they work together t days, then Jorell does $t\left(\frac{1}{20}\right)$ of the job and Ferdous does $t\left(\frac{1}{24}\right)$ of the job. We want some number t such that

$$\left(\frac{1}{20} + \frac{1}{24}\right)t = 1, \text{ or } \frac{11}{120}t = 1.$$

Carry out. We solve the equation.

$$\frac{11}{120}t = 1$$
$$\frac{120}{11} \cdot \frac{11}{120}t = \frac{120}{11} \cdot 1$$
$$t = \frac{120}{11}$$

Check. We repeat computations. The answer checks. We also expected the result to be between 8 hr. and 12 hr.

State. Working together, it takes Jorell and Ferdous 120/11 hrs, or 10 10/11 hrs.

9. **Familiarize**. The job takes Aficio 7 minutes working alone and MX 6 minutes working alone. Then in 1 minute Aficio does $\frac{1}{7}$ of the job and MX does $\frac{1}{6}$ of the job. Working together, they can do $\frac{1}{7} + \frac{1}{6}$, or $\frac{13}{42}$ of the job in 1 minute. In 2 minutes, Aficio does $2\left(\frac{1}{7}\right)$ of the job and MX does $2\left(\frac{1}{6}\right)$ of the job. Working together they can do $2\left(\frac{1}{7}\right) + 2\left(\frac{1}{6}\right)$, or $\frac{13}{21}$ of the job in 2 minutes. In 4 minutes Aficio does $4\left(\frac{1}{7}\right)$ of the job and MX does $4\left(\frac{1}{6}\right)$ of the job. Working together they can do $4\left(\frac{1}{7}\right) + 4\left(\frac{1}{6}\right)$, or $\frac{26}{21}$ of the job which is more of the job then needs to be done. The answer is somewhere between 2 minutes and 4 minutes.

Translate. If they work together t minutes, then Aficio does $t\left(\frac{1}{7}\right)$ of the job and MX does $t\left(\frac{1}{6}\right)$ of the job. We want some number t such that

$$\left(\frac{1}{7} + \frac{1}{6}\right)t = 1, \text{ or } \frac{13}{42}t = 1.$$

Carry out. We solve the equation.

$$\frac{13}{42}t = 1$$
$$\frac{42}{13} \cdot \frac{13}{42}t = \frac{42}{13} \cdot 1$$
$$t = \frac{42}{13}$$

Check. We repeat computations. The answer checks. We also expected the result to be between 2 minutes and 4 minutes.

State. Working together, it takes Aficio and MX 42/13 mins, or 3 3/13 mins.

11. **Familiarize**. We complete the table shown in the text.

	Distance	Speed	Time
	d	= r	$\cdot$ t
B & M	330	$r - 14$	$\dfrac{330}{r-14}$
AMTRAK	400	r	$\dfrac{400}{r}$

Translate. Since the time must be the same for both trains, we have the equation

$$\frac{330}{r - 14} = \frac{400}{r}.$$

Carry out. We first multiply by the LCD, $r(r - 14)$.

$$r(r - 14) \cdot \frac{330}{r - 14} = r(r - 14) \cdot \frac{400}{r}$$
$$330r = 400(r - 14)$$
$$330r = 400r - 5600$$
$$-70r = -5600$$
$$r = 80$$

If the speed of the AMTRAK train is 80 km/h, then the speed of the B & M train is $80 - 14$, or 66 km/h.

Check. The speed of the B&M train is 14 km/h slower than the speed of the AMTRAK train. At 66 km/h the B&M train travels 330 km in 330/66, or 5 hr. At 80 km/h the AMTRAK train travels 400 km in 400/80, or 5 hr. The times are the same, so the answer checks.

State. The speed of the AMTRAK train is 80 km/h, and the speed of the B&M freight train is 66 km/h.

13. ***Familiarize***. Let $r =$ the speed of Rita's Harley, in mph. Then $r + 15 =$ the speed of Sean's Camaro. We organize the information in a table using the formula time = distance/rate to fill in the last column.

	Distance	Speed	Time
Harley	120	r	$\dfrac{120}{r}$
Camaro	156	$r + 15$	$\dfrac{156}{r + 15}$

Translate. Since the times must be the same, we have the equation

$$\frac{120}{r} = \frac{156}{r + 15}.$$

Carry out. We first multiply by the LCD, $r(r + 15)$.

$$r(r + 15) \cdot \frac{120}{r} = r(r + 15) \cdot \frac{156}{r + 15}$$
$$120(r + 15) = 156r$$
$$120r + 1800 = 156r$$
$$1800 = 36r$$
$$50 = r$$

Then $r + 15 = 50 + 15 = 65$.

Check. At 50 mph, the Harley travels 120 mi in 120/50, or 2.4 hr. At 65 mph, the Camaro travels 156 mi in 156/65, or 2.4 hr. The times are the same, so the answer checks.

State. The speed of Rita's Harley is 50 mph, and the speed of Sean's Camaro is 65 mph.

15. ***Familiarize***. Let $r =$ Tau's speed, in km/h. Then Baruti's speed is $r + 4$. We organize the information in a table.

	Distance	Speed	Time
Tau	7.5	r	$\dfrac{7.5}{r}$
Baruti	13.5	$r + 4$	$\dfrac{13.5}{r + 4}$

Translate. Since the times must be the same, we have the equation

$$\frac{7.5}{r} = \frac{13.5}{r + 4}.$$

Carry out. We first multiply by the LCD, $r(r + 4)$.

$$r(r + 4) \cdot \frac{7.5}{r} = r(r + 4) \cdot \frac{13.5}{r + 4}$$
$$7.5(r + 4) = 13.5r$$
$$7.5r + 30 = 13.5r$$
$$30 = 6r$$
$$5 = r$$

Then $r + 4 = 5 + 4 = 9$.

Check. At 5 km/h, Tau traveled 7.5 km in $\dfrac{7.5}{5} = 1.5$hr. At 9 km'hr, Baruti tarvels 13.5 km in $\dfrac{13.5}{9} = 1.5$hr. Since the times are the same, the answer checks.

State. Tau's speed is 5 km/h, and Baruti's speed is 9 km/h.

17. ***Familiarize***. Let $t =$ the time it takes Caledonia to drive to town and organize the given information in a table.

	Distance	Speed	Time
Caledonia	15	r	t
Manley	20	r	$t + 1$

Translate. We can replace the r's in the table above using the formula $r = d/t$.

	Distance	Speed	Time
Caledonia	15	$\dfrac{15}{t}$	t
Manley	20	$\dfrac{20}{t + 1}$	$t + 1$

Since the speeds are the same for both riders, we have the equation

$$\frac{15}{t} = \frac{20}{t + 1}.$$

Carry out. We multiply by the LCD, $t(t + 1)$.

$$t(t + 1) \cdot \frac{15}{t} = t(t + 1) \cdot \frac{20}{t + 1}$$
$$15(t + 1) = 20t$$
$$15t + 15 = 20t$$
$$15 = 5t$$
$$3 = t$$

If $t = 3$, then $t + 1 = 3 + 1$, or 4.

Check. If Caledonia's time is 3 hr and Manley's time is 4 hr, then Manley's time is 1 hr more than Caledonia's. Caledonia's speed is 15/3, or 5 mph. Manley's speed is 20/4, or 5 mph. Since the speeds are the same, the answer checks.

State. It takes Caledonia 3 hr to drive to town.

19. We write a proportion and then solve it.

$$\frac{b}{6} = \frac{7}{4}$$

$$b = \frac{7}{4} \cdot 6$$

$$b = \frac{42}{4}, \text{ or } 10.5$$

$\left(\text{Note that the proportions } \dfrac{6}{b} = \dfrac{4}{7}, \dfrac{b}{7} = \dfrac{6}{4}, \text{ or } \dfrac{7}{b} = \dfrac{4}{6} \text{ could also be used.}\right)$

21. We write a proportion and then solve it.

$$\frac{4}{f} = \frac{6}{4}$$

$$4f \cdot \frac{4}{f} = 4f \cdot \frac{6}{4}$$

$$16 = 6f$$

$$\frac{8}{3} = f \qquad \text{Simplifying}$$

$\left(\text{One of the following proportions could also be used: } \dfrac{f}{4} = \dfrac{4}{6}, \dfrac{4}{f} = \dfrac{9}{6}, \dfrac{f}{4} = \dfrac{6}{9}, \dfrac{4}{9} = \dfrac{f}{6}, \dfrac{9}{4} = \dfrac{6}{f}\right)$

23. From the blueprint we see that 9 in. represents 36 ft and that p in. represent 15 ft. We use a proportion to find p.

$$\frac{9}{36} = \frac{p}{15}$$

$$180 \cdot \frac{9}{36} = 180 \cdot \frac{p}{15}$$

$$45 = 12p$$

$$\frac{15}{4} = p, \text{ or}$$

$$3\frac{3}{4} = p$$

The length of p is $3\frac{3}{4}$ in.

25. From the blueprint we see that 9 in. represents 36 ft and that 5 in. represents r ft. We use a proportion to find r.

$$\frac{9}{36} = \frac{5}{r}$$

$$36r \cdot \frac{9}{36} = 36r \cdot \frac{5}{r}$$

$$9r = 180$$

$$r = 20$$

The length of r is 20 ft.

27. Consider the two similar right triangles in the drawing. One has legs 4 ft and 6 ft. The other has legs 10ft and l ft. We use a proportion to find l.

$$\frac{4}{6} = \frac{10}{l}$$

$$6l \cdot \frac{4}{6} = 6l \cdot \frac{10}{l}$$

$$4l = 60$$

$$l = 15ft$$

29.
$$\frac{a}{b} = \frac{c}{d}$$

$$\frac{10}{6} = \frac{5}{d}$$

$$6d \cdot \frac{10}{6} = 6d \cdot \frac{5}{d}$$

$$10d = 30$$

$$d = 3 \text{ cm}$$

31. Consider the two similar right triangles in the drawing. One has legs 5 and 7. The other has legs 9 and r. We use a proportion to find r.

$$\frac{5}{7} = \frac{9}{r}$$

$$7r \cdot \frac{5}{7} = 7r \cdot \frac{9}{r}$$

$$5r = 63$$

$$r = \frac{63}{5}, \text{ or } 12.6$$

33. *Familiarize*. Brett had 384 text messages in 8 days. Let $n =$ the number of text messages in 30 days.

***Translate*.**

$$\text{Messages} \rightarrow \frac{384}{8} = \frac{n}{30} \leftarrow \text{Messages}$$
$$\text{Days} \rightarrow \qquad\qquad\qquad \leftarrow \text{Days}$$

***Carry out*.** We solve the proportion.

$$120 \cdot \frac{384}{8} = 120 \cdot \frac{n}{30}$$

$$5760 = 4n$$

$$1440 = n$$

***Check*.** $\dfrac{384}{8} = 48, \dfrac{1440}{30} = 48$

The ratios are the same so the answer checks.

***State*.** He will send or receive 1440 messages in 30 days.

35. *Familiarize*. Persons caught on 295 mi stretch is 12,334. Let $n =$ the number caught on 5525 mi border.

***Translate*.**

$$\text{Persons} \rightarrow \frac{12334}{295} = \frac{n}{5525} \leftarrow \text{Persons}$$
$$\text{distance} \rightarrow \qquad\qquad\qquad \leftarrow \text{distance}$$

***Carry out*.** We solve the proportion.

$$325,975 \cdot \frac{12334}{295} = 325,975 \cdot \frac{n}{5525}$$

$$13,629,070 = 59n$$

$$231,001 \approx n$$

***Check*.** $\dfrac{12334}{295} \approx 41.81, \dfrac{231001}{5525} \approx 41.81$

The ratios are the same, so the answer checks.

***State*.** Over the entire border about 231,001 people may be caught.

37. *Familiarize*. Let g = the number if gal of gas for a 810 mi trip. We can use a proportion to solve for g.

***Translate*.**

$$\text{gal} \to \frac{4}{180} = \frac{g}{810} \leftarrow \text{gal} \atop \leftarrow \text{miles}$$

***Carry out*.** We solve the proportion.

We multiply by 1620 to get g alone.

$$1620 \cdot \frac{4}{180} = 1620 \cdot \frac{g}{810}$$

$$36 = 2g$$

$$18 = g$$

***Check*.**

$$\frac{4}{180} \approx 0.02, \ \frac{18}{810} \approx 0.02$$

The ratios are the same, so the answer checks.

***State*.** For a trip of 810 mi, 18 gal of gas are needed.

39. *Familiarize*. Let w = the wing width of a stork, in cm. We can use a proportion.

***Translate*.** We translate to a proportion.

$$\begin{array}{c} \text{wing} \\ \text{span} \\ \text{wing width} \end{array} \begin{array}{c} \to \\ \to \end{array} \frac{180}{24} = \frac{200}{w} \begin{array}{c} \leftarrow \\ \leftarrow \end{array} \begin{array}{c} \text{wing} \\ \text{span} \\ \text{wing width} \end{array}$$

***Carry out*.** We solve the proportion.

$$24w \cdot \frac{180}{24} = 24w \cdot \frac{200}{w}$$

$$180w = 4800$$

$$w = \frac{80}{3} = 26\frac{2}{3}\text{cm}$$

***Check*.**

$$\frac{180}{24} = 7.5, \ \frac{200}{26\frac{2}{3}} = 7.5$$

The ratios are the same, so the answer checks.

***State*.** The wing width of a stork is $26\frac{2}{3}$cm.

41. *Familiarize*. Let H = the hat size corresponding to a 56-cm hat. We can use a proportion to solve H.

***Translate*.**

$$\text{U.S. size} \to \frac{7\frac{1}{2}}{60} = \frac{H}{56} \leftarrow \text{U.S. size} \atop \leftarrow \text{European size}$$

***Carry out*.** We solve the proportion.

$$840 \cdot \frac{7\frac{1}{2}}{60} = 840 \cdot \frac{H}{56}$$

$$105 = 15H$$

$$7 = H$$

***Check*.** $\frac{7\frac{1}{2}}{60} = 0.125, \ \frac{7}{56} = 0.125$

Since the ratios are the same, the answers check.

***State*.** The U.S. size corresponding to a 56-cm hat is size 7.

43. *Familiarize*. Let D = the number of defective bulbs in a batch of 1430 bulbs. We can use a proportion to find D.

***Translate*.**

$$\text{defective bulbs} \to \frac{8}{220} = \frac{D}{1430} \leftarrow \text{defective bulbs} \atop \leftarrow \text{batch size}$$

***Carry out*.** We solve the proportion.

$$2860 \cdot \frac{8}{220} = 2860 \cdot \frac{D}{1430}$$

$$104 = 2D$$

$$52 = D$$

***Check*.** $\frac{8}{220} = 0.0\overline{36}, \ \frac{52}{1430} = 0.0\overline{36}$

The ratios are the same, so the answer checks.

***State*.** In a batch of 1430 bulbs, 52 defective bulbs can be expected.

45. *Familiarize*. Let z = the number of ounces of water needed by a Bolognese. We can use a proportion to solve for z.

***Translate*.** We translate to a proportion.

$$\begin{array}{c} \text{dog weight} \\ \text{water} \end{array} \begin{array}{c} \longrightarrow \\ \longrightarrow \end{array} \frac{8}{12} = \frac{5}{z} \begin{array}{c} \longleftarrow \\ \longleftarrow \end{array} \begin{array}{c} \text{dog weight} \\ \text{water} \end{array}$$

***Carry out*.** We solve the proportion.

$$12z \cdot \frac{8}{12} = 12z \cdot \frac{5}{z}$$

$$8z = 60$$

$$z = \frac{60}{8} = \frac{15}{2} = 7\frac{1}{2}\text{oz}$$

***Check*.**

$$\frac{8}{12} = 0.\overline{6}, \ \frac{5}{7\frac{1}{2}} = 0.\overline{6}$$

The ratios are the same, so the answer checks.

***State*.** For a 5-lb Bolognese, approximately $7\frac{1}{2}$oz of water is required per day.

47. *Familiarize*. The ratio of moose tagged to the total number of moose in the park, M, is $\frac{69}{M}$. Of the 40 moose caught later, 15 are tagged. The ratio of tagged moose to moose caught is $\frac{15}{40}$.

***Translate*.** We translate to a proportion.

$$\begin{array}{c} \text{Moose originally} \\ \text{tagged} \\ \text{Moose} \\ \text{in forest} \end{array} \begin{array}{c} \to \\ \to \end{array} \frac{69}{M} = \frac{15}{40} \begin{array}{c} \leftarrow \\ \leftarrow \end{array} \begin{array}{c} \text{Tagged moose} \\ \text{caught later} \\ \text{Moose} \\ \text{caught later} \end{array}$$

Carry out. We solve the proportion. We multiply by the LCD, $40M$.

$$40M \cdot \frac{69}{M} = 40M \cdot \frac{15}{40}$$

$$40 \cdot 69 = M \cdot 15$$

$$2760 = 15M$$

$$\frac{2760}{15} = M$$

$$184 = M$$

Check.

$$\frac{69}{184} = 0.375, \quad \frac{15}{40} = 0.375$$

The ratios are the same, so the answer checks.

State. We estimate that there are 184 moose in the park.

49. Familiarize. Let p = the number of Whale in the pod. We use a proportion to solve for p.

Translate.

$$\begin{array}{l} \text{sighted} \rightarrow \frac{12}{27} = \frac{40}{p} \leftarrow \text{sighted} \\ \text{pod} \rightarrow \end{array}$$

Carry out. We solve the proportion.

$$27p \cdot \frac{12}{27} = 27p \cdot \frac{40}{p}$$

$$12p = 1080$$

$$p = 90$$

Check. $\frac{12}{27} = \frac{4}{9}, \frac{40}{90} = \frac{4}{p}$

The ratios are the same, so the answer checks.

State. There are 90 whales in the pod, when 40 whales are sighted.

51. Familiarize. The ratio of the weight of an object on the moon to the weight of an object on Earth is 0.16 to 1.

a) We wish to find how much a 12-ton rocket would weigh on the moon.

b) We wish to find how much a 180-lb astronaut would weigh on the moon.

Translate. We translate to proportions.

a)
$$\begin{array}{l} \text{Weight} \\ \text{on the moon} \rightarrow \frac{0.16}{1} = \frac{T}{12} \leftarrow \text{on the moon} \\ \text{Weight} \rightarrow \\ \text{on Earth} \qquad\qquad\qquad \text{on Earth} \end{array}$$

b)
$$\begin{array}{l} \text{Weight} \\ \text{on the moon} \rightarrow \frac{0.16}{1} = \frac{P}{180} \leftarrow \text{on the moon} \\ \text{Weight} \rightarrow \\ \text{on Earth} \qquad\qquad\qquad\quad \text{on Earth} \end{array}$$

Carry out. We solve each proportion.

a) $\quad \frac{0.16}{1} = \frac{T}{12} \qquad$ b) $\quad \frac{0.16}{1} = \frac{P}{180}$

$\quad 12(0.16) = T \qquad\qquad 180(0.16) = P$

$\quad\quad 1.92 = T \qquad\qquad\quad 28.8 = P$

Check. $\frac{0.16}{1} = 0.16, \frac{1.92}{12} = 0.16, \frac{28.8}{180} = 0.16$

The ratios are the same, so the answer checks.

State.

a) A 12-ton rocket would weigh 1.92 tons on the moon.

b) A 180-lb astronaut would weigh 28.8 lb on the moon.

53. Writing Exercise. No. If the workers work at different rates, two workers will complete a task in more than half the time of the faster person working alone but in less than half the slower person's time. This is illustrated in Example 1.

55. Graph: $y = 2x - 6$.

We select some x-values and compute y-values.

If $x = 1$, then $y = 2 \cdot 1 - 6 = -4$.

If $x = 3$, then $y = 2 \cdot 3 - 6 = 0$.

If $x = 5$, then $y = 2 \cdot 5 - 6 = 4$.

x	y	(x, y)
1	−4	$(1, -4)$
3	0	$(3, 0)$
5	4	$(5, 4)$

57. Graph: $3x + 2y = 12$.

We can replace either variable with a number and then calculate the other coordinate. We will find the intercepts and one other point.

If $y = 0$, we have:

$$3x + 2 \cdot 0 = 12$$

$$3x = 12$$

$$x = 4$$

The x-intercept is $(4, 0)$.

If $x = 0$, we have:

$$3 \cdot 0 + 2y = 12$$

$$2y = 12$$

$$y = 6$$

The y-intercept is $(0, 6)$.

If $y = -3$, we have:

$$3x + 2(-3) = 12$$

$$3x - 6 = 12$$

$$3x = 18$$

$$x = 6$$

The point $(6, -3)$ is on the graph.

We plot these points and draw a line through them.

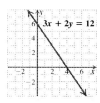

59. Graph: $y = -\dfrac{3}{4}x + 2$

We select some x-values and compute y-values. We use multiples of 4 to avoid fractions.

If $x = -4$, then $y = -\dfrac{3}{4}(-4) + 2 = 5$.

If $x = 0$, then $y = -\dfrac{3}{4} \cdot 0 + 2 = 2$.

If $x = 4$, then $y = -\dfrac{3}{4} \cdot 4 + 2 = -1$.

x	y	(x, y)
-4	5	$(-4, 5)$
0	2	$(0, 2)$
4	-1	$(4, -1)$

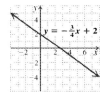

61. *Writing Exercise.* Answers may vary.

Casey can paint a room in 7 hr. It takes Lee 5 hr to paint the same room. How long would it take them to paint the room working together?

63. Equation 1: $\dfrac{1}{a} \cdot t + \dfrac{1}{b} \cdot t = 1$;

Equation 2: $\left(\dfrac{1}{a} + \dfrac{1}{b} \right) t = 1$;

Equation 3: $\dfrac{t}{a} + \dfrac{t}{b} = 1$;

Equation 4: $\dfrac{1}{a} + \dfrac{1}{b} = \dfrac{1}{t}$;

$$\dfrac{1}{a} \cdot t + \dfrac{1}{b} \cdot t = 1 \qquad \text{Equation 1}$$

$$t\left(\dfrac{1}{a} + \dfrac{1}{b} \right) = 1 \qquad \begin{array}{l}\text{Factoring out } t; \\ \text{equation1} = \text{equation 2}\end{array}$$

$$t \cdot \dfrac{1}{a} + t \cdot \dfrac{1}{b} = 1 \qquad \text{Using the distributive law}$$

$$\dfrac{t}{a} + \dfrac{t}{b} = 1 \qquad \begin{array}{l}\text{Multiplying; equation 2 =} \\ \text{equation 3}\end{array}$$

$$\dfrac{1}{t} \cdot \left(\dfrac{t}{a} + \dfrac{t}{b} \right) = \dfrac{1}{t} \cdot 1 \qquad \text{Multiplying both sides by } \dfrac{1}{t}$$

$$\dfrac{1}{t} \cdot \dfrac{t}{a} + \dfrac{1}{t} \cdot \dfrac{t}{b} = \dfrac{1}{t} \cdot 1 \qquad \text{Using the distributive law}$$

$$\dfrac{1}{a} + \dfrac{1}{b} = \dfrac{1}{t} \qquad \begin{array}{l}\text{Multiplying; equation 3 =} \\ \text{equation 4}\end{array}$$

65. *Familiarize*. Let $t =$ the time it takes Michelle working alone. Then $\dfrac{t}{2} =$ Sal's time working alone and $t - 2 =$ Kristen's time working alone. The entire job working together is 1 hr 20 min, or $\dfrac{4}{3}$hr. In $\dfrac{4}{3}$hr, Michelle does $\dfrac{4}{3} \cdot \dfrac{1}{t}$ of the job, Sal does $\dfrac{4}{3} \cdot \dfrac{1}{\frac{t}{2}}$ of the job, and Kristen does $\dfrac{4}{3} \cdot \dfrac{1}{t-2}$.

Translate. We use the information above to write an equation.

$$\dfrac{4}{3} \cdot \dfrac{1}{t} + \dfrac{4}{3} \cdot \dfrac{1}{\frac{t}{2}} + \dfrac{4}{3} \cdot \dfrac{1}{t-2}$$

Carry out. We solve the equation.

$$\dfrac{4}{3} \cdot \dfrac{1}{t} + \dfrac{4}{3} \cdot \dfrac{1}{\frac{t}{2}} + \dfrac{4}{3} \cdot \dfrac{1}{t-2} = 1, \text{ LCD} = 3t(t-2)$$

$$3t(t-2)\left(\dfrac{4}{3t} + \dfrac{8}{3t} + \dfrac{4}{3(t-2)} \right) = 3t(t-2) \cdot 1$$

$$4(t-2) + 8(t-2) + 4t = 3t(t-2)$$

$$4t - 8 + 8t - 16 + 4t = 3t^2 - 6t$$

$$0 = 3t^2 - 22t + 24$$

$$0 = (3t - 4)(t - 6)$$

$$3t - 4 = 0 \text{ or } t - 6 = 0$$

$$t = \dfrac{4}{3} \text{ or } t = 6$$

Check. Since Kristen's time, $t - 2$, is negative when $t = \dfrac{4}{3}$, and time cannot be negative in this application, so we check only 6.

$$= \dfrac{4}{3}\left(\dfrac{1}{6} + \dfrac{2}{6} + \dfrac{1}{4} \right)$$

$= 1$ complete job

The answer checks.

State. Thus, working alone it would take Michelle 6 hr, Sal 3 hr and Kristen 4 hr to wax the car.

67. Russ can reshingle the roof in 12 hr and it takes him half that time when Joan works with him. Thus, Russ and Joan have each done half of the job in 6 hr, so it would take Joan 12 hr to reshingle the roof, working alone.

69. *Familiarize*. Let t = the number of seconds for a net gain of one person. The rate of birth is $\frac{1}{7}$, the rate of death is $-\frac{1}{13}$ and the rate of new migrant is $\frac{1}{27}$.

Translate. We use the information above to write an equation.
$$t\left(\frac{1}{7} - \frac{1}{13} + \frac{1}{27}\right) = 1.$$

Carry out. We solve the equation.
$$t\left(\frac{1}{7} - \frac{1}{13} + \frac{1}{27}\right) = 1; \text{ LCD} = 2457$$
$$2457t \cdot \frac{1}{7} - 2457t \cdot \frac{1}{13} + 2457t \cdot \frac{1}{27} = 2457t \cdot 1$$
$$351t - 189t + 91t = 2457$$
$$253t = 2457$$
$$t \approx 9.7$$

Check. $\frac{9.7}{7} - \frac{9.7}{13} + \frac{9.7}{27} \approx 1$. The answer checks.

State. It will take approximately 9.7 sec. for a net gain of one person.

71.

	Distance	Speed	Time
Upstream	24	$10 - r$	t
Downstream	24	$10 + r$	$5 - t$

From the rows of the table we get two equations:
$$24 = (10 - r)t$$
$$24 = (10 + r)(5 - t)$$

We solve each equation for t and set the results equal:

Solving $24 = (10 - r)t$ for t: $t = \dfrac{24}{10 - r}$

Solving $24 = (10 + r)(5 - t)$ for t: $t = 5 - \dfrac{24}{10 + r}$

Then $\dfrac{24}{10 - r} = 5 - \dfrac{24}{10 + r}.$

$r = -2 \ \ or \ \ r = 2$

Only 2 checks in the original problem. The speed of the current is 2 mph.

73. Let p = the width of the quarry. We have similar triangles:

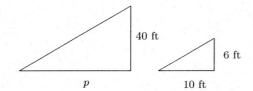

We write a proportion and solve it.

$$\frac{p}{10} = \frac{40}{6}$$
$$p = \frac{40 \cdot 10}{6} \qquad \text{Multiplying by 10}$$
$$p = \frac{400}{6}$$
$$p = \frac{200}{3}, \text{ or } 66\frac{2}{3}$$

The quarry is $66\frac{2}{3}$ ft wide.

75. Let t = the number of minutes after 5:00 at which the hands will first be together. When the minute hand moves through t minutes, the hour hand moves through $t/12$ minutes. At 5:00 the hour hand is on the 25-minute mark, so at t minutes after 5:00 it is at $25 + t/12$.

We equate the positions of the minute hand and the hour hand.
$$t = 25 + \frac{t}{12}$$
$$t = \frac{300}{11}, \text{ or } 27\frac{3}{11}$$

After 5:00 the hands on a clock will first be together in $27\frac{3}{11}$ minutes or at $27\frac{3}{11}$ minutes after 5:00.

77. *Writing Exercise*. No; consider similar right triangles with sides a, b, c and ka, kb, kc, for example, where $k \neq 1$. Their areas are $\frac{1}{2}ab$ and $\frac{1}{2} \cdot ka \cdot kb$, or $\frac{1}{2}k^2 ab$. The perimeters of these triangles are $a + b + c$ and $ka + kb + kc$, or $k(a + b + c)$, respectively. The ratio of the areas is $\dfrac{\frac{1}{2}ab}{\frac{1}{2}k^2 ab}$, or $\dfrac{1}{k^2}$, but the ratio of the perimeters is $\dfrac{a + b + c}{k(a + b + c)}$, or $\dfrac{1}{k}$. Since the ratios are not equal, the areas and perimeters are not proportional.

Chapter 6 Review

1. False; some rational expressions like $\dfrac{y^2 + 4}{y + 2}$ cannot be simplified.

3. False; when $t = 3$, then $\dfrac{t - 3}{t^2 - 4} = \dfrac{3 - 3}{3^2 - 4} = \dfrac{0}{5} = 0.$

5. True; see page 372 in the text.

7. False; see page 378 in the text.

9. $\dfrac{17}{-x^2}$

Set the denominator equal to 0 and solve for x.
$$-x^2 = 0$$
$$x = 0$$
The expression is undefined for $x = 0$.

11. $\dfrac{x-5}{x^2-36}$

Set the denominator equal to 0 and solve for x.
$$x^2-36=0$$
$$(x+6)(x-6)=0$$
$$x+6=0 \text{ or } x-6=0$$
$$x=-6 \text{ or } x=6$$
The expression is undefined for $x=-6$ and $x=6$.

13. $\dfrac{-6}{(t+2)^2}$

Set the denominator equal to 0 and solve for t.

$$(t+2)^2=0$$
$$t+2=0$$
$$t=-2$$
The expression is undefined for $t=-2$.

15. $\dfrac{14x^2-x-3}{2x^2-7x+3}=\dfrac{(2x-1)(7x+3)}{(2x-1)(x-3)}=\dfrac{7x+3}{x-3}$

17. $\dfrac{5x^2-20y^2}{2y-x}=\dfrac{-5\left(4y^2-x^2\right)}{(2y-x)}$
$$=\dfrac{-5(2y+x)(2y-x)}{(2y-x)}=-5(2y+x)$$

19. $\dfrac{6y-12}{2y^2+3y-2}\cdot\dfrac{y^2-4}{8y-8}=\dfrac{6(y-2)(y-2)(y+2)}{(2y-1)(y+2)(8)(y-1)}$
$$=\dfrac{3(y-2)^2}{4(2y-1)(y-1)}\cdot\dfrac{2(y+2)}{2(y+2)}$$
$$=\dfrac{3(y-2)^2}{4(2y-1)(y-1)}$$

21. $\dfrac{4x^4}{x^2-1}\div\dfrac{2x^3}{x^2-2x+1}=\dfrac{4x^4}{(x+1)(x-1)}\cdot\dfrac{(x-1)(x-1)}{2x^3}$
$$=\dfrac{2x(x-1)}{x+1}\cdot\dfrac{2x^3(x-1)}{2x^3(x-1)}$$
$$=\dfrac{2x(x-1)}{x+1}$$

23. $\left(t^2+3t-4\right)\div\dfrac{t^2-1}{t+4}=(t+4)(t-1)\cdot\dfrac{(t+4)}{(t+1)(t-1)}$
$$=\dfrac{(t+4)^2}{t+1}\cdot\dfrac{t-1}{t-1}$$
$$=\dfrac{(t+4)^2}{t+1}$$

25. $x^2-x=x(x-1)$
$x^5-x^3=x^3\left(x^2-1\right)=x\cdot x\cdot x\cdot(x+1)(x-1)$
$x^4=x\cdot x\cdot x\cdot x$
$\text{LCM}=x\cdot x\cdot x\cdot x\cdot(x+1)(x-1)=x^4(x+1)(x-1)$

27. $\dfrac{x+6}{x+3}+\dfrac{9-4x}{x+3}=\dfrac{x+6+9-4x}{x+3}=\dfrac{-3x+15}{x+3}$

29. $\dfrac{3x-1}{2x}-\dfrac{x-3}{x}=\dfrac{3x-1}{2x}-\dfrac{x-3}{x}\cdot\dfrac{2}{2}$
$$\text{LCD}=2x$$
$$=\dfrac{3x-1-2(x-3)}{2x}$$
$$=\dfrac{3x-1-2x+6}{2x}$$
$$=\dfrac{x+5}{2x}$$

31. $\dfrac{y^2}{y-2}+\dfrac{6y-8}{2-y}=\dfrac{y^2}{y-2}+\dfrac{8-6y}{y-2}$
$$=\dfrac{y^2-6y+8}{y-2}$$
$$=\dfrac{(y-4)(y-2)}{y-2}$$
$$=y-4$$

33. $\dfrac{d^2}{d-2}+\dfrac{4}{2-d}=\dfrac{d^2}{d-2}-\dfrac{4}{d-2}$
$$=\dfrac{d^2-4}{d-2}$$
$$=\dfrac{(d+2)(d-2)}{d-2}$$
$$=d+2$$

35. $\dfrac{3x}{x+2}-\dfrac{x}{x-2}+\dfrac{8}{x^2-4}$
$$=\dfrac{3x}{x+2}\cdot\dfrac{x-2}{x-2}-\dfrac{x}{x-2}\cdot\dfrac{x+2}{x+2}+\dfrac{8}{(x+2)(x-2)}$$
$$\text{LCD}=(x+2)(x-2)$$
$$=\dfrac{3x^2-6x-x^2-2x+8}{(x+2)(x-2)}$$
$$=\dfrac{2x^2-8x+8}{(x+2)(x-2)}=\dfrac{2\left(x^2-4x+4\right)}{(x+2)(x-2)}$$
$$=\dfrac{2(x-2)(x-2)}{(x+2)(x-2)}$$
$$=\dfrac{2(x-2)}{x+2}$$

37. $\dfrac{\dfrac{1}{z}+1}{\dfrac{1}{z^2}-1}=\dfrac{\dfrac{1}{z}+1}{\dfrac{1}{z^2}-1}\cdot\dfrac{z^2}{z^2}$
$$\text{LCD}=z^2$$
$$=\dfrac{z+z^2}{1-z^2}$$
$$=\dfrac{z(1+z)}{(1-z)(1+z)}$$
$$=\dfrac{z}{1-z}$$

39. $\dfrac{\dfrac{c}{d}-\dfrac{d}{c}}{\dfrac{1}{c}+\dfrac{1}{d}}=\dfrac{\dfrac{c}{d}-\dfrac{d}{c}}{\dfrac{1}{c}+\dfrac{1}{d}}\cdot\dfrac{cd}{cd}$
$$\text{LCD}=cd$$
$$=\dfrac{c^2-d^2}{d+c}$$
$$=\dfrac{(c-d)(c+d)}{c+d}$$
$$=c-d$$

41. $\dfrac{3}{x+4}=\dfrac{1}{x-1}$ Note $x\neq1,-4$
$$(x-1)(x+4)\dfrac{3}{x+4}=(x-1)(x+4)\dfrac{1}{x-1}$$
$$\text{LCD}=(x-1)(x+4)$$
$$3(x-1)=x+4$$
$$3x-3=x+4$$
$$3x=x+7$$
$$2x=7$$
$$x=\dfrac{7}{2}$$

43. Familiarize. The job takes Jackson 12 hours working alone and Charis 9 hours working alone. Then in 1 hour Jackson does $\frac{1}{12}$ of the job and Charis does $\frac{1}{9}$ of the job. Working together, they can do $\frac{1}{9} + \frac{1}{12}$, or $\frac{7}{36}$ of the job in 1 hour.

Translate. If they work together t hours, then Jackson does $t\left(\frac{1}{9}\right)$ of the job and Charis does $t\left(\frac{1}{12}\right)$ of the job. We want some number t such that
$$\left(\frac{1}{9} + \frac{1}{12}\right)t = 1, \text{ or } \frac{7}{36}t = 1.$$

Carry out. We solve the equation.
$$\frac{7}{36}t = 1$$
$$\frac{36}{7} \cdot \frac{7}{36}t = \frac{36}{7} \cdot 1$$
$$t = \frac{36}{7} \text{ or } 5\frac{1}{7}$$

Check. The check can be done by repeating the computation.

State. Working together, it takes them $5\frac{1}{7}$ hrs to complete the job.

45. Familiarize. The ratio of seal tagged to the total number of seal in the harbor, T, is $\frac{33}{T}$. Of the 40 seals caught later, there were 24 are tagged. The ratio of tagged seals to seals caught is $\frac{24}{40}$.

Translate. We translate to a proportion.

Seals originally tagged → $\frac{33}{T}$ = $\frac{24}{40}$ ← Tagged seals caught later / Seals caught later
Seals in harbor

Carry out. We solve the proportion.
$$40T \cdot \frac{33}{T} = 40T \cdot \frac{24}{40}$$
$$1320 = 24T$$
$$55 = T$$

Check.
$$\frac{33}{55} = 0.6, \frac{24}{40} = 0.6$$
The ratios are the same, so the answer checks.

State. We estimate that there are 55 seals in the harbor.

47. Familiarize. Let D = the number of defective radios you would expect in a sample of 540 radios. We use a proportion to solve for D.

Translate.

Defective → $\frac{4}{30}$ = $\frac{D}{540}$ ← Defective Radios → Radios

Carry out. We solve the proportion. We multiply by the LCD, 540.
$$540 \cdot \frac{4}{30} = 540 \cdot \frac{D}{540}$$
$$72 = D$$

Check. $\frac{4}{30} = \frac{2}{15}, \frac{72}{540} = \frac{2}{15}$
The ratios are the same, so the answer checks.

State. You would expect 72 defective radios in a batch of 540.

49. Writing Exercise. Although multiplying the denominators of the expressions being added results in a common denominator, it is often not the *least* common denominator. Using a common denominator other than the LCD makes the expressions more complicated, requires additional simplifying after the addition has been performed, and leaves more room for error.

51. $\frac{12a}{(a-b)(b-c)} - \frac{2a}{(b-a)(c-b)}$
$= \frac{12a - 2a}{(a-b)(b-c)}$
Since $(b-a)(c-b) = (a-b)(b-c)$
$\frac{10a}{(a-b)(b-c)}$

53. We write a proportion and solve it.
total hits → $\frac{153+x}{395+125} = \frac{4}{10}$ ← total average at-bats
total at-bats →
$$\frac{153+x}{520} = \frac{4}{10}$$
$$520 \cdot \frac{153+x}{520} = 520 \cdot \frac{4}{10}$$
$$153 + x = 208$$
$$x = 55 \text{ more hits}$$
new hits → $\frac{55}{125} = 0.44$ or 44%
new at-bats →
He must hit at 44% of the 125 at-bats.

Chapter 6 Test

1. $\frac{2-x}{5x}$
We find the number which makes the denominator 0.
$$5x = 0$$
$$x = 0$$
The expression is undefined for $x = 0$.

3. $\frac{x-7}{x^2-1}$
We find the number which makes the denominator 0.
$$x^2 - 1 = 0$$
$$(x+1)(x-1) = 0$$
$$x + 1 = 0 \text{ or } x - 1 = 0$$
$$x = -1 \text{ or } x = 1$$
The expression is undefined for $x = -1$ and $x = 1$.

5. $\frac{6x^2+17x+7}{2x^2+7x+3} = \frac{(3x+7)(2x+1)}{(x+3)(2x+1)} = \frac{(3x+7)\cancel{(2x+1)}}{(x+3)\cancel{(2x+1)}}$
$= \frac{3x+7}{x+3}$

7. $\dfrac{25y^2-1}{9y^2-6y} \div \dfrac{5y^2+9y-2}{3y^2+y-2}$

$= \dfrac{25y^2-1}{9y^2-6y} \cdot \dfrac{3y^2+y-2}{5y^2+9y-2}$

$= \dfrac{(5y+1)(5y-1)}{3y(3y-2)} \cdot \dfrac{(3y-2)(y+1)}{(5y-1)(y+2)}$

$= \dfrac{(5y+1)\cancel{(5y-1)}\cancel{(3y-2)}(y+1)}{3y\cancel{(3y-2)}\cancel{(5y-1)}(y+2)}$

$= \dfrac{(5y+1)(y+1)}{3y(y+2)}$

9. $(x^2+6x+9) \cdot \dfrac{(x-3)^2}{x^2-9}$

$= \dfrac{(x+3)(x+3)}{1} \cdot \dfrac{(x-3)(x-3)}{(x+3)(x-3)}$

$= \dfrac{\cancel{(x+3)}(x+3)\cancel{(x-3)}(x-3)}{\cancel{(x+3)}\cancel{(x-3)}}$

$= (x+3)(x-3)$

11. $\dfrac{2+x}{x^3} + \dfrac{7-4x}{x^3} = \dfrac{2+x+7-4x}{x^3} = \dfrac{-3x+9}{x^3}$

13. $\dfrac{2x-4}{x-3} + \dfrac{x-1}{3-x} = \dfrac{2x-4}{-1(3-x)} + \dfrac{x-1}{-1(3-x)}$

$= \dfrac{-(2x-4)}{3-x} + \dfrac{x-1}{3-x}$

$= \dfrac{-2x+4+x-1}{3-x}$

$= \dfrac{3-x}{3-x}$

$= 1$

15. $\dfrac{7}{t-2} + \dfrac{4}{t}$ LCD is $t(t-2)$

$= \dfrac{7}{t-2} \cdot \dfrac{t}{t} + \dfrac{4}{t} \cdot \dfrac{t-2}{t-2}$

$= \dfrac{7t}{t(t-2)} + \dfrac{4(t-2)}{t(t-2)}$

$= \dfrac{7t+4t-8}{t(t-2)}$

$= \dfrac{11t-8}{t(t-2)}$

17. $\dfrac{1}{x-1} + \dfrac{4}{x^2-1} - \dfrac{2}{x^2-2x+1} =$

$\dfrac{1}{x-1} + \dfrac{4}{(x+1)(x-1)} - \dfrac{2}{(x-1)(x-1)}$

LCD is $(x-1)(x-1)(x+1)$

$= \dfrac{1}{x-1} \cdot \dfrac{(x-1)(x+1)}{(x-1)(x+1)} +$

$\dfrac{4}{(x-1)(x+1)} \cdot \dfrac{x-1}{x-1} - \dfrac{2(x+1)}{(x+1)(x-1)^2}$

$= \dfrac{(x-1)(x+1)}{(x+1)(x-1)^2} + \dfrac{4(x-1)}{(x+1)(x-1)^2} - \dfrac{2(x+1)}{(x+1)(x-1)^2}$

$= \dfrac{(x-1)(x+1) + 4(x-1) - 2(x+1)}{(x+1)(x-1)^2}$

$= \dfrac{x^2-1+4x-4-2x-2}{(x+1)(x-1)^2}$

$= \dfrac{x^2+2x-7}{(x+1)(x-1)^2}$

19. $\dfrac{\dfrac{x}{8} - \dfrac{8}{x}}{\dfrac{1}{8} + \dfrac{1}{x}}$ LCD is $8x$

$= \dfrac{8x}{8x} \cdot \dfrac{\dfrac{x}{8} - \dfrac{8}{x}}{\dfrac{1}{8} + \dfrac{1}{x}} = \dfrac{\dfrac{8x^2}{8} - \dfrac{64x}{x}}{\dfrac{8x}{8} + \dfrac{8x}{x}}$

$= \dfrac{x^2-64}{x+8} = \dfrac{(x+8)(x-8)}{x+8} = \dfrac{\cancel{(x+8)}(x-8)}{\cancel{x+8}}$

$= x-8$

21. To avoid division by 0, we must have $x \neq 0$ and $x-2 \neq 0$, or $x \neq 0$ and $x \neq 2$.

$\dfrac{15}{x} - \dfrac{15}{x-2} = -2$ LCD $= x(x-2)$

$x(x-2)\left(\dfrac{15}{x} - \dfrac{15}{x-2}\right) = x(x-2)(-2)$

$15(x-2) - 15x = -2x(x-2)$

$15x - 30 - 15x = -2x^2 + 4x$

$2x^2 - 4x - 30 = 0$

$2(x^2 - 2x - 15) = 0$

$2(x-5)(x+3) = 0$

$x - 5 = 0$ or $x + 3 = 0$

$x = 5$ or $x = -3$

The solutions are -3 and 5.

23. Familiarize. Burning 320 calories corresponds to walking 4 mi, and we wish to find the number of miles m that correspond to burning 100 calories. We can use a proportion.

Translate.

calories burned → $\dfrac{320}{4} = \dfrac{100}{m}$ ← calories burned
miles walked → ← miles walked

Carry out. We solve the proportion.

$4m \cdot \dfrac{320}{4} = 4m \cdot \dfrac{100}{m}$

$320m = 400$

$m = \dfrac{5}{4} = 1\dfrac{1}{4}$

Check. $\dfrac{320}{4} = 80$, $\dfrac{100}{5/4} = 80$

The ratios are the same sot he answer checks.

State. Walking $1\dfrac{1}{4}$ mi corresponds to burning 100 calories.

25. Familiarize. Let $t = $ the number of hours it would take Rema to mulch the flower beds, working alone. Then $t+6$ = the number of hours it would take Perez, working alone. Note that $2\dfrac{6}{7} = \dfrac{20}{7}$ In $\dfrac{20}{7}$ hr Rema does $\dfrac{20}{7} \cdot \dfrac{1}{t}$ of the job and Perez does $\dfrac{20}{7} \cdot \dfrac{1}{t+6}$ of the job, and together they do 1 complete job.

Translate. We use the information above to write an equation. $\dfrac{20}{7} \cdot \dfrac{1}{t} + \dfrac{20}{7} \cdot \dfrac{1}{t+6} = 1$

Carry out. We solve the equation.

$$\frac{20}{7} \cdot \frac{1}{t} + \frac{20}{7} \cdot \frac{1}{t+6} = 1, \ \text{LCD} = 7t(t+6)$$

$$7t(t+6) \cdot \left(\frac{20}{7} \cdot \frac{1}{t} + \frac{20}{7} \cdot \frac{1}{t+6} \right) = 7t(t+6) \cdot 1$$

$$20(t+6) + 20t = 7t(t+6)$$

$$20t + 120 + 20t = 7t^2 + 42t$$

$$0 = 7t^2 + 2x - 120$$

$$0 = (7t+30)(t-4)$$

$$7t + 30 = 0 \ \text{or} \ t - 4 = 0$$
$$t = -\frac{30}{7} \ \text{or} \ t = 4$$

Check. Time cannot be negative in this application, so we check only 4. If Rema can mulch the flower beds working alone in 4 hr, then it would take Perez $4 + 6$, or 10 hr, working alone. In $\frac{20}{7}$ hr they would do $\frac{20}{7} \cdot \frac{1}{4} + \frac{20}{7} \cdot \frac{1}{10} = \frac{5}{7} + \frac{2}{7} = 1$ complete job. The answer checks.

State. It would take Rema 4 hr to mulch the flower beds working alone, and it would take Perez 10 hr working alone.

27. **Familiarize.** Let x = the number. Then $-\frac{1}{x}$ is the opposite of the numbers reciprocal.

Translate. The square of the number, x^2 is equivalent to $-\frac{1}{x}$, so we write a proportion. $x^2 = -\frac{1}{x}$

Carry out. We solve the equation.

$$x \cdot x^2 = x \cdot \left(-\frac{1}{x} \right)$$

$$x^3 = -1$$

$$x = \sqrt[3]{-1} = -1$$

Check. $(-1)^2 = 1$, $-\frac{1}{-1} = 1$, so the ratios are equivalent.

State. The number is -1.

Chapter 7

Systems and More Graphing

1. both; see page 430 in the text.

3. consistent; see page 431 in the text.

5. We check by substituting alphabetically 2 for x and 5 for y.

$$\frac{2x + 3y = 19}{\begin{array}{c|c} 2 \cdot 2 + 3 \cdot 5 & 19 \\ 4 + 15 & \\ & \overset{?}{19 = 19} \text{ TRUE} \end{array}} \qquad \frac{3x - y = 1}{\begin{array}{c|c} 3 \cdot 2 - 5 & 1 \\ 6 - 5 & \\ & \overset{?}{1 = 1} \text{ TRUE} \end{array}}$$

The ordered pair $(2, 5)$ is a solution of each equation. Therefore it is a solution of the system of equations.

7. We check by substituting alphabetically 3 for a and 2 for b.

$$\frac{3b - 2a = 0}{\begin{array}{c|c} 3 \cdot 2 - 2 \cdot 3 & 0 \\ 6 - 6 & \\ & \overset{?}{0 = 0} \text{ TRUE} \end{array}} \qquad \frac{b + 2a = 15}{\begin{array}{c|c} 2 + 2 \cdot 3 & 15 \\ 2 + 6 & \\ & \overset{?}{8 = 15} \text{ FALSE} \end{array}}$$

The ordered pair $(3, 2)$ is not a solution of $b + 2a = 15$. Therefore it is not a solution of the system of equations.

9. We check by substituting -15 for x and 20 for y.

$$\frac{3x + 2y = -5}{\begin{array}{c|c} 3 \cdot (-15) + 2 \cdot 20 & -5 \\ -45 + 40 & \\ & \overset{?}{-5 = -5} \text{ TRUE} \end{array}}$$

$$\frac{4y + 5x = 5}{\begin{array}{c|c} 4(20) + 5(-15) & 5 \\ 80 - 75 & \\ & \overset{?}{5 = 5} \text{ TRUE} \end{array}}$$

The ordered pair $(-15, 20)$ is a solution of each equation. Therefore it is a solution of the system of equations.

11. We graph the equations.

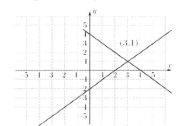

The "apparent" solution of the system, $(3, 1)$, should be checked in both equations.

Check:

$$\frac{x + y = 4}{\begin{array}{c|c} 3 + 1 & 4 \\ & \overset{?}{4 = 4} \text{ TRUE} \end{array}} \qquad \frac{x - y = 2}{\begin{array}{c|c} 3 - 1 & 2 \\ & \overset{?}{2 = 2} \text{ TRUE} \end{array}}$$

The solution is $(3, 1)$.

13. We graph the equations.

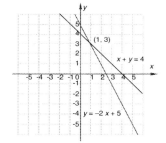

The "apparent" solution of the system, $(1, 3)$, should be checked in both equations.

Check:

$$\frac{y = -2x + 5}{\begin{array}{c|c} 3 & -2 \cdot 1 + 5 \\ & -2 + 5 \\ \overset{?}{3 = 3} & \text{ TRUE} \end{array}} \qquad \frac{x + y = 4}{\begin{array}{c|c} 1 + 3 & 4 \\ & \overset{?}{4 = 4} \text{ TRUE} \end{array}}$$

The solution is $(1, 3)$.

15. We graph the equations.

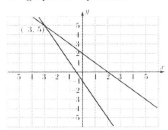

The "apparent" solution of the system, $(-3, 5)$, should be checked in both equations.

Check:

$$\frac{y = -2x - 1}{\begin{array}{c|c} 5 & -2(-3) - 1 \\ & 6 - 1 \\ \overset{?}{5 = 5} & \text{ TRUE} \end{array}} \qquad \frac{y = 2 - x}{\begin{array}{c|c} 5 & -2 - (-3) \\ & 2 + 3 \\ \overset{?}{5 = 5} & \text{ TRUE} \end{array}}$$

The solution is $(-3, 5)$.

17. We graph the equations.

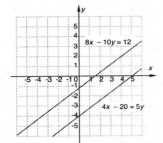

The lines are parallel. There is no solution.

19. We graph the equations.

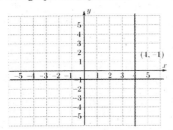

The "apparent" solution of the system, $(4, -1)$, should be checked in both equations.

Check:

$$\begin{array}{c|c} x = 4 \\ \hline 4 \overset{?}{=} 4 & \text{TRUE} \end{array} \qquad \begin{array}{c|c} y = -1 \\ \hline -1 \overset{?}{=} -1 & \text{TRUE} \end{array}$$

The solution is $(4, -1)$.

21. We graph the equations.

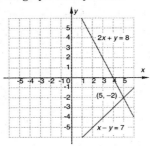

The "apparent" solution of the system, $(5, -2)$, should be checked in both equations.

Check:

$$\begin{array}{rcl} 2x + y & = & 8 \\ \hline 2 \cdot 5 + (-2) & \big| & 8 \\ 10 - 2 & \big| & \\ & \overset{?}{8} \overset{?}{=} 8 & \text{TRUE} \end{array}$$

$$\begin{array}{rcl} x - y & = & 7 \\ \hline 5 - (-2) & \big| & 7 \\ 5 + 2 & \big| & \\ & 7 \overset{?}{=} 7 & \text{TRUE} \end{array}$$

The solution is $(5, -2)$.

23. We graph the equations.

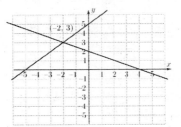

The "apparent" solution of the system, $(-2, 3)$, should be checked in both equations.

Check:

$$\begin{array}{rcl} y - x & = & 5 \\ \hline 3 - (-2) & \big| & 5 \\ 3 + 2 & \big| & \\ & 5 \overset{?}{=} 5 & \text{TRUE} \end{array}$$

$$\begin{array}{rcl} x + 2y & = & 4 \\ \hline 2 \cdot 3 + (-2) & \big| & 4 \\ 6 - 2 & \big| & \\ & 4 \overset{?}{=} 4 & \text{TRUE} \end{array}$$

The solution is $(-2, 3)$.

25. We graph the equations.

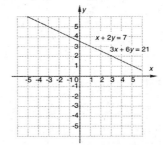

We see that the equations represent the same line. This means that any solution of one equation is a solution of the other equation as well. Thus, there is an infinite number of solutions.

27. We graph the equations.

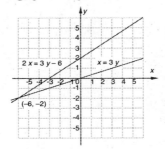

The "apparent" solution of the system, $(-6, -2)$, should be checked in both equations.

Check:

$$\begin{array}{c|c} 2x = 3y - 6 \\ \hline 2(-6) & 3(-2) - 6 \\ -12 & -6 - 6 \\ & \overset{?}{-12 = -12} \quad \text{TRUE} \end{array}$$

$$\begin{array}{c|c} x = 3y \\ \hline -6 & 3(-2) \\ & \overset{?}{-6 = -6} \quad \text{TRUE} \end{array}$$

The solution is $(-6, -2)$.

29. $y = \dfrac{1}{5}x + 4,$

$2y = \dfrac{2}{5}x + 8$

Observe that we can obtain the second equation by multiplying both sides of the first equation by 2. Thus, the equations are dependent and there is an infinite number of solutions.

31. We graph the equations.

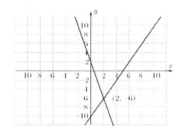

The "apparent" solution of the system, $(2, -6)$, should be checked in both equations.

$$\begin{array}{c|c} 4x + y = 2 \\ \hline 4(2) - 6 & 2 \\ 8 - 6 & \\ & \overset{?}{2 = 2} \quad \text{TRUE} \end{array}$$

$$\begin{array}{c|c} x = \frac{1}{2}y + 5 \\ \hline 2 & \frac{1}{2}(-6) + 5 \\ & -3 + 5 \\ & \overset{?}{2 = 2} \quad \text{TRUE} \end{array}$$

The solution is $(2, -6)$.

33. We graph the equations.

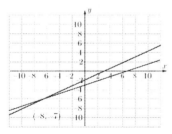

The "apparent" solution of the system, $(-8, -7)$, should be checked in both equations.

Check:

$$\begin{array}{c|c} 2x - 3y = 5 \\ \hline 2(-8) - 3(-7) & 5 \\ -16 + 21 & \\ & \overset{?}{5 = 5} \quad \text{TRUE} \end{array}$$

$$\begin{array}{c|c} x - 2y = 6 \\ \hline -8 - 2(-7) & 6 \\ -8 + 14 & \\ & \overset{?}{6 = 6} \quad \text{TRUE} \end{array}$$

The solution is $(-8, -7)$.

35. We graph the equations.

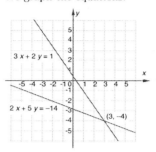

The "apparent" solution of the system, $(3, -4)$, should be checked in both equations.

Check:

$$\begin{array}{c|c} 3x + 2y = 1 \\ \hline 3 \cdot 3 + 2(-4) & 1 \\ 9 - 8 & \\ & \overset{?}{1 = 1} \quad \text{TRUE} \end{array}$$

$$\begin{array}{c|c} 2x + 5y = -14 \\ \hline 2 \cdot 3 + 5(-4) & -14 \\ 6 - 20 & \\ & \overset{?}{-14 = -14} \quad \text{TRUE} \end{array}$$

The solution is $(3, -4)$.

37. We graph the equations.

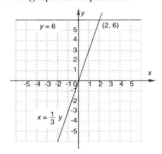

The "apparent" solution of the system, $(2, 6)$, should be checked in both equations.

Check:

$$\begin{array}{c|c} x = \frac{1}{3}y \\ \hline 2 & \frac{1}{3} \cdot 6 \\ & \overset{?}{2 = 2} \quad \text{TRUE} \end{array}$$

$$\begin{array}{c} y = 6 \\ \hline \overset{?}{6 = 6} \quad \text{TRUE} \end{array}$$

The solution is $(2, 6)$.

39. Graph $y = 2x - 1$ and $y = 3$.

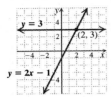

The graphs intersect at $(2, 3)$, indicating that when $x = 2$ we have $2x - 1 = 3$. The solution is 2.

41. Graph $y = x - 4$ and $y = 6 - x$.

The graphs intersect at $(5, 1)$, indicating that when $x = 5$ we have $x - 4 = 6 - x$. The solution is 5.

43. Graph $y = 2x - 1$ and $y = -x + 5$.

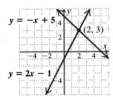

The graphs intersect at $(2, 3)$, indicating that when $x = 2$ we have $2x - 1 = -x + 5$. The solution is 2.

45. Graph $y = \dfrac{1}{2}x + 3$ and $y = -\dfrac{1}{2}x - 1$.

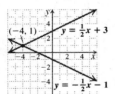

The graphs intersect at $(-4, 1)$, indicating that when $x = -4$ we have $\dfrac{1}{2}x + 3 = -\dfrac{1}{2}x - 1$. The solution is -4.

47. *Writing Exercise.* No; given a pair of linear equations there are only three possibilities: they intersect in exactly one point, they have no points of intersection, or they represent the same line and hence have an infinite number of points of intersection.

49.
$$3x - (4 - 2x) = 9$$
$$3x - 4 + 2x = 9$$
$$5x - 4 = 9$$
$$5x = 13$$
$$x = \frac{13}{5}$$

51.
$$2(8 - 5y) - y = 4$$
$$16 - 10y - y = 4$$
$$16 - 11y = 4$$
$$-11y = -12$$
$$y = \frac{12}{11}$$

53.
$$3x - 4y = 2$$
$$-4y = -3x + 2$$
$$y = \frac{3}{4}x - \frac{1}{2}$$

55. *Writing Exercise.* Yes; since the graphs of the equations are the same, there is an infinite number of solutions and the system is consistent.

57. Systems in which the graphs of the equations coincide contain dependent equations. This is the case in Exercises 18, 25, 26, and 29.

59. Systems in which the graphs of the equations are parallel are inconsistent. This is the case in Exercises 17 and 30.

61. Answers may vary. Any equation with $(-1, 4)$ as a solution that is independent of $5x + 2y = 3$ will do. One such equation is $2x + y = 2$.

63. $(2, -3)$ is a solution of $Ax - 3y = 13$. Substitute 2 for x and -3 for y and solve for A.
$$Ax - 3y = 13$$
$$A \cdot 2 - 3(-3) = 13$$
$$2A + 9 = 13$$
$$2A = 4$$
$$A = 2$$

$(2, -3)$ is a solution of $x - By = 8$. Substitute 2 for x and -3 for y and solve for B.
$$x - By = 8$$
$$2 - B(-3) = 8$$
$$2 + 3B = 8$$
$$3B = 6$$
$$B = 2$$

65. a) Let $x =$ the number of copies, up to 500, and $y =$ the cost. Then the cost equation for the copy card method of payment is $y = 20$. The cost equation for the method of paying per page is $y = 0.06x$.

b)

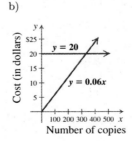

c) We see that the graphs intersect at approximately $(333, 20)$ and that for x-values greater than 333 the graph of $y = 20$ lies below the graph of $y = 0.06x$. Thus, Shelby must make more than 333 copies if the card is to be more economical.

67. a) The number of faculty in thousands, and $t =$ the number of years after 2003

$n = 8t + 632$

The equation for part time faculty is

$n = 13t + 543.$

The rate or slope is 13, the intercept is $(0, 543)$ thousand.

b) Using a graphing calculator, we can read the point of intersection as about $(18,775)$.

That is, about 2021, the number of full time faculty will equal the number of part time faculty, as about 775,500.

69. Graph $y_1 = 1.3x - 4.9$ and $y_2 = 6.3 - 3.7x$ and use the INTERSECT feature from the CALC menu to find the first coordinate of the point of intersection of the graphs. It is 2.24.

Exercise Set 7.2

1. The statement is false. See Example 2.

3. The statement is true. See Example 4(a).

5. $x + y = 9, \quad$ (1)

$\quad y = x + 1 \quad$ (2)

Substitute $x + 1$ for y in Equation (1) and solve for x.

$x + y = 9 \quad$ (1)

$x + (x + 1) = 9 \quad$ Substituting

$2x + 1 = 9$

$2x = 8$

$x = 4$

Next we substitute 4 for x in either equation of the original system and solve for y.

$y = x + 1 \quad$ (2)

$y = 4 + 1 \quad$ Substituting

$y = 5$

We check the ordered pair $(4, 5)$.

$$\begin{array}{c|c}
x + y = 9 & \\
\hline
4 + 5 \ | \ 9 & \\
9 \overset{?}{=} 9 \ \text{TRUE} &
\end{array} \qquad
\begin{array}{c|c}
y = x + 1 & \\
\hline
5 \ | \ 4 + 1 & \\
5 \overset{?}{=} 5 \qquad \text{TRUE} &
\end{array}$$

7. $x = y + 1, \quad$ (1)

$x + 2y = 4 \quad$ (2)

Substitute $y + 1$ for x in Equation (1) and solve for y.

$x + 2y = 4 \quad$ (2)

$(y + 1) + 2y = 4 \quad$ Substituting

$3y + 1 = 4$

$3y = 3$

$y = 1$

Next we substitute 1 for y in either equation of the original system and solve for x.

$x = y + 1 \quad$ (1)

$x = 1 + 1 \quad$ Substituting

$x = 2$

We check the ordered pair $(2, 1)$.

$$\begin{array}{c|c}
x = y + 1 & \\
\hline
2 \ | \ 1 + 1 & \\
2 \overset{?}{=} 2 \qquad \text{TRUE} &
\end{array} \qquad
\begin{array}{c|c}
x + 2y = 4 & \\
\hline
2 + 2 \cdot 1 \ | \ 4 & \\
2 + 2 & \\
4 \overset{?}{=} 4 \ \text{TRUE} &
\end{array}$$

Since $(2, 1)$ checks in both equations, it is the solution.

9. $y = 5x - 1, \quad$ (1)

$y - 3x = 1 \qquad$ (2)

Substitute $5x - 1$ for y in Equation (2) and solve for x.

$(5x - 1) - 3x = 1 \quad$ (2)

$5x - 1 - 3x = 1 \quad$ Substituting

$2x - 1 = 1$

$2x = 2$

$x = 1$

Next we substitute 1 for x in either equation of the original system and solve for y.

$y = 5x - 1 \quad$ (1)

$y = 5 \cdot 1 - 1 \quad$ Substituting

$y = 5 - 1$

$y = 4$

We check the ordered pair $(1, 4)$.

$$\begin{array}{c|c}
y = 5x - 1 & \\
\hline
4 \ | \ 5 \cdot 1 - 1 & \\
\ | \ 5 - 1 & \\
4 \overset{?}{=} 4 \qquad \text{TRUE} &
\end{array} \qquad
\begin{array}{c|c}
y - 3x = 1 & \\
\hline
4 - 3 \cdot 1 \ | \ 1 & \\
4 - 3 & \\
1 \overset{?}{=} 1 \ \text{TRUE} &
\end{array}$$

Since $(1, 4)$ checks in both equations, it is the solution.

11. $a = -4b \quad$ (1)

$a + 5b = 5 \qquad$ (2)

Substitute $-4b$ for a in (2)

$a + 5b = 5 \quad$ (2)

$-4b + 5b = 5 \quad$ Substituting

$b = 5$

Next we substitute 5 for b in (1)

$a = -4 \cdot 5 \quad$ (1)

$a = -20$

We check the ordered pair $(-20, 5)$.

$$\begin{array}{c|c} a = -4b \\ \hline -20 & -4 \cdot 5 \\ & ? \\ -20 \overset{?}{=} -20 & \text{TRUE} \end{array}$$

$$\begin{array}{c|c} a + 5b = 5 \\ \hline -20 + 5 \cdot 5 & 5 \\ -20 + 25 & \\ & ? \\ & 5 \overset{?}{=} 5 \quad \text{TRUE} \end{array}$$

Since $(-20, 5)$ checks in both equations, it is the solution.

13. $x = y - 5,$ (1)

$2x + 5y = 4$ (2)

Substitute $y - 5$ for x in Equation (2) and solve for y.

$$2x + 5y = 4 \quad (1)$$
$$2(y - 5) + 5y = 4 \quad \text{Substituting}$$
$$2y - 10 + 5y = 4$$
$$7y - 10 = 4$$
$$7y = 14$$
$$y = 2$$

Next we substitute 2 for y in either equation of the original system and solve for x.

$$x = 2 - 5 \quad (2)$$
$$x = -3$$

We check the ordered pair $(-3, 2)$.

$$\begin{array}{c|c} x = y - 5 \\ \hline -3 & 2 - 5 \\ & ? \\ -3 \overset{?}{=} -3 & \text{TRUE} \end{array}$$

$$\begin{array}{c|c} 2x + 5y = 4 \\ \hline 2(-3) + 5(2) & 4 \\ -6 + 10 & \\ & ? \\ & 4 \overset{?}{=} 4 \quad \text{TRUE} \end{array}$$

Since $(-3, 2)$ checks in both equations, it is the solution.

15. $x = 2y + 1,$ (1)

$3x - 6y = 2$ (2)

We substitute $2y + 1$ for x in Equation (2) and solve for y.

$$3x - 6y = 2 \quad (2)$$
$$3(2y + 1) - 6y = 2$$
$$6y + 3 - 6y = 2$$
$$3 = 2$$

We obtain a false equation, so the system has no solution.

17. $s + t = -5,$ (1)

$s - t = 3$ (2)

We solve Equation (2) for s.

$$s - t = 3 \quad (2)$$
$$s = t + 3 \quad (3)$$

We substitute $t + 3$ for s in Equation (1) and solve for t.

$$s + t = -5 \quad (1)$$
$$(t + 3) + t = -5 \quad \text{Substituting}$$
$$2t + 3 = -5$$
$$2t = -8$$
$$t = -4$$

Now we substitute -4 for t in either of the original equations or in Equation (3) and solve for s. It is easiest to use (3).

$$s = -4 + 3 = -1$$

We check the ordered pair $(-1, -4)$.

$$\begin{array}{c|c} s + t = -5 \\ \hline -1 + (-4) & -5 \\ & ? \\ -5 \overset{?}{=} -5 & \text{TRUE} \end{array}$$

$$\begin{array}{c|c} s - t = 3 \\ \hline -1 - (-4) & 3 \\ -1 + 4 & \\ & ? \\ 3 \overset{?}{=} 3 & \text{TRUE} \end{array}$$

Since $(-1, -4)$ checks in both equations, it is the solution.

19. $x - y = 5,$ (1)

$x + 2y = 7$ (2)

Solve Equation (1) for x.

$$x - y = 5 \quad (1)$$
$$x = y + 5 \quad (3)$$

Substitute $y + 5$ for x in Equation (2) and solve for y.

$$x + 2y = 7 \quad (2)$$
$$(y + 5) + 2y = 7 \quad \text{Substituting}$$
$$3y + 5 = 7$$
$$3y = 2$$
$$y = \frac{2}{3}$$

Substitute $\frac{2}{3}$ for y in Equation (3) and compute x.

$$x = y + 5 = \frac{2}{3} + 5 = \frac{2}{3} + \frac{15}{3} = \frac{17}{3}$$

The ordered pair $\left(\frac{17}{3}, \frac{2}{3}\right)$ checks in both equations. It is the solution.

21. $x - 2y = 7,$ (1)

$3x - 21 = 6y$ (2)

Solve Equation (1) for x.

$$x - 2y = 7$$
$$x = 2y + 7$$

Substitute $2y + 7$ for x in Equation (2) and solve for y.

$$3x - 21 = 6y \quad (2)$$
$$3(2y + 7) - 21 = 6y \quad \text{Substituting}$$
$$6y + 21 - 21 = 6y$$
$$6y = 6y$$

The last equation is true for any choice of y, so there is an infinite number of solutions.

23. $y = 2x + 5,$ (1)

$-2y = -4x - 10$ (2)

We substitute $2x + 5$ for y in Equation (2) and solve for x.
$$-2y = -4x - 10 \quad (2)$$
$$-2(2x + 5) = -4x - 10 \quad \text{Substituting}$$
$$-4x - 10 = -4x - 10$$

The last equation is true for any choice of x, so there is an infinite number of solutions.

25. $4x - y = -3, \quad (1)$
$2x + 5y = 2 \quad (2)$

Solve Equation (1) for y.
$$4x - y = -3 \quad (1)$$
$$4x + 3 = y$$

Substitute $4x + 3$ for y in Equation (2) and solve for x.
$$2x + 5y = 2 \quad (2)$$
$$2x + 5(4x + 3) = 2 \quad \text{Substituting}$$
$$2x + 20x + 15 = 2$$
$$22x + 15 = 2$$
$$22x = -13$$
$$x = \frac{-13}{22}$$

Now substitute $\frac{-13}{22}$ for x in Equation (3) and compute y.
$$y = 4\left(\frac{-13}{22}\right) + 3 = \frac{-26}{11} + \frac{33}{11} = \frac{7}{11}$$

The ordered pair $\left(\frac{-13}{22}, \frac{7}{11}\right)$ checks in both equations. It is the solution.

27. $a - b = 6, \quad (1)$
$3a - 2b = 12 \quad (2)$

Solve Equation (1) for a.
$$a - b = 6 \quad (1)$$
$$a = b + 6 \quad (3)$$

Substitute $b + 6$ for a in Equation (2) and solve for b.
$$3a - 2b = 12 \quad (2)$$
$$3(b + 6) - 2b = 12 \quad \text{Substituting}$$
$$3b + 18 - 2b = 12$$
$$b + 18 = 12$$
$$b = -6$$

Now substitute -6 for b in Equation (3) and compute a.
$$a = -6 + 6 = 0$$

The ordered pair $(0, -6)$ checks in both equations. It is the solution.

29. $s = \frac{1}{2}r \quad (1)$
$3r - 4s = 10 \quad (2)$

Substitute $\frac{1}{2}r$ for s in Equation (2) and solve for r.
$$3r - 4s = 10 \quad (2)$$
$$3r - 4\left(\frac{1}{2}r\right) = 10 \quad \text{Substituting}$$
$$3r - 2r = 10$$
$$r = 10$$

Next we substitute 10 for r in (1)
$$s = \frac{1}{2}(10) = 5$$
The ordered pair $(10, 5)$ checks in both equations. It is the solution.

31. $x - 3y = 7, \quad (1)$
$-4x + 12y = 28 \quad (2)$

Solve Equation (1) for x.
$$x - 3y = 7 \quad (1)$$
$$x = 3y + 7 \quad (3)$$

Substitute $3y + 7$ for x in Equation (2).
$$-4x + 12y = 28 \quad (2)$$
$$-4(3y + 7) + 12y = 28 \quad \text{Substituting}$$
$$-12y - 28 + 12y = 28$$
$$-28 = 28$$

We obtain a false equation, so the system has no solution.

33. $x - 2y = 5, \quad (1)$
$2y - 3x = 1 \quad (2)$

Solve Equation (1) for x.
$$x - 2y = 5$$
$$x = 2y + 5 \quad (3)$$

Substitute $2y + 5$ for x in Equation (2) and solve for y.
$$2y - 3x = 1 \quad (2)$$
$$2y - 3(2y + 5) = 1 \quad \text{Substituting}$$
$$2y - 6y - 15 = 1$$
$$-4y - 15 = 1$$
$$-4y = 16$$
$$y = -4$$

Next substitute -4 for y in Equation (3) and compute x.
$$x = 2y + 5 = 2(-4) + 5 = -8 + 5 = -3$$

The ordered pair $(-3, -4)$ checks in both equations. It is the solution.

35. $2x - y = 0, \quad (1)$
$2x - y = -2 \quad (2)$

Solve Equation (1) for y.
$$2x - y = 0 \quad (1)$$
$$2x = y \quad (3)$$

Substitute $2x$ for y in Equation (2) and solve for x.
$$2x - y = -2 \quad (2)$$
$$2x - 2x = -2 \quad \text{Substituting}$$
$$0 = -2$$

We obtain a false equation, so the system has no solution.

37. Familiarize. We let $x =$ the larger number and $y =$ the smaller number.

Translate.

The sum of two numbers is 63.
$$x + y = 63$$

$\underbrace{\text{One number}}_{\downarrow} \underbrace{\text{is}}_{\downarrow} \underbrace{7}_{\downarrow} \underbrace{\text{more than}}_{\downarrow} \underbrace{\text{the other.}}_{\downarrow}$

$\quad\quad x \quad\quad = 7 \quad + \quad\quad y$

The resulting system is

$$x + y = 63, \quad (1)$$
$$x = 7 + y. \quad (2)$$

Carry out. We solve the system of equations. We substitute $7 + y$ for x in Equation (1) and solve for y.

$$x + y = 63 \quad (1)$$
$$(7 + y) + y = 63 \quad \text{Substituting}$$
$$7 + 2y = 63$$
$$2y = 56$$
$$y = 28$$

Next we substitute 28 for y in either equation of the original system and solve for x.

$$x + y = 63 \quad (1)$$
$$x + 28 = 63 \quad \text{Substituting}$$
$$x = 35$$

Check. The sum of 35 and 28 is 63. The number 35 is 7 more than 28. These numbers check.

State. The numbers are 35 and 28.

39. Familiarize. Let x = the larger number and y = the smaller number.

Translate.

$\underbrace{\text{The sum of two numbers}}_{\downarrow} \underbrace{\text{is}}_{\downarrow} \underbrace{51.}_{\downarrow}$

$\quad\quad x + y \quad\quad\quad = 51$

$\underbrace{\text{The difference of two numbers}}_{\downarrow} \underbrace{\text{is}}_{\downarrow} \underbrace{13.}_{\downarrow}$

$\quad\quad x - y \quad\quad\quad = 13$

The resulting system is

$$x + y = 51, \quad (1)$$
$$x - y = 13. \quad (2)$$

Carry out. We solve the system.

We solve Equation (2) for x.

$$x - y = 13 \quad\quad (2)$$
$$x = y + 13 \quad (3)$$

We substitute $y + 13$ for x in Equation (1) and solve for y.

$$x + y = 51 \quad (1)$$
$$(y + 13) + y = 51 \quad \text{Substituting}$$
$$2y + 13 = 51$$
$$2y = 38$$
$$y = 19$$

Now we substitute 19 for y in Equation (3) and compute x.

$$x = y + 13 = 19 + 13 = 32$$

Check. The sum of 32 and 19 is 51. The difference between 32 and 19 is 13. The numbers check.

State. The numbers are 32 and 19.

41. Familiarize. Let x = the larger number and y = the smaller number.

Translate.

$\underbrace{\text{The difference between two numbers}}_{\downarrow} \underbrace{\text{is}}_{\downarrow} \underbrace{2.}_{\downarrow}$

$\quad\quad x - y \quad\quad\quad = 2$

$\underbrace{\text{Three times the larger number}}_{\downarrow} \underbrace{\text{plus}}_{\downarrow} \underbrace{\text{one-half the smaller}}_{\downarrow} \underbrace{\text{is}}_{\downarrow} \underbrace{34}_{\downarrow}$

$\quad\quad 3x \quad\quad\quad + \quad\quad \frac{1}{2}y \quad\quad = 34$

The resulting system is

$$x - y = 2, \quad\quad (1)$$
$$3x + \frac{1}{2}y = 34. \quad (2)$$

Carry out. We solve the system.

We solve Equation (1) for x.

$$x - y = 2 \quad\quad (1)$$
$$x = y + 2 \quad (3)$$

We substitute $y + 2$ for x in Equation (2) and solve for y.

$$3(y + 2) + \frac{1}{2}y = 34 \quad\quad (2)$$
$$3y + 6 + \frac{1}{2}y = 34 \quad\quad \text{Substituting}$$
$$\frac{7}{2}y = 28$$
$$\frac{2}{7} \cdot \frac{7}{2}y = \frac{2}{7} \cdot 28$$
$$y = 8$$

Next we substitute 8 for y in Equation (3) and compute x.

$$x = y + 2 = 8 + 2 = 10$$

Check. The difference between 10 and 8 is 2. Three times the larger, $3 \cdot 10$ or 30 plus one-half the smaller, $\frac{1}{2} \cdot 8$, or 4 is 34. The numbers check.

State. The numbers are 10 and 8.

43. Familiarize. Let x = one angle and y = the other angle.

Translate. Since the angles are supplementary, we have one equation.

$$x + y = 180$$

The second sentence can be translated as follows:

$\underbrace{\text{One angle}}_{\downarrow} \underbrace{\text{is}}_{\downarrow} \underbrace{15° \text{ more than twice the other.}}_{\downarrow}$

$\quad\quad x \quad\quad = \quad\quad 2y + 15$

The resulting system is

$$x + y = 180, \quad (1)$$
$$x = 2y + 15. \quad (2)$$

Carry out. We solve the system.

We substitute $2y + 15$ for x in Equation (1) and solve for y.

$$x + y = 180 \quad (1)$$
$$2y + 15 + y = 180$$
$$3y + 15 = 180$$
$$3y = 165$$
$$y = 55$$

Next we substitute 55 for y in Equation (2) and find x.

$$x = 2y + 15 = 2 \cdot 55 + 15 = 110 + 15 = 125$$

Check. The sum of the angles is $55° + 125°$, or $180°$, so the angles are supplementary. If $15°$ is added to twice $55°$, we have $2 \cdot 55° + 15°$, or $125°$, which is the other angle. The answer checks.

State. One angle is $55°$, and the other is $125°$.

45. *Familiarize*. We let $x = $ the larger angle and $y = $ the smaller angle.

Translate. Since the angles are complementary, we have one equation.

$$x + y = 90$$

We reword and translate the second statement.

$$\underbrace{\text{One angle}}_{\downarrow \atop y} \; \underbrace{\text{is}}_{\downarrow \atop =} \; \underbrace{\text{one-half the other}}_{\downarrow \atop \frac{1}{2}x} \; \underbrace{\text{less}}_{\downarrow \atop -} \; \underbrace{3°.}_{\downarrow \atop 3}$$

The resulting system is

$$x + y = 90 \quad (1)$$
$$y = \tfrac{1}{2}x - 3 \quad (2)$$

Carry out. We solve the system.

$$x + y = 90 \qquad (1)$$
$$y = \tfrac{1}{2}x - 3 \quad (2)$$

We substitute $\frac{1}{2}x - 3$ for y in Equation (1) and solve for x.

$$x + y = 90 \quad (1)$$
$$x + \left(\tfrac{1}{2}x - 3\right) = 90$$
$$\tfrac{3}{2}x - 3 = 90$$
$$\tfrac{3}{2}x = 93$$
$$x = 62$$

Next we substitute 62 for x in Equation (2) and solve for x and compute y.

$$y = \frac{1}{2}(62) - 3 = 28$$

Check. The sum of the angles is $62° + 28°$, or $90°$, so the angles are complementary. One half 62 minus $3°$ is $28°$. These numbers check.

State. The angles are $62°$ and $28°$.

47. *Familiarize*. We let $l = $ the length and $w = $ the width.

Translate. The perimeter is $2l + 2w$. We translate the first statement.

$$\underbrace{\text{The perimeter}}_{\downarrow \atop 2l + 2w} \; \underbrace{\text{is}}_{\downarrow \atop =} \; \underbrace{452 \text{ ft}}_{\downarrow \atop 452}$$

We translate the second statement.

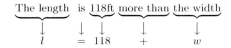

$$\underbrace{\text{The length}}_{\downarrow \atop l} \; \underbrace{\text{is}}_{\downarrow \atop =} \; \underbrace{118\text{ft}}_{\downarrow \atop 118} \; \underbrace{\text{more than}}_{\downarrow \atop +} \; \underbrace{\text{the width}}_{\downarrow \atop w}$$

The resulting system is

$$2l + 2w = 452 \quad (1)$$
$$l = 118 + w. \quad (2)$$

Solve. We solve the system. We substitute $118 + w$ for l in Equation (1) and solve for w.

$$2(118 + w) + 2w = 452$$
$$236 + 4w = 452$$
$$4w = 216$$
$$w = 54$$

Now we substitute 54 for w in Equation (2) and solve for l.

$$l = 118 + w \text{ Equation (2)}$$
$$l = 118 + 54 \text{Substituting}$$
$$l = 172$$

Check. The perimeter would be $2 \cdot 172 + 2 \cdot 54$, or 452 ft. Also, the width 54 plus 118 is 172 which is the length. These numbers check.

State. The length is 172 ft., and the width is 54 ft.

49. *Familiarize*. Recall that the perimeter of a rectangle with length l and width w is given by $2l + 2w$.

Translate.

$$\underbrace{\text{The perimeter}}_{\downarrow \atop 2l + 2w} \; \underbrace{\text{is}}_{\downarrow \atop =} \; \underbrace{1280 \text{ mi.}}_{\downarrow \atop 1280}$$

$$\underbrace{\text{The width}}_{\downarrow \atop w} \; \underbrace{\text{is}}_{\downarrow \atop =} \; \underbrace{\text{the length}}_{\downarrow \atop l} \; \underbrace{\text{less}}_{\downarrow \atop -} \; \underbrace{90\text{mi.}}_{\downarrow \atop 90}$$

The resulting system is

$$2l + 2w = 1280, \quad (1)$$
$$w = l - 90. \quad (2)$$

Carry out. We solve the system.

Substitute $l - 90$ in Equation (1) and solve for l.

$$2l + 2w = 1280 \quad (1)$$
$$2l + 2(l - 90) = 1280 \quad \text{Substituting}$$
$$2l + 2l - 180 = 1280$$
$$4l = 1460$$
$$l = 365$$

Now substitute 365 for l in Equation (2).

$$w = l - 90 \qquad (2)$$
$$w = 365 - 90 \quad \text{Substituting}$$
$$w = 275$$

Check. If the length is 365 mi and the width is 275 mi, the perimeter would be $2 \cdot 365 + 2 \cdot 275 = 1280$ mi. Also, the width is 90 mi less than the length. These numbers check.

State. The length is 365 mi, and the width is 275 mi.

51. Familiarize. Recall that the perimeter of a rectangle with length l and width w is given by $2l + 2w$.

Translate.

$$\underbrace{\text{The perimeter}}_{\downarrow \atop 2l + 2w} \; \underbrace{\text{is}}_{\downarrow \atop =} \; \underbrace{280 \text{ ft.}}_{\downarrow \atop 280}$$

$$\underbrace{\text{The width}}_{\downarrow \atop w} \; \underbrace{\text{is}}_{\downarrow \atop =} \; \underbrace{5}_{\downarrow \atop 5} \; \underbrace{\text{more than}}_{\downarrow \atop +} \; \underbrace{\text{half the length.}}_{\downarrow \atop \frac{1}{2}l}$$

The resulting system is

$$2l + 2w = 280, \quad (1)$$
$$w = 5 + \frac{1}{2}l. \quad (2)$$

Carry out. We solve the system.

Substitute $5 + \frac{1}{2}l$ for w in Equation (1) and solve for l.

$$2l + 2\left(5 + \frac{1}{2}l\right) = 280 \quad (1)$$
$$2l + 10 + l = 280$$
$$3l + 10 = 280$$
$$3l = 270$$
$$l = 90$$

Now substitute 90 for l in Equation (2) and compute w.

$$w = 5 + \frac{1}{2}l = 5 + \frac{1}{2} \cdot 90 = 5 + 45 = 50$$

Check. If the length is 90 yd and the width is 50 yd, then the perimeter is $2 \cdot 90 + 2 \cdot 50$, or $180 + 100$, or 280 yd. Also, 5 more than half the length, $5 + \frac{1}{2} \cdot 90$, or $5 + 45$, or 50 yd, is the width. The answer checks.

State. The length is 90 yd, and the width is 50 yd.

53. Familiarize. Let h = the height of the front wall and w = the width of the service zone.

Translate.

$$\underbrace{\text{The height}}_{\downarrow \atop h} \; \underbrace{\text{is}}_{\downarrow \atop =} \; \underbrace{4 \text{ times}}_{\downarrow \, \downarrow \atop 4 \quad \cdot} \; \underbrace{\text{the width.}}_{\downarrow \atop w}$$

$$\underbrace{\text{The height}}_{\downarrow \atop h} \; \underbrace{\text{plus}}_{\downarrow \atop +} \; \underbrace{\text{the width}}_{\downarrow \atop w} \; \underbrace{\text{is}}_{\downarrow \atop =} \; \underbrace{25 \text{ ft.}}_{\downarrow \atop 25}$$

The resulting system is

$$h = 4w, \quad (1)$$
$$h + w = 25. \quad (2)$$

Carry out. We solve the system.

Substitute $4w$ for h in Equation (2) and solve for w.

$$h + w = 25 \quad (2)$$
$$4w + w = 25 \quad \text{Substituting}$$
$$5w = 25$$
$$w = 5$$

Now substitute 5 for w in Equation (1) to find h.

$$h = 4w = 4 \cdot 5 = 20$$

Check. If the height is 20 ft and the width of the service zone is 5 ft, then the height is 4 times the width and the sum of the height and the width is 25. The answer checks.

State. The height of the court is 20 ft, and the width of the service zone is 5 ft.

55. Writing Exercise. This is not the best approach, in general. If the first equation has x alone on one side, for instance, or if the second equation has a variable alone on one side, solving for y in the first equation is inefficient. This procedure could also introduce fractions in the computations unnecessarily.

57.
$$3(4x + 2y) - 5(2x + y)$$
$$= 12x + 6y - 10x - 5y$$
$$= 2x + y$$

59.
$$4(5x + 6y) - 5(4x + 7y)$$
$$= 20x + 24y - 20x - 35y$$
$$= -11y$$

61.
$$2(3x - 4y) + 4(5x + 2y)$$
$$= 6x - 8y + 20x + 8y$$
$$= 26x$$

63. Writing Exercise. The equations have the same coefficients of x and y but different constant terms. This means their graphs have the same slope but different y-intercepts. Thus, the system of equations has no solution.

65.
$$\frac{1}{6}(a + b) = 1, \quad (1)$$
$$\frac{1}{4}(a - b) = 2 \quad (2)$$

Observe that $\frac{1}{6}(a + b) = 1$, so $a + b = 6$. Also, $\frac{1}{4}(a - b) = 2$, so $a - b = 8$. We need to find two numbers whose sum is 6 and whose difference is 8. The numbers are 7 and -1, so the solution of the system of equations is $(7, -1)$.

We could also solve this system of equations using the substitution method. We first clear the fractions.

$$a + b = 6 \quad (1a)$$
$$a - b = 8 \quad (2a)$$

We solve Equation (2a) for a.

$$a - b = 8 \quad (2a)$$
$$a = b + 8$$

We substitute $b + 8$ for a in Equation (1a) and solve for b.

$$(b + 8) + b = 6$$
$$2b + 8 = 6$$
$$2b = -2$$
$$b = -1$$

Next we substitute -1 for b in Equation (2a) and solve for a.

$$a - b = 8$$
$$a - (-1) = 8$$
$$a + 1 = 8$$
$$a = 7$$

Since $(7, -1)$ checks in both equations, it is the solution.

67. Graph the equations and use the INTERSECT feature from the CALC menu to find the coordinates of the point of intersection. (It might be necessary to solve each equation for y before entering them on a graphing calculator.) The solution is approximately $(4.38, 4.33)$.

69. *Familiarize*. Let t and d represent the ages Trudy and Dennis will be when the age requirement is met.

Translate. Dennis will be 20 years older than Trudy, so we have one equation:

$$d = t + 20.$$

Trudy's age will be 7 more than half of Dennis' age, so we have a second equation:

$$t = \frac{1}{2}d + 7.$$

The resulting system is

$$d = t + 20, \quad (1)$$
$$t = \frac{1}{2}d + 7. \quad (2)$$

Carry out. We solve the system.

First we substitute $t + 20$ for d in Equation (2) and solve for t.

$$t = \frac{1}{2}d + 7 \qquad (2)$$
$$t = \frac{1}{2}(t + 20) + 7$$
$$t = \frac{1}{2}t + 10 + 7$$
$$t = \frac{1}{2}t + 17$$
$$\frac{1}{2}t = 17$$
$$t = 34$$

We are asked to find only Trudy's age but we will find Dennis' as well so that we can check the answer. Substitute 34 for t in Equation (1) and find d.

$$d = t + 20 = 34 + 20 = 54$$

Check. If Trudy is 34 and Dennis is 54, then Trudy is 20 yr younger than Dennis. Since $\frac{1}{2} \cdot 54 + 7 = 27 + 7 = 34$, we see that Trudy's age is 7 more than half of Dennis' age. The answer checks.

State. The youngest age at which Trudy can marry Dennis is 34 yr.

71. $x + y + z = 180, \quad (1)$
      $x = z - 70, \qquad\quad (2)$
      $2y - z = 0 \qquad\quad (3)$

Substitute $z - 70$ for x in Equation (1).

$$(z - 70) + y + z = 180$$
$$y + 2z = 250 \quad (4)$$

We now have a system of two equations in two variables.

$$2y - z = 0 \qquad (3)$$
$$y + 2z = 250 \quad (4)$$

Solve Equation (3) for z.

$$2y - z = 0 \quad (3)$$
$$2y = z \quad (5)$$

Substitute $2y$ for z in Equation (4).

$$y + 2(2y) = 250$$
$$5y = 250$$
$$y = 50$$

Substitute 50 for y in Equation (5).

$$z = 2y = 2 \cdot 50 = 100$$

Substitute 100 for z in Equation (2).

$$x = z - 70 = 100 - 70 = 30$$

The triple $(30, 50, 100)$ checks in all three equations. It is the solution.

73. *Writing Exercise.*

$$x - 2y = 6, \quad (1)$$
$$3x + 2y = 4 \quad (2)$$

Solve (1) for $2y$.

$$x - 2y = 6$$
$$x - 6 = 2y \quad (3)$$

Substitute $x - 6$ for $2y$ in (2).

$$3x + x - 6 = 4$$
$$4x = 10$$
$$x = \frac{5}{2}$$

Substitute $\frac{5}{2}$ for x in (3).

$$\frac{5}{2} - 6 = 2y$$
$$-\frac{7}{2} = 2y$$
$$-\frac{7}{4} = y$$

The solution is $\left(\frac{5}{2}, -\frac{7}{4} \right)$.

It can be argued that neither procedure is easier to use than the other. The procedure in Example 3 requires the use of the distributive law after the first substitution while the procedure above does not. However, the procedure above requires the use of the multiplication principle in solving for the second variable while the procedure in Example 3 does not. Elsewhere, the two procedures require the same number of steps.

Exercise Set 7.3

1. The statement is false. See the introductory paragraph for this section on page 443 in the text.

3. The statement is true. See Example 6.

5.

$$\begin{array}{rcl} x - y &=& 3 \quad (1) \\ x + y &=& 13 \quad (2) \\ \hline 2x &=& 16 \quad \text{Adding} \\ x &=& 8 \end{array}$$

Substitute 8 for x in one of the original equations and solve for y.

$$\begin{array}{rcl} x + y &=& 13 \quad (2) \\ 8 + y &=& 13 \quad \text{Substituting} \\ y &=& 5 \end{array}$$

Check:

$$\begin{array}{c|c} x - y = 3 \\ \hline 8 - 5 \;\big|\; 3 \\ \overset{?}{=} \\ 3 \overset{?}{=} 3 \quad \text{TRUE} \end{array} \qquad \begin{array}{c|c} x + y = 13 \\ \hline 8 + 5 \;\big|\; 13 \\ \overset{?}{=} \\ 13 \overset{?}{=} 13 \quad \text{TRUE} \end{array}$$

Since $(8, 5)$ checks, it is the solution.

7.

$$\begin{array}{rcl} x + y &=& 6 \quad (1) \\ -x + 3y &=& -2 \quad (2) \\ \hline 4y &=& 4 \quad \text{Adding} \\ y &=& 1 \end{array}$$

Substitute 1 for y in one of the original equations and solve for x.

$$\begin{array}{rcl} x + y &=& 6 \quad (1) \\ x + 1 &=& 6 \quad \text{Substituting} \\ x &=& 5 \end{array}$$

Check:

$$\begin{array}{c|c} x + y = 6 \\ \hline 5 + 1 \;\big|\; 6 \\ \overset{?}{=} \\ 6 \overset{?}{=} 6 \quad \text{TRUE} \end{array}$$

$$\begin{array}{c|c} -x + 3y = -2 \\ \hline -5 + 3 \cdot 1 \;\big|\; -2 \\ -5 + 3 \\ \overset{?}{=} \\ -2 \overset{?}{=} -2 \quad \text{TRUE} \end{array}$$

Since $(5, 1)$ checks, it is the solution.

9.

$$\begin{array}{rcl} 4x + y &=& 5 \quad (1) \\ 2x - y &=& 7 \quad (2) \\ \hline 6x &=& 12 \quad \text{Adding} \\ x &=& 2 \end{array}$$

Substitute 2 for x in one of the original equations and solve for y.

$$\begin{array}{rcl} 4x + y &=& 5 \quad (2) \\ 4 \cdot 2 + y &=& 5 \quad \text{Substituting} \\ 8 + y &=& 5 \\ y &=& -3 \end{array}$$

Check:

$$\begin{array}{c|c} 4x + y = 5 \\ \hline 4 \cdot 2 + (-3) \;\big|\; 5 \\ 8 - 3 \\ \overset{?}{=} \\ 5 \overset{?}{=} 5 \quad \text{TRUE} \end{array}$$

$$\begin{array}{c|c} 2x - y = 7 \\ \hline 2 \cdot 2 - (-3) \;\big|\; 7 \\ 4 + 3 \\ \overset{?}{=} \\ 7 \overset{?}{=} 7 \quad \text{TRUE} \end{array}$$

Since $(2, -3)$ checks, it is the solution.

11.

$$\begin{array}{rcl} 5a + 4b &=& 7 \quad (1) \\ -5a + b &=& 8 \quad (2) \\ \hline 5b &=& 15 \\ b &=& 3 \end{array}$$

Substitute 3 for b in one of the original equations and solve for a.

$$\begin{array}{rcl} 5a + 4b &=& 7 \quad (1) \\ 5a + 4 \cdot 3 &=& 7 \\ 5a + 12 &=& 7 \\ 5a &=& -5 \\ a &=& -1 \end{array}$$

Check:

$$\begin{array}{c|c} 5a + 4b = 7 \\ \hline 5(-1) + 4 \cdot 3 \;\big|\; 7 \\ -5 + 12 \\ \overset{?}{=} \\ 7 \overset{?}{=} 7 \quad \text{TRUE} \end{array}$$

$$\begin{array}{c|c} -5a + b = 8 \\ \hline -5(-1) + 3 \;\big|\; 8 \\ 5 + 3 \\ \overset{?}{=} \\ 8 \overset{?}{=} 8 \quad \text{TRUE} \end{array}$$

Since $(-1, 3)$ checks, it is the solution.

13.

$$\begin{array}{rcl} 8x - 5y &=& -9 \quad (1) \\ 3x + 5y &=& -2 \quad (2) \\ \hline 11x &=& -11 \quad \text{Adding} \\ x &=& -1 \end{array}$$

Substitute -1 for x in either of the original equations and solve for y.

$$\begin{array}{rcl} 3x + 5y &=& -2 \quad \text{Equation (2)} \\ 3(-1) + 5y &=& -2 \quad \text{Substituting} \\ -3 + 5y &=& -2 \\ 5y &=& 1 \\ y &=& \dfrac{1}{5} \end{array}$$

Check:

$$\begin{array}{c|c} 8x - 5y = -9 \\ \hline 8(-1) - 5\left(\dfrac{1}{5}\right) & -9 \\ -8 - 1 & \\ & \overset{?}{-9 = -9} \quad \text{TRUE} \end{array}$$

$$\begin{array}{c|c} 3x + 5y = -2 \\ \hline 3(-1) + 5\left(\dfrac{1}{5}\right) & -2 \\ -3 + 1 & \\ & \overset{?}{-2 = -2} \quad \text{TRUE} \end{array}$$

Since $\left(-1, \dfrac{1}{5}\right)$ checks, it is the solution.

15. $\quad 3a - 6b = 8,$
$$\underline{-3a + 6b = -8}$$
$$0 = 0 \qquad \text{Adding}$$

The equation $0 = 0$ is always true, so the system has an infinite number of solutions.

17. $\quad -x - y = 3, \quad (1)$
$$2x - y = -3 \quad (2)$$

We multiply by -1 on both sides of Equation (1) and then add.

$$\begin{array}{ll} x + y = -3 & \text{Multiplying by } -1 \\ \underline{2x - y = -3} & \\ 3x = -6 & \text{Adding} \\ x = -2 & \end{array}$$

Substitute -2 for x in one of the original equations and solve for y.

$$\begin{aligned} 2(-2) - y &= -3 \\ -4 - y &= -3 \\ -y &= 1 \\ y &= -1 \end{aligned}$$

Check:

$$\begin{array}{c|c} -x - y = 3 \\ \hline -(-2) - (-1) & 3 \\ 2 + 1 & \\ & \overset{?}{3 = 3} \qquad \text{TRUE} \end{array}$$

$$\begin{array}{c|c} 2x - y = -3 \\ \hline 2(-2) - (-1) & -1 \\ -4 + 1 & \\ & \overset{?}{-3 = -3} \qquad \text{TRUE} \end{array}$$

Since $(-2, -1)$ checks, it is the solution.

19. $\quad x + 3y = 19,$
$$x - y = -1$$

We multiply by -1 on both sides of Equation (2) and then add.

$$\begin{array}{ll} x + 3y = 19 & \\ \underline{-x + y = 1} & \text{Multiplying by } -1 \\ 4y = 20 & \text{Adding} \\ y = 5 & \end{array}$$

Substitute 5 for y in one of the original equations and solve for x.

$$\begin{array}{ll} x - y = -1 & (2) \\ x - 5 = -1 & \text{Substituting} \\ x = 4 & \end{array}$$

Check:

$$\begin{array}{c|c} x + 3y = 19 \\ \hline 4 + 3 \cdot 5 & 19 \\ 4 + 15 & \\ & ? \\ 19 = 19 & \text{TRUE} \end{array} \qquad \begin{array}{c|c} x - y = -1 \\ \hline 4 - 5 & -1 \\ & ? \\ -1 = -1 & \text{TRUE} \end{array}$$

Since $(4, 5)$ checks, it is the solution.

21. $\quad 8x - 3y = -6 \quad (1)$
$$5x + 6y = 75 \quad (2)$$

We multiply by 2 on both sides of Equation (1) and then add.

$$\begin{array}{ll} 16x - 6y = -12 & \text{Multiplying by } 3 \\ \underline{5x + 6y = 75} & \\ 21x = 63 & \\ x = 3 & \end{array}$$

Substitute 3 for x in one of the original equations and solve for y.

$$\begin{aligned} 5x + 6y &= 75 \quad (2) \\ 5 \cdot 3 + 6y &= 75 \\ 15 + 6y &= 75 \\ 6y &= 60 \\ y &= 10 \end{aligned}$$

Check:

$$\begin{array}{c|c} 8x - 3y = -6 \\ \hline 8 \cdot 3 - 3 \cdot 10 & -6 \\ 24 - 30 & \\ & ? \\ -6 = -6 & \text{TRUE} \end{array} \qquad \begin{array}{c|c} 5x + 6y = 75 \\ \hline 5 \cdot 3 + 6 \cdot 10 & 75 \\ 15 + 60 & \\ & ? \\ 75 = 75 & \text{TRUE} \end{array}$$

Since $(3, 10)$ checks, it is the solution.

23. $\quad 2w - 3z = -1, \quad (1)$
$$-4w + 6z = 5 \quad (2)$$

We multiply by 2 on both sides of Equation (1) and then add.

$$\begin{array}{l} 4w - 6z = -2 \\ \underline{-4w + 6z = 5} \\ 0 = 3 \end{array}$$

We get a false equation, so there is no solution.

25. $\quad 4a + 6b = -1, \quad (1)$
$$a - 3b = 2 \quad (2)$$

We multiply by 2 on both sides of Equation (2) and then add.

$$\begin{array}{ll} 4a + 6b = -1 & \\ \underline{2a - 6b = 4} & \text{Multiplying by } 2 \\ 6a = 3 & \text{Adding} \\ a = \tfrac{1}{2} & \end{array}$$

Substitute $\frac{1}{2}$ for a in one of the original equations and solve for b.

$$4 \cdot \tfrac{1}{2} + 6b = -1$$
$$2 + 6b = -1$$
$$6b = -3$$
$$b = -\tfrac{1}{2}$$

Check:

$$\begin{array}{c|c} 4a + 6b = -1 \\ \hline 4(\tfrac{1}{2}) + 6(-\tfrac{1}{2}) & -1 \\ 2 - 3 & \\ & \overset{?}{-1 = -1} \quad \text{TRUE} \end{array}$$

$$\begin{array}{c|c} a - 3b = 2 \\ \hline \tfrac{1}{2} - 3(-\tfrac{1}{2}) & 2 \\ \tfrac{1}{2} + \tfrac{3}{2} & \\ & \overset{?}{2 = 2} \quad \text{TRUE} \end{array}$$

Since $\left(\dfrac{1}{2}, -\dfrac{1}{2}\right)$ checks, it is the solution.

27.
$$3y = x, \quad (1)$$
$$5x + 14 = y \quad (2)$$

We first get each equation in the form $Ax + By = C$.

$$\begin{array}{ll} x - 3y = 0, & (1\text{a}) \quad \text{Adding } -3y \\ 5x - y = -14 & (2\text{a}) \quad \text{Adding } -y - 14 \end{array}$$

We multiply by -5 on both sides of Equation (1a) and add.

$$\begin{array}{rl} -5x + 15y = \ \ 0 & \text{Multiplying by } -5 \\ \underline{5x - \ \ y = -14} & \\ 14y = -14 & \text{Adding} \\ y = \ -1 & \end{array}$$

Substitute -1 for y in Equation (1a) and solve for x.

$$x - 3y = 0$$
$$x - 3(-1) = 0 \quad \text{Substituting}$$
$$x + 3 = 0$$
$$x = -3$$

Check:

$$\begin{array}{c|c} x = 3y \\ \hline -3 & 3(-1) \\ & \overset{?}{-3 = -3} \quad \text{TRUE} \end{array}$$

$$\begin{array}{c|c} 5x + 14 = y \\ \hline 5(-3) + 14 & -1 \\ -15 + 14 & \\ & \overset{?}{-1 = -1} \quad \text{TRUE} \end{array}$$

Since $(-3, -1)$ checks, it is the solution.

29.
$$4x - 10y = 13, \quad (1)$$
$$-2x + 5y = 8 \quad (2)$$

We multiply by 2 on both sides of Equation (2) and then add.

$$\begin{array}{rl} 4x - 10y = 13 & \\ \underline{-4x + 10y = 16} & \text{Multiplying by 2} \\ 0 = 29 & \end{array}$$

The equation $0 = 29$ is false for any pair (x, y), so there is no solution.

31.
$$2n - 15 - 10m = 40, \quad (1)$$
$$28 = n - 4m \quad (2)$$

We first get each equation in the form $Am + Bn = C$.

$$\begin{array}{ll} -10m + 2n = 55, & (1) \quad \text{Adding 15} \\ -4m + n = 28 & (2) \end{array}$$

We multiply by -2 on both sides of Equation (2) and add.

$$\begin{array}{rl} -10m + 2n = \ \ 55 & \\ \underline{8m - 2n = -56} & \\ -2m \ \ \ \ \ = -1 & \\ m = \ \ \tfrac{1}{2} & \end{array}$$

Substitute $\frac{1}{2}$ for m in Equation (2) and solve for n.

$$-4m + n = 32$$
$$-4\left(\frac{1}{2}\right) + n = 28$$
$$-2 + n = 28$$
$$n = 30$$

Check:

$$\begin{array}{c|c} 2n - 15 - 10m = 40 \\ \hline 2(30) - 15 - 10\left(\dfrac{1}{2}\right) & 40 \\ 60 - 15 - 5 & \\ & \overset{?}{40 = 40} \quad \text{TRUE} \end{array}$$

$$\begin{array}{c|c} 28 = n - 4m \\ \hline 28 & 30 - 4\left(\dfrac{1}{2}\right) \\ & 30 - 2 \\ \overset{?}{28 = 28} & \quad \text{TRUE} \end{array}$$

Since $\left(\frac{1}{2}, 30\right)$ checks, it is the solution.

33.
$$3x + 5y = 4, \quad (1)$$
$$-2x + 3y = 10 \quad (2)$$

We use the multiplication principle with both equations and then add.

$$\begin{array}{rl} 6x + 10y = \ \ 8 & \text{Multiplying (1) by 2} \\ \underline{-6x + 9y = 30} & \text{Multiplying (2) by 3} \\ 19y = 38 & \text{Adding} \\ y = \ \ 2 & \end{array}$$

Substitute 2 for y in one of the original equations and solve for x.

$$3x + 5y = 4 \quad (1)$$
$$3x + 5 \cdot 2 = 4$$
$$3x + 10 = 4$$
$$3x = -6$$
$$x = -2$$

Check:

$$\frac{3x + 5y = 4}{\begin{array}{c|c} 3(-2) + 5 \cdot 2 & 4 \\ -6 + 10 & \end{array}}$$

$$4 \overset{?}{=} 4 \qquad \text{TRUE}$$

$$\frac{-2x + 3y = 10}{\begin{array}{c|c} -2(-2) + 3 \cdot 2 & 10 \\ 4 + 6 & \end{array}}$$

$$10 \overset{?}{=} 10 \qquad \text{TRUE}$$

Since $(-2, 2)$ checks, it is the solution.

35. $0.06x + 0.05y = 0.07,$

$0.04x - 0.03y = 0.11$

We first multiply each equation by 100 to clear the decimals.

$$6x + 5y = 7, \quad (1)$$
$$4x - 3y = 11 \quad (2)$$

We use the multiplication principle with both equations of the resulting system.

$$\begin{array}{ll} 18x + 15y = 21 & \text{Multiplying (1) by 3} \\ \underline{20x - 15y = 55} & \text{Multiplying (2) by 5} \\ 38x = 76 & \text{Adding} \\ x = 2 \end{array}$$

Substitute 2 for x in Equation (1) and solve for y.

$$6x + 5y = 7$$
$$6 \cdot 2 + 5y = 7$$
$$12 + 5y = 7$$
$$5y = -5$$
$$y = -1$$

Check:

$$\frac{0.06x + 0.05y = 0.07}{\begin{array}{c|c} 0.06(2) + 0.05(-1) & 0.07 \\ 0.12 - 0.05 & \end{array}}$$

$$0.07 \overset{?}{=} 0.07 \qquad \text{TRUE}$$

$$\frac{0.04x - 0.03y = 0.11}{\begin{array}{c|c} 0.04(2) - 0.03(-1) & 0.11 \\ 0.08 + 0.03 & \end{array}}$$

$$0.11 \overset{?}{=} 0.11 \qquad \text{TRUE}$$

Since $(2, -1)$ checks, it is the solution.

37. $x + \dfrac{9}{2}y = \dfrac{15}{4},$

$\dfrac{9}{10}x - y = \dfrac{9}{20}$

First we clear fractions. We multiply both sides of the first equation by 4 and both sides of the second equation by 20.

$$4\left(x + \frac{9}{2}y\right) = 4 \cdot \frac{15}{4}$$
$$4x + 4 \cdot \frac{9}{2}y = 15$$
$$4x + 18y = 15$$

$$20\left(\frac{9}{10}x - y\right) = 20 \cdot \frac{9}{20}$$
$$20 \cdot \frac{9}{10}x - 20y = 9$$
$$18x - 20y = 9$$

The resulting system is

$$4x + 18y = 15, \quad (1)$$
$$18x - 20y = 9. \quad (2)$$

We use the multiplication principle with both equations.

$$\begin{array}{ll} 72x + 324y = 270 & \text{Multiplying (1) by 18} \\ \underline{-72x + 80y = -36} & \text{Multiplying (2) by } -4 \\ 404y = 234 & \\ y = \dfrac{234}{404}, \text{ or } \dfrac{117}{202} \end{array}$$

Substitute $\dfrac{117}{202}$ for y in (1) and solve for x.

$$4x + 18\left(\frac{117}{202}\right) = 15$$
$$4x + \frac{1053}{101} = 15$$
$$4x = \frac{462}{101}$$
$$x = \frac{1}{4} \cdot \frac{462}{101}$$
$$x = \frac{231}{202}$$

The ordered pair $\left(\dfrac{231}{202}, \dfrac{117}{202}\right)$ checks in both equations. It is the solution.

39. *Familiarize*. We let $m = $ the number of miles driven and $c = $ the total cost of the truck rental.

***Translate*.** We reword and translate the first statement, using \$0.22 for 22¢.

$$\underset{\downarrow}{\text{\$27}} \underset{\downarrow}{\text{ plus }} \underset{\downarrow}{\text{22¢}} \underset{\downarrow}{\text{ times }} \underbrace{\text{the number of}}_{\underset{\downarrow}{\text{miles driven}}} \underset{\downarrow}{\text{ is }} \underset{\downarrow}{\text{cost.}}$$
$$\underset{}{27} \underset{}{+} \underset{}{0.22} \underset{}{\cdot} \underset{}{m} \underset{}{=} \underset{}{c}$$

We reword and translate the second statement using \$0.17 for 17¢.

$$\underset{\downarrow}{\text{\$29}} \underset{\downarrow}{\text{ plus }} \underset{\downarrow}{\text{17¢}} \underset{\downarrow}{\text{ times }} \underbrace{\text{the number of}}_{\underset{\downarrow}{\text{miles driven}}} \underset{\downarrow}{\text{ is }} \underset{\downarrow}{\text{cost.}}$$
$$\underset{}{29} \underset{}{+} \underset{}{0.17} \underset{}{\cdot} \underset{}{m} \underset{}{=} \underset{}{c}$$

We have a system of equations:

$$27 + 0.22m = c,$$
$$29 + 0.17m = c$$

Carry out. To solve the system of equations, we multiply the second equation by -1 and add to eliminate c.

$$27 + 0.22m = c$$
$$\underline{-29 - 0.17m = -c}$$
$$-2 + 0.05m = 0$$
$$0.05m = 2$$
$$m = 40$$

Check. For 40 mi, the cost of the cargo van is $27 + $0.22(40)$, or \$35.80. For 40 mi, the cost of the pickup truck is $29 + $0.17(40)$, or \$35.80. The cost is the same for 40 mi.

State. When driven 40 mi, the cost of the cargo van and the pickup truck are the same.

41. ***Familiarize***. We let $x =$ the larger angle and $y =$ the smaller angle.

Translate. We reword and translate the first statement.

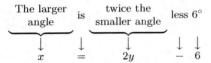

We reword and translate the second statement.

We have a system of equations:

$$x + y = 90,$$
$$x = 2y - 6$$

Carry out. We solve the system. We will use the elimination method, although we could also easily use the substitution method. First we get the second equation in the form $Ax + By = C$.

$$x + y = 90 \quad (1)$$
$$x - 2y = -6 \quad (2) \quad \text{Adding } -2y$$

Now we multiply Equation (1) by 2 and add.

$$2x + 2y = 180$$
$$\underline{x - 2y = -6}$$
$$3x \quad\quad = 174$$
$$x = 58$$

Then we substitute 58 for x in Equation (1) and solve for y.

$$x + y = 90 \quad (1)$$
$$58 + y = 90 \quad \text{Substituting}$$
$$y = 32$$

Check. The sum of the angles is $58° + 32°$, or $90°$, so the angles are complementary. The larger angle, $58°$, is $6°$ less than twice the smaller angle, $32°$. These numbers check.

State. The angles are $58°$ and $32°$.

43. ***Familiarize***. Let $m =$ the number of long distance minutes used in a month and $c =$ the cost of the calls, in cents.

Translate. We reword the problem and translate. We will express \$2.99 as 299¢.

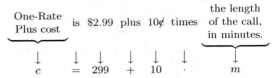

We reword the second statement. We will express \$4.99 as 499¢.

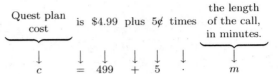

We have a system of equations:

$$c = 299 + 10m,$$
$$c = 499 + 5m$$

Carry out. To solve the system, we multiply the second equation by -1 and add to eliminate c.

$$c = 299 \quad + 10m$$
$$\underline{-c = -499 - 5m}$$
$$0 = -200 + 5m$$
$$200 = 5m$$
$$40 = m$$

Check. For 40 min, the cost of the One-Rate plan is $299 + 10 \cdot 40$, or 699¢ and the cost of the Quest plan is $499 + 5 \cdot 40$, or 699¢. Thus the costs are the same for 40 long distance minutes.

State. For 40 minutes of long distance calls, the monthly costs are the same.

45. ***Familiarize***. Let $x =$ the measure of one angle and $y =$ the measure of the other angle.

Translate. We reword the problem.

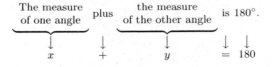

The resulting system is

$$x + y = 180,$$
$$x = 4y - 5.$$

Carry out. We solve the system. We will use the elimination method although we could also easily use the substitution method. First we get the second equation in the form $Ax + By = C$.

$$x + y = 180 \quad (1)$$
$$x - 4y = -5 \quad (2) \text{ Adding } -4y$$

Now we multiply Equation (2) by -1 and add.

$$x + y = 180$$
$$\underline{-x + 4y = 5}$$
$$5y = 185$$
$$y = 37$$

Then we substitute 37 for y in Equation (1) and solve for x.

$$x + y = 180$$
$$x + 37 = 180$$
$$x = 143$$

Check. The sum of the angle measures is $37° + 143°$, or $180°$, so the angles are supplementary. Also, $5°$ less than four times the $37°$ angle is $4 \cdot 37° - 5°$, or $148° - 5°$, or $143°$, the measure of the other angle. These numbers check.

State. The measures of the angles are $37°$ and $143°$.

47. Familiarize. Let $C =$ the number of acre of Chardonnay and $R =$ the number of acres of Riesling.

Translate.

The number of acres of Chardonnay and Riesling is 820 acres

$$C + R = 820$$

The number of acres of Chardonnay is 140 more.

The number of acres of Chardonnay is 140 more than Riesling

$$C = 140 + R$$

The resulting system is

$$C + R = 820$$
$$C = 140 + R$$

Carry out. We solve the system. We will use the elimination method, although we could also easily use the substitution method. First we get the second equation in the form $Ax + By = C$. Then we add the equations.

$$C + R = 820 \quad (1)$$
$$\underline{C - R = 140} \quad (2) \text{ Adding } -R$$
$$2C = 960$$
$$C = 480$$

Substitute 480 for C in Equation (1) and solve for R.

$$C + R = 820$$
$$480 + R = 820$$
$$R = 340$$

Check. The total number of acres is $480 + 340$, or 820. the number of acres of Chardonnay, 480 is 140 more than 340, the number of acres of Riesling. These numbers check.

State. South Wind Vineyards plants 480 acres of Chardonnay and 340 acres of Riesling.

49. Familiarize. Let $l =$ the length of the frame and $w =$ the width, in feet.

Translate.

The perimeter is 18 ft.

$$2l + 2w = 18$$

The length is twice the width.

$$l = 2w$$

The resulting system is

$$2l + 2w = 18,$$
$$l = 2w.$$

Carry out. We solve the system. We will use the elimination method, although we could also easily use the substitution method. First we get the second equation in the form $Al + Bw = C$. Then we add the equations.

$$2l + 2w = 18 \quad (1)$$
$$\underline{l - 2w = 0} \quad (2)$$
$$3l = 18$$
$$l = 6$$

Substitute 6 for l in Equation (2) and solve for w.

$$l - 2w = 0$$
$$6 - 2w = 0$$
$$6 = 2w$$
$$3 = w$$

Check. The perimeter is $2 \cdot 6 + 2 \cdot 3$, or 18 ft. Twice the width is $2 \cdot 3$, or 6 ft, which is the length. These numbers check.

State. The length of the mirror is 6 ft, and the width is 3 ft.

51. Writing Exercise. Write an equation $Ax + By = C$. Then write a second equation $kAx + kBy = kC$.

53. $12.2\% = 12.2 \times 0.01$ Replacing % by $\times 0.01$
$$= 0.122$$

55. Translate. Let $x =$ the percent.

What percent of 65 is 26?

$$x \cdot 65 = 26$$

We solve the equation.

$$65x = 26$$
$$x = 0.4 = 40\%$$

This is equivalent to moving the decimal point two places to the right and inserting the percent symbol.

57. Let $n =$ the number of liters

12 % of the number of liters

$$= 0.12 \cdot n$$
$$= 0.12n$$

59. Writing Exercise. No; only ordered pairs that satisfy the equation(s) of the system are solutions.

61. $y = 3x + 4,$ (1)
$3 + y = 2(y - x)$ (2)

Substitute $3x + 4$ for y in Equation (2) and solve for x.

$$3 + y = 2(y - x) \qquad (1)$$
$$3 + (3x + 4) = 2((3x + 4) - x) \quad \text{Substituting}$$
$$3x + 7 = 2(2x + 4)$$
$$3x + 7 = 4x + 8$$
$$7 = x + 8$$
$$-1 = x$$

Now substitute -1 for x in Equation (1) and compute y.

$$y = 3(-1) + 4 = -3 + 4 = 1$$

The ordered pair $(-1, 1)$ checks, so it is the solution.

63. $0.05x + y = 4,$

$\dfrac{x}{2} + \dfrac{y}{3} = 1\dfrac{1}{3}$

Multiply the first equation by 100 to clear the decimal. Also, multiply the second equation by 6 to clear the fractions.

$$5x + 100y = 400, \quad (1)$$
$$3x + 2y = 8 \quad (2)$$

Multiply Equation (2) by -50 and add.

$$\begin{aligned} 5x + 100y &= 400 \\ -150x - 100y &= -400 \\ \hline -145x &= 0 \\ x &= 0 \end{aligned}$$

Now substitute 0 for x in Equation (1) or (2) and solve for y.

$$3x + 2y = 8 \quad (2)$$
$$3 \cdot 0 + 2y = 8$$
$$2y = 8$$
$$y = 4$$

The ordered pair $(0, 4)$ checks, so it is the solution.

65. $y = -\dfrac{2}{7}x + 3,$

$y = \dfrac{4}{5}x + 3$

Observe that these equations represent lines with different slopes and the same y-intercept. Thus, their point of intersection is the y-intercept, $(0, 3)$ and this is the solution of the system of equations.

67. $ax + by + c = 0,$

$ax + cy + b = 0$

Put both equations in the form $Ax + By = C$.

$$ax + by = -c, \quad (1)$$
$$ax + cy = -b \quad (2)$$

$$\begin{aligned} ax + by &= -c \qquad (1) \\ -ax - cy &= b \qquad \text{Multiplying (2) by } -1 \\ \hline by - cy &= b - c \\ y(b - c) &= b - c \\ y &= 1 \end{aligned}$$

Substitute 1 for y in (1).

$$ax + by = -c$$
$$ax + b \cdot 1 = -c$$
$$ax = -b - c$$
$$x = \frac{-b - c}{a}$$

The solution is $\left(\dfrac{-b - c}{a}, 1 \right)$.

69. *Familiarize*. Let x represent the number of rabbits and y the number of pheasants in the cage. Each rabbit has one head and four feet. Thus, there are x rabbit heads and $4x$ rabbit feet in the cage. Each pheasant has one head and two feet. Thus, there y pheasant heads and $2y$ pheasant feet in the cage.

Translate. We reword the problem.

Rabbit heads	plus	pheasant heads	is	35.
x	$+$	y	$=$	35

Rabbit feet	plus	pheasant feet	is	94.
$4x$	$+$	$2y$	$=$	94

The resulting system is

$$x + y = 35, \quad (1)$$
$$4x + 2y = 94. \quad (2)$$

Carry out. We solve the system of equations. We multiply Equation (1) by -2 and then add.

$$\begin{aligned} -2x - 2y &= -70 \\ 4x + 2y &= 94 \\ \hline 2x &= 24 \quad \text{Adding} \\ x &= 12 \end{aligned}$$

Substitute 12 for x in one of the original equations and solve for y.

$$x + y = 35 \quad (1)$$
$$12 + y = 35 \quad \text{Substituting}$$
$$y = 23$$

Check. If there are 12 rabbits and 23 pheasants, the total number of heads in the cage is $12 + 23$, or 35. The total number of feet in the cage is $4 \cdot 12 + 2 \cdot 23$, or $48 + 46$, or 94. The numbers check.

State. There are 12 rabbits and 23 pheasants.

71. *Familiarize*. Let $x =$ the man's age and $y =$ his daughter's age. Five years ago their ages were $x - 5$ and $y - 5$.

Translate.

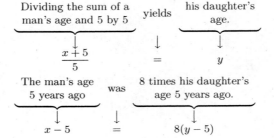

We have a system of equations:
$$\frac{x+5}{5} = y$$
$$x - 5 = 8(y - 5)$$

Carry out. Solve the system.

Multiply the first Equation by 5 to clear the fraction.
$$x + 5 = 5y$$
$$x - 5y = -5$$

Simplify the second equation.
$$x - 5 = 8(y - 5)$$
$$x - 5 = 8y - 40$$
$$x - 8y = -35$$

The resulting system is
$$x - 5y = -5, \quad (1)$$
$$x - 8y = -35. \quad (2)$$

Multiply Equation (2) by -1 and add.

$$\begin{array}{ll} x - 5y = -5 & \\ \underline{-x + 8y = 35} & \text{Multiplying by } -1 \\ \quad\;\; 3y = 30 & \text{Adding} \\ \quad\;\;\; y = 10 & \end{array}$$

Substitute 10 for y in Equation (1) and solve for x.
$$x - 5y = -5$$
$$x - 5 \cdot 10 = -5 \quad \text{Substituting}$$
$$x - 50 = -5$$
$$x = 45$$

Possible solution: Man is 45, daughter is 10.

Check. If 5 is added to the man's age, $5 + 45$, the result is 50. If 50 is divided by 5, the result is 10, the daughter's age. Five years ago the father and daughter were 40 and 5, respectively, and $40 = 8 \cdot 5$. The numbers check.

State. The man is 45 years old; his daughter is 10 years old.

Chapter 7 Connecting the Concepts

1.
$$x = y \quad (1)$$
$$x + y = 4 \quad (2)$$

Use the substitution method.
$$x + (x) = 4 \quad \text{Substituting Equation (1) and (2)}$$
$$2x = 4$$
$$x = 2$$
From Equation (1) if $x = 2$, then $y = 2$. The solution is $(2, 2)$.

3.
$$y = x + 1, \quad (1)$$
$$2x + y = 6 \quad (2)$$

Use the substitution method.
$$2x + (x + 1) = 6 \quad (1)$$
$$3x + 1 = 6$$
$$3x = 5$$
$$x = \frac{5}{3}$$
Substituting $\frac{5}{3}$ for x in Equation (1).

$$y = \frac{5}{3} + 1 = \frac{8}{3}$$
The solution is $\left(\frac{5}{3}, \frac{8}{3} \right)$.

5.
$$y = 2x + 1 \quad (1)$$
$$y = \frac{1}{2}x - 2 \quad (2)$$

Use the substitution method.

$$2x + 1 = \frac{1}{2}x - 2 \quad \text{Substituting Equation (1) into (2)}$$
$$4x + 2 = x - 4 \quad \text{Clearing fractions; Multiplying by 2}$$
$$4x = x - 6$$
$$3x = -6$$
$$x = -2$$
Substituting -2 for x in Equation (1).
$$y = 2(-2) + 1$$
$$y = -3$$
The solution is $(-2, -3)$.

7.
$$3x - 4y = 1 \quad (1)$$
$$2x + 2y = 5 \quad (2)$$

We multiply by 2 on both sides of equation (2).

$$\begin{array}{l} 3x - 4y = \;\; 1 \\ \underline{4x + 4y = 10} \\ 7x \quad\quad = 11 \\ \quad\;\; x = \dfrac{11}{7} \end{array}$$

Substituting $\frac{11}{7}$ for x in Equation (1).

$$2\left(\frac{11}{7}\right) + 2y = 5 \quad (1)$$
$$\frac{22}{7} + 2y = 5$$
$$2y = \frac{13}{7}$$
$$y = \frac{13}{14}$$
The solution is $\left(\frac{11}{7}, \frac{13}{14} \right)$.

9.
$$x = -1, \quad (1)$$
$$y = 5 \quad (2)$$
The solution is $(-1, 5)$.

11.
$$y - 3x = 1 \quad (1)$$
$$y = 3x - 5 \quad (2)$$

Solving Equation (1) for y.
$$y = 3x + 1 \quad (1)$$
$$y = 3x - 5$$
We observe the equations, they have the same slope but different y-intercepts. the lines never intersect. So there is no solution.

13.
$$x + y = 3 \quad (1)$$
$$2x + 2y = 6 \quad (2)$$

Multiplying equation (1) by 2.
$$2x + 2y = 6 \quad (1)$$
$$2x + 2y = 6 \quad (2)$$
We observe that the equations are the same. There is an infinite number of solutions.

15. Use the elimination method.
$$\begin{array}{rl} 8x - 11y = 12 & \\ \underline{3x + 11y = 10} & \text{(2)} \\ 11x \qquad\quad = 22 & \\ x = 2 & \end{array}$$

Substituting 2 for x in Equation (2).
$$\begin{aligned} 3(2) + 11y &= 10 \quad \text{(1)} \\ 6 + 11y &= 10 \\ 11y &= 4 \\ y &= \frac{4}{11} \end{aligned}$$

The solution is $\left(2, \dfrac{4}{11}\right)$.

17. $\quad 2x + 7y = 0 \quad$ (1)
$\quad\ 9x - 10y = 0 \quad$ (2)

Solving Equation (1) for x.
$$\begin{aligned} 2x &= -7y \\ x &= -\frac{7}{2}y \end{aligned}$$

Substituting $-\dfrac{7}{2}y$ for x in Equation (2).
$$\begin{aligned} 9\left(-\frac{7}{2}y\right) - 10y &= 0 \quad \text{(1)} \\ \frac{-63}{2}y - 10y &= 0 \\ -\frac{83}{2}y &= 0 \\ y &= 0 \end{aligned}$$

Substituting 0 for y in Equation (1)
$$\begin{aligned} 2x + 7 \cdot 0 &= 0 \\ 2x &= 0 \\ x &= 0 \end{aligned}$$

Solution is $(0, 0)$.

19. Use the elimination method.
$$\begin{aligned} \frac{1}{2}x + \frac{1}{5}y &= 9 \quad \text{(1)} \\ \frac{1}{3}x - \frac{4}{15}y &= -2 \quad \text{(2)} \end{aligned}$$

Multiply Equation (1) by 2 and Equation (2) by -3.
$$\begin{array}{rl} x + \frac{2}{5}y = 18 & \\ \underline{-x + \frac{4}{5}y = 6} & \\ \frac{6}{5}y = 24 & \\ y = 20 & \end{array}$$

Substitute 20 for y in Equation (2).
$$\begin{aligned} \frac{1}{3}x - \frac{4}{15}(20) &= -2 \quad \text{(1)} \\ \frac{1}{3}x - \frac{16}{3} &= -2 \\ \frac{1}{3}x &= \frac{10}{3} \\ x &= 10 \\ &= \end{aligned}$$

The solution is $(10, 20)$.

---

1. *Familiarize*. Let $x =$ the number of two-point shots that were made and $y =$ the number of three-pointers made. Then Stephen scored $2x$ points on two-point shots and $3y$ points on three-pointers.

***Translate*.** We reword the problem and translate.
$$\underbrace{\text{Total number of shots}}_{x+y} \ \underset{=}{\text{is}} \ \underset{10}{10.}$$
$$\underbrace{\text{Total number of points}}_{2x+3y} \ \underset{=}{\text{is}} \ \underset{27}{27.}$$

The resulting system is
$$\begin{aligned} x + y &= 10, \quad \text{(1)} \\ 2x + 3y &= 27. \quad \text{(2)} \end{aligned}$$

***Carry out*.** We solve using the elimination method.
$$\begin{array}{rll} -2x - 2y = -20 & \text{Multiplying (1) by } -2 \\ \underline{2x + 3y = \quad 27} & \text{(2)} \\ y = \quad 7 & \text{Adding} \end{array}$$

Substitute 7 for y in Equation (1) and solve for x.
$$\begin{aligned} x + y &= 10 \quad \text{(1)} \\ x + 7 &= 10 \\ x &= 3 \end{aligned}$$

***Check*.** If Stephen made 3 two-pointers and 7 three-pointers, then he made $3 + 7$, or 10 shots for a total of $2 \cdot 3 + 3 \cdot 7$, or $6 + 21$, or 27 points. The numbers check.

***State*.** Stephen Jackson made 3 two-point shots and 7 three-point shots.

3. *Familiarize*. Let $x =$ the number of two-point shots that were made and $y =$ the number of three-pointers made. Then the Miami Heat scored $2x$ points on two-point shots and $3y$ points on three-pointers.

***Translate*.** We reword the problem and translate.
$$\underbrace{\text{Total number of shots}}_{x+y} \ \underset{=}{\text{is}} \ \underset{39}{39.}$$
$$\underbrace{\text{Total number of points}}_{2x+3y} \ \underset{=}{\text{is}} \ \underset{84}{84.}$$

The resulting system is
$$\begin{aligned} x + y &= 39, \quad \text{(1)} \\ 2x + 3y &= 84. \quad \text{(2)} \end{aligned}$$

***Carry out*.** We solve using the elimination method.
$$\begin{array}{rll} -2x - 2y = -78 & \text{Multiplying (1) by } -2 \\ \underline{2x + 3y = \quad 84} & \text{(2)} \\ y = \quad 6 & \text{Adding} \end{array}$$

Substitute 6 for y in Equation (1) and solve for x.

$$x + y = 39 \quad (1)$$
$$x + 6 = 39$$
$$x = 33$$

Check. If the Miami Heat made 33 two-pointers and 6 three-pointers, then they made $33 + 6$, or 39 shots and scored $2 \cdot 33 + 3 \cdot 6$, or $66 + 18$, or 84 points. The numbers check.

State. The Miami Heat made 33 two-point shots and 6 three-point shots.

5. Familiarize. Let $x =$ the number of 3-credit courses and $y =$ the number of 4-credit courses. Then the 3-credit courses account for $3x$ credits and the 4-credit courses account for $4y$ credits.

Translate.

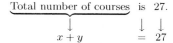

Total number of courses is 27.

$$x + y = 27$$

Total number of credits is 89.

$$3x + 4y = 89$$

The resulting system is

$$x + y = 27, \quad (1)$$
$$3x + 4y = 89. \quad (2)$$

Carry out. We solve using the elimination method.

$$\begin{array}{rl} -3x - 3y = -81 & \text{Multiplying (1) by } -3 \\ \underline{3x + 4y = 89} & (2) \\ y = 8 & \text{Adding} \end{array}$$

Substitute 8 for y in Equation (1) and solve for x.

$$x + y = 27$$
$$x + 8 = 27$$
$$x = 19$$

Check. If there are 19 3-credit courses and 8 4-credit courses, the total number of courses is $19 + 8$, or 27. The total number of credits is $3 \cdot 19 + 4 \cdot 8$, or $57 + 32$, or 89. The answer checks.

State. 19 3-credit courses and 8 4-credit courses are being taken.

7. Familiarize. Let $x =$ the number of 5-cent bottles or cans collected and $y =$ the number of 10-cent bottles or cans collected.

Translate. We organize the given information in a table.

	$0.05	$0.10	Total
Number	x	y	430
Total Value	$0.05x$	$0.10y$	26.20

A system of two equations can be formed using the rows of the table.

$$x + y = 430,$$
$$0.05x + 0.10y = 26.20$$

Carry out. First we multiply on both sides of the second equation by 100 to clear the decimals.

$$x + y = 430, \quad (1)$$
$$5x + 10y = 2620 \quad (2)$$

Now multiply Equation (1) by -5 and add.

$$\begin{array}{rl} -5x - 5y = -2150 \\ \underline{5x + 10y = 2620} \\ 5y = 470 \\ y = 94 \end{array}$$

Substitute 94 for y in Equation (1) and solve for x.

$$x + y = 430$$
$$x + 94 = 430$$
$$x = 336$$

Check. If 336 5-cent bottles and cans and 94 10-cent bottles and cans were collected, then a total of $336 + 94$, or 430 bottles and cans were collected. Their total value is $\$0.05(336) + \$0.10(94)$, or $\$16.80 + \$9.40$, or $\$26.20$. These numbers check.

State. 336 5-cent bottles and cans and 94 10-cent bottles and cans were collected.

9. Familiarize. Let $c =$ the number of cars and $m =$ the number of motorcycles that enter the park on a typical day. Then the cars account for payments of $25c$ and the motorcycles for $20m$.

Total number of cars and motorcycles is 5950.

$$c + m = 5950$$

Total payment is $\$137,650$.

$$25c + 20m = 137,650$$

The resulting system is

$$c + m = 5950, \quad (1)$$
$$25c + 20m = 137,650 \quad (2)$$

Carry out. We use the elimination method.

$$\begin{array}{rl} -20c - 20m = -119,000 & \text{Multiplying (1) by } -15 \\ \underline{25c + 20m = 137,650} & (2) \\ 5c = 18,650 & \text{Adding} \\ c = 3730 \end{array}$$

Since the problem asks only for the number of motorcycles, we could have solved for m first and stopped there, but we solve for both c and m so that we can check the solution.

Substitute 3730 for c in Equation (1) and solve for m.

$$c + m = 5950$$
$$3730 + m = 5950$$
$$m = 2220$$

Check. If there are 3730 cars and 2220 motorcycles, the total number of cars and motorcycles is $3730 + 2220$, or 5950. The total payments are $25(3730) + 20(2220) = 93,250 + 44,400$, or $\$137,650$. The answer checks.

State. On a typical day 2220 motorcycles enter the park.

11. _Familiarize_. Let x = the number of adult admissions and y = the number of child admissions.

Translate. We organize the information in a table.

	Adult	Children	Total
Admission	17	14	
Number	x	y	642
Money Taken In	$17x$	$14y$	9369

We use the last two rows of the table to form a system of equations.

$$x + \quad y = 642, \quad (1)$$
$$17x + 14y = 9369 \quad (2)$$

Carry out. We solve using the elimination method. We first multiply Equation (1) by -14 and add.

$$
\begin{array}{r}
-14x - 14y = -8988 \\
17x + 14y = 9369 \\
\hline
3x = 381 \\
x = 127
\end{array}
$$

Since the problem asks only for the number of adult admissions, this is the number that we needed to find. We will find y also, however, in order to be able to check the solution.

Substitute 127 for x in Equation (1) and solve for y.

$$x + y = 642$$
$$127 + y = 642$$
$$y = 515$$

Check. If $x = 127$ and $y = 515$, then there were $127 + 515$, or 642 admissions sold. The amount collected for adults is $17 \cdot 127$, or \$2159, and the amount collected for children is $14 \cdot 515$, or \$7210. The total amount collected was $\$2159 + \$7210$, or \$9369. The numbers check.

State. There were 127 adult admissions.

13. _Familiarize_. Let x = the number of students receiving private lessons and y = the number of students receiving group lessons.

Translate. We present the information in a table.

	Private	Group	Total
Price	\$25	\$18	
Number	x	y	12
Money Earned	$25x$	$18y$	265

The last two rows of the table give us a system of equations.

$$x + \quad y = 12, \quad (1)$$
$$25x + 18y = 265 \quad (2)$$

Carry out. We solve using the elimination method. First we multiply Equation (1) by -18 and add.

$$
\begin{array}{r}
-18x - 18y = -216 \\
25x + 18y = 265 \\
\hline
7x = 49 \\
x = 7
\end{array}
$$

Substitute 7 for x in Equation (1) and solve for y.

$$x + y = 12$$
$$7 + y = 12$$
$$y = 5$$

Check. If $x = 7$ and $y = 5$, then a total of $7 + 5$, or 12 students received lessons. Alice earned $\$25 \cdot 7$, or \$175 teaching private lessons and $\$18 \cdot 5$, or \$90 teaching group lessons. Thus, she earned a total of $\$175 + \$90$, or \$265. The numbers check.

State. Jillian gave private lessons to 7 students and group lessons to 5 students.

15. _Familiarize_. Let x = the number of Xbox 360 games Jessie bought and y = the number of PS3 games. Then the Xbox games cost a total of $9.99x$ dollars and the PS3 games cost $17.99y$ dollars.

Translate.

Total Xbox games and PS3 games is 11.

$$x + y \qquad\qquad = 11$$

Total purchase price is \$125.89.

$$9.99x + 17.99y \qquad = 125.89$$

The resulting system is

$$x + \qquad y = \qquad 11, \quad (1)$$
$$9.99x + 17.99y = 125.89. \quad (2)$$

Carry out. We use the elimination method. First we multiply Equation (1) by -9.99 and add.

$$
\begin{array}{r}
-9.99x - 9.99y = -109.89 \\
9.99x + 17.99y = 125.89 \\
\hline
8y = 16 \\
y = 2
\end{array}
$$

Substitute 2 for y in Equation (1) and solve for x.

$$x + y = 11$$
$$x + 2 = 11$$
$$x = 9$$

Check. If Jessie bought 9 Xbox 360 games and 2 PS3 games, she bought a total of $9 + 2$, or 11 items. She paid a total of $\$9.99(9) + \$17.99(2)$, or $\$89.91 + 35.98$, or \$125.89. The answer checks.

State. Jessie bought 9 Xbox 360 games and 2 PS3 games.

17. _Familiarize_. Let x = the number of kg of Brazilian coffee to be used and y = the number of kg of Turkish coffee to be used.

Translate. Organize the given information in a table.

Type of coffee	Brazilian	Turkish	Mixture
Cost of coffee	\$19	\$22	\$20
Amount (in kg)	x	y	300
Value	$19x$	$22y$	\$20(300); or \$6000

The last two rows of the table give us two equations. Since the total amount of the mixture is 300 lb, we have

$$x + y = 300.$$

The value of the Brazilian coffee is $19x$ (x lb at \$19 per pound), the value of the Turkish coffee is $22y$ (y lb at \$22 per pound), and the value of the mixture is \$20(300) or \$6000. Thus we have

$$19x + 22y = 6000.$$

The resulting system is

$$x + y = 300, \quad (1)$$
$$19x + 22y = 6000. \quad (2)$$

Carry out. We use the elimination method. We multiply on both sides of Equation (1) by -19 and then add.

$$-19x - 19y = -5700$$
$$\underline{19x + 22y = 6000}$$
$$3y = 300$$
$$y = 100$$

Now substitute 100 for y in Equation (1) and solve for x.

$$x + y = 300$$
$$x + 100 = 300$$
$$x = 200$$

Check. The sum of 100 and 200 is 300. The value of the mixture is \$19(200) + \$22(100), or \$3800 + \$2200, or \$6000. These numbers check.

State. 200 kg of Brazilian coffee and 100 kg of Turkish coffee should be used.

19. **Familiarize**. Let x and y represent the number of kilograms of cashews and pecans to be used, respectively.

Translate. We organize the given information in a table.

Type of nut	Cashews	Pecans	Mixture
Cost per kilogram	8.00	9.00	8.40
Amount	x	y	10
Value	$8.00x$	$9.00y$	8.40(10); or 84.00

The last two rows of the table form a system of equations.

$$x + y = 10,$$
$$8.00x + 9.00y = 84.00$$

Carry out. First we clear decimals in the second equation.

$$x + y = 10, \quad (1)$$
$$8x + 9y = 84 \quad (2)$$

Now multiply Equation (1) by -8 and add.

$$-8x - 8y = -80$$
$$\underline{8x + 9y = 84}$$
$$y = 4$$

Substitute 4 for y in Equation (1) and solve for x.

$$x + y = 10$$
$$x + 4 = 10$$
$$x = 6$$

Check. The sum of 6 and 4 is 10. The value of the mixture is \$8.00(6) + \$9.00(4), or \$48.00 + \$36.00, or \$84.00. These numbers check.

State. 6 kg of cashews and 4 kg of Pecans should be used.

21. **Familiarize**. From the table in the text, note that x represents the number of milliliters of solution A to be used and y represents the number of milliliters of solution B.

Translate. We complete the table in the text.

Type of solution	50%-acid	80%-acid	68%-acid
Amount of solution	x	y	200
Percent acid	50%	80%	68%
Amount of acid in solution	$0.5x$	$0.8y$	0.68×200, or 136

Since the total amount of solution is 200 mL, we have

$$x + y = 200.$$

The amount of acid in the mixture is to be 68% of 200 mL, or 136 mL. The amounts of acid from the two solutions are $50\%x$ and $80\%y$. Thus

$$50\%x + 80\%y = 136,$$
$$\text{or} \quad 0.5x + 0.8y = 136,$$
$$\text{or} \quad 5x + 8y = 1360 \quad \text{Clearing decimals}$$

Carry out. We use the elimination method.

$$x + y = 200, \quad (1)$$
$$5x + 8y = 1360 \quad (2)$$

We multiply Equation (1) by -5 and then add.

$$-5x - 5y = -1000$$
$$\underline{5x + 8y = 1360}$$
$$3y = 360$$
$$y = 120$$

Next we substitute 120 for y in one of the original equations and solve for x.

$$x + y = 200 \quad (1)$$
$$x + 120 = 200 \quad \text{Substituting}$$
$$x = 80$$

Check. The sum of 80 and 120 is 200. Now 50% of 80 is 40 and 80% of 120 is 96. These add up to 136. The numbers check.

State. 80 mL of the 50%-acid solution and 120 mL of the 80%-acid solution should be used.

23. **Familiarize**. Let x and y represent the number of liters of 80%-base solution and 30%-base solution to be used in the mixture, respectively.

Translate. We organize the given information in a table.

Type of solution	80%	30%	62%
Amount of solution	x	y	200
Percent base	80%	30%	62%
Amount of base in solution	$0.80x$	$0.30y$	0.62(200), or 124

We get a system of equations from the first and third rows of the table.

$$x + \quad y = 200,$$
$$0.8x + 0.3y = 124$$

Clearing decimals we have

$$x + \quad y = 200, \quad (1)$$
$$8x + 3y = 1240 \quad (2)$$

Carry out. We use the elimination method. Multiply Equation (1) by -3 and add.

$$
\begin{array}{r}
-3x - 3y = -600 \\
8x + 3y = 1240 \\
\hline
5x = 640 \\
x = 128
\end{array}
$$

Now substitute 128 for x in Equation (1) and solve for y.

$$x + y = 200$$
$$128 + y = 200$$
$$y = 72$$

Check. The sum of 128 and 72 is 200. The amount of base in the mixture is $0.8(128) + 0.3(72)$, or $102.4 + 21.6$, or 124. These numbers check.

State. 128 L of the 80%-base solution and 72 L of the 30%-base solution should be used.

25. *Familiarize*. Let x and y represent the number of gallons of 87-octane gas and 93-octane gas to be blended, respectively. We organize the given information in a table.

Type of gasoline	87-octane	93-octane	91-octane
Amount of gas	x	y	12
Octane rating	87	93	91
Mixture	$87x$	$93y$	$91 \cdot 12$, or 1092

Translate. We get a system of equations from the first and third rows of the table.

$$x + \quad y = 12, \quad (1)$$
$$87x + 93y = 1092 \quad (2)$$

Carry out. We use the elimination method. First we multiply Equation (1) by -87 and add.

$$
\begin{array}{r}
-87x - 87y = -1044 \\
87x + 93y = 1092 \\
\hline
6y = 48 \\
y = 8
\end{array}
$$

Now substitute 8 for y in Equation (1) and solve for x.

$$x + y = 12$$
$$x + 8 = 12$$
$$x = 4$$

Check. The sum of 4 and 8 is 12. The mixture is $87(4) + 93(8)$, or $348 + 744$, or 1092. These numbers check.

State. 4 gal of 87-octane gas and 8 gal of 93-octane gas should be blended.

27. *Familiarize*. Let $x =$ the number of 1300-word pages that were filled and $y =$ the number of 1850-word pages that were filled. Then the number of words on the x pages that hold 1300 words each is $1300x$ and on the y pages that hold 1850 words each is $1850y$.

Translate. We reword the problem and translate.

$$\underbrace{\text{Total number of pages}}_{x + y} \; \underbrace{\text{is}}_{=} \; \underbrace{12.}_{12}$$

$$\underbrace{\text{Total number of words}}_{1300x + 1850y} \; \underbrace{\text{is}}_{=} \; \underbrace{18,350.}_{18,350}$$

The resulting system of equations is

$$x + \quad y = 12, \quad (1)$$
$$1300x + 1850y = 18,350. \quad (2)$$

Carry out. We solve using the elimination method. First we multiply Equation (1) by -1300 and add.

$$
\begin{array}{r}
-1300x - 1300y = -15,600 \\
1300x + 1850y = 18,350 \\
\hline
550y = 2750 \\
y = 5
\end{array}
$$

Substitute 5 for y in Equation (1) and solve for x.

$$x + y = 12$$
$$x + 5 = 12$$
$$x = 7$$

Check. If $x = 7$ and $y = 5$, then $7 + 5$, or 12 pages are filled. The 1300-word pages contain $1300 \cdot 7$, or 9100 words and the 1850-word pages contain $1850 \cdot 5$ or 9250 words. The total number of words is $9100 + 9250$, or 18,350. The numbers check.

State. The typesetter used 7 1300-word pages and 5 1850-word pages.

29. *Familiarize*. Let $f =$ the number of foul shots made and $t =$ the number of two point shots made. Then Chamberlain scored f points from foul shots and $2t$ points from two-pointers.

Translate. We reword the problem and translate.

$$\underbrace{\text{Total number of shots}}_{f + t} \; \underbrace{\text{is}}_{=} \; \underbrace{64.}_{64}$$

$$\underbrace{\text{Total number of points}}_{f + 2t} \; \underbrace{\text{is}}_{=} \; \underbrace{100.}_{100}$$

The resulting system of equations is

$$f + t = 64, \quad (1)$$
$$f + 2t = 100. \quad (2)$$

Carry out. We solve using the elimination method. First we multiply Equation (1) by -1 and add.

$$\begin{array}{r} -f - t = -64 \\ \underline{f + 2t = 100} \\ t = 36 \end{array}$$

Substitute 36 for t in Equation (1) and solve for f.

$$f + t = 64$$
$$f + 36 = 64$$
$$f = 28$$

Check. If Chamberlain made 28 foul shots and 36 two point shots, he made $28 + 36$, or 64 shots for a total of $28 + 2 \cdot 36$, or $28 + 72$, or 100 points. The numbers check.

State. Chamberlain made 28 foul shots and 36 two point shots.

31. **Familiarize**. Let $x =$ the number of fluid ounces of Kinney's suntan lotion that should be used and $y =$ the number of fluid ounces of Coppertone that should be used.

Translate. We present the information in a table.

	Kinney's	Coppertone	Mixture
spf Rating	15	30	20
Amount	x	y	50
spf Value	$15x$	$30y$	$20 \cdot 50$, or 1000

The last two rows of the table give us a system of equations.

$$x + y = 50, \quad (1)$$
$$15x + 30y = 1000. \quad (2)$$

Carry out. We solve using the elimination method. First we multiply Equation (1) by -15 and add.

$$\begin{array}{r} -15x - 15y = -750 \\ \underline{15x + 30y = 1000} \\ 15y = 250 \\ y = 16\frac{2}{3} \end{array}$$

Substitute $16\frac{2}{3}$ for y in Equation (1) and solve for x.

$$x + y = 50$$
$$x + 16\frac{2}{3} = 50$$
$$x = 33\frac{1}{3}$$

Check. If $x = 33\frac{1}{3}$ and $y = 16\frac{2}{3}$, then the total amount of suntan lotion is $33\frac{1}{3} + 16\frac{1}{3}$, or 50 fluid ounces. The spf value of the mixture is $15\left(33\frac{1}{3}\right) + 30\left(16\frac{2}{3}\right)$, or $500 + 500$, or 1000. The numbers check.

State. The mixture should contain $33\frac{1}{3}$ fluid ounces of Kinney's and $16\frac{2}{3}$ fluid ounces of Coppertone.

33. Observe that 27.5 is midway between 20 and 35. Thus the mixture would contain equal parts of the cereals that get 20% and 35% of their calories from fat. Since a 40-lb mixture is desired, it should contain 20 lb of New England Natural Bakers Muesli and 20 lb of Breadshop Supernatural granola.

35. *Writing Exercise*. Although it makes sense to consider fractional parts of pounds or gallons, it does not make sense to talk about fractional parts of baskets or rolls of film.

37. $x + 2 < 5$

   $x < 3$

   The solution set is $\{x | x < 3\}$.

39. $5 - 3x > 20$

   $-3x > 15$

   $x < -5$

   The solution set is $\{x | x < -5\}$.

41. $4 \le -\frac{1}{3}x + 5$

   $-1 \le -\frac{1}{3}x$

   $3 \ge x$, or $x \le 3$

   The solution set is $\{x | x \le 3\}$.

43. *Writing Exercise*. No; we would need more information to solve the problem. We have only enough information to write two equations, each containing three variables.

45. **Familiarize**. Let $k =$ the number of pounds of pure Kona beans that should be added to the Columbian beans and $m =$ the total weight of the final mixture, in pounds.

Translate. We present the information in a table.

	Kona	Columbian	Total
Amount	k	45	m
Percent of Kona	100%	0%	30%
Amount of Kona	$1 \cdot k$, or k	$0 \cdot 45$, or 0	$0.3m$

The table gives us two equations.

$$k + 45 = m, \quad (1)$$
$$k = 0.3m \quad (2)$$

Carry out. We use the substitution method. First substitute $k + 45$ for m in Equation (2) and solve for k.

$$k = 0.3(x + 45)$$
$$k = 0.3x + 13.5$$
$$0.7k = 13.5$$
$$k = \frac{13.5}{0.7} = \frac{135}{7} = 19\frac{2}{7}$$

This is the number the problem asks for. We will also find m so that we can check the answer. Substitute $19\frac{2}{7}$ for k in Equation (1) and compute m.

$$k + 45 = m$$
$$19\frac{2}{7} + 45 = m$$
$$64\frac{2}{7} = m$$

Check. If a coffee mixture that weighs $64\frac{2}{7}$ lb contains $19\frac{2}{7}$ lb of Kona coffee, then the percent of Kona coffee in the mixture is $\frac{19\frac{2}{7}}{64\frac{2}{7}}$, or 0.3, or 30%. The answer checks.

State. $19\frac{2}{7}$ lb of Kona coffee should be added to 45 lb of Columbian coffee to obtain the desired mixture.

47. Familiarize. In a table we arrange the information regarding the solution *after* some of the 30% solution is drained and replaced with pure antifreeze. We let x represent the amount of the original (30%) solution remaining, and we let y represent the amount of the 30% mixture that is drained and replaced with pure antifreeze.

Type of solution	Original (30%)	Pure anti-freeze	Mixture
Amount of solution	x	y	6.3
Percent of antifreeze	30%	100%	50%
Amount of antifreeze in solution	$0.3x$	$1 \cdot y$, or y	0.5(6.3), or 3.15

Translate. The table gives us two equations.

Amount of solution: $x + y = 6.3$

Amount of antifreeze in solution: $0.3x + y = 3.15$, or $30x + 100y = 315$

The resulting system is

$$x + y = 63, \qquad (1)$$
$$30x + 100y = 315. \quad (2)$$

Carry out. We multiply Equation (1) by -30 and then add.

$$-30x - 30y = -189$$
$$\underline{\ 30x + 100y = 315\ }$$
$$70y = 126$$
$$y = \frac{126}{70} = 1.8$$

Then we substitute 1.8 for y in Equation (1) and solve for x.

$$x + y = 6.3$$
$$x + 1.8 = 6.3$$
$$x = 4.5$$

Check. When $x = 4.5$ L and $y = 1.8$ L, the total is 6.3 L. The amount of antifreeze in the mixture is $0.3(4.5) + 1.8$, or

1.35 + 1.8, or 3.15 L. This is 50% of 6.3 L, so the numbers check.

State. 1.8 L of the original mixture should be drained and replaced with pure antifreeze.

49. Familiarize. Let x represent the number of gallons of 91-octane gas to be added to the tank and let y represent the total number of gallons in the tank after the 91-octane gas is added. We organize the given information in a table.

Type of gasoline	85-octane	91-octane	Mixture
Amount of gas	5	x	y
Octane rating	85	91	87
Mixture	$85 \cdot 5$, or 425	$91x$	$87y$

Translate. We get a system of equations from the first and third rows of the table.

$$5 + x = y, \qquad (1)$$
$$425 + 91x = 87y \quad (2)$$

Carry out. Substitute $5 + x$ for y in Equation (2) and solve for x.

$$425 + 91x = 87y$$
$$425 + 91x = 87(5 + x)$$
$$425 + 91x = 435 + 87x$$
$$425 + 4x = 435$$
$$4x = 10$$
$$x = 2.5$$

Although the original problem asks us to find only x, we will find y also in order to check the answer. Substitute 2.5 for x in Equation (1) and compute y.

$$y = 5 + 2.5 = 7.5$$

Check. The mixture is $425 + 91(2.5)$, or 652.5. This is equal to $87(7.5)$, so the answer checks.

State. Kim should add 2.5 gal of 91-octane gas to her tank.

51. Familiarize. Let x = Juanita's regular hourly pay rate and let y = her overtime pay rate. Since $55 - 40 = 15$, she worked 15 hr of overtime. For the first 40 hr she earned $40x$ dollars and for the 15 overtime hours she earned $15y$ dollars.

Translate.

Total pay is $812.50.
$$40x + 15y = 812.50$$

Overtime rate is 1.5 times regular rate.
$$y = 1.5 \cdot x$$

The resulting system is

$$40x + 15y = 812.50, \quad (1)$$
$$y = 1.5x. \qquad (2)$$

Carry out. It will be most efficient to use the substitution method. First substitute $1.5x$ for y in Equation (1) and solve for x.

$$40x + 15(1.5x) = 812.50$$
$$40x + 22.5x = 812.50$$
$$62.5x = 812.50$$
$$x = 13$$

Since we are asked to find only the regular hourly rate, we could stop here but we will also find the overtime rate so that we can check our work.

Substitute 13 for x in Equation (2) and compute y.

$$y = 1.5x = 1.5(13) = 19.5$$

Check. \$19.50 is one and a half times \$13. Also, Juanita would earn $40(\$13) + 15(\$19.50)$, or $\$520 + \$292.50$, or \$812.50. The answer checks.

State. Juanita's regular hourly pay rate is \$13.

53. *Familiarize*. Let $x =$ the original number of \$20 per hour workers and $y =$ the original number of \$25 per hour workers. When the number of \$20 workers is increased by 50%, the are $x + 0.50x$, or $1.5x$ workers. When the number of \$25 per hour workers is decreased by 20% there are $y - 0.20y$, or $0.8y$ workers.

Translate.

Original total pay is \$325
$$20x + 25y \quad = \quad 325$$

New total pay is \$400
$$20\,(x + 0.5y) + 25\,(y - 0.2y) \quad = \quad 400$$

The resulting system is

$$20x + 25y = 325 \quad (1)$$
$$30x + 20y = 400 \quad (2)$$

Carry out. We use the elimination method. Multiplying Equation (1) by -4 and Equation (2) by 5 and adding.

$$\begin{array}{r} -80x - 100y = -1300 \\ 150x + 100y = 2000 \\ \hline 70x = 700 \\ x = 10 \end{array}$$

We substitute 10 for x in the Equation (1) and compute y.

$$20x + 25y = 325$$
$$20(10) + 25y = 325$$
$$200 + 25y = 325$$
$$25y = 125$$
$$y = 5$$

Check. With 10 workers at \$20 and 5 workers at \$25, the total is $10(20) + 5(25)$, or $200 + 125$, or \$325/hr. With 15 workers at \$20 and 4 workers at \$25, the total is $15 \cdot 20 + 4 \cdot 25$, or $300 + 100$, or \$400/hr. The numbers check.

State. Ace Engineering employed 10 workers at \$20/hr and 5 workers at \$25/hr.

55. *Familiarize*. Let x represent the ten's digit and y the one's digit. Then the number is $10x + y$.

Translate.

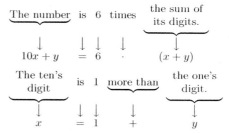

We simplify the first equation.

$$10x + y = 6(x + y)$$
$$10x + y = 6x + 6y$$
$$4x - 5y = 0$$

The system of equations is

$$4x - 5y = 0, \quad (1)$$
$$x = 1 + y. \quad (2)$$

Carry out. We use the substitution method. We substitute $1 + y$ for x in Equation (1) and solve for y.

$$4(1 + y) - 5y = 0$$
$$4 + 4y - 5y = 0$$
$$4 - y = 0$$
$$4 = y$$

Then we substitute 4 for y in Equation (2) and compute x.

$$x = 1 + y = 1 + 4 = 5$$

Check. We consider the number 54. The number is 6 times the sum of the digits, 9. The ten's digit is 1 more than the one's digit. This number checks.

State. The number is 54.

57. *Familiarize*. Let $x =$ Tweedledum's weight and $y =$ Tweedledee's weight, in pounds.

Translate.

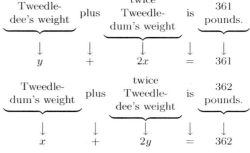

We have a system of equations.

$$y + 2x = 361,$$
$$x + 2y = 362, \text{ or}$$

$$2x + y = 361, \quad (1)$$
$$x + 2y = 362 \quad (2)$$

Carry out. We use elimination. First multiply Equation (2) by -2 and add.

$$2x + y = 361$$
$$\underline{-2x - 4y = -724}$$
$$-3y = -363$$
$$y = 121$$

Now substitute 121 for y in Equation (2) and solve for x.

$$x + 2y = 362 \quad (2)$$
$$x + 2 \cdot 121 = 362 \quad \text{Substituting}$$
$$x + 242 = 362$$
$$x = 120$$

Check. If Tweedledum weighs 120 lb and Tweedledee weighs 121 lb, then the sum of Tweedledee's weight and twice Tweedledum's is $121 + 2 \cdot 120$, or $121 + 240$, or 361 lb. The sum of Tweedledum's weight and twice Tweedledee's is $120 + 2 \cdot 121$, or $120 + 242$, or 362 lb. The answer checks.

State. Tweedledum weighs 120 lb, and Tweedledee weighs 121 lb.

Exercise Set 7.5

1. True; see the box on page 463 in the text.

3. True; see the box on page 463 in the text.

5. We use alphabetical order of variables. We replace x by -2 and y by -6.

$$\begin{array}{c|c} \multicolumn{2}{c}{2x + y < -10} \\ \hline 2(-2) + (-6) & -10 \\ -4 - 6 & \\ & \overset{?}{} \\ -10 \overset{?}{<} -10 & \text{FALSE} \end{array}$$

Since $-10 < -10$ is false, $(-2, -6)$ is not a solution.

7. We use alphabetical order of variables. We replace x by $\frac{1}{3}$ and y by $\frac{9}{10}$.

$$\begin{array}{c|c} \multicolumn{2}{c}{2y + 3x \geq -1} \\ \hline 2\left(\dfrac{9}{10}\right) + 3\left(\dfrac{1}{3}\right) & -1 \\ \dfrac{9}{5} + 1 & \\ \dfrac{14}{5} \overset{?}{\geq} -1 & \text{TRUE} \end{array}$$

Since $\frac{14}{5} \geq -1$ is true, $\left(\frac{1}{3}, \frac{9}{10}\right)$ is a solution.

9. Graph $y \leq x - 1$.

First graph the line $y = x - 1$. The intercepts are $(0, -1)$ and $(1, 0)$. We draw a solid line since the inequality symbol is $\leq$. Then we pick a test point that is not on the line We try $(0, 0)$.

$$\begin{array}{c|c} \multicolumn{2}{c}{y \leq x - 1} \\ \hline 0 & 0 - 1 \\ & \overset{?}{} \\ 0 \overset{?}{\leq} -1 & \text{FALSE} \end{array}$$

We see that $(0, 0)$ is not a solution of the inequality, so we shade the region that does not contain $(0, 0)$.

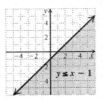

11. Graph $y < x + 4$.

First graph the line $y = x + 4$. The intercepts are $(0, 4)$ and $(-4, 0)$. We draw a dashed line since the inequality symbol is $<$. Then we pick a test point that is not on the line. We try $(0, 0)$.

$$\begin{array}{c|c} \multicolumn{2}{c}{y < x + 4} \\ \hline 0 & 0 + 4 \\ & \overset{?}{} \\ 0 \overset{?}{<} 4 & \text{TRUE} \end{array}$$

Since $(0, 0)$ is a solution of the inequality, we shade the region that contains $(0, 0)$.

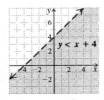

13. Graph $y \geq x - 1$.

First graph the line $y = x - 1$. The intercepts are $(0, -1)$ and $(1, 0)$. We draw a solid line since the inequality symbol is $\geq$. Then we test the point $(0, 0)$.

$$\begin{array}{c|c} \multicolumn{2}{c}{y \geq x - 1} \\ \hline 0 & 0 - 1 \\ & \overset{?}{} \\ 0 \overset{?}{\geq} -1 & \text{TRUE} \end{array}$$

Since $(0, 0)$ is a solution of the inequality, we shade the region containing $(0, 0)$.

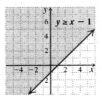

15. Graph $y \leq 3x + 2$.

First graph the line $y = 3x + 2$. The intercepts are $(0, 2)$ and $\left(-\frac{2}{3}, 0\right)$. We draw a solid line since the inequality symbol is $\leq$. Then we test the point $(0, 0)$.

$$\frac{y \le 3x + 2}{0 \ \bigg| \ 3 \cdot 0 + 2}$$
$$0 \overset{?}{\le} 2 \qquad \text{TRUE}$$

Since $(0, 0)$ is a solution of the inequality, we shade the region that contains $(0, 0)$.

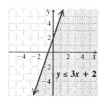

17. Graph $x + y \le 4$.

First graph the line $x + y = 4$. The intercepts are $(0, 4)$ and $(4, 0)$. We draw a solid line since the inequality symbol is $\le$. Then we test the point $(0, 0)$.

$$\frac{x + y \le 4}{0 + 0 \ \bigg| \ 4}$$
$$0 \overset{?}{\le} 4 \quad \text{TRUE}$$

Since $(0, 0)$ is a solution of the inequality, we shade the region that contains $(0, 0)$.

19. Graph $x - y > 7$.

First graph the line $x - y = 7$. The intercepts are $(0, -7)$ and $(7, 0)$. We draw a dashed line since the inequality symbol is $>$. Then we test the point $(0, 0)$.

$$\frac{x - y > 7}{0 - 0 \ \bigg| \ 7}$$
$$0 \overset{?}{>} 7 \quad \text{FALSE}$$

Since $(0, 0)$ is not a solution of the inequality, we shade the region that does not contain $(0, 0)$.

21. Graph $y \ge 1 - 2x$.

First graph the line $y = 1 - 2x$. The intercepts are $(0, 1)$ and $\left(\frac{1}{2}, 0\right)$. We draw a solid line since the inequality symbol is $\ge$. Then we test the point $(0, 0)$.

$$\frac{y \ge 1 - 2x}{0 \ \bigg| \ 1 - 2 \cdot 0}$$
$$0 \overset{?}{\ge} 1 \qquad \text{FALSE}$$

Since $(0, 0)$ is not a solution of the inequality, we shade the region that does not contain $(0, 0)$.

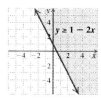

23. Graph $y + 2x > 0$.

First graph the line $y + 2x = 0$. Two points on the line are $(0, 0)$ and $(1, -2)$. We draw a dashed line, since the inequality symbol is $>$. Then we test the point $(1, 1)$, which is not a point on the line.

$$\frac{y + 2x > 0}{1 + 2 \cdot 1 \ \bigg| \ 0}$$
$$3 \overset{?}{>} 0 \quad \text{TRUE}$$

Since $(1, 1)$ is a solution of the inequality, we shade the region that contains $(1, 1)$.

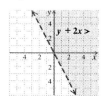

25. Graph $x \ge 4$.

First graph the line $x = 4$ using a solid line since the inequality symbol is $\ge$. Then use $(5, -3)$ as a test point. We can write the inequality as $x + 0y \ge 4$.

$$\frac{x + 0y \ge 4}{5 + 0(-3) \ \bigg| \ 4}$$
$$5 \overset{?}{\ge} 4 \quad \text{TRUE}$$

Since $(5, -3)$ is a solution of the inequality, we shade the region containing $(5, -3)$.

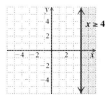

27. Graph $y > -1$.

Graph the line $y = -1$ using a dashed line since the inequality symbol is $>$. Then use $(2, 1)$ as a test point. We can write the inequality as $0x + y > -1$.

$$\frac{0x + y > -1}{0 \cdot 2 + 1 \mid -1}$$

$$1 \overset{?}{>} -1 \quad \text{TRUE}$$

Since $(2, 1)$ is a solution of the inequality, we shade the region containing $(2, 1)$.

29. Graph $y < 0$.

Graph the line $y = 0$ using a dashed line since the inequality symbol is $<$. Then use $(2, -2)$ as a test point. We can write the inequality as $0x + y < 0$.

$$\frac{0x + y < 0}{0 \cdot 2 + (-2) \mid 0}$$

$$-2 \overset{?}{<} 0 \quad \text{TRUE}$$

Since $(2, -2)$ is a solution of the inequality, we shade the region containing $(2, -2)$.

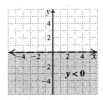

31. Graph $x \le -2$.

Graph the line $x = -2$ using a solid line since the inequality symbol is $\le$. Then use $(-1, 2)$ as a test point. We can write the inequality as $x + 0y \le -2$.

$$\frac{x + 0y \le -2}{-1 + 0 \cdot 2 \mid -2}$$

$$-1 \overset{?}{\le} -2 \quad \text{FALSE}$$

Since $(-1, 2)$ is not a solution of the inequality, we shade the region that does not contain $(-1, 2)$.

33. Graph $\frac{1}{3}x - y < -5$.

Graph the line $\frac{1}{3}x - y < -5$. The intercepts are $(0, 5)$ and $(-15, 0)$. We draw a dashed line since the inequality symbol is $<$. Then we test the point $(0,0)$.

$$\frac{\frac{1}{3}x - y < -5}{\frac{1}{3}0 - 0 \mid -5}$$

$$0 \overset{?}{<} -5 \quad \text{FALSE}$$

Since $(0, 0)$ is not a solution of the inequality, we shade the region that does not contain $(0, 0)$.

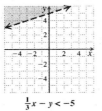

$$\frac{1}{3}x - y < -5$$

35. Graph $2x + 3y \le 12$.

First graph the line $2x + 3y = 12$. The intercepts are $(0, 4)$ and $(6, 0)$. We draw a solid line since the inequality symbol is $\le$. Then we test the point $(0, 0)$.

$$\frac{2x + 3y \le 12}{2 \cdot 0 + 3 \cdot 0 \mid 12}$$

$$0 \overset{?}{\le} 12 \quad \text{TRUE}$$

Since $(0, 0)$ is a solution of the inequality, we shade the region containing $(0, 0)$.

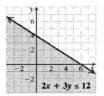

37. Graph $3x - 2y \ge -6$.

Graph the line $3x - 2y = -6$. The intercepts are $(0, -3)$ and $(2, 0)$. We draw a solid line since the inequality symbol is $\ge$. Then we test the point $(0,0)$.

$$\frac{3x - 2y \ge -6}{3 \cdot 0 - 2 \cdot 0 \mid -6}$$

$$0 \overset{?}{\ge} -6 \quad \text{TRUE}$$

Since $(0, 0)$ is a solution of the inequality, we shade the region that contains $(0, 0)$.

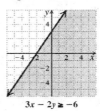

$$3x - 2y \ge -6$$

39. *Writing Exercise.* The point $(4.5, -1)$ is on the boundary line rather than in one of the regions created by the boundary line.

41.
$$2x - 7 = 3x + 8$$
$$2x - 15 = 3x$$
$$-15 = x$$

43.
$$\frac{2}{3}(x-1) = -12$$
$$\frac{3}{2}\cdot\frac{2}{3}(x-1) = \frac{3}{2}\cdot(-12)$$
$$x-1 = -18$$
$$x = -17$$

The solution is -17.

45.
$$\frac{1}{2}-\frac{1}{4} = \frac{1}{t} \qquad \text{Note } t \neq 0$$
$$\frac{1}{4} = \frac{1}{t}$$
$$4t\left(\frac{1}{4}\right) = 4t\left(\frac{1}{t}\right) \quad \text{LCD} = 4t$$
$$t = 4$$

The solution is 4.

47.
$$x^2 + 64 = 16x$$
$$x^2 - 16x + 64 = 0$$
$$(x-8)(x-8) = 0$$
$$x-8=0 \ \text{ or } \ x-8=0$$
$$x=8 \ \text{ or } \ x=8$$

The solution is 8.

49. *Writing Exercise.* When a test point on an axis is chosen, the either x is 0 or y is 0. Then the computation is simpler.

51. The c children weigh $75c$ lb, and the a adults weigh $150a$ lb. Together, the children and adults weigh $75c + 150a$ lb. When this total is more than 1000 lb the elevator is overloaded, so we have $75c + 150a > 1000$. (Of course, c and a would also have to be nonnegative, so we show only the portion of the graph that is in the first quadrant.)

To graph $75c + 150a > 1000$, we first graph $75c + 150a = 1000$ using a dashed line. (Remember to use alphabetical order of variables.) Then we test the point $(0,0)$.

$$\frac{75c + 150a > 1000}{75 \cdot 0 + 150 \cdot 0 \ \bigg| \ 1000}$$
$$0 \overset{?}{>} 1000 \quad \text{FALSE}$$

Since $(0,0)$ is not a solution of the inequality, we shade the region that does not contain $(0,0)$.

53. First find the equation of the line containing the points $(2,0)$ and $(0,-2)$. The slope is
$$\frac{-2-0}{0-2} = \frac{-2}{-2} = 1.$$

We know that the y-intercept is $(0,-2)$, so we write the equation using the slope-intercept form.
$$y = mx + b$$
$$y = 1\cdot x + (-2)$$
$$y = x - 2$$

Since the line is dashed, the inequality symbol will be $<$ or $>$. To determine which, we substitute the coordinates of a point in the shaded region. We will use $(0,0)$.

$$\frac{y \quad\quad x-2}{0 \ \bigg| \ 0-2}$$
$$0 \overset{?}{=} -2$$

Since $0 > -2$ is true, the correct symbol is $>$. The inequality is $y > x - 2$.

55. Graph $xy \leq 0$.

From the principle of zero products, we know that $xy = 0$ when $x = 0$ or $y = 0$. Therefore, the graph contains the lines $x = 0$ and $y = 0$, or the y- and x-axes. Also, $xy < 0$ when x and y have different signs. This is the case for all points in the second quadrant (x is negative and y is positive) and in the fourth quadrant (x is positive and y is negative). Thus, we shade the second and fourth quadrants.

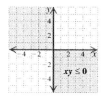

57. First solve the equation for y.
$$y + 3x \leq 4.9$$
$$y \leq -3x + 4.9$$

Then graph the line $y = -3x + 4.9$ and shade below the line.

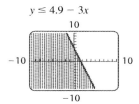

Exercise Set 7.6

1. $x + y \leq 3,$
$x - y \leq 5$

We graph the lines $x + y = 3$ and $x - y = 5$ using solid lines. We indicate the region for each inequality by the arrows at the ends of the lines. We shade the area where the regions overlap.

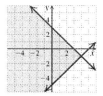

3. $x + y < 6$,

$x + y > 0$

We graph the lines $x + y = 6$ and $x + y = 0$ using dashed lines. We indicate the region for each inequality by the arrows at the ends of the lines. We shade the area where the regions overlap.

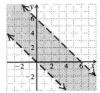

5. $x > 3$,

$x + y \leq 4$

We graph the line $x = 3$ using a dashed line and the line $x + y = 4$ using a solid line. We indicate the region for each inequality by the arrows at the ends of the lines. We shade the area where the regions overlap.

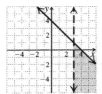

7. $y \geq x$,

$y \leq 1 - x$

We graph the lines $y = x$ and $y = 1 - x$ using solid lines. We indicate the region for each inequality by the arrows at the ends of the lines. We shade the area where the regions overlap.

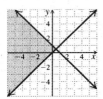

9. $x \geq -3$,

$y \leq 2$

We graph the lines $x = -3$ and $y = 2$ using solid lines. We indicate the region for each inequality by the arrows at the ends of the lines. We shade the area where the regions overlap.

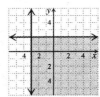

11. $x \leq 0$,

$y \leq 0$

We graph the lines $x = 0$ and $y = 0$ using solid lines. We indicate the region for each inequality by the arrows at the ends of the lines. We shade the area where the regions overlap.

13. $2x - 3y \geq 9$,

$2y + x > 6$

We graph the line $2x - 3y = 9$ using a solid line and the line $2y + x = 6$ using a dashed line. We indicate the region for each inequality by the arrows at the ends of the lines. We shade the area where the regions overlap.

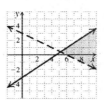

15. $y > \frac{1}{2}x - 2$,

$x + y \leq 1$

We graph the line $y = \frac{1}{2}x - 2$ using a dashed line and the line $x + y = 1$ using a solid line. We indicate the region for each inequality by the arrows at the ends of the lines. We shade the area where the regions overlap.

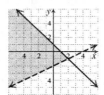

17. $x + y \leq 5$,

$x \geq 0$,

$y \geq 0$,

$y \leq 3$

We graph the lines $x + y = 5$, $x = 0$, $y = 0$, and $y = 3$ using solid lines. We indicate the region for each inequality by the arrows at the ends of the lines. We shade the area where the regions overlap.

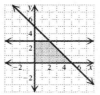

19. $y - x \geq 1,$

   $y - x \leq 3,$

   $x \leq 5,$

   $x \geq 2$

We graph the lines $y - x = 1$, $y - x = 3$, $x = 5$, and $x = 2$ using solid lines. We indicate the region for each inequality by the arrows at the ends of the lines. We shade the area where the regions overlap.

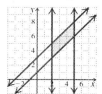

21. $y \leq x,$

   $x \geq -2,$

   $x \leq -y$

We graph the lines $y = x$, $x = -2$ and $x = -y$ using solid lines. We indicate the region for each inequality by the arrows at the ends of the lines. We shade the area where the regions overlap.

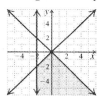

23. *Writing Exercise.* No, a system of linear inequalities will not always have a solution. One condition under which a system of two linear inequalities will have no solution is that boundary lines of the graphs are parallel and the graphs have no points in common. Another is that the boundary lines of the graphs coincide, but at most one graph includes the boundary line, and the graphs have no non-boundary line points in common.

25. $210 = k \cdot 10$

   $$\frac{210}{10} = \frac{k \cdot 10}{10}$$

   $21 = k$

The solution is 21.

27. $0.4 = k \cdot 0.5$

   $$\frac{0.4}{0.5} = \frac{k \cdot 0.5}{0.5}$$

   $$\frac{4}{5} = k$$

The solution is $\frac{4}{5}$.

29. $5 = \dfrac{k}{8}$

   $8 \cdot 5 = 8 \cdot \dfrac{k}{8}$

   $40 = k$

The solution is 40.

31. *Writing Exercise.* The solution of a system of linear inequalities would be a line if both inequalities had the same boundary line, the boundary line is included in the graph of each inequality, and opposite sides of the boundary line are shaded. Such a system could be written in the form

   $y \geq mx + b,$

   $y \leq mx + b.$

33. *Writing Exercise.* The solution of a system of inequalities could be a line segment if the conditions in Exercise 31 exist and if there is an additional restriction of the form $a \leq x \leq c$ or $m \leq y \leq n$.

35. $3r + 6t \geq 36,$

   $2r + 3t \geq 21,$

   $5r + 3t \geq 30,$

   $t \geq 0,$

   $r \geq 0$

We graph the boundary lines, find the region for each inequality, and shade the area where the regions overlap.

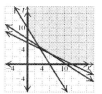

37. $x + 3y \leq 6,$

   $2x + y \geq 4,$

   $3x + 9y \geq 18$

We graph the boundary lines and find the region for each inequality. We find that the solution set consists of a single point, $\left(\dfrac{6}{5}, \dfrac{8}{5} \right)$, or $(1.2, 1.6)$.

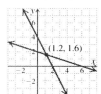

39. Note that the graphs of the related equations for the first two inequalities are parallel lines since $4x + 6y = 2(2x + 3y)$ and $9 \neq 2 \cdot 1$. Also observe that we would shade below the first line when graphing the solution set of the first inequality and we would shade above the second line when graphing the solution set of the second inequality. Thus there is no region of overlap for these two inequalities and hence the system of inequalities has no solution.

41. let l = the length and w = the width, in inches of the quilt. Then the perimeter is $2l + 2w$. The maximum perimeter is 200 in,. so we can form an inequality $2l + 2w \leq 200$. Also since length and width must be positive we can also include the inequalities $l > 0$ and $w > 0$. Thus we have the system of inequalities

$$2l + 2w \leq 200$$
$$l > 0,$$
$$w > 0.$$

We can graph this system using the horizontal axis for l and the vertical axis for w.

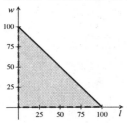

43.

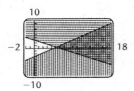

Exercise Set 7.7

1. As the number of copiers increases, the time required to complete the job decreases, so the situation reflects inverse variation.

3. As the number of cabinets increases, the number of flaws increases, so the situation reflects direct variation.

5. As the number of heaters increases, the time required to heat the room decreases, so the situation reflects inverse variation.

7. We substitute to find k.
$$y = kx \quad \text{y varies directly as x.}$$
$$40 = k \cdot 8 \quad \text{Substituting 40 for y and 8 for x}$$
$$\frac{40}{8} = k$$
$$5 = k \quad \text{k is the constant of variation.}$$
The equation of variation is $y = 5x$.

9. We substitute to find k.
$$y = kx \quad \text{y varies directly as x.}$$
$$1.75 = k \cdot 0.25 \quad \text{Substituting 1.75 for y and 0.25 for x}$$
$$\frac{1.75}{0.25} = k$$
$$7 = k$$
The equation of variation is $y = 7x$.

11. We substitute to find k.
$$y = kx$$
$$0.3 = k \cdot 0.5 \quad \text{Substituting 0.3 for y and 0.5 for x}$$
$$\frac{0.3}{0.5} = k$$
$$\frac{3}{5} = k$$

The equation of variation is $y = \frac{3}{5}x$.

13. We substitute to find k.
$$y = kx$$
$$200 = k \cdot 300 \quad \text{Substituting 200 for y and 300 for x}$$
$$\frac{200}{300} = k$$
$$\frac{2}{3} = k$$
The equation of variation is $y = \frac{2}{3}x$.

15. We substitute to find k.
$$y = \frac{k}{x}$$
$$10 = \frac{k}{12} \quad \text{Substituting 10 for y and 12 for x}$$
$$120 = k$$
The equation of variation is $y = \frac{120}{x}$.

17. We substitute to find k.
$$y = \frac{k}{x}$$
$$0.25 = \frac{k}{4} \quad \text{Substituting 0.25 for y and 4 for x}$$
$$1 = k$$
The equation of variation is $y = \frac{1}{x}$.

19. We substitute to find k.
$$y = \frac{k}{x}$$
$$50 = \frac{k}{0.4} \quad \text{Substituting 50 for y and 0.4 for x}$$
$$20 = k$$
The equation of variation is $y = \frac{20}{x}$.

21. We substitute to find k.
$$y = \frac{k}{x}$$
$$42 = \frac{k}{5} \quad \text{Substituting 42 for y and 5 for x}$$
$$210 = k$$
The equation of variation is $y = \frac{210}{x}$.

23. *Familiarize and Translate*. The problem states that we have direct variation between the variables P and H. Thus, an equation $P = kH$ applies.

Carry out. First find an equation of variation.
$$P = kH$$
$$135 = k \cdot 15 \quad \text{Substituting 135 for P and 15 for H}$$
$$\frac{135}{15} = k$$
$$9 = k$$

The equation of variation is $P = 9H$. When $H = 23$, we have:

$$P = 9H$$

$$P = 9(23) \quad \text{Substituting 23 for } H$$

$$P = 207$$

Check. This check might be done by repeating the computations. We might also do some reasoning about the answer. The paycheck increased from \$135 to \$207. Similarly, the hours increased from 15 to 23. The ratios 15/135 and 23/207 are the same value: $0.\overline{1}$.

State. For 23 hours work, the paycheck is \$207.

25. Familiarize and Translate. The problem states that we have direct variation between the variables n and c. Thus, an equation $n = kc$ applies.

Carry out. First find an equation of variation.

$$n = kc$$

$$18.5 = k \cdot 2 \quad \text{Substituting 18.5 for } n \text{ and 2 for } c.$$

$$\frac{18.5}{2} = k$$
$$9.25 = k$$

The equation of variation is $n = 9.25c$. When $c = 3$, we have:

$$n = 9.25c$$

$$n = 9.25(3) \quad \text{Substituting 3 for } c$$

$$n = 27.75$$

Check. A check might be done by repeating the computations.

State. A 3-ton air conditioner uses 27.75 kWh in 10 hr.

27. Familiarize and Translate. The problem states that we have direct variation between the variables R and P. Thus, an equation $R = kP$ applies.

Carry out. First find an equation of variation.

$$R = kP$$

$$14 = k \cdot 0.5825 \quad \text{Substituting 14 for } R \text{ and } 0.5825 \text{ for } P$$

$$\frac{14}{0.5825} = k$$

$$24.034 \approx k$$

The equation of variation is $R = 24.034P$.

When $R = 24$, we have

$$R = 24.034P$$

$$24 = 24.034P \quad \text{Substituting 24 for } R$$

$$0.9986 \approx P$$
$$0.9986 = 99.86\%$$

Check. A check might be done by repeating the computation.

State. A 24-karat gold chain is 99.86% gold.

29. Familiarize and Translate. The problem states that we have inverse variation between the variables n and t. Thus, an equation $n = \dfrac{k}{t}$ applies.

Carry out. First find an equation of variation.

$$n = \frac{k}{t}$$

$$20 = \frac{k}{4} \quad \text{Substituting 20 for } n \text{ and } 4 \text{ for } t$$

$$80 = k$$

The equation of variation is $n = \dfrac{80}{t}$. When $n = 25$, we have

$$n = \frac{80}{t}$$

$$25 = \frac{80}{t}$$

$$25t = 80$$

$$t = 3.2$$

Check. A check might be done by repeating the computations.

State. It would take 25 people 3.2 hr to complete the job.

31. Familiarize and Translate. The problem states that we have inverse variation between the variables P and W. Thus, an equation $P = \dfrac{k}{W}$ applies.

Carry out. First find an equation of variation.

$$P = \frac{k}{W}$$

$$440 = \frac{k}{2.4} \quad \text{Substituting 440 for } P \text{ and } 2.4 \text{ for } W$$

$$1056 = k$$

The equation of variation is $P = \dfrac{1056}{W}$. When $P = 660$, we have

$$P = \frac{1056}{W}$$

$$660 = \frac{1056}{W} \quad \text{Substituting 660 for } P$$

$$660W = 1056$$

$$W = \frac{1056}{660} = 1.6$$

Check. A check might be done by repeating the computations. We can also do some reasoning about the answer. Note that as the pitch increases, the wavelength decreases as expected. Also note that 2.4(440) and 1.6(660) are both 1056.

State. The wavelength of a trumpet's E above concert A is 1.6 ft.

33. Familiarize and Translate. The problem states that we have inverse variation between the variables c and n. Thus, an equation $c = \dfrac{k}{n}$ applies.

Carry out. First find an equation of variation.

$$c = \frac{k}{n}$$

$$70 = \frac{k}{40} \quad \text{Substituting 70 for } c \text{ and } 40 \text{ for } n$$

$$2800 = k$$

The equation of variation is $c = \dfrac{2800}{n}$. When $c = 140$, we have

$$c = \frac{2800}{n}$$

$$140 = \frac{2800}{n}$$

$$140n = 2800$$

$$n = 20$$

Check. A check might be done by repeating the computations.

State. When costs are \$140 per person, 20 people chartered the boat.

35. *Writing Exercise*. Neither; the cost of mailing a package in the United States is not affected by the distance it travels.

37. *Writing Exercise*. Direct variation; the greater the weight, the longer the cooking time.

39. $10^2 = (10)(10) = 100$

41. $(-5x)^2 = (-5x)(-5x) = 25x^2$

43. $(ab^2)^2 = (a)^2(b^2)^2 = a^2b^4$

45. $(x+y)^2 = (x+y)(x+y)$
$$\qquad\quad = x^2 + 2xy + y^2 \qquad (A+B)^2 = A^2 + 2AB + B^2$$

47. *Writing Exercise*. If $x = \dfrac{k}{y}$ and $y = \dfrac{m}{z}$, then $x = \dfrac{k}{m/z} = k \cdot \dfrac{z}{m} = \dfrac{k}{m} \cdot z$, or $x = Kz$, where $K = \dfrac{k}{m}$, so x varies directly as z.

49. $P = kv^3$

51. $S = kv^6$

53. $N = k \cdot \dfrac{T}{P}$

55. $P = kS$

Since an octagon has 8 sides, $k = 8$, and we have $P = 8S$.

57. $A = kr^2$

From the formula for the area of a circle we know that $k = \pi$, and we have $A = \pi r^2$.

Chapter 7 Review

1. ordered

3. parallel

5. coefficients

7. intersection

9. We check

$$\begin{array}{c|c} x - 2y = 12 & \\ \hline 2 - 2(-5) & 12 \\ 2 + 10 & \\ & \overset{?}{12 = 12} \quad \text{TRUE} \end{array}$$

$$\begin{array}{c|c} 2x - y = 1 & \\ \hline 2(2) - (-5) & 1 \\ 4 + 5 & \\ & \overset{?}{9 = 1} \quad \text{FALSE} \end{array}$$

$(2, -5)$ is not a solution.

11. We graph the equations

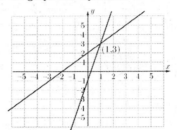

The apparent solution of the system, $(1, 3)$ should be checked in both equations.

We check

$$\begin{array}{c|c} y = 4x - 1 & \\ \hline 3 & 4(1) - 1 \\ & \\ \overset{?}{3 = 3} & \text{TRUE} \end{array}$$

$$\begin{array}{c|c} y = x + 2 & \\ \hline 3 & 1 + 2 \\ & \\ & \overset{?}{3 = 3} \quad \text{TRUE} \end{array}$$

The solution is $(1, 3)$.

13. We graph the equations.

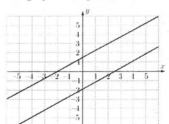

The lines are parallel, There is no solution.

15.
$$\begin{aligned} y &= 4 - x \quad (1) \\ 3x + 4y &= 21 \quad (2) \end{aligned}$$

Substitute $4 - x$ for y in Equation (2) and solve for x.

$$\begin{aligned} 3x + 4(4 - x) &= 21 \quad \text{Substituting} \\ 3x + 16 - 4x &= 21 \\ -x + 16 &= 21 \\ -x &= 5 \\ x &= -5 \end{aligned}$$

Substitute -5 for x in Equation (1).

$$\begin{aligned} y &= 4 - (-5) \\ y &= 9 \end{aligned}$$

The solution is $(-5, 9)$.

17. $x + y = 5$ (1)
$y = 2 - x$ (2)

Substitute $2 - x$ for y in Equation (1) and solve for x.

$x + (2 - x) = 5$ (1)
$2 = 5$

We obtain a false solution, so the system has no solution.

19. $x - 2y = 5$ (1)
$3x + 4y = 10$ (2)

Solve Equation (1) for x.

$x = 2y + 5$ (3)

Substitute $2y + 5$ for x in Equation (2) and solve for y.

$3(2y + 5) + 4y = 10$ Substituting
$6y + 15 + 4y = 10$
$10y + 15 = 10$
$10y = -5$
$y = -\dfrac{1}{2}$

Substitute $-\dfrac{1}{2}$ for y in Equation (3).

$x = 2\left(-\dfrac{1}{2}\right) + 5$
$x = -1 + 5$
$x = 4$

The solution is $\left(4, -\dfrac{1}{2}\right)$.

21. $3x - 2y = 0$ (1)
$\underline{2x + 2y = 50}$ (2)
$5x = 50$ Adding
$x = 10$

Substitute 10 for x in Equation (2) and solve for y.

$2(10) + 2y = 50$ (1)
$20 + 2y = 50$
$2y = 30$
$y = 15$

The solution is $(10, 15)$.

23. $x - \dfrac{1}{3}y = -\dfrac{13}{3},$ (1)
$3x - y = -13$ (2)

Multiply Equation (1) by -3 and then add.

$-3x + y = 13$
$\underline{3x - y = -13}$
$0 = 0$

The equation $0 = 0$ is always true so the system has an infinite number of equations.

25. $5x - 2y = 7,$ (1)
$4x - 3y = 14$ (2)

Multiply Equation (1) by 3 and Equation (2) by -2.

$15x - 6y = 21$
$\underline{-8x + 6y = -28}$
$7x = -7$
$x = -1$

Substitute -1 for x in Equation (2).

$4(-1) - 3y = 14$ (1)
$-4 - 3y = 14$
$-3y = 18$
$y = -6$

The solution is $(-1, -6)$.

27. $2x + 5y = 8,$ (1)
$3x + 4y = 10$ (2)

Multiply Equation (1) by 3 and Equation (2) by -2.

$6x + 15y = 24$
$\underline{-6x - 8y = -20}$
$7y = 4$
$y = \dfrac{4}{7}$

Substitute $\dfrac{4}{7}$ for y in Equation (2).

$3x + 4\left(\dfrac{4}{7}\right) = 10$ (1)
$3x + \dfrac{16}{7} = 10$
$3x = \dfrac{54}{7}$
$x = \dfrac{18}{7}$

The solution is $\left(\dfrac{18}{7}, \dfrac{4}{7}\right)$.

29. *Familiarize*. Recall that the perimeter of a rectangle with length l and width w is given by $2l + 2w$.

Translate.

$$\underbrace{\text{The perimeter}}_{} \;\; \text{is} \;\; \underbrace{\text{66 ft.}}_{}$$
$$\downarrow \downarrow \;\;\; \downarrow$$
$$2l + 2w = 66$$

$$\underbrace{\text{The length}}_{} \; \text{is} \; \text{1 ft.} \; \underbrace{\text{more than}}_{} \; \underbrace{\text{three}}_{} \, \underbrace{\text{times}}_{} \; \underbrace{\text{the width}}_{}$$
$$\downarrow \downarrow \downarrow \downarrow \downarrow \downarrow \downarrow$$
$$l = 1 + 3 \cdot w$$

The resulting system is

$2l + 2w = 66,$ (1)
$l = 1 + 3w$ (2)

Carry out. We solve the system.

Substitute $1 + 3w$ for l in Equation (1) and solve for w.

$2l + 2w = 66$ (1)

$2(l + 3w) + 2w = 66$ Substituting
$2 + 6w + 2w = 66$
$2 + 8w = 66$
$8w = 64$
$w = 8$

Now substitute 8 for w in Equation (2)

$l = 1 + 3w$ (2)

$l = 1 + 3 \cdot 8$

$l = 25$

Check. If the length is 25 ft and the width is 8 ft, then the perimeter would be $2 \cdot 25 + 2 \cdot 8$, or $50 + 16$, or 66 ft. Also, the length is 1 ft more than $3 \cdot 8$, or 25. These numbers check.

State. The length is 25 ft and the width is 8 ft.

31. **Familiarize**. Let x = the number of turkey subs and and y = the number of veggie subs.

Translate. We organize the given information in a table.

Type of sub	Turkey	Veggie	Total
Price	6.49	5.09	
Number	x	y	50
Value	$6.49x$	$5.09y$	303.50

The last two rows of the table give us a system of equations.

$$x + y = 50, \quad (1)$$
$$6.49x + 5.09y = 303.50 \quad (2)$$

Carry out. We solve using the elimination method. First we multiply Equation (1) by -5.09 and add.

$$\begin{array}{rcr} -5.09x - 5.09y &=& -254.50 \\ 6.49x + 5.09y &=& 303.50 \\ \hline 1.4x &=& 49 \\ x &=& 35 \end{array}$$

Substitute 35 for x in Equation (1) and solve for y.

$$x + y = 50$$
$$35 + y = 50$$
$$y = 15$$

Check. If $x = 35$ and $y = 15$, then a total of $35 + 15$, or 50 subs were ordered. Their total value was $\$6.49(35) + \$5.09(15)$, or $\$227.15 + \$76.35$, or $\$303.50$. The numbers check.

State. 35 turkey subs and 15 veggie subs were delivered.

33. Graph $x \leq y$.

First graph the line $x = y$ using a solid line since the inequality symbol is $\leq$. Then use $(1, 2)$ as a test point.

$$\frac{x \leq y}{1 \mid 2}$$
$$1 \overset{?}{\leq} 2 \quad \text{TRUE}$$

Since $(1, 2)$ is a solution of the inequality, we shade the region that contains $(1, 2)$.

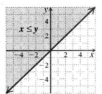

35. Graph $y > -2$.

First graph the line $y = -2$ using a dashed line since the inequality symbol is $>$. Then use $(1, 1)$ as a test point. We can write the inequality as $0x + y > -2$.

$$\frac{0x + y > -2}{0 \cdot 1 + 1 \mid -2}$$
$$1 \overset{?}{>} -2 \quad \text{TRUE}$$

Since $(1, 1)$ is a solution of the inequality, we shade the region that containing $(1, 1)$.

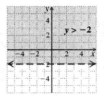

37. Graph $2x + y < 1$.

Graph the line $2x + y = 1$ using a dashed line since the inequality symbol is $<$. The intercepts are $(0, 1)$ and $(\frac{1}{2}, 0)$. Then we test the point $(0, 0)$.

$$\frac{2x + y < 1}{2 \cdot 0 + 0 \mid 1}$$
$$0 \overset{?}{<} 1 \quad \text{TRUE}$$

Since $(0, 0)$ is a solution of the inequality, we shade the region containing $(0, 0)$.

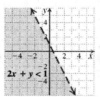

39. $x \geq 1,$

$y \leq -1$

We graph the lines $x = 1$ and $y = -1$ using solid lines. We indicate the region for each inequality by the arrows at the ends of the lines. We shade the area where the regions overlap.

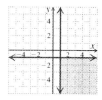

41. We substitute to find k.

$y = kx$ y varies directly as x

$81 = k \cdot 3$ Substituting
$27 = k$

The equation of variation is $y = 27x$.

43. *Writing Exercise* A solution of a system of two equations is an ordered pair that makes both equations true. The graph of an equation represents all ordered pairs that make that equation true. So in order for an ordered pair to make <u>both</u> equations true, it must be on both graphs.

45. $2x - Dy = 6$ (1)
$Cx + 4y = 14$ (2)

The solution is $(6, 2)$. Substitute $(6, 2)$ in Equation (1).

$2 \cdot 6 - D \cdot 2 = 6$ (1)
$12 - 2D = 6$
$-2D = 6$
$D = 3$

Substitute $(6, 2)$. in Equation (2).

$C \cdot 6 + 4 \cdot 2 = 14$ (1)
$6C + 8 = 14$
$6C = 6$
$C = 1$

The solution is $C = 1$ and $D = 3$.

47. $3(x - y) = 4 + x$ (1)
$x = 5y + 2$ (2)

Substitute $5y + 2$ for x in Equation (1).

$3(5y + 2 - y) = 4 + 5y + 2$ (1)
$3(4y + 2) = 5y + 6$
$12y + 6 = 5y + 6$
$12y = 5y$
$7y = 0$
$y = 0$

Substitute 0 for y in Equation (2).

$x = 5 \cdot 0 + 2$ (1)
$x = 2$

The solution is $(2, 0)$.

49. Let $x =$ the value of the computer and $c =$ the compensation. Compensation is \$42,000 plus a computer translates to $c = 42,000 + x$. (1) Prorated 7 months compensation is \$23,750 and the computer translates to

$\frac{7}{12}c = 23,750 + x.$ (2)

Substitute $42,000 + x$ for c in Equation (2).

$\frac{7}{12}(42,000 + x) = 23,750 + x$
$294,000 + 7x = 285,000 + 12x$ Multiplying by 12
$900 = 5x$
$1800 = x$

The value of the computer is \$1800.

Chapter 7 Test

1. We check by substituting alphabetically 3 for x and -2 for y.

$$\begin{array}{c|c} 2x + y = 4 \\ \hline 2 \cdot 3 + (-2) & 4 \\ 6 - 2 & 4 \\ 4 \stackrel{?}{=} 4 & \text{TRUE} \end{array}$$

$$\begin{array}{c|c} 5x + 6y = 27 \\ \hline 5 \cdot 3 - 6(-2) & 27 \\ 15 + 12 & \\ 27 \stackrel{?}{=} 27 & \text{TRUE} \end{array}$$

The ordered pair $(3, 2)$ is a solution of each equation. Therefore it is a solution of the system of equations.

3. We graph the equations.

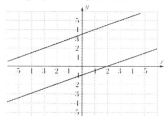

The lines are parallel. There is no solution.

5. $4x + y = 5$ (1)
$2x + y = 4$ (2)

We solve Equation (1) for y.

$4x + y = 5$ (1)
$y = -4x + 5$ (3)

We substitute $-4x + 5$ for y in Equation (2) and solve for x.

$2x + y = 4$ (2)
$2x + (-4x + 5) = 4$ Substituting
$-2x + 5 = 4$
$-2x = -1$
$x = \frac{1}{2}$

Now we substitute $\frac{1}{2}$ for x in either of the original equations or in Equation (3) and solve for y. It is easiest to use (3).

$y = -4x + 5 = -4\left(\frac{1}{2}\right) + 5 = -2 + 5 = 3$

We check the ordered pair $\left(\frac{1}{2}, 3\right)$.

$$\frac{4x+y=5}{4(\frac{1}{2})+3 \mid 5}$$
$$2+3$$
$$5 \overset{?}{=} 5 \quad \text{TRUE}$$

$$\frac{2x+y=4=4}{2\left(\frac{1}{2}\right)+3 \mid 4}$$
$$1+3$$
$$4 \overset{?}{=} 4 \quad \text{TRUE}$$

Since $(\frac{1}{2}, 3)$ checks in both equations, it is the solution.

7. $x - y = 1$ (1)
$$\frac{2x+y=8 \quad (2)}{3x \quad\quad = 9}$$
$$x = 3$$

Substitute 3 for x in one of the original equations and solve for y.
$$x - y = 1 \quad (1)$$
$$3 - y = 1$$
$$-y = -2$$
$$y = 2$$

$$\frac{x-y=1}{3-2 \mid 1}$$
$$1$$
$$1 \overset{?}{=} 1 \quad \text{TRUE}$$

$$\frac{2x+y=8}{2(3)+2 \mid 8}$$
$$6+2$$
$$8 \overset{?}{=} 8 \quad \text{TRUE}$$

Since $(3, 2)$ checks, it is the solution.

9. $4x + 5y = 5$,   (1)
$6x + 7y = 7$ (2)

We use the multiplication principle with both equations and then add.
$$\frac{\begin{array}{ll} 12x + 15y = 15 & \text{Multiplying (1) by 3} \\ -12x - 14y = -14 & \text{Multiplying (2) by -2} \end{array}}{y = 1}$$

Substitute 1 for y in one of the original equations and solve for x.
$$4x + 5y = 5 \quad (1)$$
$$4x + 5(1) = 5$$
$$4x + 5 = 5$$
$$4x = 0$$
$$x = 0$$

Check.
$$\frac{4x+5y=5}{4(0)+5(1) \mid 5}$$
$$5$$
$$5 \overset{?}{=} 5 \quad \text{TRUE}$$

$$\frac{6x+7y=7}{6(0)+7(1) \mid 7}$$
$$7$$
$$7 \overset{?}{=} 7 \quad \text{TRUE}$$

Since $(0, 1)$ checks, it is the solution.

11. *Familiarize*. Let x = one angle and y = the other angle.

Translate. Since the angles are complementary, we have one equation.
$$x + y = 90$$

The second sentence can be translated as follows:

One angle is 2° less than three times the other
$$x \quad = \quad 3y - 2$$

The resulting system is
$$x + y = 90 \quad (1)$$
$$x = 3y - 2. \quad (2)$$

Carry out. We solve the system. We substitute $3y - 2$ for x in Equation (1) and solve for y.
$$x + y = 90 \quad (1)$$
$$3y - 2 + y = 90$$
$$4y - 2 = 90$$
$$4y = 92$$
$$y = 23$$

Next we substitute 23 for y in Equation (2) and solve for x.
$$x = 3y - 2 = 3(23) - 2 = 69 - 2 = 67$$

Check.The sum of the angles is $23° + 67°$, $90°$, so the angles are complementary. If 2 is subtracted from three times $23°$, we have $3(23°) - 2°$ or $67°$, which is the other angle. The answers checks.

State. One angle is $23°$, and the other $67°$.

13. *Familiarize*. We let m = the number of half-miles driven and c = the total cost of the taxi.

Translate. We reword and translate the first statement.
$2.50 plus $1.00 per half mile is cost
$$2.50 + 1.00 \cdot m = c$$

We reword and translate the second statement.
$1.75 plus $1.20 per half mile is cost
$$1.75 + 1.20 \cdot m = c$$

We have a system of equations:
$$2.50 + m = c, \quad (1)$$
$$1.75 + 1.20m = c \quad (2)$$

Carry out. We solve the system of equations, we multiply the second equation by -1 and add to eliminate c.

$$2.50 + m = c \qquad (1)$$

$$\frac{-1.75 - 1.2m = -c}{0.75 - 0.2m = 0}$$

$$-0.2m = -0.75$$

$$m = 3.75$$

Check. For 3.75 half-miles, the cost of the taxi in New York City is $2.50 + 1.00 \cdot 3.75$, or \$6.25. For 3.75 half miles, the cost of the taxi in Boston is $1.75 + 1.2 \cdot 3.75$, or \$6.25. The cost is same for 3.75 half-miles, or 3.75/2 or 1 7/8 miles.

State. When the taxis are driving 1 7/8 mi, the cost is the same.

15. Graph $2x - y \leq 4$.

First graph the line $2x - y = 4$. The intercepts are $(0, -4)$ and $(2, 0)$. We draw a solid line since the inequality symbol is $\leq$. Then we test the point $(0, 0)$.

$$\frac{2x - y \leq 4}{2(0) - 0 \mid 4}$$

$$0 \overset{?}{\leq} 4 \quad \text{TRUE}$$

Since $(0, 0)$ is a solution of the inequality, we shade the region that contains $(0, 0)$.

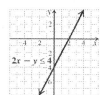

17. $y \geq x - 5,$

$y < \frac{1}{2}x$

We graph the line $y = x - 5$ using a solid line and the line $y = \frac{1}{2}x$ using a dashed line. We indicate the region for each inequality by the arrows at the ends of the lines. We shade the area where the regions overlap.

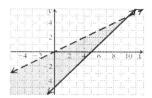

19. We substitute to find k.

$$y = \frac{k}{x}$$

$$9 = \frac{k}{2} \qquad \begin{array}{l}\text{Substituting 9 for } y \text{ and} \\ 2 \text{ for } x\end{array}$$

$$18 = k$$

The equation of variation is $y = \dfrac{18}{x}$.

21. Familiarize. We let $x =$ the number of people behind. The y is the number of people ahead. All the people in line is $x + y + 1$, or behind + ahead + you.

Translate. Two more people ahead than behind you: $y = x + 2$

Three times as many people as there are behind you: $3x = x + y + 1$.

The resulting system is

$$y = x + 2 \qquad (1)$$
$$3x = x + y + 1 \qquad (2)$$

Carry out. We solve the system of equations. We substitute $x + 2$ for y in Equation (2) and Solve for x.

$$3x = x + y + 1 \qquad (2)$$
$$3x = x + (x + 2) + 1 \qquad \text{Substituting}$$
$$3x = 2x + 3$$
$$x = 3$$

Next we substitute 3 for x in either equation of the original system and solve for y.

$$y = x + 2$$
$$y = 3 + 2$$
$$y = 5$$

All the people in line: $x + y + 1 = 3 + 5 + 1 = 9$

State. There are 9 people in line.

23. $(-2, 3)$ is a solution of $Cx - 4y = 4$. Substitute -2 for x and 3 for y and solve for C.

$$Cx - 4y = 7$$
$$C(-2) - 4(3) = 7$$
$$-2C - 12 = 7$$
$$-2C = 19$$
$$C = \frac{-19}{2}$$

$(-2, 3)$ is a solution of $3x + Dy = 8$. Substitute -2 for x and 3 for y and solve for D.

$$3x + Dy = 8$$
$$3(-2) + D(3) = 8$$
$$-6 + 3D = 8$$
$$3D = 14$$
$$D = \frac{14}{3}$$

Chapter 8

Radical Expressions and Equations

Exercise Set 8.1

1. The name for an expression written under a radical is a radicand, so choice (c) is correct.

3. The name for a number that is real but not rational is an irrational number, so choice (a) is correct.

5. When x represents a negative number, then $3x$ is the product of a positive number and a negative number so the sign of $3x$ is negative. Choice (b) is correct.

7. $6^2 = 36$, $7^2 = 49$, and 37 is between 36 and 49, so it is true that $\sqrt{37}$ is between 6 and 7.

9. $11^2 = 121$, $12^2 = 144$, and 150 is not between 121 and 144, so it is false that $\sqrt{150}$ is between 11 and 12.

11. The square roots of 100 are 10 and -10, since $10^2 = 100$ and $(-10)^2 = 100$.

13. The square roots of 36 are 6 and -6, since $6^2 = 36$ and $(-6)^2 = 36$.

15. The square roots of 1 are 1 and -1, because $1^2 = 1$ and $(-1)^2 = 1$.

17. The square roots of 144 are 12 and -12, because $12^2 = 144$ and $(-12)^2 = 144$.

19. $\sqrt{100} = 10$, taking the principal square root.

21. $\sqrt{1} = 1$, so $-\sqrt{1} = -1$.

23. $\sqrt{0} = 0$

25. $\sqrt{121} = 11$, so $-\sqrt{121} = -11$.

27. $\sqrt{900} = 30$

29. $\sqrt{144} = 12$, so $-\sqrt{144} = -12$

31. The radicand is the expression under the radical, $10a$.

33. The radicand is the expression under the radical, $t^3 - 2$.

35. The radicand is the expression under the radical, $\dfrac{7}{x+y}$.

37. $\sqrt{4}$ is rational, since 4 is a perfect square.

39. $\sqrt{11}$ is irrational, since 11 is not a perfect square.

41. $\sqrt{32}$ is irrational, since 32 is not a perfect square.

43. $-\sqrt{16}$ is rational, since 16 is a perfect square.

45. Since the radicand is expressed as a number squared, we know that it is a perfect square and thus the number in the exercise is rational.

47. 2.828

49. 3.873

51. 9.110

53. For any real number A, $\sqrt{A^2} = |A|$, so $x^2 = |x|$. Use either one.

55. For any real number A, $\sqrt{A^2} = |A|$, so $\sqrt{(10x)^2} = |10x|$, or $10|x|$.

57. For any real number A, $\sqrt{A^2} = |A|$, so $\sqrt{(x+7)^2} = |x+7|$.

59. For any real number A, $\sqrt{A^2} = |A|$, so
$\sqrt{(5-2x)^2} = |5-2x|$.

61. $\sqrt{x^2} = x$ Since x is assumed to be nonnegative

63. $\sqrt{(5y)^2} = 5y$ Since y is assumed to be nonnegative

65. $\sqrt{16t^2} = \sqrt{(4t)^2} = 4t$ Since t is assumed to be nonnegative

67. $\sqrt{(n+7)^2} = n+7$ Since n is assumed to be nonnegative

69. a) We substitute 36 in the formula.
$N = 2.5\sqrt{36} = 2.5(6) = 15$

For an average of 36 arrivals, 15 spaces are needed.

b) We substitute 29 in the formula. We use a calculator or Table 2 to find an approximation.
$N = 2.5\sqrt{29} \approx 2.5(5.385) \approx 13.463 \approx 14$

For an average of 29 arrivals, 14 spaces are needed.

71. Substitute 39 in the formula.
$T = 0.144\sqrt{39} \approx 0.899$ sec

73. *Writing Exercise.* $\sqrt{12}$ is more exact, since 3.464101615 is an approximation of $\sqrt{12}$ while $\sqrt{12}$ denotes the exact value of the principal square root of 12.

75. $(t^{10})^2 = t^{10 \cdot 2} = t^{20}$

77. $2a \cdot 25a^{12} = 2 \cdot 25 \cdot a \cdot a^{12} = 50a^{1+12} = 50a^{13}$

79. $3m \cdot (10m^4)^2 = 3m \cdot 10^2 m^{4 \cdot 2} = 3 \cdot 100 \cdot m \cdot m^8 = 300m^9$

81. *Writing Exercise.* $\sqrt{A^2}$ denotes the principal, or positive, square root of A^2. Thus, when A is negative $\sqrt{A^2} \neq A$.

83. $\sqrt{\dfrac{1}{100}} = \sqrt{\dfrac{1}{10^2}} = \dfrac{\sqrt{1}}{\sqrt{10^2}} = \dfrac{1}{10}$

85. $\sqrt{A^2} = |A|$, so $\sqrt{(-7)^2} = 7$

87. $\sqrt{3^2 + 4^2} = \sqrt{9+16} = \sqrt{25} = 5$

89. $-\sqrt{36} < -\sqrt{33} < -\sqrt{25}$, or $-6 < -\sqrt{33} < -5$, so $-\sqrt{33}$ is between -6 and -5.

91. If $\sqrt{t^2} = 4$, then $t^2 = 4^2$, or 16. Thus, $t = 4$ or $t = -4$. The solutions are 4 and -4.

93. If $-\sqrt{x^2} = -3$, then $\sqrt{x^2} = 3$ and $x^2 = 3^2$, or 9. Thus, $x = 3$ or $x = -3$. The solutions are 3 and -3.

95. The values of x for which $\sqrt{(x-10)}$ is false are the values of x for which $x - 10 < 0$, so we have $x < 10$, or $(-\infty, 10)$, or $\{x | x < 10\}$.

97. $\sqrt{\dfrac{144x^8}{36y^6}} = \sqrt{\dfrac{4x^8}{y^6}} = \sqrt{\left(\dfrac{2x^4}{y^3}\right)^2} = \dfrac{2x^4}{y^3}$

99. $\sqrt{\dfrac{400}{m^{16}}} = \sqrt{\left(\dfrac{20}{m^8}\right)^2} = \dfrac{20}{m^8}$

101. a) Locate 3 on the x-axis, move up vertically to the graph, and then move left horizontally to the y-axis to read the approximation.

     $\sqrt{3} \approx 1.7$ (Answers may vary.)

b) Locate 5 on the x-axis, move up vertically to the graph, and then move left horizontally to the y-axis to read the approximation.

     $\sqrt{5} \approx 2.2$ (Answers may vary.)

c) Locate 7 on the x-axis, move up vertically to the graph, and then move left horizontally to the y-axis to read the approximation.

     $\sqrt{7} \approx 2.6$ (Answers may vary.)

103. We substitute 5 in the formula.

$V = \dfrac{1087\sqrt{273+5}}{16.52}$

$V = \dfrac{1087\sqrt{278}}{16.52}$

$V \approx \dfrac{1087(16.673)}{16.52}$

$V \approx 1097.1$ ft/sec

105.

$y_1 = \sqrt{x} - 2;\ y_2 = \sqrt{x} + 7;$
$y_3 = 5 + \sqrt{x};\ y_4 = -4 + \sqrt{x}$

107. The values of a for which $\sqrt{(a-1)^2} = a - 1$ is false are the values fo a for which $a - 1 < 0$. So we have $a < 1$, or $(-\infty, 1)$, or $\{a | a < 1\}$.

Exercise Set 8.2

1. $\sqrt{3} \cdot \sqrt{7} = \sqrt{3 \cdot 7} = \sqrt{21}$, so choice (h) is correct.

3. $\sqrt{3 \cdot 49} = \sqrt{3} \cdot \sqrt{49} = \sqrt{3} \cdot 7$, so choice (f) is correct.

5. $\sqrt{a^2} = a$, so choice (i) is correct.

7. $\sqrt{a^6} = \sqrt{(a^3)^2}$, so choice (b) is correct.

9. $\sqrt{ab} \cdot \sqrt{bc} = \sqrt{ab \cdot bc} = \sqrt{ab^2c} = \sqrt{b^2ac}$, so choice (c) is correct.

11. $\sqrt{2}\sqrt{5} = \sqrt{2 \cdot 5} = \sqrt{10}$

13. $\sqrt{4}\sqrt{3} = \sqrt{12}$, or
$\sqrt{4}\sqrt{3} = 2\sqrt{3}$ Taking the square root of 4

15. $\sqrt{\dfrac{3}{5}}\sqrt{\dfrac{7}{8}} = \sqrt{\dfrac{3 \cdot 7}{5 \cdot 8}} = \sqrt{\dfrac{21}{40}}$

17. $\sqrt{10}\sqrt{10} = \sqrt{10 \cdot 10} = \sqrt{10^2} = 10$

19. $\sqrt{25}\sqrt{3} = \sqrt{75}$, or
$\sqrt{25}\sqrt{3} = 5\sqrt{3}$ Taking the square root of 25

21. $\sqrt{2}\sqrt{x} = \sqrt{2 \cdot x} = \sqrt{2x}$

23. $\sqrt{7}\sqrt{2a} = \sqrt{7 \cdot 2a} = \sqrt{14a}$

25. $\sqrt{3x}\sqrt{7y} = \sqrt{3x \cdot 7y} = \sqrt{21xy}$

27. $\sqrt{3a}\sqrt{2bc} = \sqrt{3a \cdot 2bc} = \sqrt{6abc}$

29. $\sqrt{12} = \sqrt{4 \cdot 3}$ 4 is a perfect square.
$= \sqrt{4}\sqrt{3}$ Factoring into a product of radicals
$= 2\sqrt{3}$

31. $\sqrt{75} = \sqrt{25 \cdot 3}$ 25 is a perfect square.
$= \sqrt{25}\sqrt{3}$ Factoring into a product of radicals
$= 5\sqrt{3}$

33. $\sqrt{500} = \sqrt{100 \cdot 5} = \sqrt{100}\sqrt{5} = 10\sqrt{5}$

35. $\sqrt{16t} = \sqrt{16 \cdot t} = \sqrt{16}\sqrt{t} = 4\sqrt{t}$

37. $\sqrt{20z} = \sqrt{4 \cdot 5z} = \sqrt{4}\sqrt{5z} = 2\sqrt{5z}$

39. $\sqrt{100y^2} = \sqrt{100}\sqrt{y^2} = 10y$, or $\sqrt{100y^2} = \sqrt{(10y)^2} = 10y$

41. $\sqrt{13x^2} = \sqrt{13}\sqrt{x^2} = \sqrt{13} \cdot x$, or $x\sqrt{13}$

43. $\sqrt{27b^2} = \sqrt{9 \cdot b^2 \cdot 3} = \sqrt{9}\sqrt{b^2}\sqrt{3} = 3b\sqrt{3}$

45. $\sqrt{144x^2y} = \sqrt{144 \cdot x^2 \cdot y} = \sqrt{144}\sqrt{x^2}\sqrt{y} = 12x\sqrt{y}$

47. $\sqrt{b^2 - 4ac} = \sqrt{4^2 - 4(2)(-1)}$
$= \sqrt{16 - 8(-1)}$
$= \sqrt{16 + 8}$
$= \sqrt{24}$
$= \sqrt{4 \cdot 6}$
$= 2\sqrt{6}$

49. $\sqrt{b^2 - 4ac} = \sqrt{4^2 - 4\,(1)\,(4)}$

$= \sqrt{16 - 4\,(4)}$

$= \sqrt{16 - 16}$

$= \sqrt{0}$

$= 0$

51. $\sqrt{b^2 - 4ac} = \sqrt{(-6)^2 - 4 \cdot 3(-4)}$

$= \sqrt{36 - 12(-4)}$

$= \sqrt{36 + 48}$

$= \sqrt{84}$

$= \sqrt{2 \cdot 2 \cdot 3 \cdot 7}$

$= \sqrt{4} \cdot \sqrt{21}$

$= 2\sqrt{21}$

53. $\sqrt{a^{18}} = \sqrt{(a^9)^2} = a^9$

55. $\sqrt{x^{16}} = \sqrt{(x^8)^2} = x^8$

57. $\sqrt{r^5} = \sqrt{r^4 \cdot r}$

$= \sqrt{r^4}\sqrt{r}$

$= r^2\sqrt{r}$

59. $\sqrt{t^{15}} = \sqrt{t^{14} \cdot t} = \sqrt{t^{14}}\sqrt{t} = t^7\sqrt{t}$

61. $\sqrt{40a^3} = \sqrt{4 \cdot a^2 \cdot 10 \cdot a} = \sqrt{4}\sqrt{a^2}\sqrt{10a} = 2a\sqrt{10a}$

63. $\sqrt{45x^5} = \sqrt{9x^4 \cdot 5x} = \sqrt{9x^4}\sqrt{5x} = 3x^2\sqrt{5x}$

65. $\sqrt{200p^{25}} = \sqrt{100p^{24} \cdot 2p} = 10p^{12}\sqrt{2p}$

67. $\sqrt{2}\sqrt{10} = \sqrt{2 \cdot 10}$ Multiplying

$= \sqrt{2 \cdot 2 \cdot 5}$ Writing the prime
factorization

$= \sqrt{2^2}\sqrt{5}$

$= 2\sqrt{5}$

69. $\sqrt{3} \cdot \sqrt{27} = \sqrt{3 \cdot 27}$ Multiplying

$= \sqrt{3 \cdot 3 \cdot 3 \cdot 3}$ Writing the prime
factorization

$= \sqrt{3^4}$

$= 3^2$

$= 9$

71. $\sqrt{3x}\sqrt{12y} = \sqrt{3x \cdot 12y}$

$= \sqrt{3 \cdot x \cdot 2 \cdot 2 \cdot 3 \cdot y}$

$= \sqrt{2^2}\sqrt{3^2}\sqrt{xy}$

$= 2 \cdot 3\sqrt{xy}$

$= 6\sqrt{xy}$

73. $\sqrt{17}\sqrt{17x} = \sqrt{17 \cdot 17x} = \sqrt{17 \cdot 17 \cdot x} = \sqrt{17^2}\sqrt{x} = 17\sqrt{x}$

75. $\sqrt{10b}\sqrt{50b} = \sqrt{10b \cdot 50b}$

$= \sqrt{10 \cdot b \cdot 5 \cdot 10 \cdot b}$

$= \sqrt{10^2}\sqrt{b^2}\sqrt{5}$

$= 10b\sqrt{5}$

77. Since the radicands are indentical, the product will be the radicand, $12t$. We could also do this problem as follows.
$\sqrt{12t}\sqrt{12t} = \sqrt{12t \cdot 12t} = \sqrt{(12t)^2} = 12t$

79. $\sqrt{ab}\sqrt{ac} = \sqrt{a^2bc} = \sqrt{a^2}\sqrt{bc} = a\sqrt{bc}$

81. $\sqrt{x^7}\sqrt{x^{10}} = \sqrt{x^7 \cdot x^{10}} = \sqrt{x^{16} \cdot x} = x^8\sqrt{x}$

83. $\sqrt{7m^5}\sqrt{14m} = \sqrt{7m^5 \cdot 14m} = \sqrt{7 \cdot 2 \cdot 7m^6}$
$= 7m^3\sqrt{2}$

85. $\sqrt{x^2y^3}\sqrt{xy^4}$

$= \sqrt{x^2y^3}\sqrt{x(y^2)^2}$ x^2 and $(y^2)^2$ are
perfect squares

$= \sqrt{x^2} \cdot \sqrt{y^3} \cdot \sqrt{x} \cdot \sqrt{(y^2)^2}$

$= x \cdot \sqrt{y^3} \cdot \sqrt{x} \cdot y^2$

$= xy^2\sqrt{y^3 \cdot x}$

$= xy^2\sqrt{x \cdot y^2 \cdot y}$

$= xy^2\sqrt{y^2}\sqrt{xy}$

$= xy^2 \cdot y \cdot \sqrt{xy}$

$= xy^3\sqrt{xy}$

87. $\sqrt{6ab}\sqrt{12a^2b^5} = \sqrt{6ab \cdot 12a^2b^5} = \sqrt{6 \cdot 6 \cdot 2 \cdot a^2 \cdot a \cdot b^6}$
$= 6ab^3\sqrt{2a}$

89. Substitute 20 (thousand) for p in the formula:
$f = 400\sqrt{p} = 400\sqrt{20}$
The required flow is about 1789 gal/min.

91. First we substitute 20 for L in the formula:
$r = 2\sqrt{5L} = 2\sqrt{5 \cdot 20} = 2\sqrt{100} = 2 \cdot 10 = 20$ mph

Then we substitute 150 for L:
$r = 2\sqrt{5 \cdot 150} = 2\sqrt{750} = 2\sqrt{25 \cdot 30} = 2\sqrt{25}\sqrt{30}$
$= 2 \cdot 5\sqrt{30} = 10\sqrt{30} \approx 10(5.477) \approx 54.77$ mph, or 54.8 mph
(rounded to the nearest tenth)

93. *Writing Exercise.* Consider a replacement for x, say 8. Then $\sqrt{8^2 - 49} = \sqrt{64 - 49} = \sqrt{15}$, but $\sqrt{8^2} - \sqrt{49}$
$= 8 - 7 = 1$ and $\sqrt{15} \neq 1$.

95. $\dfrac{x^5y^6}{x^2y} = x^{5-2}y^{6-1} = x^3y^5$

97. $\dfrac{7a}{8b} \cdot \dfrac{3a}{2b} = \dfrac{7a \cdot 3a}{8b \cdot 2b} = \dfrac{21a^2}{16b^2}$

99. $\dfrac{2r^3}{7t} \cdot \dfrac{rt}{rt} = \dfrac{2r^3}{7t} \cdot 1 = \dfrac{2r^3}{7t}$

101. *Writing Exercise.* $\sqrt{16x^4} = \sqrt{(4x^2)^2} = 4x^2$, but
$\sqrt{4x^{16}} = \sqrt{(2x^8)^2} = 2x^8 \neq 2x^4$.

103. $\sqrt{0.01} = \sqrt{(0.1)^2} = 0.1$

105. $\sqrt{0.0625} = \sqrt{(0.25)^2} = 0.25$

107. $4\sqrt{14} = \sqrt{16 \cdot 14} = \sqrt{224}$ and $15 = \sqrt{225}$, so
$4\sqrt{14} < 15$.

109. $\sqrt{450} = \sqrt{225 \cdot 2} = 15\sqrt{2}$, so $\sqrt{450} = 15\sqrt{2}$

111. $8^2 = 64$

$(\sqrt{15} + \sqrt{17})^2 = 15 + 2\sqrt{255} + 17 = 32 + 2\sqrt{255}$

Now $\sqrt{255} < \sqrt{256}$, or $\sqrt{255} < 16$, so
$2\sqrt{255} < 2 \cdot 16 = 32$.   Then $32 + 2\sqrt{255} < 32 + 32$, or
$(\sqrt{15} + \sqrt{17})^2 < 64$. Thus, $8 > \sqrt{15} + \sqrt{17}$.

113. $\quad \sqrt{54(x+1)} \, \sqrt{6y(x+1)^2}$

$= \sqrt{54(x+1) \cdot 6y(x+1)^2}$

$= \sqrt{9 \cdot 6 \cdot (x+1) \cdot 6 \cdot y \cdot (x+1)^2}$

$= \sqrt{9 \cdot 6 \cdot 6 \cdot (x+1)^2 \cdot y(x+1)}$

$= \sqrt{9} \, \sqrt{6 \cdot 6} \, \sqrt{(x+1)^2} \, \sqrt{y(x+1)}$

$= 3 \cdot 6 \cdot (x+1)\sqrt{y(x+1)}$

$= 18(x+1)\sqrt{y(x+1)}$

115. $\sqrt{x^9} \, \sqrt{2x} \, \sqrt{10x^5} = \sqrt{x^9 \cdot 2x \cdot 10x^5}$

$= \sqrt{x^8 \cdot x \cdot 2 \cdot x \cdot 2 \cdot 5 \cdot x^4 \cdot x}$

$= \sqrt{x^8 \cdot x \cdot x \cdot 2 \cdot 2 \cdot x^4 \cdot 5 \cdot x}$

$= \sqrt{x^8} \, \sqrt{x \cdot x} \, \sqrt{2 \cdot 2} \, \sqrt{x^4} \, \sqrt{5x}$

$= x^4 \cdot x \cdot 2 \cdot x^2 \sqrt{5x} = 2x^7 \sqrt{5x}$

117. $7x^{14}\sqrt{6x^7} = \sqrt{(7x^{14})^2} \cdot \sqrt{6x^7} = \sqrt{49x^{28}} \cdot \sqrt{6x^7}$

$= \sqrt{49x^{28} \cdot 6x^7} = \sqrt{294x^{35}} = \sqrt{21x^9 \cdot 14x^{26}}$

$= \sqrt{21x^9} \cdot \sqrt{14x^{26}}$

119. $\sqrt{x^{16n}} = \sqrt{(x^{8n})^2} = x^{8n}$

121. $\sqrt{y^n} = (y^n)^{\frac{1}{2}} = y^{n/2}$

Exercise Set 8.3

1. The statement is true by the quotient rule for square roots.

3. The statement is false. See Example 4 and 5.

5. $\dfrac{\sqrt{500}}{\sqrt{5}} = \sqrt{\dfrac{500}{5}} = \sqrt{100} = 10$

7. $\dfrac{\sqrt{40}}{\sqrt{10}} = \sqrt{\dfrac{40}{10}} = \sqrt{4} = 2$

9. $\dfrac{\sqrt{55}}{\sqrt{5}} = \sqrt{\dfrac{55}{5}} = \sqrt{11}$

11. $\dfrac{\sqrt{5}}{\sqrt{20}} = \sqrt{\dfrac{5}{20}} = \sqrt{\dfrac{1}{4}} = \dfrac{1}{2}$

13. $\dfrac{\sqrt{18}}{\sqrt{32}} = \sqrt{\dfrac{18}{32}} = \sqrt{\dfrac{9}{16}} = \dfrac{3}{4}$

15. $\dfrac{\sqrt{8x}}{\sqrt{2x}} = \sqrt{\dfrac{8x}{2x}} = \sqrt{4} = 2$

17. $\dfrac{\sqrt{63y^3}}{\sqrt{7y}} = \sqrt{\dfrac{63y^3}{7y}} = \sqrt{9y^2} = 3y$

19. $\dfrac{\sqrt{500a^{10}}}{\sqrt{5a^2}} = \sqrt{\dfrac{500a^{10}}{5a^2}} = \sqrt{100a^8} = 10a^4$

21. $\dfrac{\sqrt{21a^9}}{\sqrt{7a^3}} = \sqrt{\dfrac{21a^9}{7a^3}} = \sqrt{3a^6} = a^3\sqrt{3}$

23. $\sqrt{\dfrac{4}{25}} = \dfrac{\sqrt{4}}{\sqrt{25}} = \dfrac{2}{5}$

25. $\sqrt{\dfrac{49}{16}} = \dfrac{\sqrt{49}}{\sqrt{16}} = \dfrac{7}{4}$

27. $-\sqrt{\dfrac{25}{81}} = -\dfrac{\sqrt{25}}{\sqrt{81}} = -\dfrac{5}{9}$

29. $\sqrt{\dfrac{2a^5}{50a}} = \sqrt{\dfrac{a^4}{25}} = \dfrac{\sqrt{a^4}}{\sqrt{25}} = \dfrac{a^2}{5}$

31. $\sqrt{\dfrac{6x^7}{32x}} = \sqrt{\dfrac{3x^6}{16}} = \dfrac{\sqrt{3x^6}}{\sqrt{16}} = \dfrac{x^3\sqrt{3}}{4}$

33. $\sqrt{\dfrac{21t^9}{28t^3}} = \sqrt{\dfrac{3t^6}{4}} = \dfrac{\sqrt{3t^6}}{\sqrt{4}} = \dfrac{t^3\sqrt{3}}{2}$

35. $\dfrac{1}{\sqrt{3}} = \dfrac{1}{\sqrt{3}} \cdot \dfrac{\sqrt{3}}{\sqrt{3}} = \dfrac{\sqrt{3}}{3}$

37. $\dfrac{5}{\sqrt{7}} = \dfrac{5}{\sqrt{7}} \cdot \dfrac{\sqrt{7}}{\sqrt{7}} = \dfrac{5\sqrt{7}}{7}$

39. $\dfrac{\sqrt{16}}{\sqrt{27}} = \dfrac{\sqrt{16}}{\sqrt{9}\sqrt{3}} = \dfrac{4}{3\sqrt{3}} \cdot \dfrac{\sqrt{3}}{\sqrt{3}} = \dfrac{4\sqrt{3}}{3 \cdot 3} = \dfrac{4\sqrt{3}}{9}$

41. $\dfrac{\sqrt{6}}{\sqrt{5}} = \dfrac{\sqrt{6}}{\sqrt{5}} \cdot \dfrac{\sqrt{5}}{\sqrt{5}} = \dfrac{\sqrt{30}}{5}$

43. $\dfrac{\sqrt{3}}{\sqrt{50}} = \dfrac{\sqrt{3}}{\sqrt{25}\sqrt{2}} = \dfrac{\sqrt{3}}{5\sqrt{2}} = \dfrac{\sqrt{3}}{5\sqrt{2}} \cdot \dfrac{\sqrt{2}}{\sqrt{2}}$

$= \dfrac{\sqrt{6}}{5 \cdot 2} = \dfrac{\sqrt{6}}{10}$

45. $\dfrac{\sqrt{2a}}{\sqrt{45}} = \dfrac{\sqrt{2a}}{\sqrt{9}\sqrt{5}} = \dfrac{\sqrt{2a}}{3\sqrt{5}} = \dfrac{\sqrt{2a}}{3\sqrt{5}} \cdot \dfrac{\sqrt{5}}{\sqrt{5}}$

$= \dfrac{\sqrt{10a}}{3 \cdot 5} = \dfrac{\sqrt{10a}}{15}$

47. $\sqrt{\dfrac{12}{5}} = \dfrac{\sqrt{4}\sqrt{3}}{\sqrt{5}} = \dfrac{2\sqrt{3}}{\sqrt{5}} = \dfrac{2\sqrt{3}}{\sqrt{5}} \cdot \dfrac{\sqrt{5}}{\sqrt{5}} = \dfrac{2\sqrt{15}}{5}$

49. $\sqrt{\dfrac{7}{z}} = \dfrac{\sqrt{7}}{\sqrt{z}} \cdot \dfrac{\sqrt{z}}{\sqrt{z}} = \dfrac{\sqrt{7z}}{z}$

51. $\sqrt{\dfrac{a}{200}} = \dfrac{\sqrt{a}}{\sqrt{200}} = \dfrac{\sqrt{a}}{\sqrt{100}\sqrt{2}} = \dfrac{\sqrt{a}}{10\sqrt{2}} = \dfrac{\sqrt{a}}{10\sqrt{2}} \cdot \dfrac{\sqrt{2}}{\sqrt{2}}$

$= \dfrac{\sqrt{2a}}{10 \cdot 2} = \dfrac{\sqrt{2a}}{20}$

53. $\sqrt{\dfrac{x}{90}} = \dfrac{\sqrt{x}}{\sqrt{90}} = \dfrac{\sqrt{x}}{\sqrt{9}\sqrt{10}} = \dfrac{\sqrt{x}}{3\sqrt{10}} = \dfrac{\sqrt{x}}{3\sqrt{10}} \cdot \dfrac{\sqrt{10}}{\sqrt{10}}$

$= \dfrac{\sqrt{10x}}{3 \cdot 10} = \dfrac{\sqrt{10x}}{30}$

55. Since the denominator, 25, is a perfect square we need only to simplify the expression in order to rationalize the denominator.

$$\sqrt{\dfrac{3a}{25}} = \dfrac{\sqrt{3a}}{\sqrt{25}} = \dfrac{\sqrt{3a}}{5}$$

57. $\sqrt{\dfrac{5x^3}{12x}} = \sqrt{\dfrac{5x^2}{12}} = \dfrac{\sqrt{5x^2}}{\sqrt{12}} = \dfrac{\sqrt{5x^2}}{\sqrt{4}\sqrt{3}} = \dfrac{x\sqrt{5}}{2\sqrt{3}}$

$= \dfrac{x\sqrt{5}}{2\sqrt{3}} \cdot \dfrac{\sqrt{3}}{\sqrt{3}} = \dfrac{x\sqrt{15}}{2\cdot 3} = \dfrac{x\sqrt{15}}{6}$

59. $v = \dfrac{18,500}{\sqrt{t+1.0565}} = \dfrac{18,500}{\sqrt{3+1.0565}} = \dfrac{18,500}{\sqrt{4.0565}} \approx \$9185$

61. $v = \dfrac{18,500}{\sqrt{t+1.0565}} = \dfrac{18,500}{\sqrt{1+1.0565}} \approx \$12,901$

63. 32 ft: $T \approx 2(3.14)\sqrt{\dfrac{32}{32}} \approx 6.28\sqrt{1} \approx 6.28(1)$

≈ 6.28 sec

50 ft: $T \approx 2(3.14)\sqrt{\dfrac{50}{32}} \approx 6.28\sqrt{\dfrac{25}{16}} \approx 6.28\dfrac{\sqrt{25}}{\sqrt{16}}$

$\approx 6.28\left(\dfrac{5}{4}\right) \approx 7.85$ sec

65. Substitute $\dfrac{2}{\pi^2}$ for L in the formula.

$T = 2\pi\sqrt{\dfrac{L}{32}} = 2\pi\sqrt{\dfrac{\frac{2}{\pi^2}}{32}} = 2\pi\sqrt{\dfrac{2}{\pi^2}\cdot\dfrac{1}{32}}$

$= 2\pi\sqrt{\dfrac{2}{32\pi^2}} = 2\pi\sqrt{\dfrac{1}{16\pi^2}} = 2\pi\cdot\dfrac{\sqrt{1}}{\sqrt{16\pi^2}} = 2\pi\cdot\dfrac{1}{4\pi} = \dfrac{2\pi}{4\pi}$

$= 0.5$ sec

It takes 0.5 sec to move from one side to the other and back.

67. Substitute 72 for L in the formula.

$T \approx 2(3.14)\sqrt{\dfrac{72}{32}} \approx 6.28\sqrt{2.25} \approx 9.42$ sec

69. *Writing Exercise.* If division requires rationalizing the denominator, it is necessary to know how to multiply radical expressions.

71. $9x + 6x = 15x$

73. $9y - z + 8y = 9y + 8y - z = 17y - z$

75. $9x(2x - 7) = 9x\cdot 2x - 9x\cdot 7 = 18x^2 - 63x$

77. $(2x-3)(2x+3) = (2x)^2 - 3^2$
$= 4x^2 - 9$

Using $(A+B)(A-B) = A^2 - B^2$

79. *Writing Exercise.* $\sqrt{2} \approx 1.414213562$; the long division $1.414213562/2$ is easier to perform than the long division $1/1.414213562$.

81. $\sqrt{\dfrac{7}{1000}} = \dfrac{\sqrt{7}}{\sqrt{1000}} = \dfrac{\sqrt{7}}{\sqrt{100}\sqrt{10}} = \dfrac{\sqrt{7}}{10\sqrt{10}}$

$= \dfrac{\sqrt{7}}{10\sqrt{10}}\cdot\dfrac{\sqrt{10}}{\sqrt{10}} = \dfrac{\sqrt{70}}{10\cdot 10} = \dfrac{\sqrt{70}}{100}$

83. $\sqrt{\dfrac{3x^2}{8x^7y^3}} = \sqrt{\dfrac{3}{8x^5y^3}} = \dfrac{\sqrt{3}}{\sqrt{8x^5y^3}}$

$= \dfrac{\sqrt{3}}{\sqrt{4x^4y^2}\sqrt{2xy}} = \dfrac{\sqrt{3}}{2x^2y\sqrt{2xy}}$

$= \dfrac{\sqrt{3}}{2x^2y\sqrt{2xy}}\cdot\dfrac{\sqrt{2xy}}{\sqrt{2xy}} = \dfrac{\sqrt{6xy}}{2x^2y\cdot 2xy} = \dfrac{\sqrt{6xy}}{4x^3y^2}$

85. $\sqrt{\dfrac{1}{5zw^2}} = \dfrac{1}{w\sqrt{5z}} = \dfrac{1}{w\sqrt{5z}}\cdot\dfrac{\sqrt{5z}}{\sqrt{5z}} = \dfrac{\sqrt{5z}}{5zw}$

87. $\dfrac{2}{\sqrt{\sqrt{5}}} = \dfrac{2}{\sqrt{\sqrt{5}}}\cdot\dfrac{\sqrt{5\sqrt{5}}}{\sqrt{5\sqrt{5}}} = \dfrac{2\sqrt{5\sqrt{5}}}{\sqrt{5}\cdot\sqrt{5}} = \dfrac{2\sqrt{5\sqrt{5}}}{5}$

89. $\sqrt{\dfrac{1}{x^2} - \dfrac{2}{xy} + \dfrac{1}{y^2}}$, LCD is x^2y^2

$= \sqrt{\dfrac{1}{x^2}\cdot\dfrac{y^2}{y^2} - \dfrac{2}{xy}\cdot\dfrac{xy}{xy} + \dfrac{1}{y^2}\cdot\dfrac{x^2}{x^2}}$

$= \sqrt{\dfrac{y^2 - 2xy + x^2}{x^2y^2}}$

$= \sqrt{\dfrac{(y-x)^2}{x^2y^2}}$

$= \dfrac{\sqrt{(y-x)^2}}{\sqrt{x^2y^2}}$

$= \dfrac{y-x}{xy}$

An alternate method of simplifying this expression is shown below.

$\sqrt{\dfrac{1}{x^2} - \dfrac{2}{xy} + \dfrac{1}{y^2}} = \sqrt{\left(\dfrac{1}{x} - \dfrac{1}{y}\right)^2}$

$= \dfrac{1}{x} - \dfrac{1}{y}$

The two answers are equivalent.

91. $\sqrt{\dfrac{2x-3}{8}} = \dfrac{5}{2}$

If this equation is true, we have:

$\dfrac{2x-3}{8} = \left(\dfrac{5}{2}\right)^2$

$\dfrac{2x-3}{8} = \dfrac{25}{4}$

$2x - 3 = 50 \qquad$ Multiplying by 8

$2x = 53$

$x = \dfrac{53}{2}$

The number $\dfrac{53}{2}$ checks in the original equation, so the solution is $\dfrac{53}{2}$.

Exercise Set 8.4

---

1. (b); see page 503 in the text.

3. (d); see page 503 in the text.

5. $3\sqrt{10} + 8\sqrt{10} = (3+8)\sqrt{10}$
$= 11\sqrt{10}$

7. $4\sqrt{2} - \sqrt{2} = (4-1)\sqrt{2}$
$= 3\sqrt{2}$

9. $4\sqrt{t} + 9\sqrt{t} = (4+9)\sqrt{t} = 13\sqrt{t}$

11. $7\sqrt{x} - 8\sqrt{x} = (7-8)\sqrt{x} = -\sqrt{x}$

13. $5\sqrt{2a} + 3\sqrt{2a} = (5+3)\sqrt{2a} = 8\sqrt{2a}$

15. $9\sqrt{10y} - \sqrt{10y} = (9-1)\sqrt{10y} = 8\sqrt{10y}$

17. $6\sqrt{7} + 2\sqrt{7} + 4\sqrt{7} = (6+2+4)\sqrt{7} = 12\sqrt{7}$

19. $5\sqrt{2} - 9\sqrt{2} + 8\sqrt{2} = (5-9+8)\sqrt{2} = 4\sqrt{2}$

21. $\begin{aligned}5\sqrt{3} + \sqrt{8} &= 5\sqrt{3} + \sqrt{4\cdot 2} \qquad \text{Factoring 8}\\ &= 5\sqrt{3} + \sqrt{4}\sqrt{2}\\ &= 5\sqrt{3} + 2\sqrt{2}\end{aligned}$

$5\sqrt{3} + \sqrt{8}$, or $5\sqrt{3} + 2\sqrt{2}$, cannot be simplified further.

23. $\begin{aligned}\sqrt{x} - \sqrt{16x} &= \sqrt{x} - \sqrt{16}\sqrt{x}\\ &= \sqrt{x} - 4\sqrt{x}\\ &= (1-4)\sqrt{x}\\ &= -3\sqrt{x}\end{aligned}$

25. $\begin{aligned}2\sqrt{3} - 4\sqrt{75} &= 2\sqrt{3} - 4\sqrt{25\cdot 3}\\ &= 2\sqrt{3} - 4\sqrt{25}\sqrt{3}\\ &= 2\sqrt{3} - 4\cdot 5\sqrt{3}\\ &= 2\sqrt{3} - 20\sqrt{3}\\ &= -18\sqrt{3}\end{aligned}$

27. $\begin{aligned}6\sqrt{18} + 5\sqrt{8} &= 6\sqrt{9\cdot 2} + 5\sqrt{4\cdot 2}\\ &= 6\sqrt{9}\sqrt{2} + 5\sqrt{4}\sqrt{2}\\ &= 6\cdot 3\sqrt{2} + 5\cdot 2\sqrt{2}\\ &= 18\sqrt{2} + 10\sqrt{2}\\ &= 28\sqrt{2}\end{aligned}$

29. $\begin{aligned}\sqrt{72} + \sqrt{98} &= \sqrt{36\cdot 2} + \sqrt{49\cdot 2}\\ &= 6\sqrt{2} + 7\sqrt{2}\\ &= 13\sqrt{2}\end{aligned}$

31. $9\sqrt{8} + \sqrt{72} - 9\sqrt{8}$

Observe that $9\sqrt{8} - 9\sqrt{8} = 0$, so we need only to simplify $\sqrt{72}$.
$$\sqrt{72} = \sqrt{36\cdot 2} = \sqrt{36}\sqrt{2} = 6\sqrt{2}$$

33. $\begin{aligned}&5\sqrt{18} - 2\sqrt{32} - \sqrt{50}\\ &= 5\sqrt{9\cdot 2} - 2\sqrt{16\cdot 2} - \sqrt{25\cdot 2}\\ &= 5\cdot 3\sqrt{2} - 2\cdot 4\sqrt{2} - 5\sqrt{2}\\ &= 15\sqrt{2} - 8\sqrt{2} - 5\sqrt{2}\\ &= 2\sqrt{2}\end{aligned}$

35. $\begin{aligned}\sqrt{16a} - 4\sqrt{a} + \sqrt{25a} &= 4\sqrt{a} - 4\sqrt{a} + 5\sqrt{a}\\ &= 5\sqrt{a}\end{aligned}$

37. $\begin{aligned}\sqrt{3}\left(\sqrt{2} + \sqrt{7}\right) &= \sqrt{3}\sqrt{2} + \sqrt{3}\sqrt{7}\\ &= \sqrt{6} + \sqrt{21}\end{aligned}$

39. $\begin{aligned}\sqrt{5}(\sqrt{6} - \sqrt{10}) &= \sqrt{5}\sqrt{6} - \sqrt{5}\sqrt{10}\\ &= \sqrt{30} - \sqrt{50}\\ &= \sqrt{30} - \sqrt{25\cdot 2}\\ &= \sqrt{30} - 5\sqrt{2}\end{aligned}$

41. $\begin{aligned}&(3+\sqrt{2})(4+\sqrt{2})\\ &= 3\cdot 4 + 3\cdot\sqrt{2} + \sqrt{2}\cdot 4 + \sqrt{2}\cdot\sqrt{2} \quad\text{Using FOIL}\\ &= 12 + 3\sqrt{2} + 4\sqrt{2} + 2\\ &= 14 + 7\sqrt{2}\end{aligned}$

43. $\begin{aligned}&(\sqrt{7} - 2)(\sqrt{7} - 5)\\ &= \sqrt{7}\cdot\sqrt{7} - \sqrt{7}\cdot 5 - 2\cdot\sqrt{7} + 2\cdot 5 \quad\text{Using FOIL}\\ &= 7 - 5\sqrt{7} - 2\sqrt{7} + 10\\ &= 17 - 7\sqrt{7}\end{aligned}$

45. $\begin{aligned}\left(\sqrt{6} + 5\right)\left(\sqrt{6} - 5\right) &= \left(\sqrt{6}\right)^2 - 5^2\\ &\qquad \text{Using } (A+B)(A-B)\\ &\qquad\qquad = A^2 - B^2\\ &= 6 - 25 = -19\end{aligned}$

47. $\begin{aligned}\left(\sqrt{7} - \sqrt{3}\right)\left(\sqrt{7} + \sqrt{3}\right) &= \left(\sqrt{7}\right)^2 - \left(\sqrt{3}\right)^2\\ &\text{Using } (A+B)(A-B) = A^2 - B^2\\ &= 7 - 3\\ &= 4\end{aligned}$

49. $\begin{aligned}&(2+3\sqrt{2})(3-\sqrt{2})\\ &= 2\cdot 3 - 2\cdot\sqrt{2} + 3\sqrt{2}\cdot 3 - 3\sqrt{2}\cdot\sqrt{2} \quad\text{Using FOIL}\\ &= 6 - 2\sqrt{2} + 9\sqrt{2} - 3\cdot 2\\ &= 6 - 2\sqrt{2} + 9\sqrt{2} - 6\\ &= 7\sqrt{2}\end{aligned}$

51. $\begin{aligned}&(7+\sqrt{3})^2\\ &= 7^2 + 2\cdot 7\cdot\sqrt{3} + \left(\sqrt{3}\right)^2 \quad\text{Using } (A+B)^2 =\\ &\qquad\qquad\qquad\qquad\qquad\quad A^2 + 2AB + B^2\\ &= 49 + 14\sqrt{3} + 3\\ &= 52 + 14\sqrt{3}\end{aligned}$

53. $\begin{aligned}&(1-2\sqrt{3})^2\\ &= 1^2 - 2\cdot 1\cdot 2\sqrt{3} + (2\sqrt{3})^2 \quad\text{Using } (A-B)^2 =\\ &\qquad\qquad\qquad\qquad\qquad\qquad A^2 - 2AB + B^2\\ &= 1 - 4\sqrt{3} + 4\cdot 3\\ &= 1 - 4\sqrt{3} + 12\\ &= 13 - 4\sqrt{3}\end{aligned}$

55. $\begin{aligned}(\sqrt{x} - \sqrt{10})^2 &= (\sqrt{x})^2 - 2\sqrt{x}\sqrt{10} + (\sqrt{10})^2\\ &= x - 2\sqrt{10x} + 10\end{aligned}$

57. $\begin{aligned}\frac{2}{5+\sqrt{2}} &= \frac{2}{5+\sqrt{2}}\cdot\frac{5-\sqrt{2}}{5-\sqrt{2}}\\ &= \frac{2\left(5-\sqrt{2}\right)}{\left(5+\sqrt{2}\right)\left(5-\sqrt{2}\right)}\\ &= \frac{10-2\sqrt{2}}{25-2} = \frac{10-2\sqrt{2}}{23}\end{aligned}$

59. $\dfrac{2}{1-\sqrt{3}} = \dfrac{2}{1-\sqrt{3}} \cdot \dfrac{1+\sqrt{3}}{1+\sqrt{3}}$

$= \dfrac{2\left(1+\sqrt{3}\right)}{\left(1-\sqrt{3}\right)\left(1+\sqrt{3}\right)}$

$= \dfrac{2\left(1+\sqrt{3}\right)}{1-3}$

$= \dfrac{2\left(1+\sqrt{3}\right)}{-2} = -1 - \sqrt{3}$

61. $\dfrac{2}{\sqrt{7}+5}$

$= \dfrac{2}{\sqrt{7}+5} \cdot \dfrac{\sqrt{7}-5}{\sqrt{7}-5}$

$= \dfrac{2(\sqrt{7}-5)}{(\sqrt{7})^2 - 5^2}$

$= \dfrac{2\sqrt{7}-10}{7-25}$

$= \dfrac{2\sqrt{7}-10}{-18}$ Since 2 is a common factor, we simplify.

$= \dfrac{\cancel{2}(\sqrt{7}-5)}{\cancel{2}(-9)}$ Factoring and removing a factor equal to 1

$= \dfrac{\sqrt{7}-5}{-9}$

$= \dfrac{-(\sqrt{7}-5)}{9}$

$= \dfrac{5-\sqrt{7}}{9}$

63. $\dfrac{\sqrt{10}}{\sqrt{10}+4} = \dfrac{\sqrt{10}}{\sqrt{10}+4} \cdot \dfrac{\sqrt{10}-4}{\sqrt{10}-4}$

$= \dfrac{\sqrt{10}\left(\sqrt{10}-4\right)}{\left(\sqrt{10}+4\right)\left(\sqrt{10}-4\right)}$

$= \dfrac{10 - 4\sqrt{10}}{10-16}$

$= \dfrac{-2\left(-5+2\sqrt{10}\right)}{-6}$

$= \dfrac{-5+2\sqrt{10}}{3}$

65. $\dfrac{\sqrt{7}}{\sqrt{7}-\sqrt{3}} = \dfrac{\sqrt{7}}{\sqrt{7}-\sqrt{3}} \cdot \dfrac{\sqrt{7}+\sqrt{3}}{\sqrt{7}+\sqrt{3}} = \dfrac{\sqrt{7}\sqrt{7}+\sqrt{7}\sqrt{3}}{(\sqrt{7})^2-(\sqrt{3})^2}$

$= \dfrac{7+\sqrt{21}}{7-3} = \dfrac{7+\sqrt{21}}{4}$

67. $\dfrac{\sqrt{3}}{\sqrt{5}-\sqrt{3}} = \dfrac{\sqrt{3}}{\sqrt{5}-\sqrt{3}} \cdot \dfrac{\sqrt{5}+\sqrt{3}}{\sqrt{5}+\sqrt{3}}$

$= \dfrac{\sqrt{3}\sqrt{5}+\sqrt{3}\sqrt{3}}{(\sqrt{5})^2-(\sqrt{3})^2} = \dfrac{\sqrt{15}+3}{5-3} = \dfrac{\sqrt{15}+3}{2}$

69. $\dfrac{2}{\sqrt{7}+\sqrt{2}} = \dfrac{2}{\sqrt{7}+\sqrt{2}} \cdot \dfrac{\sqrt{7}-\sqrt{2}}{\sqrt{7}-\sqrt{2}}$

$= \dfrac{2\sqrt{7}-2\sqrt{2}}{(\sqrt{7})^2-(\sqrt{2})^2} = \dfrac{2\sqrt{7}-2\sqrt{2}}{7-2} = \dfrac{2\sqrt{7}-2\sqrt{2}}{5}$

71. $\dfrac{\sqrt{6}-\sqrt{x}}{\sqrt{6}+\sqrt{x}} = \dfrac{\sqrt{6}-\sqrt{x}}{\sqrt{6}+\sqrt{x}} \cdot \dfrac{\sqrt{6}-\sqrt{x}}{\sqrt{6}-\sqrt{x}}$

$= \dfrac{\left(\sqrt{6}-\sqrt{x}\right)^2}{\left(\sqrt{6}+\sqrt{x}\right)\left(\sqrt{6}-\sqrt{x}\right)}$

$= \dfrac{(\sqrt{6})^2 - 2\sqrt{6}\sqrt{x} + (\sqrt{x})^2}{(\sqrt{6})^2 - (\sqrt{x})^2}$

$= \dfrac{6 - 2\sqrt{6x} + x}{6 - x}$

73. *Writing Exercise.* Since the inner and outer products of the multiplication are opposites, the product of conjugates contain no radicals.

75. $3x + 5 + 2(x-3) = 4 - 6x$

$3x + 5 + 2x - 6 = 4 - 6x$

$5x - 1 = 4 - 6x$

$11x - 1 = 4$

$11x = 5$

$x = \dfrac{5}{11}$

The solution is $\dfrac{5}{11}$.

77. $x^2 - 5x = 6$

$x^2 - 5x - 6 = 0$

$(x+1)(x-6) = 0$

$x + 1 = 0 \quad or \quad x - 6 = 0$

$x = -1 \quad or \qquad x = 6$

The solutions are -1 and 6.

79. **Familiarize**. Let x the number of minutes of use in one month.

Translate.

pay-per-use 15¢ per minute	is same cost as	unlimited $9.99
↓	↓	↓
$0.15x$	$=$	9.99

Carry out. We solve the equation.

$0.15x = 9.99$

$x = \dfrac{9.99}{0.15}$

$x = 66.6$

Check. If 66.6 min used in the pay-per-use plan, then the cost is $0.15(66.6)$, or $9.99. The answer checks.

State. The costs are the same when 66.6 minutes are used.

81. *Writing Exercise.* She is not correct. For example, $\left(\sqrt{x}+\sqrt{y}\right)^2 \neq x+y$, since $\left(\sqrt{x}+\sqrt{y}\right)\left(\sqrt{x}+\sqrt{y}\right) = x + 2\sqrt{xy} + y$.

83. $\sqrt{\dfrac{25}{x}} + \dfrac{\sqrt{x}}{2x} - \dfrac{5}{\sqrt{2}}$

$= \dfrac{5}{\sqrt{x}} + \dfrac{\sqrt{x}}{2x} - \dfrac{5}{\sqrt{2}}$

$= \dfrac{5}{\sqrt{x}} \cdot \dfrac{\sqrt{x}}{\sqrt{x}} + \dfrac{\sqrt{x}}{2x} - \dfrac{5}{\sqrt{2}} \cdot \dfrac{\sqrt{2}}{\sqrt{2}}$

$= \dfrac{5\sqrt{x}}{x} + \dfrac{\sqrt{x}}{2x} - \dfrac{5\sqrt{2}}{2}$

$= \dfrac{5\sqrt{x}}{x} \cdot \dfrac{2}{2} + \dfrac{\sqrt{x}}{2x} - \dfrac{5\sqrt{2}}{2} \cdot \dfrac{x}{x}$

$= \dfrac{10\sqrt{x}}{2x} + \dfrac{\sqrt{x}}{2x} - \dfrac{5x\sqrt{2}}{2x}$

$= \dfrac{11\sqrt{x} - 5x\sqrt{2}}{2x}$

85. $\sqrt{8x^6 y^3} - x\sqrt{2y^7} - \dfrac{x}{3}\sqrt{18x^2 y^9}$

$= \sqrt{4x^6 y^2 \cdot 2y} - x\sqrt{y^6 \cdot 2y} - \dfrac{x}{3}\sqrt{9x^2 y^8 \cdot 2y}$

$= 2x^3 y\sqrt{2y} - xy^3\sqrt{2y} - \dfrac{x}{3} \cdot 3xy^4\sqrt{2y}$

$= 2x^3 y\sqrt{2y} - xy^3\sqrt{2y} - x^2 y^4\sqrt{2y}$

$= xy\sqrt{2y}(2x^2 - y^2 - xy^3)$

87. $7x\sqrt{12xy^2} - 9y\sqrt{27x^3} + 5\sqrt{300x^3 y^2}$

$= 7x\sqrt{4y^2 \cdot 3x} - 9y\sqrt{9x^2 \cdot 3x} + 5\sqrt{100x^2 y^2 \cdot 3x}$

$= 7x \cdot 2y\sqrt{3x} - 9y \cdot 3x\sqrt{3x} + 5 \cdot 10xy\sqrt{3x}$

$= 14xy\sqrt{3x} - 27xy\sqrt{3x} + 50xy\sqrt{3x}$

$= (14xy - 27xy + 50xy)\sqrt{3x}$

$= 37xy\sqrt{3x}$

89. $\sqrt{10} + \sqrt{50} = \sqrt{10} + \sqrt{10}\,\sqrt{5} = \sqrt{10}(1 + \sqrt{5})$

$\sqrt{10} + \sqrt{50} = \sqrt{10} + \sqrt{25 \cdot 2} = \sqrt{10} + 5\sqrt{2}$

$\sqrt{10} + \sqrt{50} = \sqrt{2}\,\sqrt{5} + \sqrt{2}\,\sqrt{25} =$

$\sqrt{2}(\sqrt{5} + \sqrt{25}) = \sqrt{2}(\sqrt{5} + 5),$ or $\sqrt{2}(5 + \sqrt{5})$

All three are correct.

Exercise Set 8.5

1. The statement is true by the principle of squaring.

3. The statement is false. See Example 1(b).

5. $\sqrt{x} = 6$

$\left(\sqrt{x}\right)^2 = 6^2$ Simplifying both sides

$x = 36$ Simplifying

Check: $\dfrac{\sqrt{x} = 6}{\sqrt{36}\ \bigg|\ 6}$

$\qquad\qquad 6 \overset{?}{=} 6$ TRUE

The solution is 36.

7. $\sqrt{x} - 3 = 9$

$\sqrt{x} = 12$ Adding 3

$\left(\sqrt{x}\right)^2 = 12^2$

$x = 144$

Check: $\dfrac{\sqrt{x} - 3 = 9}{\sqrt{144} - 3\ \bigg|\ 9}$

$\qquad\qquad 12 - 3\ \bigg|$

$\qquad\qquad\quad 9 \overset{?}{=} 9$ TRUE

The solution is 144.

9. $\sqrt{3x + 1} = 8$

$\left(\sqrt{3x + 1}\right)^2 = 8^2$

$3x + 1 = 64$

$3x = 63$

$x = 21$

Check: $\dfrac{\sqrt{3x + 1} = 8}{\sqrt{3 \cdot 21 + 1}\ \bigg|\ 8}$

$\qquad\quad \sqrt{63 + 1}\ ?$

$\qquad\qquad\qquad 8 = 8$ TRUE

The solution is 21.

11. $2 + \sqrt{3 - y} = 9$

$\sqrt{3 - y} = 7$ Subtracting 2

$\left(\sqrt{3 - y}\right)^2 = 7^2$ Squaring both sides

$3 - y = 49$

$-y = 46$

$y = -46$

Check: $\dfrac{2 + \sqrt{3 - y} = 9}{2 + \sqrt{3 - (-46)}\ \bigg|\ 9}$

$\qquad\qquad 2 + \sqrt{49}\ \bigg|$

$\qquad\qquad\quad 2 + 7\ \bigg|$

$\qquad\qquad\qquad 9 \overset{?}{=} 9$ TRUE

The solution is -46.

13. $10 - 2\sqrt{3n} = 0$

$10 = 2\sqrt{3n}$ Adding $2\sqrt{3n}$

$5 = \sqrt{3n}$ Dividing by 2

$5^2 = \left(\sqrt{3n}\right)^2$ Squaring both sides

$25 = 3n$

$\dfrac{25}{3} = n$

Check:

$$\frac{10 - 2\sqrt{3n} = 0}{}$$

$$10 - 2\sqrt{3 \cdot \frac{25}{3}} \mid 0$$

$$10 - 2\sqrt{25}$$

$$10 - 2 \cdot 5$$

$$10 - 10$$

$$0 \stackrel{?}{=} 0 \quad \text{TRUE}$$

The solution is $\frac{25}{3}$.

15.
$$\sqrt{8t + 3} = \sqrt{6t + 7}$$
$$\left(\sqrt{8t + 3}\right)^2 = \left(\sqrt{6t + 7}\right)^2$$
$$8t + 3 = 6t + 7$$
$$2t + 3 = 7$$
$$2t = 4$$
$$t = 2$$

Check:
$$\frac{\sqrt{8t + 3} = \sqrt{6t + 7}}{}$$
$$\sqrt{8 \cdot 2 + 3} \mid \sqrt{6 \cdot 2 + 7}$$
$$\sqrt{16 + 3} \mid \sqrt{12 + 7}$$
$$\sqrt{19} \stackrel{?}{=} \sqrt{19} \quad \text{TRUE}$$

The solution is 2.

17. $5\sqrt{y} = -2$
$$\sqrt{y} = -\frac{2}{5}$$

Since the principle square roof of a number cannot be negative, we see that this equation has no solution.

19.
$$\sqrt{6 - 4t} = \sqrt{2 - 5t}$$
$$\left(\sqrt{6 - 4t}\right)^2 = \left(\sqrt{2 - 5t}\right)^2$$
$$6 - 4t = 2 - 5t$$
$$6 + t = 2$$
$$t = -4$$

Check:
$$\frac{\sqrt{6 - 4t} = \sqrt{2 - 5t}}{}$$
$$\sqrt{6 - 4(-4)} \mid \sqrt{2 - 5(-4)}$$
$$\sqrt{6 + 16} \mid \sqrt{2 + 20}$$
$$\sqrt{22} \stackrel{?}{=} \sqrt{22} \quad \text{TRUE}$$

The solution is -4.

21.
$$\sqrt{3x + 1} = x - 3$$
$$(\sqrt{3x + 1})^2 = (x - 3)^2$$
$$3x + 1 = x^2 - 6x + 9$$
$$0 = x^2 - 9x + 8$$
$$0 = (x - 1)(x - 8)$$
$$x - 1 = 0 \quad or \quad x - 8 = 0$$
$$x = 1 \quad or \quad x = 8$$

Check:
$$\frac{\sqrt{3x + 1} = x - 3}{}$$
$$\sqrt{3 \cdot 1 + 1} \mid 1 - 3$$
$$\sqrt{4} \mid -2$$
$$2 \stackrel{?}{=} -2 \quad \text{FALSE}$$

$$\frac{\sqrt{3x + 1} = x - 3}{}$$
$$\sqrt{3 \cdot 8 + 1} \mid 8 - 3$$
$$\sqrt{25} \mid 5$$
$$5 \stackrel{?}{=} 5 \quad \text{TRUE}$$

The number 1 does not check, but 8 does. The solution is 8.

23.
$$a - 9 = \sqrt{a - 3}$$
$$(a - 9)^2 = (\sqrt{a - 3})^2$$
$$a^2 - 18a + 81 = a - 3$$
$$a^2 - 19a + 84 = 0$$
$$(a - 12)(a - 7) = 0$$
$$a - 12 = 0 \quad or \quad a - 7 = 0$$
$$a = 12 \quad or \quad a = 7$$

Check:
$$\frac{a - 9 = \sqrt{a - 3}}{}$$
$$12 - 9 \mid \sqrt{12 - 3}$$
$$3 \mid \sqrt{9}$$
$$3 \stackrel{?}{=} 3 \quad \text{TRUE}$$

$$\frac{a - 9 = \sqrt{a - 3}}{}$$
$$7 - 9 \mid \sqrt{7 - 3}$$
$$-2 \mid \sqrt{4}$$
$$-2 \stackrel{?}{=} 2 \quad \text{FALSE}$$

The number 12 checks, but 7 does not. The solution is 12.

25.
$$x + 1 = 6\sqrt{x - 7}$$
$$(x + 1)^2 = (6\sqrt{x - 7})^2$$
$$x^2 + 2x + 1 = 36(x - 7)$$
$$x^2 + 2x + 1 = 36x - 252$$
$$x^2 - 34x + 253 = 0$$
$$(x - 11)(x - 23) = 0$$
$$x - 11 = 0 \quad or \quad x - 23 = 0$$
$$x = 11 \quad or \quad x = 23$$

Check:
$$\frac{x + 1 = 6\sqrt{x - 7}}{}$$
$$11 + 1 \mid 6\sqrt{11 - 7}$$
$$12 \mid 6\sqrt{4}$$
$$\mid 6 \cdot 2$$
$$12 \stackrel{?}{=} 12 \quad \text{TRUE}$$

$$x + 1 = 6\sqrt{x - 7}$$

$23 + 1$	$6\sqrt{23 - 7}$
24	$6\sqrt{16}$
	$6 \cdot 4$
$24 \overset{?}{=} 24$	TRUE

The solutions are 11 and 23.

27.
$$\sqrt{5x + 21} = x + 3$$
$$(\sqrt{5x + 21})^2 = (x + 3)^2$$
$$5x + 21 = x^2 + 6x + 9$$
$$0 = x^2 + x - 12$$
$$0 = (x + 4)(x - 3)$$
$$x + 4 = 0 \quad or \quad x - 3 = 0$$
$$x = -4 \quad or \quad x = 3$$

Check:

$\sqrt{5x + 21} = x + 3$	
$\sqrt{5(-4) + 21}$	$-4 + 3$
$\sqrt{1}$	-1
$1 \overset{?}{=} -1$	FALSE

$\sqrt{5x + 21} = x + 3$	
$\sqrt{5 \cdot 3 + 21}$	$3 + 3$
$\sqrt{36}$	6
$6 \overset{?}{=} 6$	TRUE

The number 3 checks, but -4 does not. The solution is 3.

29.
$$t + 4 = 4\sqrt{t + 1}$$
$$(t + 4)^2 = (4\sqrt{t + 1})^2$$
$$t^2 + 8t + 16 = 16(t + 1)$$
$$t^2 + 8t + 16 = 16t + 16$$
$$t^2 - 8t = 0$$
$$t(t - 8) = 0$$
$$t = 0 \quad or \quad t - 8 = 0$$
$$t = 0 \quad or \quad t = 8$$

Check:

$t + 4 = 4\sqrt{t + 1}$	
$0 + 4$	$4\sqrt{0 + 1}$
4	$4\sqrt{1}$
	$4 \cdot 1$
$4 \overset{?}{=} 4$	TRUE

$t + 4 = 4\sqrt{t + 1}$	
$8 + 4$	$4\sqrt{8 + 1}$
12	$4\sqrt{9}$
	$4 \cdot 3$
$12 \overset{?}{=} 12$	TRUE

The solutions are 0 and 8.

31.
$$\sqrt{x^2 + 6} - x + 3 = 0$$
$$\sqrt{x^2 + 6} = x - 3 \quad \text{Isolating the radical}$$
$$(\sqrt{x^2 + 6})^2 = (x - 3)^2$$
$$x^2 + 6 = x^2 - 6x + 9$$
$$-3 = -6x \quad \text{Adding } -x^2 \text{ and } -9$$
$$\frac{1}{2} = x$$

Check:

$\sqrt{x^2 + 6} - x + 3 = 0$	
$\sqrt{\left(\frac{1}{2}\right)^2 + 6} - \frac{1}{2} + 3$	0
$\sqrt{\frac{25}{4}} - \frac{1}{2} + 3$	
$\frac{5}{2} - \frac{1}{2} + 3$	
$2 + 3$	
$5 \overset{?}{=} 0$ FALSE	

The number $\frac{1}{2}$ does not check. There is no solution.

33.
$$\sqrt{(4x + 5)(x + 4)} = 2x + 5$$
$$\left(\sqrt{(4x + 5)(x + 4)}\right)^2 = (2x + 5)^2$$
$$(4x + 5)(x + 4) = 4x^2 + 20x + 25$$
$$4x^2 + 21x + 20 = 4x^2 + 20x + 25$$
$$x = 5$$

The number 5 checks. It is the solution.

35.
$$\sqrt{8 - 3x} = \sqrt{13 + x}$$
$$\left(\sqrt{8 - 3x}\right)^2 = \left(\sqrt{13 + x}\right)^2$$
$$8 - 3x = 13 + x$$
$$-5 = 4x$$
$$-\frac{5}{4} = x$$

The number $\frac{-5}{4}$ checks. It is the solution.

37.
$$x = 1 + \sqrt{1 - x}$$
$$x - 1 = \sqrt{1 - x}$$
$$(x - 1)^2 = \left(\sqrt{1 - x}\right)^2$$
$$x^2 - 2x + 1 = 1 - x$$
$$x^2 - x = 0$$
$$x(x - 1) = 0$$
$$x = 0 \quad or \quad x - 1 = 0$$
$$x = 0 \quad or \quad x = 1$$

Only 1 checks. The number 0 does not check. The solution is 1.

39. $r = 2\sqrt{5L}$

$40 = 2\sqrt{5L}$ Substituting 40 for r

$20 = \sqrt{5L}$

$20^2 = (\sqrt{5L})^2$

$400 = 5L$

$80 = L$

The car will skid 80 ft at 40 mph.

$60 = 2\sqrt{5L}$ Substituting 60 for r

$30 = \sqrt{5L}$

$30^2 = (\sqrt{5L})^2$

$900 = 5L$

$180 = L$

The car will skid 180 ft at 60 mph.

41. Familiarize. We will use the formula $s = 21.9\sqrt{5t + 2457}$, where t is in degrees Fahrenheit and s is in feet per second.

Translate. We substitute 1113 for s in the formula.

$1113 = 21.9\sqrt{5t + 2457}$

Carry out. We solve for t.

$1113 = 21.9\sqrt{5t + 2457}$

$\dfrac{1113}{21.9} = \sqrt{5t + 2457}$

$\left(\dfrac{1113}{21.9}\right)^2 = (\sqrt{5t + 2457})^2$

$2582.9 \approx 5t + 2457$

$125.9 \approx 5t$

$25.2 \approx t$

Check. We can substitute 25.2 for t in the formula.

$21.9\sqrt{5(25.2) + 2457} = 21.9\sqrt{2583} \approx 1113$

The answer checks.

State. The temperature was about $25.2°$F.

43. Familiarize and Translate. We substitute 56 for V in the equation $V = 3.5\sqrt{h}$.

Carry out.
$56 = 3.5\sqrt{h}$

$16 = \sqrt{h}$

$16^2 = \left(\sqrt{h}\right)^2$

$256 = h$

If h is 256, then $3.5\sqrt{256}$ is 56. The answer checks.

State. The attitude of the scout's eyes is 256 m.

45. Familiarize and Translate. We substitute 62 for v in the equation $v = 3.1\sqrt{d}$.

$62 = 3.1\sqrt{d}$

Carry out.
$62 = 3.1\sqrt{d}$

$20 = \sqrt{d}$

$20^2 = \left(\sqrt{d}\right)^2$

$400 = d$

Check. If d is 400, then $3.1\sqrt{400}$ is 62. The answer checks.

State. The water depth is 400 m.

47. $T = 2\pi\sqrt{\dfrac{L}{32}}$

$4.4 = 2(3.14)\sqrt{\dfrac{L}{32}}$ Substituting 4.4 for T and 3.14 for π

$4.4 = 6.28\sqrt{\dfrac{L}{32}}$

$\dfrac{4.4}{6.28} = \sqrt{\dfrac{L}{32}}$

$\left(\dfrac{4.4}{6.28}\right)^2 = \left(\sqrt{\dfrac{L}{32}}\right)^2$

$0.4909 \approx \dfrac{L}{32}$

$15.71 \approx L$

The pendulum is about 15.71 ft long.

49. *Writing Exercise.* Yes; if $a = b$, then by the multiplication principle we can multiply on the left by a n times and on the right by b n times to produce an equivalent equation.

51. Graph $y = 2x - 1$. We make a table of values and plot these points to find the graph.

x	y
-1	-3
0	-1
1	1
2	3

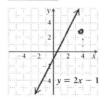

53. Graph $x = 3$

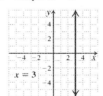

55. Graph $y < \dfrac{1}{3}x - 2$.

First graph the line $y = \dfrac{1}{3}x - 2$. The intercepts are $(0, -2)$ and $(6, 0)$. We draw a dashed line since the inequality symbol is $<$. Then we test the point $(0, 0)$.

Check: $y < \dfrac{1}{3}x - 2$

$$
\begin{array}{c|c}
0 & \dfrac{1}{3} \cdot 0 - 2 \\[2mm]
  & -2 \\[2mm]
  & \overset{?}{} \\
0 & < -2 \qquad \text{FALSE}
\end{array}
$$

Since $(0,0)$ is not a solution of the inequality, we shade the region that does not contain $(0,0)$.

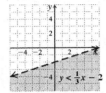

57. No he is not correct. The principle square root of a number cannot be negative. This is different from the solution of a radical equation. For example, in Exercise 11, the solution is -46.

59. *Familiarize*. Let x represent the number. Then three times its square root is $3\sqrt{x}$ and the opposite of three times its square root is $-3\sqrt{x}$.

Translate. We reword the problem.

$$
\underbrace{\text{The opposite of three times}}_{\downarrow \atop -3\sqrt{x}} \text{ is } -33.
$$

The opposite of three times the square root of a number   is -33.

$$-3\sqrt{x} \qquad = \quad -33$$

Carry out. We solve the equation.

$$
\begin{aligned}
-3\sqrt{x} &= -33 \\
\sqrt{x} &= 11 \qquad \text{Dividing by } -3 \\
(\sqrt{x})^2 &= 11^2 \\
x &= 121
\end{aligned}
$$

Check. $\sqrt{121} = 11$ and $-3 \cdot 11 = -33$. The answer checks.

State. The number is 121.

61.
$$
\begin{aligned}
1 + \sqrt{x} &= \sqrt{x + 9} \\
(1 + \sqrt{x})^2 &= (\sqrt{x + 9})^2 \\
1 + 2\sqrt{x} + x &= x + 9 \\
1 + 2\sqrt{x} &= 9 \qquad \text{Adding } -x \\
2\sqrt{x} &= 8 \\
\sqrt{x} &= 4 \\
(\sqrt{x})^2 &= 4^2 \\
x &= 16
\end{aligned}
$$

The number 16 checks. It is the solution.

63.
$$
\begin{aligned}
\sqrt{t + 4} &= 1 - \sqrt{3t + 1} \\
(\sqrt{t + 4})^2 &= (1 - \sqrt{3t + 1})^2 \\
t + 4 &= 1 - 2\sqrt{3t + 1} + 3t + 1 \\
t + 4 &= 2 - 2\sqrt{3t + 1} + 3t \\
-2t + 2 &= -2\sqrt{3t + 1} \\
& \qquad \text{Isolating the radical} \\
t - 1 &= \sqrt{3t + 1} \\
& \qquad \text{Multiplying by } -\dfrac{1}{2} \\
(t - 1)^2 &= (\sqrt{3t + 1})^2 \\
t^2 - 2t + 1 &= 3t + 1 \\
t^2 - 5t &= 0 \\
t(t - 5) &= 0 \\
t = 0 \quad &or \quad t - 5 = 0 \\
t = 0 \quad &or \qquad t = 5
\end{aligned}
$$

Check:

$$
\begin{array}{c|c}
\multicolumn{2}{c}{\sqrt{t + 4} = 1 - \sqrt{3t + 1}} \\
\hline
\sqrt{0 + 4} & 1 - \sqrt{3 \cdot 0 + 1} \\
\sqrt{4} & 1 - \sqrt{1} \\
2 & 1 - 1 \\
 & \overset{?}{} \\
2 &= 0 \qquad \text{FALSE}
\end{array}
$$

$$
\begin{array}{c|c}
\multicolumn{2}{c}{\sqrt{t + 4} = 1 - \sqrt{3t + 1}} \\
\hline
\sqrt{5 + 4} & 1 - \sqrt{3 \cdot 5 + 1} \\
\sqrt{9} & 1 - \sqrt{16} \\
3 & 1 - 4 \\
 & \overset{?}{} \\
3 &= -3 \qquad \text{FALSE}
\end{array}
$$

Neither number checks. There is no solution.

65.
$$
\begin{aligned}
\sqrt{y + 1} - \sqrt{y - 2} &= \sqrt{2y - 5} \\
(\sqrt{y + 1} - \sqrt{y - 2})^2 &= (\sqrt{2y - 5})^2 \\
y + 1 - 2\sqrt{(y + 1)(y - 2)} + y - 2 &= 2y - 5 \\
2y - 1 - 2\sqrt{(y + 1)(y - 2)} &= 2y - 5 \\
-2\sqrt{(y + 1)(y - 2)} &= -4 \quad \text{Adding } -2y \text{ and } 1 \\
\sqrt{(y + 1)(y - 2)} &= 2 \quad \text{Dividing by } -2 \\
\left(\sqrt{(y + 1)(y - 2)}\right)^2 &= 2^2 \\
(y + 1)(y - 2) &= 4 \\
y^2 - y - 2 &= 4 \\
y^2 - y - 6 &= 0 \\
(y - 3)(y + 2) &= 0 \\
y - 3 = 0 \quad &or \quad y + 2 = 0 \\
y = 3 \quad &or \qquad y = -2
\end{aligned}
$$

The number 3 checks, but -2 does not. The solution is 3.

67.

$$2\sqrt{x-1} - \sqrt{x-9} = \sqrt{3x-5}$$
$$(2\sqrt{x-1} - \sqrt{x-9})^2 = (\sqrt{3x-5})^2$$
$$4(x-1) - 4\sqrt{(x-1)(x-9)} + x - 9 = 3x - 5$$
$$4x - 4 - 4\sqrt{x^2 - 10x + 9} + x - 9 = 3x - 5$$
$$5x - 13 - 4\sqrt{x^2 - 10x + 9} = 3x - 5$$
$$-4\sqrt{x^2 - 10x + 9} = -2x + 8$$
$$2\sqrt{x^2 - 10x + 9} = x - 4$$
$$\text{Multiplying by } -\frac{1}{2}$$
$$(2\sqrt{x^2 - 10x + 9})^2 = (x-4)^2$$
$$4(x^2 - 10x + 9) = x^2 - 8x + 16$$
$$4x^2 - 40x + 36 = x^2 - 8x + 16$$
$$3x^2 - 32x + 20 = 0$$
$$(3x - 2)(x - 10) = 0$$
$$3x - 2 = 0 \quad or \quad x - 10 = 0$$
$$3x = 2 \quad or \quad x = 10$$
$$x = \frac{2}{3} \quad or \quad x = 10$$

The number 10 checks, but $\frac{2}{3}$ does not. The solution is 10.

69. Familiarize. We will use the formula $V = 3.5\sqrt{h}$. We present the information in a table.

	Height	Distance to the horizon
First sighting	h	V
Second sighting	$h + 100$	$V + 20$

Translate. The rows of the table give us two equations.

$$V = 3.5\sqrt{h}, \qquad (1)$$
$$V + 20 = 3.5\sqrt{h + 100} \quad (2)$$

Carry out. We substitute $3.5\sqrt{h}$ for V in Equation (2) and solve for h.

$$3.5\sqrt{h} + 20 = 3.5\sqrt{h + 100}$$
$$(3.5\sqrt{h} + 20)^2 = (3.5\sqrt{h + 100})^2$$
$$12.25h + 140\sqrt{h} + 400 = 12.25(h + 100)$$
$$12.25h + 140\sqrt{h} + 400 = 12.25h + 1225$$
$$140\sqrt{h} = 825$$
$$28\sqrt{h} = 165 \quad \text{Multiplying by } \frac{1}{5}$$
$$(28\sqrt{h})^2 = (165)^2$$
$$784h = 27,225$$
$$h = 34\frac{569}{784}, \text{ or}$$
$$h \approx 34.726$$

Check. When $h \approx 34.726$, then $V \approx 3.5\sqrt{34.726} \approx 20.625$ km. When $h \approx 100 + 34.726$, or 134.726, then $V \approx 3.5\sqrt{134.726} \approx 40.625$ km. This is 20 km more than 20.625. The answer checks.

State. The climber was at a height of $34\frac{569}{784}$ m, or about 34.726 m when the first computation was made.

71. Graph $y = \sqrt{x}$.

We make a table of values. Note that we must choose nonnegative values of x in order to have a nonnegative radicand.

x	y
0	0
1	1
2	1.414
4	2
5	2.236

We plot these points and connect them with a smooth curve.

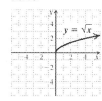

73. Graph $y = \sqrt{x + 2}$.

We make a table of values. Note that we must choose values for x that are greater than or equal to -2 in order to have a nonnegative radicand.

x	y
4	0
5	1
6	1.414
7	1.732
8	2

We plot these points and connect them with a smooth curve.

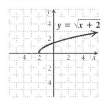

75. Given a car that is 15 ft long and a speed r, in mph, one car length per 10 mph of speed is represented by $15 \cdot \frac{r}{10}$, or $\frac{3r}{2}$. Substitute $\frac{3r}{2}$ for L in the formula and solve for r.

$$r = 2\sqrt{5L}$$

$$r = 2\sqrt{5 \cdot \frac{3r}{2}}$$

$$r = 2\sqrt{\frac{15r}{2}}$$

$$r^2 = \left(2\sqrt{\frac{15r}{2}}\right)^2$$

$$r^2 = 4 \cdot \frac{15r}{2}$$

$$r^2 = 30r$$

$$r^2 - 30r = 0$$

$$r(r - 30) = 0$$

$$r = 0 \ \ or \ \ r - 30 = 0$$

$$r = 0 \ \ or \ \ \ \ \ \ \ r = 30$$

The number 0 has no meaning in this problem. The number 30 checks, so the answer is 30 mph.

77. First we make a table of values for each equation.

For $y = 1 + \sqrt{x}$:   For $y = \sqrt{x + 9}$:

x	y
0	1
1	2
4	3
9	4
16	5

x	y
-9	0
-8	1
-5	2
0	3
7	4

We graph the equations.

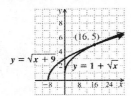

The graphs intersect at $(16, 5)$, so the solution of $1 + \sqrt{x} = \sqrt{x + 9}$ is 16.

79. Graph $y_1 = -\sqrt{x + 3}$ and $y_2 = 2x - 1$ and then find the first coordinate(s) of the point(s) of intersection. The solution is about -0.32.

Chapter 8 Connecting the Concepts

1. $\sqrt{100t^2} = 10t$

3. $\sqrt{x} + 1 = 7$

$\ \ \ \ \sqrt{x} = 6$

$\ \ \ \ (\sqrt{x})^2 = 6^2$ Squaring both sides

$\ \ \ \ \ \ \ \ x = 36$

Check:

$$\frac{\sqrt{x} + 1 = 7}{\begin{array}{c|c} \sqrt{36} + 1 & 7 \\ 6 + 1 & \\ & \overset{?}{} \\ 7 \overset{?}{=} 7 & \text{TRUE} \end{array}}$$

The solution is 36.

5. $\sqrt{15t} + 4\sqrt{15t} = (1 + 4)\sqrt{15t} = 5\sqrt{15t}$

7. $\sqrt{6}\left(\sqrt{10} - \sqrt{33}\right) = \sqrt{60} - \sqrt{198}$

$= \sqrt{4 \cdot 15} - \sqrt{9 \cdot 22}$

$= 2\sqrt{15} - 3\sqrt{22}$

9. $2\sqrt{3} - 5\sqrt{12} = 2\sqrt{3} - 10\sqrt{3} = -8\sqrt{3}$

11. $\sqrt{4n - 5} = \sqrt{n + 1}$

$\left(\sqrt{4n - 5}\right)^2 = \left(\sqrt{n + 1}\right)^2$

$4n - 5 = n + 1$

$4n = n + 6$

$3n = 6$

$n = 2$

Check:

$$\frac{\sqrt{4n - 5} = \sqrt{n + 1}}{\begin{array}{c|c} \sqrt{4 \cdot 2 - 5} & \sqrt{2 + 1} \\ \sqrt{8 - 5} & \sqrt{3} \\ & \overset{?}{} \\ \sqrt{3} \overset{?}{=} \sqrt{3} & \text{TRUE} \end{array}}$$

The solution is 2.

13. $4 = 3\sqrt{2x}$

$\frac{4}{3} = \sqrt{2x}$

$\left(\frac{4}{3}\right)^2 = \left(\sqrt{2x}\right)^2$

$\frac{16}{9} = 2x$

$\frac{8}{9} = x$

Check:

$$\frac{4 = 3\sqrt{2x}}{\begin{array}{c|c} 4 & 3\sqrt{2 \cdot \frac{8}{9}} \\ & 3\sqrt{\frac{16}{9}} \\ & 3 \cdot \frac{4}{3} \\ 4 \overset{?}{=} 4 & \text{TRUE} \end{array}}$$

The solution is $\frac{8}{9}$.

15. $\left(\sqrt{5} + 1\right)\left(\sqrt{10} + 3\right) = \sqrt{50} + 3\sqrt{5} + \sqrt{10} + 3$

$= 5\sqrt{2} + 3\sqrt{5} + \sqrt{10} + 3$

17.
$$\sqrt{5x+6} = x+2$$
$$\left(\sqrt{5x+6}\right)^2 = (x+2)^2$$
$$5x+6 = x^2+4x+4$$
$$0 = x^2-x-2$$
$$0 = (x-2)(x+1)$$
$$x-2=0 \text{ or } x+1=0$$
$$x=2 \text{ or } x=-1$$

Check:

$\sqrt{5x+6} = x+2$	
$\sqrt{5\cdot 2+6}$	$2+2$
$\sqrt{10+6}$	4
$\sqrt{16}$	
$4 \overset{?}{=} 4$	TRUE

Check:

$\sqrt{5x+6} = x+2$	
$\sqrt{5(-1)+6}$	$-1+2$
$\sqrt{-5+6}$	1
$\sqrt{1}$	
$1 \overset{?}{=} 1$	TRUE

The solution are -1 and 2.

19.
$$x-1 = 2\sqrt{x-2}$$
$$(x-1)^2 = \left(2\sqrt{x-2}\right)^2$$
$$x^2-2x+1 = 4(x-2)$$
$$x^2-2x+1 = 4x-8$$
$$x^2-6x+9 = 0$$
$$(x-3)(x-3) = 0$$
$$x-3=0 \text{ or } x-3=0$$
$$x=3 \text{ or } x=3$$

Check:

$x-1 = 2\sqrt{x-2}$	
$3-1$	$2\sqrt{3-2}$
2	$2\sqrt{1}$
$2 \overset{?}{=} 2$	TRUE

The solution is 3.

Exercise Set 8.6

1. hypotenuse; see page 512 in the text.

3. Pythagorean; see page 512 in the text.

5.
$$a^2+b^2 = c^2$$
$$8^2+15^2 = x^2 \quad \text{Substituting}$$
$$64+225 = x^2$$
$$289 = x^2$$
$$\sqrt{289} = x$$
$$17 = x$$

7.
$$a^2+b^2 = c^2$$
$$7^2+7^2 = x^2$$
$$49+49 = x^2$$
$$98 = x^2$$
$$\sqrt{98} = x \qquad \text{Exact answer}$$
$$9.899 \approx x \qquad \text{Approximation}$$

9.
$$a^2+b^2 = c^2$$
$$5^2+x^2 = 13^2$$
$$25+x^2 = 169$$
$$x^2 = 144$$
$$x = 12$$

11.
$$a^2+b^2 = c^2$$
$$\left(\sqrt{5}\right)^2+b^2 = 6^2$$
$$5+x^2 = 36$$
$$x^2 = 31$$
$$x = \sqrt{31}$$
$$x \approx 5.568$$

13.
$$a^2+b^2 = c^2$$
$$12^2+5^2 = c^2$$
$$144+25 = c^2$$
$$169 = c^2$$
$$13 = c$$

15.
$$a^2+b^2 = c^2$$
$$9^2+b^2 = 15^2$$
$$81+b^2 = 225$$
$$b^2 = 144$$
$$b = 12$$

17.
$$a^2+b^2 = c^2$$
$$a^2+1^2 = \left(\sqrt{10}\right)^2$$
$$a^2+1 = 10$$
$$a^2 = 9$$
$$a = 3$$

19.
$$a^2+b^2 = c^2$$
$$1^2+b^2 = (\sqrt{3})^2$$
$$1+b^2 = 3$$
$$b^2 = 2$$
$$b = \sqrt{2} \qquad \text{Exact answer}$$
$$b \approx 1.414 \quad \text{Approximation}$$

21.
$$a^2+b^2 = c^2$$
$$a^2+(5\sqrt{3})^2 = 10^2$$
$$a^2+25\cdot 3 = 100$$
$$a^2+75 = 100$$
$$a^2 = 25$$
$$a = 5$$

23. Familiarize. Referring to the drawing in the text, let l represent the length of the string of lights.

Translate. We use the Pythagorean theorem, substituting 8 for a, 15 for b, and l for c.

$$8^2 + 15^2 = l^2.$$

Carry out. We solve the equation.

$$8^2 + 15^2 = l^2$$
$$64 + 225 = l^2$$
$$289 = l^2$$
$$17 = l$$

Check. We check by substituting 8, 15, and 17 in the Pythagorean theorem.

$a^2 + b^2 = c^2$	
$8^2 + 15^2$	17^2
$64 + 225$	289

$$289 \overset{?}{=} 289 \quad \text{TRUE}$$

State. The string of lights needs to be 17 ft long.

25. Familiarize. Let h = the height of the back of the jump, in inches.

Translate. We use the Pythagorean theorem, substituting 30 for a, h for b, and 33 for c.

$$30^2 + h^2 = 33^2$$

Carry out. We solve the equation.

$$30^2 + h^2 = 33^2$$
$$900 + h^2 = 1089$$
$$h^2 = 189$$
$$h = \sqrt{189} = \sqrt{9 \cdot 21}$$
$$h = 3\sqrt{21} \quad \text{Exact answer}$$
$$h \approx 13.748$$

Check. We check by substituting 30, $\sqrt{189}$, and 33 in the Pythagorean theorem.

$a^2 + b^2 = c^2$	
$30^2 + (\sqrt{189})^2$	33^2
$900 + 189$	1089

$$1089 \overset{?}{=} 1089 \quad \text{TRUE}$$

State. The back of the jump should be $3\sqrt{21}$ in., or about 13.748 in. high.

27. Familiarize. We first make a drawing. We label the unknown length w.

w 12 ft

8 ft

Translate. We use the Pythagorean theorem, substituting 8 for a, 12 for b, and w for c.

$$8^2 + 12^2 = w^2$$

Carry out. We solve the equation.

$$8^2 + 12^2 = w^2$$
$$64 + 144 = w^2$$
$$208 = w^2$$
$$\sqrt{208} = w$$
$$\sqrt{16 \cdot 13} = w$$
$$4\sqrt{13} = w \quad \text{Exact answer}$$
$$14.422 \approx w \quad \text{Approximation}$$

Check. We check by substituting 8, 12, and $\sqrt{208}$ into the Pythagorean theorem:

$a^2 + b^2 = c^2$	
$8^2 + 12^2$	$(\sqrt{208})^2$
$64 + 144$	208

$$208 \overset{?}{=} 208 \quad \text{TRUE}$$

State. The pipe should be $4\sqrt{13}$ feet or about 14.422 feet long.

29. Familiarize. Let d = the distance from first base to third base, in feet.

Translate. We use the Pythagorean theorem, substituting 90 for a, 90 for b, and d for c.

$$90^2 + 90^2 = d^2$$

Carry out. We solve the equation.

$$90^2 + 90^2 = d^2$$
$$8100 + 8100 = d^2$$
$$16,200 = d^2$$
$$\sqrt{16,200} = d$$
$$\sqrt{8100 \cdot 2} = d$$
$$90\sqrt{2} = d \quad \text{Exact answer}$$
$$127.279 \approx d$$

Check. We check by substituting 90, 90, and $\sqrt{16,200}$ into the Pythagorean theorem.

$a^2 + b^2 = c^2$	
$90^2 + 90^2$	$(\sqrt{16,200})^2$
$8100 + 8100$	$16,200$

$$16,200 \overset{?}{=} 16,200 \quad \text{TRUE}$$

State. It is $90\sqrt{2}$ ft, or about 127.279 ft, from first base to third base.

31. Familiarize: Let d = the length of the diagonal in yds.

Translate. We use the Pythagorean theorem, substituting in $a^2 + b^2 = c^2$
Carry out. We solve the equation.

$$60^2 + 110^2 = c^2$$
$$3600 + 12,100 = c^2$$
$$15,700 = c^2$$
$$\sqrt{15,700} = c$$
$$125.300 \text{ yd} \approx c$$

Check:

$$\frac{a^2 + b^2 = c^2}{60^2 + 110^2 \quad \left(\sqrt{15,700}\right)^2}$$
$$3600 + 12,100 \quad 15,700$$
$$15,700 \overset{?}{=} 15,700 \qquad \text{TRUE}$$

State. The length of the diagonal is $\sqrt{15,700}$ yds or about 125.300 yds long.

33. Familiarize: Let d = the length of the diagonal in cm.

Translate: We use the Pythagorean theorem, substituting in $a^2 + b^2 = c^2$
Carry out. We solve the equation.

$$30^2 + 60^2 = c^2$$
$$900 + 3600 = c^2$$
$$4500 = c^2$$
$$\sqrt{4500} \text{ cm} = c$$
$$67.082 \text{ cm} \approx c$$

Check. We check by substituting 30, 60 and $\sqrt{4500}$ in the Pythagorean theorem.

Check:

$$\frac{a^2 + b^2 = c^2}{30^2 + 60^2 \quad \left(\sqrt{4500}\right)}$$
$$900 + 3600 \quad 4500$$
$$4500 \overset{?}{=} 4500 \qquad \text{TRUE}$$

State. The length of a diagonal is $\sqrt{4500}$ cm, or about 67.082 cm.

35. Familiarize. We make a drawing. We let h = the height the hose can reach.

Translate. We use the Pythagorean theorem, substituting h for a, 24 for b, and 32 for c.
$$h^2 + 24^2 = 32^2$$
Carry out. We solve the equation.
$$h^2 + 24^2 = 32^2$$
$$h^2 + 576 = 1024$$
$$h^2 = 448$$
$$h = \sqrt{448} \qquad \text{Exact answer}$$
$$h \approx 21.166$$

Check. We check by substituting $\sqrt{448}$ for a, 24 for b, and 32 for c in the Pythagorean equation.

$$\frac{a^2 + b^2 = c^2}{\left(\sqrt{448}\right)^2 + 24^2 \quad 32^2}$$
$$448 + 576 \quad 1024$$
$$1024 \overset{?}{=} 1024 \quad \text{TRUE}$$

State. The hose can reach $\sqrt{448}$ ft, or about 21.166 ft, up the far corner of the house.

37.
$$d = \sqrt{(x_2 - x_1)^2 + (y_2 - y_1)^2}$$
$$d = \sqrt{(6 - 2)^2 + (10 - 3)^2}$$
$$d = \sqrt{4^2 + 7^2}$$
$$d = \sqrt{16 + 49}$$
$$d = \sqrt{65} \qquad \text{This is exact}$$
$$d \approx 8.062 \qquad \begin{array}{l}\text{Using a calculator}\\ \text{for an approximation}\end{array}$$

39.
$$d = \sqrt{(x_2 - x_1)^2 + (y_2 - y_1)^2}$$
$$= \sqrt{(4 - 0)^2 + (0 - 3)^2} \qquad \text{Substituting}$$
$$= \sqrt{4^2 + (-3)^2}$$
$$= \sqrt{16 + 9}$$
$$= \sqrt{25}$$
$$= 5$$

41.
$$d = \sqrt{(x_2 - x_1)^2 + (y_2 - y_1)^2}$$
$$= \sqrt{(-1 - (-3))^2 + (5 - 2)^2}$$
$$= \sqrt{2^2 + 3^2}$$
$$= \sqrt{4 + 9}$$
$$= \sqrt{13} \qquad \text{This is exact}$$
$$\approx 3.606 \qquad \begin{array}{l}\text{Using a calculator}\\ \text{for an approximation}\end{array}$$

43.
$$d = \sqrt{(x_2 - x_1)^2 + (y_2 - y_1)^2}$$
$$= \sqrt{(-8 - (-2))^2 + (-4 - 4)^2} \qquad \text{Substituting}$$
$$= \sqrt{(-6)^2 + (-8)^2}$$
$$= \sqrt{36 + 64}$$
$$= \sqrt{100}$$
$$= 10$$

45. *Writing Exercise.* Yes; consider an isosceles triangle with two sides of length s. Then a triangle whose third side has length $s^2 + s^2$, or $2s^2$, is an isosceles right triangle.

47. $(-2)^5 = (-2)(-2)(-2)(-2)(-2) = -32$

49. $\left(\dfrac{2}{3}\right)^3 = \dfrac{2}{3} \cdot \dfrac{2}{3} \cdot \dfrac{2}{3} = \dfrac{8}{27}$

51. $3^4 = 3 \cdot 3 \cdot 3 \cdot 3 = 81$

53. $2^4 = 2 \cdot 2 \cdot 2 \cdot 2 = 16$

55. $(-4)^3 = (-4) \cdot (-4) \cdot (-4) = -64$

57. *Writing Exercise.* Yes; consider the ladder to be the hypotenuse of a right triangle. If the base of the ladder is positioned 2 ft from the house, then the ladder touches the house about 27.9 ft above the ground. From a position near the top of the ladder, the homeowner could make the repair.

59. *Writing Exercise.* Consider a right triangle with legs 3 units and 4 units, we can use the rope to measure the first leg 3 units and the second leg of 4 units. Completing the triangle, we have a hypotenuse of 5 units. Therefore a right triangle is formed.

61. *Familiarize.* The base of the triangle is $\frac{1}{2}$ the base of the square side or 352 ft. The sloping side is the hypotenuse of the triangle. The height h is one leg of the right triangle.

Translate. We use the Pythagorean Theorem.
$$a^2 + b^2 = c^2$$
$$352^2 + h^2 = 588^2$$

Carry out. We solve the equation.
$$352^2 + h^2 = 588^2$$
$$123904 + h^2 = 345744$$
$$h^2 = 221840$$
$$h = \sqrt{221840}$$
$$h \approx 471 ft.$$

Check. We check by substituting in Pythagorean Theorem.

$a^2 + b^2 = c^2$	
$352^2 + (\sqrt{221840})^2$	588^2
$123,904 + 221,840$	$345,744$

$$345,744 \stackrel{?}{=} 345,744 \quad \text{TRUE}$$

State. The pyramid is $\sqrt{221,840}$ ft or 471 ft high.

63. *Familiarize.* Find the vertical distance $32,000 - 21,000$ or $11,000$ ft, let d = distance traveled.

Convert 5 miles to feet
$$5 \text{ miles} = 5 \text{ mi} \times \frac{5280 \text{ ft}}{1 \text{ mi}} = 26,400 \text{ ft}$$

Translate. Using the Pythagorean theorem.
$a^2 + b^2 = c^2$, we have
$$11,000^2 + 26400^2 = d^2$$

Carry out. Solve the equation.
$$11000^2 + 26400^2 = d^2$$
$$121,000,000 + 696,960,000 = d^2$$
$$817,960,000 = d^2$$
$$\sqrt{817,960,000} = d$$
$$28600 \text{ ft} = d$$

Check. We check by substituting in Pythagorean theorem.

$a^2 + b^2 = c^2$	
$11000^2 + 26400^2$	28600^2

$$817,960,000 \stackrel{?}{=} 817,960,000 \quad \text{TRUE}$$

State. The distance traveled by the plan is 28,600 ft.

65. *Familiarize* Let s = the length of a side of the square in m.

Translate. We use the formula for the area of a square, $A = s^2$

$$\underbrace{\text{Area}}_{7} \quad \underset{=}{\text{is}} \quad \underset{s^2}{s^2}$$

Carry out. We solve the equation.
$$7 = s^2$$
$$\sqrt{7} \text{ m} = s$$
$$2.646 \text{ m} \approx s$$

Check. We substitute into the formula.

$A = s^2$	
7	$(\sqrt{7})^2$

$$7 \stackrel{?}{=} 7 \quad \text{TRUE}$$

State. The length of a side of the square is $\sqrt{7}$ m.

67. *Familiarize.* Let d = the diagonal $\overline{AC}$

Translate. We use the Pythagorean theorem.
$$\left(\frac{\sqrt{2}}{3}\right)^2 + \left(\frac{\sqrt{2}}{3}\right)^2 = d^2$$

Carry out. We solve the equation.
$$\frac{2}{9} + \frac{2}{9} = d^2$$
$$\frac{4}{9} = d^2$$
$$\frac{2}{3} = d$$

Check. We substitute in Pythagorean Theorem.

$a^2 + b^2 = c^2$	
$\left(\frac{\sqrt{2}}{3}\right)^2 + \left(\frac{\sqrt{2}}{3}\right)^2$	$\left(\frac{2}{3}\right)$
$\frac{2}{9} + \frac{2}{9}$	$\frac{4}{9}$

$$12 \stackrel{?}{=} 12 \quad \text{TRUE}$$

State. The diagonal $\overline{AC}$ is $\frac{2}{3}$.

69. *Familiarize*. If the area of square $PQRS$ is $100ft^2$, then each side measures 10 ft. If A, B, C, and D are midpoints, then each segment PB, BQ, QC, CR, RD, DS, SA and AP measures 5 ft. Let d = diagonal $AD = CD = BC = BA$

Translate. We use the Pythagorean theorem.
$$a^2 + b^2 = c^2$$
$$5^2 + 5^2 = d^2 \quad \text{substituting}$$

Carry out. We solve the equation.
$$5^2 + 5^2 = d^2$$
$$25 + 25 = d^2$$
$$50 = d^2$$
$$\sqrt{50} = d$$

Using the formula for the area of square
$A = \left(\sqrt{50}\right)^2 = 50 \text{ ft}^2$

Check. We substitute into Pythagorean theorem

$$a^2 + b^2 = c^2$$

$$\begin{array}{c|c} 5^2 + 5^2 & \left(\sqrt{50}\right)^2 \\ 25 + 25 & 50 \\ \hline & \overset{?}{50} = 50 \quad \text{TRUE} \end{array}$$

State. If a side of square $ABCD$ is $\sqrt{50}$, then the area is 50 ft^2.

71. Familiarize. Let l = diagonal in ft. of the base. Let d = diagonal ft of room.

Translate. We use the Pythagorean theorem.

$$20^2 + 40^2 = l^2$$

Carry out. We solve the equation.

$$20^2 + 40^2 = l^2$$
$$400 + 1600 = l^2$$
$$2000 = l^2$$
$$\sqrt{2000} = l$$

We solve the other equation.

$$20^2 + l^2 = d^2$$
$$20^2 + \left(\sqrt{2000}\right)^2 = d^2$$
$$400 + 2000 = d^2$$
$$2400 = d^2$$
$$\sqrt{2400} \text{ ft} = d$$
$$20\sqrt{6} \text{ ft} = d$$
$$48.990 \text{ ft} \approx d$$

State. The longest straight-line distance that can be measured is $\sqrt{2400}$ ft ≈ 48.990 ft

73. Familiarize. Let a = leg of the small triangle

Translate. We use the Pythagorean theorem.

$$a^2 + b^2 = c^2 \text{ and } (a+x)^2 + 5^2 = 13^2$$

Carry out. Solve the equation.

$$a^2 + 5^2 = 7^2$$
$$a^2 + 25 = 49$$
$$a^2 = 24$$
$$a = \sqrt{24}, \text{ or } 2\sqrt{6}$$

Now substitute and solve the second equation.

$$(a+x)^2 + 5^2 = 13^2$$
$$\left(2\sqrt{6}+x\right)^2 + 25 = 169$$
$$\left(2\sqrt{6}+x\right)^2 = 144$$
$$2\sqrt{6}+x = 12 \quad \text{taking the principal square root}$$
$$x = 12 - 2\sqrt{6}$$
$$x \approx 7.101$$

Exercise Set 8.7

1. True; see page 525 in the text.

3. True; see page 525 in the text.

5. $\sqrt[3]{-8} = -2 \qquad (-2)^3 = (-2)(-2)(-2) = -8$

7. $\sqrt[3]{-1000} = -10$
$(-10)^3 = (-10) \cdot (-10) \cdot (-10) = -1000$

9. $\sqrt[3]{125} = 5 \qquad (5)^3 = 5 \cdot 5 \cdot 5 = 125$

11. $-\sqrt[3]{216} = -6 \qquad \sqrt[3]{216} = 6$, so $-\sqrt[3]{216} = -6$.

13. $\sqrt[4]{625} = 5 \qquad 5^4 = 5 \cdot 5 \cdot 5 \cdot 5 = 625$

15. $\sqrt[5]{0} = 0 \qquad 0^5 = 0 \cdot 0 \cdot 0 \cdot 0 \cdot 0 = 0$

17. $\sqrt[5]{-1} = -1 \quad (-1)^5 = (-1)(-1)(-1)(-1)(-1) = -1$

19. $\sqrt[4]{-81}$ is not a real number, because it is an even root of a negative number.

21. $\sqrt[4]{10,000} = 10 \quad 10^4 = 10 \cdot 10 \cdot 10 \cdot 10 = 10,000$
We might also observe that $10,000 = 10^4$, so we have $\sqrt[4]{10^4} = 10$.

23. $\sqrt[3]{6^3} = 6 \qquad 6^3 = 6 \cdot 6 \cdot 6$

25. $\sqrt[8]{1} = 1 \qquad 1^8 = 1 \cdot 1 \cdot 1 \cdot 1 \cdot 1 \cdot 1 \cdot 1 \cdot 1 = 1$

27. $\sqrt[7]{a^7} = a \qquad a^7 = a \cdot a \cdot a \cdot a \cdot a \cdot a \cdot a$

29. $\sqrt[3]{54} = \sqrt[3]{27 \cdot 2} = \sqrt[3]{27}\sqrt[3]{2} = 3\sqrt[3]{2}$

31. $\sqrt[4]{48} = \sqrt[4]{16 \cdot 3} = \sqrt[4]{16}\sqrt[4]{3} = 2\sqrt[4]{3}$

33. $\sqrt[3]{\dfrac{64}{125}} = \dfrac{\sqrt[3]{64}}{\sqrt[3]{125}} = \dfrac{4}{5}$

35. $\sqrt[5]{\dfrac{32}{243}} = \dfrac{\sqrt[5]{32}}{\sqrt[5]{243}} = \dfrac{2}{3}$

37. $\sqrt[3]{\dfrac{7}{8}} = \dfrac{\sqrt[3]{7}}{\sqrt[3]{8}} = \dfrac{\sqrt[3]{7}}{2}$

39. $\sqrt[4]{\dfrac{14}{81}} = \dfrac{\sqrt[4]{14}}{\sqrt[4]{81}} = \dfrac{\sqrt[4]{14}}{3}$

41. $49^{1/2} = \sqrt{49} = 7$

43. $1000^{1/3} = \sqrt[3]{1000} = 10$

45. $16^{1/4} = \sqrt[4]{16} = 2$

47. $16^{3/4} = (16^{1/4})^3 = (\sqrt[4]{16})^3 = 2^3 = 8$

49. $16^{3/2} = (16^{1/2})^3 = (\sqrt{16})^3 = 4^3 = 64$

51. $64^{2/3} = (64^{1/3})^2 = (\sqrt[3]{64})^2 = 4^2 = 16$

53. $1000^{4/3} = (1000^{1/3})^4 = (\sqrt[3]{1000})^4 = 10^4 = 10,000$

55. $100^{5/2} = (100^{1/2})^5 = (\sqrt{100})^5 = 10^5 = 100,000$

57. $9^{-1/2} = \dfrac{1}{9^{1/2}} = \dfrac{1}{3}$

59. $256^{-1/4} = \dfrac{1}{256^{1/4}} = \dfrac{1}{\sqrt[4]{256}} = \dfrac{1}{4}$

61. $16^{-3/4} = \dfrac{1}{16^{3/4}} = \dfrac{1}{(\sqrt[4]{16})^3} = \dfrac{1}{2^3} = \dfrac{1}{8}$

63. $81^{-5/4} = \dfrac{1}{81^{5/4}} = \dfrac{1}{(\sqrt[4]{81})^5} = \dfrac{1}{3^5} = \dfrac{1}{243}$

65. $125^{-2/3} = \dfrac{1}{125^{2/3}} = \dfrac{1}{(\sqrt[3]{125})^2} = \dfrac{1}{5^2} = \dfrac{1}{25}$

67. *Writing Exercise.* You would probably use $(\sqrt[n]{a})^m$, because it is easier to compute $(\sqrt{25})^3 = 5^3 = 125$ than $\sqrt{25^3} = \sqrt{15,625} = 125$.

69. $x^2 + 7x + 12 = 0$

   $(x + 4)(x + 3) = 0$

   $x + 4 = 0 \quad or \quad x + 3 = 0$

   $\quad x = -4 \; or \qquad x = -3$

   The solutions are -4 and -3.

71. $16t^2 - 9 = 0$

   $(4t + 3)(4t - 3) = 0$

   $4t + 3 = 0 \quad or \quad 4t - 3 = 0$

   $\quad t = -\dfrac{3}{4} \;\; or \qquad t = \dfrac{3}{4}$

   The solutions are $-\dfrac{3}{4}$ and $\dfrac{3}{4}$.

73. $3x^2 - x - 10 = 0$

   $(3x + 5)(x - 2) = 0$

   $3x + 5 = 0 \quad or \quad x - 2 = 0$

   $\quad x = -\dfrac{5}{3} \;\; or \qquad x = 2$

   The solutions are $-\dfrac{5}{3}$ and 2.

75. *Writing Exercise.* Yes; we can think of a^n as the product of n factors of $a^{1/n}$ and of b^n as the product of n factors of $b^{1/n}$. Then, if $a > b$, each factor of $a^{1/n}$ must be greater than each factor of $b^{1/n}$, or $a^{1/n} > b^{1/n}$.

77. $8^{4/5} \approx 5.278$

79. $48^{5/8} \approx 11.240$

81. $a^{1/4}a^{3/2} = a^{1/4 + 6/4} = a^{7/4}$

83. $m^{-2/3}m^{1/4}m^{3/2} = m^{-2/3 + 1/4 + 3/2}$
$= m^{-8/12 + 3/12 + 18/12} = m^{13/12}$

85. Graph $y = \sqrt[3]{x}$

We make a table of values.

x	y
-8	-2
-1	-1
0	0
1	1
8	2

We plot these points and connect them with a smooth curve.

87. Substitute 16 for w in the formula and solve for c.

$c = 10w^{3/4} = 10(16)^{3/4} = 10(\sqrt[4]{16})^3 = 10(2)^3 = 10 \cdot 8 = 80$

The daily calorie requirement is 80 calories.

Chapter 8 Review

1. False; see page 502 in the text.

3. True; see page 524 in the text.

5. False; see page 523 in the text.

7. False; the Pythagorean Theorem only applies to right triangles.

9. True; see page 515 in the text.

11. The square roots of 16 are 4 and -4, since $4^2 = 16$ and $(-4)^2 = 16$.

13. The square roots of 400 are 20 and -20, since $20^2 = 400$ and $(-20)^2 = 400$.

15. $\sqrt{144} = 12$, taking the principle square root.

17. $-\sqrt{4} = -2$

19. The radicand is the expression under the radical, $5x^3y$.

21. $-\sqrt{36}$ is rational since 36 is a perfect square

23. $\sqrt{99}$ is irrational since 99 is not a perfect square

25. 2.236

27. For any real number A, $A^2 = |A|$, so $\sqrt{(5x)^2} = |5x|$ or $5|x|$.

29. $\sqrt{p^2} = p$ since p is assumed to be nonnegative.

31. $\sqrt{49n^2} = 7n$ since n is assumed to be nonnegative.

33. $\sqrt{48} = \sqrt{16 \cdot 3} = \sqrt{16}\sqrt{3} = 4\sqrt{3}$

35. $\sqrt{32p} = \sqrt{16 \cdot 2p} = 4\sqrt{2p}$

37. $\sqrt{12a^3} = \sqrt{4a^{12} \cdot 3a} = 2a^6\sqrt{3a}$

39. $\sqrt{5}\sqrt{11} = \sqrt{5 \cdot 11} = \sqrt{55}$

41. $\sqrt{3s}\sqrt{7t} = \sqrt{3s \cdot 7t} = \sqrt{21st}$

43. $\sqrt{5x}\sqrt{10xy^2} = \sqrt{50x^2y^2} = \sqrt{25x^2y^2 \cdot 2} = 5xy\sqrt{2}$

45. $\dfrac{\sqrt{35}}{\sqrt{45}} = \sqrt{\dfrac{35}{45}} = \sqrt{\dfrac{7}{9}} = \dfrac{\sqrt{7}}{3}$

47. $\sqrt{\dfrac{49}{64}} = \dfrac{7}{8}$

49. $\sqrt{\dfrac{64t}{t^7}} = \sqrt{\dfrac{64}{t^6}} = \dfrac{8}{t^3}$

51. $\sqrt{80} - \sqrt{45} = \sqrt{16 \cdot 5} - \sqrt{9 \cdot 5} = 4\sqrt{5} - 3\sqrt{5}$
$= (4 - 3)\sqrt{5} = \sqrt{5}$

53. $\left(2 + \sqrt{3}\right)^2 = 4 + 2 \cdot 2\sqrt{3} + \left(\sqrt{3}\right)^2$ Using $(A + B)^2$
$$= A^2 + 2AB + B^2$$
$$= 4 + 4\sqrt{3} + 3$$
$$= 7 + 4\sqrt{3}$$

55. $\left(1 + 2\sqrt{7}\right)\left(3 - \sqrt{7}\right)$
$$= 3 - \sqrt{7} + 6\sqrt{7} - 14 \qquad \text{Using FOIL}$$
$$= -11 + 5\sqrt{7}$$

57. $\dfrac{\sqrt{5}}{\sqrt{8}} = \dfrac{\sqrt{5}}{\sqrt{8}} \cdot \dfrac{\sqrt{8}}{\sqrt{8}} = \dfrac{\sqrt{40}}{8} = \dfrac{\sqrt{4 \cdot 10}}{8} = \dfrac{2\sqrt{10}}{8} = \dfrac{\sqrt{10}}{4}$

59. $\dfrac{2}{\sqrt{3}} = \dfrac{2}{\sqrt{3}} \cdot \dfrac{\sqrt{3}}{\sqrt{3}} = \dfrac{2\sqrt{3}}{3}$

61. $\dfrac{1 + \sqrt{5}}{2 - \sqrt{5}} = \dfrac{1 + \sqrt{5}}{2 - \sqrt{5}} \cdot \dfrac{2 + \sqrt{5}}{2 + \sqrt{5}} = \dfrac{2 + \sqrt{5} + 2\sqrt{5} + 5}{4 - 5}$
$$= \dfrac{7 + 3\sqrt{5}}{-1} = -7 - 3\sqrt{5}$$

63. $\sqrt{5x + 3} = \sqrt{2x - 1}$
$$\left(\sqrt{5x + 3}\right)^2 = \left(\sqrt{2x - 1}\right)^2 \qquad \text{Simplifying both sides}$$
$$5x + 3 = 2x - 1 \qquad \text{Simplifying}$$
$$5x = 2x - 4$$
$$3x = -4$$
$$x = \dfrac{-4}{3}$$

Check:
$$\dfrac{\sqrt{5x + 3} = \sqrt{2x - 1}}{\sqrt{5\left(-\dfrac{4}{3}\right) + 3} \;\Bigg|\; \sqrt{2\left(-\dfrac{4}{3}\right) - 1}}$$
$$\sqrt{\dfrac{-20}{3} + 3} \;\Bigg|\; \sqrt{-\dfrac{8}{3} - 1}$$
$$\sqrt{-\dfrac{11}{3}} \overset{?}{=} \sqrt{-\dfrac{11}{3}} \qquad \text{TRUE}$$

Since we cannot have a square root of a negative number, there is no solution.

65.
$$1 + x = \sqrt{1 + 5x}$$
$$(1 + x)^2 = \left(\sqrt{1 + 5x}\right)^2$$
$$1 + 2x + x^2 = 1 + 5x$$
$$x^2 - 3x = 0$$
$$x(x - 3) = 0$$
$$x = 0 \text{ or } x - 3 = 0$$
$$x = 0 \text{ or } x = 3$$

Check: For $x = 0$
$$\dfrac{1 + x = \sqrt{1 + 5x}}{\begin{array}{c|c} 1 + 0 & \sqrt{1 + 5 \cdot 0} \\ 1 & \sqrt{1} \\ 1 \overset{?}{=} 1 & \text{TRUE} \end{array}}$$

Check: For $x = 3$
$$\dfrac{1 + x = \sqrt{1 + 5x}}{\begin{array}{c|c} 1 + 3 & \sqrt{1 + 5 \cdot 3} \\ 4 & \sqrt{1 + 15} \\ & \sqrt{16} \\ 4 \overset{?}{=} 4 & \text{TRUE} \end{array}}$$

The solutions are 0 and 3.

67.
$$a^2 + b^2 = c^2$$
$$15^2 + b^2 = 25^2 \qquad \text{Substituting}$$
$$225 + b^2 = 625$$
$$b^2 = 400$$
$$b = 20$$

69. **Familiarize** Let $x =$ the distance of the wire.

Translate. We use the Pythagorean theorem. Substituting 48 for a, 48 for b, and x for c.
$$48^2 + 48^2 = x^2$$

Carry out. We solve the equation.
$$48^2 + 48^2 = x^2$$
$$2 \cdot 48^2 = x^2$$
$$\sqrt{2 \cdot 48^2} = x$$
$$48\sqrt{2} = x \qquad \text{Exact answer}$$
$$67.882 \approx x$$

Check. We check by substituting 48, 48 and $48\sqrt{2}$ in the Pythagorean Theorem.

Check:
$$\dfrac{a^2 + b^2 = c^2}{\begin{array}{c|c} 48^2 + 48^2 & \left(48\sqrt{2}\right)^2 \\ 2 \cdot 48^2 & 48^2 \cdot 2 \\ 4806 \overset{?}{=} 4806 & \text{TRUE} \end{array}}$$

State. The length of the wire should be $48\sqrt{2}$, or about 67.882 ft.

71. $\sqrt[5]{-32} = -2$
$$(-2)^5 = (-2)(-2)(-2)(-2)(-2) = -32$$

73. $\sqrt[3]{\dfrac{8}{27}} = \dfrac{\sqrt[3]{8}}{\sqrt[3]{27}} = \dfrac{2}{3}$

75. $81^{1/2} = \sqrt{81} = 9$

77. $25^{3/2} = \left(25^{1/2}\right)^3 = \left(\sqrt{25}\right)^3 = 5^3 = 125$

79. *Writing Exercise.* Absolute-value signs may be necessary when simplifying a radical expression with an even index. For n an even number, if it is possible that A is negative, then $\sqrt[n]{A^n} = |A|$.

81. $\sqrt{\sqrt{\sqrt{256}}} = \left(\left(256^{1/2}\right)^{1/2}\right)^{1/2} = 256^{\frac{1}{2} \cdot \frac{1}{2} \cdot \frac{1}{2}} = 256^{\frac{1}{8}}$
$$= \sqrt[8]{256} = 2$$

83. Factor $x^2 - 5$.

If we set this expression equal to zero and solve using the principle of square roots, we have,

$$x^2 - 5 = 0$$
$$x^2 = 5$$
$$x = \pm\sqrt{5}.$$

Then factoring the expression, $x^2 - 5$, we have

$$x^2 - 5 = \left(x + \sqrt{5}\right)\left(x - \sqrt{5}\right).$$

Chapter 8 Test

1. The square roots of 49 are 7 and -7, because $7^2 = 49$ and $(-7)^2 = 49$.

3. $-\sqrt{25} = -5$

5. $\sqrt{44}$ is irrational, since 44 is not a perfect square.

7. $\sqrt{2} \approx 1.414$

9. $\sqrt{a^2} = a$ since a is assumed to be nonnegative

11. $\sqrt{60} = \sqrt{4 \cdot 15}$ 4 is a perfect square.
$\quad = \sqrt{4}\sqrt{15}$ Factoring
$\quad = 2\sqrt{15}$

13. $\sqrt{36t^{11}} = \sqrt{36 \cdot t^{10} \cdot t}$
$\quad = \sqrt{36}\sqrt{t^{10}}\sqrt{t}$
$\quad = 6t^5\sqrt{t}$

15. $\sqrt{5}\sqrt{15} = \sqrt{5 \cdot 15}$
$\quad = \sqrt{5 \cdot 5 \cdot 3}$
$\quad = \sqrt{5^2 \cdot 3}$
$\quad = 5\sqrt{3}$

17. $\sqrt{2t}\sqrt{8t} = \sqrt{2t \cdot 8t}$
$\quad = \sqrt{2 \cdot t \cdot 2 \cdot 2 \cdot 2 \cdot t}$
$\quad = 4t$

19. $\dfrac{\sqrt{28}}{\sqrt{63}} = \sqrt{\dfrac{28}{63}} = \sqrt{\dfrac{4}{9}} = \dfrac{2}{3}$

21. $\dfrac{\sqrt{27}}{\sqrt{12}} = \sqrt{\dfrac{27}{12}} = \sqrt{\dfrac{9}{4}} = \dfrac{3}{2}$

23. $3\sqrt{18} - 5\sqrt{18} = (3 - 5)\sqrt{18}$
$\quad = -2\sqrt{18}$
$\quad = -2 \cdot 3\sqrt{2}$
$\quad = -6\sqrt{2}$

25. $\left(4 - \sqrt{5}\right)^2 = 4^2 - 2 \cdot 4\sqrt{5} + \left(\sqrt{5}\right)^2$
$\quad = 16 - 8\sqrt{5} + 5$
$\quad = 21 - 8\sqrt{5}$

27. $\sqrt{\dfrac{2}{5}} = \dfrac{\sqrt{2}}{\sqrt{5}} \cdot \dfrac{\sqrt{5}}{\sqrt{5}} = \dfrac{\sqrt{10}}{5}$

29.
$$\frac{10}{4 - \sqrt{5}} = \frac{10}{4 - \sqrt{5}} \cdot \frac{4 + \sqrt{5}}{4 + \sqrt{5}}$$
$$= \frac{10\left(4 + \sqrt{5}\right)}{4^2 - \left(\sqrt{5}\right)^2}$$
$$= \frac{10\left(4 + \sqrt{5}\right)}{16 - 5}$$
$$= \frac{10\left(4 + \sqrt{5}\right)}{11}$$
$$= \frac{40 + 10\sqrt{5}}{11}$$

31. *Familiarize.* Let w = the length of the wire.

Translate. We use the Pythagorean theorem, substituting 25 for a, 100 for b, and w for c. $25^2 + 100^2 = w^2$

Carry out. We solve the equation.

$$25^2 + 100^2 = w^2$$
$$625 + 10,000 = w^2$$
$$10,625 = w^2$$
$$\sqrt{10,625} = w$$
$$103.078 \approx w$$

Check. We check by substituting 25, 100 and $\sqrt{10,625}$ in the Pythagorean theorem.

State. The length of the wire is $\sqrt{10,625}$ ft or about 103.078 ft.

33. $\sqrt{6x + 13} = x + 3$
$\quad \left(\sqrt{6x + 13}\right)^2 = (x + 3)^2$
$\quad 6x + 13 = x^2 + 6x + 9$
$\quad 0 = x^2 - 4$
$\quad 0 = (x - 2)(x + 2)$
$\quad x - 2 = 0 \ or \ x + 2 = 0$
$\quad x = 2 \ or \ x = -2$

35. $\sqrt[4]{16} = 2$ $2 \cdot 2 \cdot 2 \cdot 2 = 16$

37. $\sqrt[3]{-64} = -4$ $(-4)(-4)(-4) = 64$

39. $9^{1/2} = \sqrt{9} = 3$

41. $100^{3/2} = \left(\sqrt{100}\right)^3 = (10)^3 = 1000$

43.
$$\sqrt{1 - x} + 1 = \sqrt{6 - x}$$
$$\left(\sqrt{1 - x} + 1\right)^2 = \left(\sqrt{6 - x}\right)^2$$
$$1 - x + 2\sqrt{1 - x} + 1 = 6 - x$$
$$2\sqrt{1 - x} = 4$$
$$\sqrt{1 - x} = 2$$
$$\left(\sqrt{1 - x}\right)^2 = 2^2$$
$$1 - x = 4$$
$$x = -3$$

The number -3 checks. It is the solution.

Chapter 9

Quadratic Equations

Exercise Set 9.1

1. True; see page 536 in the text.

3. False; see Examples 2, 3(b), and 4(b).

5.
$$t^2 = 81$$
$$t = \sqrt{81} \quad or \quad t = -\sqrt{81} \quad \text{Using the principle}$$
$$t = 9 \quad or \quad t = -9 \quad \text{of square roots}$$

We can check mentally that $9^2 = 81$ and $(-9)^2 = 81$. The solutions are 9 and -9.

7.
$$x^2 = 1$$
$$x = \sqrt{1} \quad or \quad x = -\sqrt{1} \quad \text{Using the principle}$$
$$x = 1 \quad or \quad x = -1 \quad \text{of square roots}$$

We can check mentally that $1^2 = 1$ and $(-1)^2 = 1$. The solutions are 1 and -1.

9.
$$a^2 = 11$$
$$a = \sqrt{11} \quad or \quad a = -\sqrt{11} \quad \text{Using the principle}$$
$$\text{of square roots}$$

Check: For $\sqrt{11}$:   For $-\sqrt{11}$:

$$\frac{a^2 = 11}{(\sqrt{11})^2 \mid 11} \qquad \frac{a^2 = 11}{(-\sqrt{11})^2 \mid 11}$$
$$11 \overset{?}{=} 11 \text{ TRUE} \qquad 11 \overset{?}{=} 11 \text{ TRUE}$$

The solutions are $\sqrt{11}$ and $-\sqrt{11}$.

11.
$$10x^2 = 40$$
$$x^2 = 4 \quad \text{Dividing by 3}$$
$$x = \sqrt{4} \quad or \quad x = -\sqrt{4} \quad \text{Using the principle}$$
$$\text{of square roots}$$
$$x = 2 \quad or \quad x = -2$$

Both numbers check. The solutions are 2 and -2.

13. $3t^2 = 6$
$$t^2 = 2$$
$$t = \sqrt{2} \quad or \quad t = -\sqrt{2}$$

Both numbers check. The solutions are $\sqrt{2}$ and $-\sqrt{2}$.

15. $4 - 9x^2 = 0$
$$4 = 9x^2$$
$$\frac{4}{9} = x^2$$

$$R+6 = \sqrt{-13}$$

$$x = \sqrt{\frac{4}{9}} \quad or \quad x = -\sqrt{\frac{4}{9}}$$
$$x = \frac{2}{3} \quad or \quad x = -\frac{2}{3}$$

Both numbers check. The solutions are $\frac{2}{3}$ and $-\frac{2}{3}$.

17.
$$12y^2 + 1 = 1$$
$$12y^2 = 0$$

Observe that y^2 must be 0, so $y = 0$. the solution is 0.

19.
$$15x^2 - 25 = 0$$
$$15x^2 = 25$$
$$x^2 = \frac{25}{15}$$
$$x = \sqrt{\frac{25}{15}} \quad or \quad x = -\sqrt{\frac{25}{15}}$$
$$x = \sqrt{\frac{5}{3}} \quad or \quad x = -\sqrt{\frac{5}{3}}$$
$$x = \frac{\sqrt{5}}{\sqrt{3}} \cdot \frac{\sqrt{3}}{\sqrt{3}} \quad or \quad x = -\frac{\sqrt{5}}{\sqrt{3}} \cdot \frac{\sqrt{3}}{\sqrt{3}}$$
$$x = \frac{\sqrt{15}}{3} \quad or \quad x = -\frac{\sqrt{15}}{3}$$

The solutions are $\frac{\sqrt{15}}{3}$ and $-\frac{\sqrt{15}}{3}$.

21.
$$(x - 1)^2 = 49$$
$$x - 1 = 7 \quad or \quad x - 1 = -7 \quad \text{Using the principle}$$
$$\text{of square roots}$$
$$x = 8 \quad or \quad x = -6$$

The solutions are 8 and -6.

23.
$$(t + 6)^2 = 4$$
$$t + 6 = 2 \quad or \quad t + 6 = -2 \quad \text{Using the principle}$$
$$\text{of square roots}$$
$$t = -4 \quad or \quad t = -8$$

The solutions are -8 and -4.

25.
$$(m + 3)^2 = 6$$
$$m + 3 = \sqrt{6} \quad or \quad m + 3 = -\sqrt{6}$$
$$m = -3 + \sqrt{6} \quad or \quad m = -3 - \sqrt{6}$$

The solutions are $-3 + \sqrt{6}$ and $-3 - \sqrt{6}$, or $-3 \pm \sqrt{6}$.

27. $(a - 7)^2 = 0$

Observe that $a - 7$ must be 0, so $a - 7 = 0$, or $a = 7$. The solution is 7.

29.
$$(5 - x)^2 = 14$$
$$5 - x = \sqrt{14} \quad or \quad 5 - x = -\sqrt{14}$$
$$-x = -5 + \sqrt{14} \quad or \quad -x = -5 - \sqrt{14}$$
$$x = 5 - \sqrt{14} \quad or \quad x = 5 + \sqrt{14}$$

The solutions are $5 - \sqrt{14}$ and $5 + \sqrt{14}$, or $5 \pm \sqrt{14}$.

31.
$$(t + 1)^2 = 1$$
$$t + 1 = 1 \quad or \quad t + 1 = -1$$
$$t = 0 \quad or \quad t = -2$$

The solutions are 0 and -2.

33.
$$\left(y - \frac{3}{4}\right)^2 = \frac{17}{16}$$

$$y - \frac{3}{4} = \sqrt{\frac{17}{16}} \quad or \quad y - \frac{3}{4} = -\sqrt{\frac{17}{16}}$$

$$y - \frac{3}{4} = \frac{\sqrt{17}}{4} \quad or \quad y - \frac{3}{4} = -\frac{\sqrt{17}}{4}$$

$$y = \frac{3}{4} + \frac{\sqrt{17}}{4} \quad or \quad y = \frac{3}{4} - \frac{\sqrt{17}}{4}$$

The solutions are $\frac{3}{4} + \frac{\sqrt{17}}{4}$ and $\frac{3}{4} - \frac{\sqrt{17}}{4}$, or $\frac{3}{4} \pm \frac{\sqrt{17}}{4}$.

35.
$$x^2 - 10x + 25 = 100$$
$$(x - 5)^2 = 100$$

$$x - 5 = 10 \quad or \quad x - 5 = -10$$
$$x = 15 \quad or \qquad x = -5$$

The solutions are 15 and -5.

37.
$$p^2 + 8p + 16 = 1$$
$$(p + 4)^2 = 1$$

$$p + 4 = 1 \quad or \quad p + 4 = -1$$
$$p = -3 \quad or \qquad p = -5$$

The solutions are -3 and -5.

39.
$$t^2 - 16t + 64 = 7$$
$$(t - 8)^2 = 7$$

$$t - 8 = \sqrt{7} \quad or \quad t - 8 = -\sqrt{7}$$
$$t = 8 + \sqrt{7} \quad or \qquad t = 8 - \sqrt{7}$$

The solutions are $8 + \sqrt{7}$ and $8 - \sqrt{7}$, or $8 \pm \sqrt{7}$.

41.
$$x^2 + 12x + 36 = 18$$
$$(x + 6)^2 = 18$$

$$x + 6 = \sqrt{18} \quad or \quad x + 6 = -\sqrt{18}$$
$$x + 6 = 3\sqrt{2} \quad or \quad x + 6 = -3\sqrt{2}$$
$$x = -6 + 3\sqrt{2} \quad or \qquad x = -6 - 3\sqrt{2}$$

The solutions are $-6 + 3\sqrt{2}$ and $-6 - 3\sqrt{2}$, or $-6 \pm 3\sqrt{2}$.

43. *Writing Exercise.* It is easier to use the principle of square roots than the principle of zero products to solve a quadratic equation when the equation is in the form $ax^2 = p$ or $(x + k)^2 = p$.

45. $(x - 3)^2 = x^2 - 6x + 9$

47. $\left(x + \frac{1}{2}\right)^2 = x^2 + x + \frac{1}{4}$

49. $x^2 + 2x + 1 = (x + 1)(x + 1) = (x + 1)^2$

51. $x^2 - 4x + 4 = (x - 2)(x - 2) = (x - 2)^2$

53. *Writing Exercise.* A quadratic equation has only one solution when it can be written in the form $ax^2 = 0$ or $(x + k)^2 = 0$.

55.
$$x^2 - 5x + \frac{25}{4} = \frac{13}{4}$$
$$\left(x - \frac{5}{2}\right)^2 = \frac{13}{4}$$

$$x - \frac{5}{2} = \frac{\sqrt{13}}{2} \quad or \quad x - \frac{5}{2} = -\frac{\sqrt{13}}{2}$$

$$x = \frac{5}{2} + \frac{\sqrt{13}}{2} \quad or \qquad x = \frac{5}{2} - \frac{\sqrt{13}}{2}$$

$$x = \frac{5}{2} \pm \frac{\sqrt{13}}{2}$$

57.
$$t^2 + 3t + \frac{9}{4} = \frac{49}{4}$$
$$\left(t + \frac{3}{2}\right)^2 = \frac{49}{4}$$

$$t + \frac{3}{2} = \frac{7}{2} \quad or \quad t + \frac{3}{2} = \frac{-7}{2}$$

$$t = \frac{4}{2} \quad or \qquad t = \frac{-10}{2}$$

$$t = 2 \quad or \qquad t = -5$$

The solutions are -5 and 2.

59.
$$x^2 + 2.5x + 1.5625 = 9.61$$
$$(x + 1.25)^2 = 9.61$$

$$x + 1.25 = 3.1 \quad or \quad x + 1.25 = -3.1$$
$$x = 1.85 \quad or \qquad x = -4.35$$

The solutions are 1.85 and -4.35.

61. From the graph we see that when $y = 1$, then $x = -4$ or $x = -2$. Thus, the solutions of $(x + 3)^2 = 1$ are -4 and -2.

63. From the graph we see that when $y = 9$, then $x = -6$ or $x = 0$. Thus, the solutions of $(x + 3)^2 = 9$ are -6 and 0.

65.
$$f = \frac{kMm}{d^2}$$

$$d^2 f = kMm \qquad \text{Multiplying by } d^2$$

$$d^2 = \frac{kMm}{f} \qquad \text{Dividing by } f$$

$$d = \sqrt{\frac{kMm}{f}} \qquad \text{Taking the principal square root}$$

$$d = \frac{\sqrt{kMmf}}{f} \qquad \text{Rationalizing the denominator}$$

Exercise Set 9.2

1. First complete the square for $x^2 + 6x$:

$$\left(\frac{6}{2}\right)^2 = 3^2 = 9$$

Then we have:
$$x^2 + 6x = 2$$
$$x^2 + 6x + 9 = 2 + 9$$
$$x^2 + 6x + 9 = 11$$

Choice (c) is correct.

3. Factoring on the left side, we have $(x + 3)^2 = 10$. Choice (a) is correct.

5. First complete the square for $x^2 + 8x$:

$$\left(\frac{8}{2}\right)^2 = 4^2 = 16$$

Then we have:

$$x^2 + 8x = 2$$
$$x^2 + 8x + 16 = 2 + 16$$
$$(x + 4)^2 = 18$$

Choice (d) is correct.

7. To complete the square for $x^2 + 8x$, we take half the coefficient of x and square it:

$$\left(\frac{8}{2}\right)^2 = 4^2 = 16$$

The trinomial $x^2 + 8x + 16$ is the square of $x + 4$.
Check: $(x + 4)^2 = x^2 + 8x + 16$.

9. To complete the square for $x^2 - 2x$, we take half the coefficient of x and square it:

$$\left(\frac{-2}{2}\right)^2 = 1$$

The trinomial $x^2 - 2x + 1$ is the square of $x - 1$.
Check: $(x - 1)^2 = x^2 - 2x + 1$.

11. To complete the square for $x^2 - 3x$, we take half the coefficient of x and square it:

$$\left(\frac{-3}{2}\right)^2 = \frac{9}{4}$$

The trinomial $x^2 - 3x + \frac{9}{4}$ is the square of $x - \frac{3}{2}$.
Check: $\left(x - \frac{3}{2}\right)^2 = x^2 - 3x + \frac{9}{4}$.

13. To complete the square for $t^2 + t$, we take half the coefficient of t and square it:

$$\left(\frac{1}{2}\right)^2 = \frac{1}{4}$$

The trinomial $t^2 + t + \frac{1}{4}$ is the square of $t + \frac{1}{2}$.
Check: $\left(t + \frac{1}{2}\right)^2 = t^2 + t + \frac{1}{4}$.

15. To complete the square for $x^2 + \frac{3}{2}x$, we take half the coefficient of x and square it:

$$\left(\frac{1}{2} \cdot \frac{3}{2}\right)^2 = \left(\frac{3}{4}\right)^2 = \frac{9}{16}$$

The trinomial $x^2 + \frac{3}{2}x + \frac{9}{16}$ is the square of $x + \frac{3}{4}$.
Check: $\left(x + \frac{3}{4}\right)^2 = x^2 + \frac{3}{2}x + \frac{9}{16}$.

17. To complete the square for $m^2 - \frac{8}{3}m$, we take half the coefficient of m and square it:

$$\left[\frac{1}{2}\left(-\frac{8}{3}\right)\right]^2 = \left(-\frac{4}{3}\right)^2 = \frac{16}{9}$$

The trinomial $m^2 - \frac{8}{3}m + \frac{16}{9}$ is the square of $m - \frac{4}{3}$.
Check: $\left(m - \frac{4}{3}\right)^2 = m^2 - \frac{8}{3}m + \frac{16}{9}$.

19.
$$x^2 + 7x + 10 = 0$$
$$x^2 + 7x = -10 \qquad \text{Subtracting 10}$$
$$x^2 + 7x + \frac{49}{4} = -10 + \frac{49}{4} \quad \text{Adding } \frac{49}{4}:$$
$$\left(\frac{7}{2}\right)^2 = \frac{49}{4}$$
$$\left(x + \frac{7}{2}\right)^2 = \frac{9}{4}$$
$$x + \frac{7}{2} = \frac{3}{2} \quad or \quad x + \frac{7}{2} = -\frac{3}{2} \quad \text{Principle of square roots}$$
$$x = -2 \quad or \qquad x = -5$$

The solutions are -5 and -2.

21.
$$x^2 - 24x + 21 = 0$$
$$x^2 - 24x = -21 \qquad \text{Subtracting 21}$$
$$x^2 - 24x + 144 = -21 + 144 \quad \text{Adding 144:}$$
$$\left(\frac{-24}{2}\right)^2 = (-12)^2 = 144$$
$$(x - 12)^2 = 123$$
$$x - 12 = \sqrt{123} \qquad or \quad x - 12 = -\sqrt{123}$$
$$\text{Principle of square roots}$$
$$x = 12 + \sqrt{123} \quad or \qquad x = 12 - \sqrt{123}$$

The solutions are $12 + \sqrt{123}$ and $12 - \sqrt{123}$, or $12 \pm \sqrt{123}$.

23.
$$t^2 + 10t + 12 = 0$$
$$t^2 + 10t = -12$$
$$t^2 + 10t + 25 = -12 + 25 \qquad \text{Adding 25}$$
$$(t + 5)^2 = 13$$
$$t + 5 = \sqrt{13} \quad or \quad t + 5 = -\sqrt{13} \qquad \text{Principle of square roots}$$
$$t = -5 + \sqrt{13} \quad or \quad t = -5 - \sqrt{13}$$

The solutions are $-5 + \sqrt{13}$ and $-5 - \sqrt{13}$, or $-5 \pm \sqrt{13}$.

25.
$$2x^2 - 8x = 14$$
$$\frac{1}{2}\left(2x^2 - 8x\right) = \frac{1}{2}(14)$$
$$x^2 - 4x = 7$$
$$x^2 - 4x + 4 = 7 + 4$$
$$(x - 2)^2 = 11$$
$$x - 2 = \sqrt{11} \quad or \quad x - 2 = -\sqrt{11}$$
$$x = 2 + \sqrt{11} \quad or \quad x = 2 - \sqrt{11}$$

The solutions are $2 \pm \sqrt{11}$.

27. $x^2 + 3x - 3 = 0$

$x^2 + 3x \quad = 3$

$x^2 + 3x + \dfrac{9}{4} = 3 + \dfrac{9}{4}$ Adding $\dfrac{9}{4}$:

$$\left(\dfrac{3}{2}\right)^2 = \dfrac{9}{4}$$

$$\left(x + \dfrac{3}{2}\right)^2 = \dfrac{21}{4}$$

$x + \dfrac{3}{2} = \sqrt{\dfrac{21}{4}}$ or $x + \dfrac{3}{2} = -\sqrt{\dfrac{21}{4}}$

$x + \dfrac{3}{2} = \dfrac{\sqrt{21}}{2}$ or $x + \dfrac{3}{2} = -\dfrac{\sqrt{21}}{2}$

$x = -\dfrac{3}{2} + \dfrac{\sqrt{21}}{2}$ or $x = -\dfrac{3}{2} - \dfrac{\sqrt{21}}{2}$

The solutions are $-\dfrac{3}{2} \pm \dfrac{\sqrt{21}}{2}$, or $\dfrac{-3 \pm \sqrt{21}}{2}$.

29. $2t^2 + t - 1 = 0$

$\dfrac{1}{2}\left(2t^2 + t - 1\right) = \dfrac{1}{2} \cdot 0$

$t^2 + \dfrac{1}{2}t - \dfrac{1}{2} = 0$

$t^2 + \dfrac{1}{2}t + \dfrac{1}{16} = \dfrac{1}{2} + \dfrac{1}{16}$

$\left(t + \dfrac{1}{4}\right)^2 = \dfrac{9}{16}$

$t + \dfrac{1}{4} = \dfrac{3}{4}$ or $t + \dfrac{1}{4} = -\dfrac{3}{4}$

$t = \dfrac{1}{2}$ or $t = -1$

The solutions are -1 and $\dfrac{1}{2}$.

31. $x^2 - \dfrac{3}{2}x - 2 = 0$

$x^2 - \dfrac{3}{2}x \quad = 2$

$x^2 - \dfrac{3}{2}x + \dfrac{9}{16} = 2 + \dfrac{9}{16}$ Adding $\dfrac{9}{16}$:

$$\left[\dfrac{1}{2}\left(-\dfrac{3}{2}\right)\right]^2 = \left(-\dfrac{3}{4}\right)^2 = \dfrac{9}{16}$$

$$\left(x - \dfrac{3}{4}\right)^2 = \dfrac{32}{16} + \dfrac{9}{16} = \dfrac{41}{16}$$

$x - \dfrac{3}{4} = \dfrac{\sqrt{41}}{4}$ or $x - \dfrac{3}{4} = -\dfrac{\sqrt{41}}{4}$

$x = \dfrac{3}{4} + \dfrac{\sqrt{41}}{4}$ or $x = \dfrac{3}{4} - \dfrac{\sqrt{41}}{4}$

The solutions are $\dfrac{3}{4} \pm \dfrac{\sqrt{41}}{4}$, or $\dfrac{3 \pm \sqrt{41}}{4}$.

33. $2x^2 - 5x - 10 = 0$

$\dfrac{1}{2}\left(2x^2 - 5x - 10\right) = \dfrac{1}{2} \cdot 0$

$x^2 - \dfrac{5}{2}x - 5 = 0$

$x^2 - \dfrac{5}{2}x = 5$

$x^2 - \dfrac{5}{2}x + \dfrac{25}{16} = 5 + \dfrac{25}{16}$

$\left(x - \dfrac{5}{4}\right)^2 = \dfrac{105}{16}$

$x - \dfrac{5}{4} = \dfrac{\sqrt{105}}{4}$ or $x - \dfrac{5}{4} = \dfrac{-\sqrt{105}}{4}$

$x = \dfrac{5}{4} + \dfrac{\sqrt{105}}{4}$ or $x = \dfrac{5}{4} - \dfrac{\sqrt{105}}{4}$

The solutions are $\dfrac{5}{4} \pm \dfrac{\sqrt{105}}{4}$ or $\dfrac{5 \pm \sqrt{105}}{4}$.

35. $3t^2 + 4t - 5 = 0$

$\dfrac{1}{3}\left(3t^2 + 4t - 5\right) = \dfrac{1}{3} \cdot 0$

$t^2 + \dfrac{4}{3}t - \dfrac{5}{3} = 0$

$t^2 + \dfrac{4}{3}t = \dfrac{5}{3}$

$t^2 + \dfrac{4}{3}t + \dfrac{4}{9} = \dfrac{5}{3} + \dfrac{4}{9}$

$\left(t + \dfrac{2}{3}\right)^2 = \dfrac{19}{9}$

$t + \dfrac{2}{3} = \dfrac{\sqrt{19}}{3}$ or $t + \dfrac{2}{3} = \dfrac{-\sqrt{19}}{3}$

$t = \dfrac{-2}{3} + \dfrac{\sqrt{19}}{3}$ or $t = -\dfrac{2}{3} - \dfrac{\sqrt{19}}{3}$

The solutions are $-\dfrac{2}{3} \pm \dfrac{\sqrt{19}}{3}$ or $\dfrac{-2 \pm \sqrt{19}}{3}$.

37. $2x^2 = 5 + 9x$

$2x^2 - 9x - 5 = 0$

$\dfrac{1}{2}(2x^2 - 9x - 5) = \dfrac{1}{2} \cdot 0$

$x^2 - \dfrac{9}{2}x - \dfrac{5}{2} = 0$

$x^2 - \dfrac{9}{2}x = \dfrac{5}{2}$

$x^2 - \dfrac{9}{2}x + \dfrac{81}{16} = \dfrac{5}{2} + \dfrac{81}{16}$

$\left(x - \dfrac{9}{4}\right)^2 = \dfrac{121}{16}$

$x - \dfrac{9}{4} = \dfrac{11}{4}$ or $x - \dfrac{9}{4} = -\dfrac{11}{4}$

$x = 5$ or $x = -\dfrac{1}{2}$

The solutions are 5 and $-\dfrac{1}{2}$.

39.
$$6x^2 + 11x = 10$$
$$\frac{1}{6}\left(6x^2 + 11x\right) = \frac{1}{6} \cdot 10$$
$$x^2 + \frac{11}{6}x = \frac{5}{3}$$
$$x^2 + \frac{11}{6}x + \frac{121}{144} = \frac{5}{3} + \frac{121}{144}$$
$$\left(x + \frac{11}{12}\right)^2 = \frac{361}{144}$$

$$x + \frac{11}{12} = \frac{19}{12} \quad or \quad x + \frac{11}{12} = -\frac{19}{12}$$
$$x = \frac{2}{3} \quad or \qquad x = -\frac{5}{2}$$

The solutions are $\frac{2}{3}$ and $-\frac{5}{2}$.

41. *Writing Exercise.* Given an equation that cannot be solved using the principle of zero products or the principle of square roots, completing the square enables us to express the equation in an equivalent form that can be solved using the principle of square roots.

43. $-b = -7$

45. $b^2 - 4ac = 6^2 - 4\,(2)\,(3) = 36 - 24 = 12$

47. $\dfrac{2\left(1 - \sqrt{5}\right)}{2} = \dfrac{\cancel{2}\left(1 - \sqrt{5}\right)}{\cancel{2} \cdot 1} = 1 - \sqrt{5}$

49. $\dfrac{24 - 3\sqrt{5}}{9} = \dfrac{\cancel{3}(8 - \sqrt{5})}{\cancel{3} \cdot 3} = \dfrac{8 - \sqrt{5}}{3}$

51. *Writing Exercise.* No; the solutions are of the form $x = \dfrac{g \pm \sqrt{h}}{k}$. They are opposites only when $g = 0$ (that is, when the quadratic equation has no x-term.)

53. $x^2 + bx + 25$

The trinomial is a square if the square of one-half the x-coefficient is equal to 25. Thus, we have:
$$\left(\frac{b}{2}\right)^2 = 25$$
$$\frac{b^2}{4} = 25$$
$$b^2 = 100$$
$$b = 10 \quad or \quad b = -10$$

55. $x^2 + bx + 45$

The trinomial is a square if the square of one-half the x-coefficient is equal to 45. Thus, we have:
$$\left(\frac{b}{2}\right)^2 = 45$$
$$\frac{b^2}{4} = 45$$
$$b^2 = 180$$
$$b = \sqrt{180} \quad or \quad b = -\sqrt{180}$$
$$b = 6\sqrt{5} \quad or \quad b = -6\sqrt{5}$$

57. $4x^2 + bx + 16$

The trinomial is a square if the square of one-half the x-coefficient is equal to 16. Thus, we have:
$$4\left(x^2 + \frac{b}{4}x + 4\right)$$
$$\left(\frac{b/4}{2}\right)^2 = 4$$
$$\left(\frac{b}{8}\right)^2 = 4$$
$$\frac{b^2}{64} = 4$$
$$b^2 = 256$$
$$b = 16 \quad or \quad b = -16$$

59. $-0.39, -7.61$

61. $23.09, 0.91$

63. $-2.5, 0.67$

65. *Writing Exercise.* Answers may vary. Most would consider that using the principle of zero products is the best way to solve the equations since $x^2 + 8x$ is easily factored.

Exercise Set 9.3

1. True.

3. False. Using the quadratic formula to solve for x, if $b^2 - 4ac = 0$, then $x = \dfrac{-b \pm 0}{2a} = \dfrac{-b}{2a}$, a single solution.

5.
$$x^2 - 8x = 20$$
$$x^2 - 8x - 20 = 0 \qquad \text{Standard form}$$
We can factor.
$$x^2 - 8x - 20 = 0$$
$$(x - 10)(x + 2) = 0$$
$$x - 10 = 0 \quad or \quad x + 2 = 0$$
$$x = 10 \quad or \qquad x = -2$$
The solutions are -2 and 10.

7.
$$t^2 = 2t - 1$$
$$t^2 - 2t + 1 = 0 \qquad \text{Standard form}$$
We can factor.
$$t^2 - 2t + 1 = 0$$
$$(t - 1)(t - 1) = 0$$
$$t - 1 = 0 \quad or \quad t - 1 = 0$$
$$t = 1 \quad or \qquad t = 1$$
The solution is 1.

9. $3y^2 + 7y + 4 = 0$

We can factor.
$$3y^2 + 7y + 4 = 0$$
$$(3y + 4)(y + 1) = 0$$
$$3y + 4 = 0 \quad or \quad y + 1 = 0$$
$$3y = -4 \quad or \qquad y = -1$$
$$y = -\frac{4}{3} \quad or \qquad y = -1$$
The solutions are $-\dfrac{4}{3}$ and -1.

11.
$$4x^2 - 12x = 7$$
$$4x^2 - 12x - 7 = 0$$

We can factor.
$$4x^2 - 12x - 7 = 0$$
$$(2x + 1)(2x - 7) = 0$$

$$2x + 1 = 0 \quad or \quad 2x - 7 = 0$$
$$2x = -1 \quad or \quad 2x = 7$$
$$x = -\frac{1}{2} \quad or \quad x = \frac{7}{2}$$

The solutions are $-\frac{1}{2}$ and $\frac{7}{2}$.

13.
$$p^2 = 25$$
$$p = 5 \quad or \quad p = -5 \quad \text{Principle of square roots}$$

The solutions are 5 and -5.

15. $x^2 + 4x - 7 = 0$

We use the quadratic formula.
$$a = 1, \, b = 4, \, c = -7$$
$$x = \frac{-b \pm \sqrt{b^2 - 4ac}}{2a}$$
$$x = \frac{-4 \pm \sqrt{4^2 - 4 \cdot 1 \cdot (-7)}}{2 \cdot 1}$$
$$x = \frac{-4 \pm \sqrt{16 + 28}}{2}$$
$$x = \frac{-4 \pm \sqrt{44}}{2} = \frac{-4 \pm \sqrt{4 \cdot 11}}{2}$$
$$x = \frac{-4 \pm 2\sqrt{11}}{2} = \frac{2(-2 \pm \sqrt{11})}{2 \cdot 1}$$
$$x = -2 \pm \sqrt{11}$$

The solutions are $-2 + \sqrt{11}$ and $-2 - \sqrt{11}$, or $-2 \pm \sqrt{11}$.

17. $y^2 - 10y + 19 = 0$

We use the quadratic formula.
$$a = 1, \, b = -10, \, c = 19$$
$$y = \frac{-b \pm \sqrt{b^2 - 4ac}}{2a}$$
$$y = \frac{-(-10) \pm \sqrt{(-10)^2 - 4 \cdot 1 \cdot 19}}{2 \cdot 1}$$
$$y = \frac{10 \pm \sqrt{100 - 76}}{2}$$
$$y = \frac{10 \pm \sqrt{24}}{2} = \frac{10 \pm \sqrt{4 \cdot 6}}{2}$$
$$y = \frac{10 \pm 2\sqrt{6}}{2} = \frac{2(5 \pm \sqrt{6})}{2 \cdot 1}$$
$$y = 5 \pm \sqrt{6}$$

The solutions are $5 + \sqrt{6}$ and $5 - \sqrt{6}$, or $5 \pm \sqrt{6}$.

19. $x^2 - 10x + 25 = 3$

Observe that $x^2 - 10x + 25$ is a perfect-square trinomial. Then we can use the principle of square roots.
$$x^2 - 10x + 25 = 3$$
$$(x - 5)^2 = 3$$

$$x - 5 = \sqrt{3} \quad or \quad x - 5 = -\sqrt{3}$$
$$x = 5 + \sqrt{3} \quad or \quad x = 5 - \sqrt{3}$$

The solutions are $5 + \sqrt{3}$ and $5 - \sqrt{3}$, or $5 \pm \sqrt{3}$.

21. $3t^2 + 8t + 2 = 0$

We use the quadratic formula.
$$a = 3, \, b = 8, \, c = 2$$
$$t = \frac{-b \pm \sqrt{b^2 - 4ac}}{2a}$$
$$t = \frac{-8 \pm \sqrt{8^2 - 4 \cdot 3 \cdot 2}}{2 \cdot 3}$$
$$t = \frac{-8 \pm \sqrt{64 - 24}}{6} = \frac{-8 \pm \sqrt{40}}{6}$$
$$t = \frac{-8 \pm \sqrt{4 \cdot 10}}{6} = \frac{-8 \pm 2\sqrt{10}}{6}$$
$$t = \frac{2(-4 \pm \sqrt{10})}{2 \cdot 3} = \frac{-4 \pm \sqrt{10}}{3}$$

The solutions are $\frac{-4 + \sqrt{10}}{3}$ and $\frac{-4 - \sqrt{10}}{3}$, or $\frac{-4 \pm \sqrt{10}}{3}$, or $-\frac{4}{3} \pm \frac{\sqrt{10}}{3}$.

23.
$$2x^2 - 5x = 1$$
$$2x^2 - 5x - 1 = 0 \quad \text{Standard form}$$

We use the quadratic formula.
$$a = 2, \, b = -5, \, c = -1$$
$$x = \frac{-b \pm \sqrt{b^2 - 4ac}}{2a}$$
$$x = \frac{-(-5) \pm \sqrt{(-5)^2 - 4 \cdot 2 \cdot (-1)}}{2 \cdot 2}$$
$$x = \frac{5 \pm \sqrt{25 + 8}}{4} = \frac{5 \pm \sqrt{33}}{4}$$

The solutions are $\frac{5 + \sqrt{33}}{4}$ and $\frac{5 - \sqrt{33}}{4}$, or $\frac{5 \pm \sqrt{33}}{4}$, or $\frac{5}{4} \pm \frac{\sqrt{33}}{4}$.

25. $4y^2 + 2y - 3 = 0$

We use the quadratic formula.
$$a = 4, \, b = 2, \, c = -3$$
$$y = \frac{-b \pm \sqrt{b^2 - 4ac}}{2a}$$
$$y = \frac{-2 \pm \sqrt{2^2 - 4 \cdot 4 \cdot (-3)}}{2 \cdot 4}$$
$$y = \frac{-2 \pm \sqrt{4 + 48}}{8} = \frac{-2 \pm \sqrt{52}}{8}$$
$$y = \frac{-2 \pm \sqrt{4 \cdot 13}}{8} = \frac{-2 \pm 2\sqrt{13}}{8}$$
$$y = \frac{2(-1 \pm \sqrt{13})}{2 \cdot 4} = \frac{-1 \pm \sqrt{13}}{4}$$

The solutions are $\frac{-1 + \sqrt{13}}{4}$ and $\frac{-1 - \sqrt{13}}{4}$, or $\frac{-1 \pm \sqrt{13}}{4}$, or $-\frac{1}{4} \pm \frac{\sqrt{13}}{4}$.

27. $2m^2 - m + 3 = 0$

We use the quadratic formula.

$a = 2,\ b = -1,\ c = 3$

$$m = \frac{-b \pm \sqrt{b^2 - 4ac}}{2a}$$

$$m = \frac{-(-1) \pm \sqrt{(-1)^2 - 4 \cdot 2 \cdot 3}}{2 \cdot 2}$$

$$m = \frac{1 \pm \sqrt{1 - 24}}{4} = \frac{1 \pm \sqrt{-23}}{4}$$

Since the radicand, -23, is negative, there are no real-number solutions.

29. $3x^2 - 5x = 4$

$3x^2 - 5x - 4 = 0$

We use the quadratic formula.

$a = 3,\ b = -5,\ c = -4$

$$x = \frac{-b \pm \sqrt{b^2 - 4ac}}{2a}$$

$$x = \frac{-(-5) \pm \sqrt{(-5)^2 - 4 \cdot 3 \cdot (-4)}}{2 \cdot 3}$$

$$x = \frac{5 \pm \sqrt{25 + 48}}{6} = \frac{5 \pm \sqrt{73}}{6}$$

The solutions are $\frac{5 + \sqrt{73}}{6}$ and $\frac{5 - \sqrt{73}}{6}$, or $\frac{5 \pm \sqrt{73}}{6}$, or $\frac{5}{6} \pm \frac{\sqrt{73}}{6}$.

31. $2y^2 - 6y = 10$

$2y^2 - 6y - 10 = 0$

$y^2 - 3y - 5 = 0$ Multiplying by $\frac{1}{2}$

We use the quadratic formula.

$a = 1,\ b = -3,\ c = -5$

$$y = \frac{-b \pm \sqrt{b^2 - 4ac}}{2a}$$

$$y = \frac{-(-3) \pm \sqrt{(-3)^2 - 4 \cdot 1 \cdot (-5)}}{2 \cdot 1}$$

$$y = \frac{3 \pm \sqrt{9 + 20}}{2} = \frac{3 \pm \sqrt{29}}{2}$$

The solutions are $\frac{3 + \sqrt{29}}{2}$ and $\frac{3 - \sqrt{29}}{2}$, or $\frac{3 \pm \sqrt{29}}{2}$, or $\frac{3}{2} \pm \frac{\sqrt{29}}{2}$.

33. $6t^2 + 26t = 20$

$6t^2 + 26t - 20 = 0$

$2(3t^2 + 13t - 10) = 0$

$2(3t - 2)(t + 5) = 0$

$3t - 2 = 0$ or $t + 5 = 0$

$3t = 2$ or $t = -5$

$t = \frac{2}{3}$ or $t = -5$

The solutions are $\frac{2}{3}$ and -5.

35. $5t^2 - 7t = -4$

$5t^2 - 7t + 4 = 0$ Standard form

We use the quadratic formula.

$a = 5,\ b = -7,\ c = 4$

$$t = \frac{-b \pm \sqrt{b^2 - 4ac}}{2a}$$

$$t = \frac{-(-7) \pm \sqrt{(-7)^2 - 4 \cdot 5 \cdot 4}}{2 \cdot 5}$$

$$t = \frac{7 \pm \sqrt{49 - 80}}{10} = \frac{7 \pm \sqrt{-31}}{10}$$

Since the radicand, -31, is negative, there are no real-number solutions.

37. $5y^2 = 60$

$y^2 = 12$ Dividing by 5

$y = \sqrt{12}$ or $y = -\sqrt{12}$ Principle of square roots

$y = 2\sqrt{3}$ or $y = -2\sqrt{3}$

The solutions are $2\sqrt{3}$ and $-2\sqrt{3}$, or $\pm 2\sqrt{3}$.

39. $x^2 + 3x - 2 = 0$

$a = 1,\ b = 3,\ c = -2$

$$x = \frac{-3 \pm \sqrt{3^2 - 4 \cdot 1 \cdot (-2)}}{2 \cdot 1}$$

$$x = \frac{-3 \pm \sqrt{9 + 8}}{2} = \frac{-3 \pm \sqrt{17}}{2}$$

Using a calculator, we see that $\sqrt{17} \approx 4.123$:

$$\frac{-3 + \sqrt{17}}{2} \approx \frac{-3 + 4.123}{2} \quad or \quad \frac{-3 - \sqrt{17}}{2} \approx \frac{-3 - 4.123}{2}$$
$$\approx 0.562 \qquad or \qquad \approx -3.562$$

The approximate solutions, to the nearest thousandth, are 0.562 and -3.562.

41. $y^2 - 5y - 1 = 0$

$a = 1,\ b = -5,\ c = -1$

$$y = \frac{-b \pm \sqrt{b^2 - 4ac}}{2a}$$

$$y = \frac{-(-5) \pm \sqrt{(-5)^2 - 4 \cdot 1 \cdot (-1)}}{2 \cdot 1}$$

$$y = \frac{5 \pm \sqrt{25 + 4}}{2} = \frac{5 \pm \sqrt{29}}{2}$$

Using a calculator or Table 2, we see that $\sqrt{29} \approx 5.385$:

$$\frac{5 + \sqrt{29}}{2} \approx \frac{5 + 5.385}{2} \quad or \quad \frac{5 - \sqrt{29}}{2} \approx \frac{5 - 5.385}{2}$$
$$\approx 5.193 \qquad or \qquad \approx -0.193$$

The approximate solutions, to the nearest thousandth, are 5.193 and -0.193.

43.
$$4x^2 + 4x = 1$$

$$4x^2 + 4x - 1 = 0 \qquad \text{Standard form}$$

$$a = 4, \; b = 4, \; c = -1$$

$$x = \frac{-4 \pm \sqrt{4^2 - 4 \cdot 4 \cdot (-1)}}{2 \cdot 4}$$

$$x = \frac{-4 \pm \sqrt{16 + 16}}{8} = \frac{-4 \pm \sqrt{32}}{8}$$

$$x = \frac{-4 \pm \sqrt{16 \cdot 2}}{8} = \frac{-4 \pm 4\sqrt{2}}{8}$$

$$x = \frac{4(-1 \pm \sqrt{2})}{4 \cdot 2} = \frac{-1 \pm \sqrt{2}}{2}$$

Using a calculator or Table 2, we see that $\sqrt{2} \approx 1.414$:

$$\frac{-1 + \sqrt{2}}{2} \approx \frac{-1 + 1.414}{2} \quad or \quad \frac{-1 - \sqrt{2}}{2} \approx \frac{-1 - 1.414}{2}$$

$$\approx \frac{0.414}{2} \quad or \quad \approx \frac{-2.414}{2}$$

$$\approx 0.207 \quad or \quad \approx -1.207$$

The approximate solutions, to the nearest thousandth, are 0.207 and −1.207.

45. *Familiarize*. We will use the formula
$$d = \frac{n^2 - 3n}{2},$$
where d is the number of diagonals and n is the number of sides.

***Translate*.** We substitute 35 for d.
$$35 = \frac{n^2 - 3n}{2}$$

***Carry out*.** We solve the equation.
$$\frac{n^2 - 3n}{2} = 35$$
$$n^2 - 3n = 70 \quad \text{Multiplying by 2}$$
$$n^2 - 3n - 70 = 0$$
$$(n - 10)(n + 7) = 0$$
$$n - 10 = 0 \quad or \quad n + 7 = 0$$
$$n = 10 \quad or \qquad n = -7$$

***Check*.** Since the number of sides cannot be negative, −7 cannot be a solution. To check 10, we substitute 10 for n in the original formula and determine if this yields $d = 35$. This is left to the student.

***State*.** The polygon has 10 sides.

47. *Familiarize*. We will use the formula $s = 16t^2$.

***Translate*.** We substitute 1482 for s.
$$1482 = 16t^2$$

***Carry out*.** We solve the equation.
$$1482 = 16t^2$$
$$\frac{1482}{16} = t^2$$
$$\sqrt{\frac{1482}{16}} = t \quad or \quad -\sqrt{\frac{1482}{16}} = t \quad \begin{array}{l}\text{Principle of}\\ \text{square roots}\end{array}$$
$$9.62 \approx t \quad or \qquad -9.62 \approx t$$

***Check*.** The number −9.62 cannot be a solution, because time cannot be negative in this situation. We substitute 9.62 in the original equation:
$$s = 16(9.62)^2 = 16(92.5444) \approx 1481.$$
This is close to 1482. Remember that we approximated the solution. Thus, we have a check.

***State*.** It would take about 9.62 sec for an object to fall to the ground from the top of the Petronas Towers.

49. *Familiarize*. We will use the formula $s = 16t^2$.

***Translate*.** We substitute 28 for s.
$$28 = 16t^2$$

***Carry out*.** We solve the equation.
$$28 = 16t^2$$
$$\frac{7}{4} = t^2$$
$$\sqrt{\frac{7}{4}} = t \quad or \quad -\sqrt{\frac{7}{4}} = t \quad \begin{array}{l}\text{Principle of}\\ \text{square roots}\end{array}$$
$$1.32 \approx t \quad or \quad -1.32 \approx t$$

***Check*.** The number −1.32 cannot be a solution, because time cannot be negative in this situation. We substitute 1.36 in the original equation:
$$s = 16(1.32)^2 = 16(1.7424) = 27.8784.$$
This is close to 28. Remember that we approximated the solution. Thus, we have a check.

***State*.** The free-fall portion of the jump lasted about 1.32 sec.

51. *Familiarize*. From the drawing in the text we have $s =$ the length of the shorter leg and $s + 17 =$ the length of the longer leg, in feet.

***Translate*.** We use the Pythagorean theorem.
$$s^2 + (s + 17)^2 = 25^2$$

***Carry out*.** We solve the equation.
$$s^2 + s^2 + 34s + 289 = 625$$
$$2s^2 + 34s - 336 = 0$$
$$s^2 + 17s - 168 = 0 \qquad \text{Multiplying by } \frac{1}{2}$$
$$(s - 7)(s + 24) = 0$$
$$s - 7 = 0 \quad or \quad s + 24 = 0$$
$$s = 7 \quad or \qquad s = -24$$

***Check*.** Since the length of a leg cannot be negative, −24 does not check. But 7 does check. If the smaller leg is 7, the other leg is $7 + 17$, or 24. Then, $7^2 + 24^2 = 49 + 576 = 625$, and $\sqrt{625} = 25$, the length of the hypotenuse.

***State*.** The legs measure 7 ft and 24 ft.

53. *Familiarize*. From the drawing in the text, we see that w represents the width of the rectangle and $w + 4$ represents the length, in centimeters.

***Translate*.** The area is length $\times$ width. Thus, we have two expressions for the area of the rectangle: $(w+4)w$ and 60. This gives us a translation.
$$(w + 4)w = 60$$

Carry out. We solve the equation.

$$w^2 + 4w = 60$$
$$w^2 + 4w - 60 = 0$$
$$(w + 10)(w - 6) = 0$$
$$w + 10 = 0 \quad or \quad w - 6 = 0$$
$$w = -10 \quad or \quad w = 6$$

Check. Since the length of a side cannot be negative, -10 does not check. But 6 does check. If the width is 6, then the length is $6 + 4$, or 10. The area is 10×6, or 60. This checks.

State. The length is 10 cm, and the width is 6 cm.

55. *Familiarize*. We make a drawing. We let w = the width of the yard. Then $w + 6$ = the length, in meters.

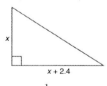

Translate. We use the Pythagorean theorem.

$$w^2 + (w + 6)^2 = 30^2$$

Carry out. We solve the equation.

$$w^2 + w^2 + 12w + 36 = 900$$
$$2w^2 + 12w - 864 = 0$$
$$w^2 + 6w - 432 = 0$$
$$(w + 24)(w - 18) = 0$$
$$w + 24 = 0 \quad or \quad w - 18 = 0$$
$$w = -24 \quad or \quad w = 18$$

Check. Since the width cannot be negative, -24 does not check. But 18 does check. If the width is 18, then the length is $18 + 6$, or 24, and $18^2 + 24^2 = 324 + 576 = 900 = 30^2$.

State. The yard is 18 m by 24 m.

57. *Familiarize*. We make a drawing. Let x = the length of the shorter leg of the right triangle. Then $x + 5$ = the length of the longer leg, in cm.

Translate. Using the formula $A = \frac{1}{2}bh$, we substitute 26 for A, $x + 5$ for b, and x for h.

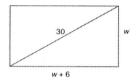

$$26 = \frac{1}{2}(x + 5)(x)$$

Carry out. We solve the equation.

$$26 = \frac{1}{2}(x + 5)(x)$$
$$52 = (x + 5)(x) \qquad \text{Multiplying by 2}$$
$$52 = x^2 + 5x$$
$$0 = x^2 + 5x - 52$$

We use the quadratic formula.

$$a = 1, \, b = 5, \, c = -52$$
$$x = \frac{-5 \pm \sqrt{5^2 - 4 \cdot 1 \cdot (-52)}}{2 \cdot 1}$$
$$x = \frac{-5 \pm \sqrt{233}}{2}$$
$$x = \frac{-5 + \sqrt{233}}{2} \quad or \quad x = \frac{-5 - \sqrt{233}}{2}$$
$$x \approx 5.13 \qquad or \quad x \approx -10.1$$

Check. Since the length of a leg cannot be negative, -10.1 does not check. But 5.13 does. If the shorter leg is 5.13 cm, then the longer leg is $5.13 + 5$, $= 10.13$ cm, and $A = \frac{1}{2}(5.13)(10.13) \approx 25.98 \approx 26$.

State. The lengths of the legs are about 5.13 cm and 10.13 cm.

59. *Familiarize*. We first make a drawing. We let x represent the width and $x + 5$ the length, in feet.

Translate. The area is length × width. We have two expressions for the area of the rectangle: $(x + 5)x$ and 25. This gives us a translation.

$$(x + 5)x = 25$$

Carry out. We solve the equation.

$$x^2 + 5x = 25$$
$$x^2 + 5x - 25 = 0$$
$$a = 1, \, b = 5, \, c = -25$$
$$x = \frac{-5 \pm \sqrt{5^2 - 4 \cdot 1 \cdot (-25)}}{2 \cdot 1}$$
$$x = \frac{-5 \pm \sqrt{125}}{2} = \frac{-5 \pm 5\sqrt{5}}{2}$$
$$x = \frac{-5 + 5\sqrt{5}}{2} \quad or \quad x = \frac{-5 - 5\sqrt{5}}{2}$$
$$x \approx 3.09 \qquad or \quad x \approx -8.09$$

Check. Since the width cannot be negative, -8.09 does not check. But 3.09 does check. If the width is 3.09 ft., then the length is $3.09 + 5$, or 8.09 ft., and the area is $3.09(8.09) \approx 25 \text{ ft}^2$.

State. The length is about 8.09 ft., and the width is about 3.09 ft.

61. *Familiarize*. We first make a drawing. We let x represent the width and $2x$ the length, in meters.

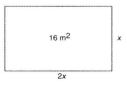

Translate. The area is length × width. We have two expressions for the area of the rectangle: $2x \cdot x$ and 16. This gives us a translation.

$$2x \cdot x = 16$$

Carry out. We solve the equation.

$$2x^2 = 16$$
$$x^2 = 8$$
$$x = \sqrt{8} \quad or \quad x = -\sqrt{8}$$
$$x = 2.83 \quad or \quad x \approx -2.83 \quad \text{Using a calculator}$$
$$\text{or Table 2}$$

Check. Since the length cannot be negative, -2.83 does not check. But 2.83 does check. If the width is $\sqrt{8}$ m, then the length is $(2\sqrt{8})$ or 5.66 m. The area is $(5.66)(2.83)$, or $16.0178 \approx 16$.

State. The length is about 5.66 m, and the width is about 2.83 m.

63. ***Familiarize***. We will use the formula $A = P(1 + r)^t$.

Translate. We substitute 2560 for P, 3610 for A, and 2 for t.

$$3610 = 2560(1 + r)^2$$

Carry out. We solve the equation.

$$3610 = 2560(1 + r)^2$$
$$\frac{3610}{2560} = (1 + r)^2$$
$$\sqrt{\frac{3610}{2560}} = 1 + r \quad or \quad -\sqrt{\frac{3610}{2560}} = 1 + r \quad \text{Principle}$$
$$\text{of square roots}$$
$$1.1875 - 1 = r \quad or \quad -1.1875 - 1 = r$$
$$0.1875 = r \quad or \quad -2.1875 = r$$

Check. Since the interest rate cannot be negative, we check only 0.1875, or 18.75%. We substitute in the formula: $2560(1 + 0.1875)^2 = 2560(1.1875)^2 = 3610$. The answer checks.

State. The interest rate is 18.75%.

65. ***Familiarize***. We will use the formula $A = P(1 + r)^t$.

Translate. We substitute 6000 for P, 6615 for A, and 2 for t.

$$6615 = 6000(1 + r)^2$$

Carry out. We solve the equation.

$$6615 = 6000(1 + r)^2$$
$$1.1025 = (1 + r)^2$$
$$\sqrt{1.1025} = 1 + r \quad or \quad -\sqrt{1.1025} = 1 + r$$
$$\text{Principle of square roots}$$
$$1.05 = 1 + r \quad or \quad -1.05 = 1 + r$$
$$0.05 = r \quad or \quad -2.05 = r$$

Check. Since the interest rate cannot be negative, we check only 0.05, or 5%. We substitute in the formula: $6000(1 + 0.05)^2 = 6000(1.05)^2 = 6615$. The answer checks.

State. The interest rate is 5%.

67. ***Familiarize***. Let d = the diameter (or width) of the oil slick, in miles. Then $d/2$ = the radius. We will use the formula for the area of the circle, $A = \pi r^2$.

Translate. We substitute 100 for A, 3.14 for π, and $d/2$ for r in the formula.

$$100 = 3.14\left(\frac{d}{2}\right)^2$$
$$100 = 3.14\left(\frac{d^2}{4}\right)$$

Carry out. We solve the equation.

$$100 = 3.14\left(\frac{d^2}{4}\right)$$
$$100 = 0.785d^2 \quad \left(\frac{3.14}{4} = 0.785\right)$$
$$\frac{100}{0.785} = d^2$$
$$d = \sqrt{\frac{100}{0.785}} \quad or \quad d = -\sqrt{\frac{100}{0.785}}$$
$$d \approx 11.28 \quad or \quad d \approx -11.28$$

Check. Since the diameter cannot be negative, -11.28 cannot be a solution. If $d = 11.28$, then

$$A = 3.14\left(\frac{11.28}{2}\right)^2 \approx 100. \text{ The answer checks.}$$

State. The oil slick was about 11.28 mi wide.

69. *Writing Exercise*. The quadratic formula would not be the easiest way to solve a quadratic equation when the equation can be solved by factoring or by using the principle of square roots.

71.
$$a = \frac{1}{3}b$$
$$3a = b$$

73.
$$t = \frac{c + d}{2}$$
$$2t = c + d$$
$$2t - c = d$$

75.
$$y = Ax + B$$
$$y - B = Ax$$
$$\frac{y - B}{A} = x$$

77. *Writing Exercise*. When using the quadratic formula, the quantity $b^2 - 4ac$ is the radicand. If $b^2 - 4ac$ is a perfect square, then $\sqrt{b^2 - 4ac}$ is rational and the solutions are rational. If $b^2 - 4ac$ is not a perfect square, then $\sqrt{b^2 - 4ac}$ is not rational, and the solutions are also not rational.

79.
$$5x = -x(x + 6)$$
$$5x = -x^2 - 6x$$
$$x^2 + 5x + 6x = 0$$
$$x^2 + 11x = 0$$
$$x(x + 11) = 0$$
$$x = 0 \quad or \quad x + 11 = 0$$
$$x = 0 \quad or \quad x = -11$$

The solutions are 0 and -11.

81. $3 - x(x - 3) = 4$

$3 - x^2 + 3x = 4$

$0 = x^2 - 3x + 1$

$a = 1,\ b = -3,\ c = 1$

$x = \dfrac{-(-3) \pm \sqrt{(-3)^2 - 4 \cdot 1 \cdot 1}}{2 \cdot 1}$

$x = \dfrac{3 \pm \sqrt{5}}{2}$

The solutions are $\dfrac{3 + \sqrt{5}}{2}$ and $\dfrac{3 - \sqrt{5}}{2}$, or $\dfrac{3 \pm \sqrt{5}}{2}$, or $\dfrac{3}{2} \pm \dfrac{\sqrt{5}}{2}$.

83. $(y + 4)(y + 3) = 15$

$y^2 + 7y + 12 = 15$

$y^2 + 7y - 3 = 0$

$a = 1,\ b = 7,\ c = -3$

$y = \dfrac{-7 \pm \sqrt{7^2 - 4 \cdot 1 \cdot (-3)}}{2 \cdot 1}$

$y = \dfrac{-7 \pm \sqrt{61}}{2}$

The solutions are $\dfrac{-7 + \sqrt{61}}{2}$ and $\dfrac{-7 - \sqrt{61}}{2}$, or $\dfrac{-7 \pm \sqrt{61}}{2}$, or $-\dfrac{7}{2} \pm \dfrac{\sqrt{61}}{2}$.

85. $\dfrac{x^2}{x + 3} = \dfrac{11}{x + 3}$,   LCD is $x + 3$

$(x + 3)\dfrac{x^2}{x + 3} = (x + 3)\dfrac{11}{x + 3}$

$x^2 = 11$

$x = \pm\sqrt{11}$ Principle of square roots

Both numbers check. The solutions are $\sqrt{11}$ and $-\sqrt{11}$, or $\pm\sqrt{11}$.

87. $\dfrac{1}{x} + \dfrac{1}{x + 1} = \dfrac{1}{3}$,   LCM is $3x(x + 1)$

$3x(x + 1)\left(\dfrac{1}{x} + \dfrac{1}{x + 1}\right) = 3x(x + 1) \cdot \dfrac{1}{3}$

$3(x + 1) + 3x = x(x + 1)$

$3x + 3 + 3x = x^2 + x$

$6x + 3 = x^2 + x$

$0 = x^2 - 5x - 3$

$x = \dfrac{-(-5) \pm \sqrt{(-5)^2 - 4 \cdot 1 \cdot (-3)}}{2 \cdot 1}$

$x = \dfrac{5 \pm \sqrt{37}}{2}$

The solutions are $\dfrac{5 + \sqrt{37}}{2}$ and $\dfrac{5 - \sqrt{37}}{2}$, or $\dfrac{5 \pm \sqrt{37}}{2}$, or $\dfrac{5}{2} \pm \dfrac{\sqrt{37}}{2}$.

89. *Familiarize.* From the drawing in the text, we see that we have a right triangle where $r =$ the length of each leg and $r + 2 =$ the length of the hypotenuse, in centimeters.

Translate. We use the Pythagorean theorem.

$r^2 + r^2 = (r + 2)^2$.

Carry out. We solve the equation.

$2r^2 = r^2 + 4r + 4$

$r^2 - 4r - 4 = 0$

$a = 1,\ b = -4,\ c = -4$

$r = \dfrac{-(-4) \pm \sqrt{(-4)^2 - 4 \cdot 1 \cdot (-4)}}{2 \cdot 1}$

$r = \dfrac{4 \pm \sqrt{16 + 16}}{2} = \dfrac{4 \pm \sqrt{32}}{2}$

$r = \dfrac{4 \pm \sqrt{16 \cdot 2}}{2} = \dfrac{4 \pm 4\sqrt{2}}{2}$

$r = \dfrac{2(2 \pm 2\sqrt{2})}{2 \cdot 1} = 2 \pm 2\sqrt{2}$

$x = 2 - 2\sqrt{2}$ *or* $x = 2 + 2\sqrt{2}$

$x \approx 2 - 2.828$ *or* $x \approx 2 + 2.828$

$x \approx -0.828$ *or* $x \approx 4.828$

$x \approx -0.83$ *or* $x \approx 4.83$ Rounding to the nearest hundredth

Check. Since the length of a leg cannot be negative, -0.83 cannot be a solution of the original equation. When $x \approx 4.83$, then $x + 2 \approx 6.83$ and $(4.83)^2 + (4.83)^2 = 23.3289 + 23.3289 = 46.6578 \approx (6.83)^2$. This checks.

State. In the figure, $r = 2 + 2\sqrt{2}$ cm ≈ 4.83 cm.

91. *Familiarize.* Let $w =$ the width of the rectangle, in meters. Then $1.6w =$ the length. Recall that the formula for the area of a rectangle is $A = l \times w$.

Translate. We substitute 9000 for A, $1.6w$ for w, and w for w in the formula.

$9000 = 1.6w(w)$

$9000 = 1.6w^2$

Carry out. We solve the equation.

$9000 = 1.6w^2$

$\dfrac{9000}{1.6} = w^2$

$w = \sqrt{\dfrac{9000}{1.6}}$ *or* $w = -\sqrt{\dfrac{9000}{1.6}}$

$w = 75$ *or* $w = -75$

Check. Since the width of the rectangle cannot be negative we will not check -75. If $w = 75$, then the length is $1.6(75)$, or 120, and the area is $120(75) = 9000$ m^2. The answer checks.

State. The length of the rectangle is 120 m, and the width is 75 m.

93. *Familiarize.* We will use the formula $A = P(1 + r)^2$ twice. The amount in the account for the \$4000 invested for 2 yr is given by $4000(1 + r)^2$. The amount for the \$2000 invested for 1 yr is $2000(1 + r)$.

Translate. The total amount in the account at the end of 2 yr is the sum of the amounts above.

$6510 = 4000(1 + r)^2 + 2000(1 + r)$

Carry out. We solve the equation. Begin by letting $x = 1 + r$.

$$6510 = 4000x^2 + 2000x$$
$$0 = 4000x^2 + 2000x - 6510$$
$$0 = 400x^2 + 200x - 651 \quad \text{Dividing by 10}$$

We will use the quadratic formula.

$$a = 400, \ b = 200, \ c = -651$$

$$x = \frac{-200 \pm \sqrt{200^2 - 4 \cdot 400(-651)}}{2 \cdot 400}$$

$$x = \frac{-200 \pm \sqrt{1,081,600}}{800} = \frac{-200 \pm 1040}{800}$$

$$x = \frac{-200 + 1040}{800} \quad or \quad x = \frac{-200 - 1040}{800}$$

$$x = 1.05 \qquad or \qquad x = -1.55$$

$$1 + r = 1.05 \qquad or \quad 1 + r = -1.55$$

$$r = 0.05 \qquad or \qquad r = -2.55$$

Check. Since the interest rate cannot be negative, we check only 0.05 or 5%. At the end of 2 yr, $4000 invested at 5% interest has grown to $4000(1 + 0.05)^2$, or $4410. At the end of 1 yr, $2000 invested at 5% interest has grown to $2000(1 + 0.05)$, or $2100. Then the total amount in the account is $4410 + $2100, or $6510. The answer checks.

State. The interest rate is 5%.

95. Familiarize. The area of the actual strike zone is $15(40)$, so the area of the enlarged zone is $15(40) + 0.4(15)(40)$, or $1.4(15)(40)$. From the drawing in the text we see that the dimensions of the enlarged strike zone are $15 + 2x$ by $40 + 2x$.

Translate. Using the formula $A = lw$, we write an equation for the area of the enlarged strike zone.

$$1.4(15)(40) = (15 + 2x)(40 + 2x)$$

Carry out. We solve the equation.

$$1.4(15)(40) = (15 + 2x)(40 + 2x)$$
$$840 = 600 + 110x + 4x^2 \quad \begin{array}{l}\text{Multiplying} \\ \text{on both sides}\end{array}$$
$$0 = 4x^2 + 110x - 240$$
$$0 = 2x^2 + 55x - 120 \quad \text{Dividing by 2}$$

$$a = 2, \ b = 55, \ c = -120$$

$$x = \frac{-55 \pm \sqrt{55^2 - 4 \cdot 2 \cdot (-120)}}{2 \cdot 2}$$

$$x = \frac{-55 \pm \sqrt{3985}}{4}$$

$$x \approx 2.03 \ \ or \ \ x \approx -29.53$$

Check. Since the measurement cannot be negative, -29.53 cannot be a solution. If $x = 2.03$, then the dimensions of the enlarged strike zone are $15 + 2(2.03)$, or 19.06, by $40 + 2(2.03)$, or 44.06, and the area is $19.06(44.06) = 839.7836 \approx 840$. The answer checks.

State. The dimensions of the enlarged strike zone are 19.06 in. by 44.06 in.

Chapter 9 Connecting the Concepts

1. $(x + 2)(x - 1) = 0$ Already factored

$$x + 2 = 0 \ \ \text{or} \ \ x - 1 = 0$$
$$x = -2 \ \ \text{or} \ \ x = 1$$

The solutions are -2 and 1.

3. $x^2 + 3x - 5 = 0$

$$x = \frac{-b \pm \sqrt{b^2 - 4ac}}{2a} \qquad \begin{array}{l}\text{Using the} \\ \text{quadratic formula}\end{array}$$

$$x = \frac{-3 \pm \sqrt{3^2 - 4 \cdot 1 \cdot (-5)}}{2(1)}$$

$$x = \frac{-3 \pm \sqrt{9 + 20}}{2}$$

$$x = \frac{-3 \pm \sqrt{29}}{2}$$

The solutions are $\dfrac{-3 \pm \sqrt{29}}{2}$ or $-\dfrac{3}{2} \pm \dfrac{\sqrt{29}}{2}$.

5. $2x^2 - 5x - 3 = 0$

$$(2x + 1)(x - 3) = 0 \qquad \text{Using factoring}$$

$$2x + 1 = 0 \ \ \text{or} \ \ x - 3 = 0$$

$$x = -\frac{1}{2} \ \ \text{or} \ \ x = 3$$

The solutions are $-\dfrac{1}{2}$ and 3.

7. $x^2 + x + 1 = 0$

Since $b^2 - 4ac = 1^2 - 4 \cdot 1 \cdot 1 = -3 < 0$

there are no real-number solutions.

9. $x^2 = 5x$

$$x^2 - 5x = 0 \qquad \text{Using factoring}$$
$$x(x - 5) = 0$$
$$x = 0 \ \ \text{or} \ \ x - 5 = 0$$
$$x = 0 \ \ \text{or} \ \ x = 5$$

The solutions are 0 and 5.

11.
$$x^2 + x = 6$$
$$x^2 + x - 6 = 0$$
$$(x + 3)(x - 2) = 0 \qquad \text{Using factoring}$$
$$x + 3 = 0 \ \ \text{or} \ \ x - 2 = 0$$
$$x = -3 \ \ \text{or} \ \ x = 2$$

The solutions are -3 and 2.

13. $x^2 + 8x + 1 = 0$

$$x^2 + 8x + 16 = -1 + 16 \qquad \begin{array}{l}\text{Using completing} \\ \text{the square}\end{array}$$

$$(x + 4)^2 = 15$$
$$x + 4 = \pm\sqrt{15}$$
$$x = -4 \pm \sqrt{15}$$

The solutions are $-4 \pm \sqrt{15}$.

15.
$$121x^2 - 1 = 0$$
$$(11x - 1)(11x + 1) = 0 \qquad \text{Using factoring}$$
$$11x - 1 = 0 \quad \text{or} \quad 11x - 1 = 0$$
$$x = \frac{1}{11} \quad \text{or} \quad x = -\frac{1}{11}$$
The solutions are $-\dfrac{1}{11}$ and $\dfrac{1}{11}$.

17.
$$(x + 1)^2 - 3 = 0$$
$$(x + 1)^2 = 3$$
$$x + 1 = \pm\sqrt{3} \qquad \text{Using principle}$$
$$\text{of square roots}$$
$$x = -1 \pm \sqrt{3}$$
The solutions are $-1 \pm \sqrt{3}$.

19.
$$(x + 2)(x - 1) = 3$$
$$x^2 + x - 2 = 3$$
$$x^2 + x - 5 = 0$$
$$x = \frac{-b \pm \sqrt{b^2 - 4ac}}{2a} \qquad \text{Using the}$$
$$\text{quadratic formula}$$
$$x = \frac{-1 \pm \sqrt{1^2 - 4(1)(-5)}}{2(1)}$$
$$x = \frac{-1 \pm \sqrt{1 + 20}}{2}$$
$$x = \frac{-1 \pm \sqrt{21}}{2}$$
The solutions are $\dfrac{-1 \pm \sqrt{21}}{2}$ or $-\dfrac{1}{2} \pm \dfrac{\sqrt{21}}{2}$.

Exercise Set 9.4

1. Since t is the radicand, choice (b) is correct.

3. Since multiplying both sides by a will isolate t, choice (c) is correct.

5. $A = \dfrac{1}{2}bh$

$$2A = bh$$
$$\frac{2A}{b} = h$$

7. $Q = \dfrac{100m}{c}$

$$cQ = 100m \qquad \text{Multiplying by } c$$
$$c = \frac{100m}{Q} \qquad \text{Dividing by } Q$$

9.
$$A = P(1 + rt)$$
$$A = P + Prt$$
$$A - P = Prt$$
$$\frac{A - P}{Pr} = t, \text{ or}$$
$$\frac{A}{Pr} - \frac{1}{r} = t$$

11. $d = c\sqrt{h}$

$$d^2 = (c\sqrt{h})^2 \qquad \text{Squaring both sides}$$
$$d^2 = c^2 \cdot h$$
$$\frac{d^2}{c^2} = h \qquad \text{Dividing by } c^2$$

13.
$$\frac{1}{R} = \frac{1}{r_1} + \frac{1}{r_2}$$
$$Rr_1r_2 \cdot \frac{1}{R} = Rr_1r_2\left(\frac{1}{r_1} + \frac{1}{r_2}\right) \begin{array}{l}\text{Multiplying by} \end{array}$$
$$\text{the LCD, } Rr_1r_2, \text{ to clear fractions}$$
$$r_1r_2 = Rr_1r_2 \cdot \frac{1}{r_1} + Rr_1r_2 \cdot \frac{1}{r_2}$$
$$r_1r_2 = Rr_2 + Rr_1$$
$$r_1r_2 = R(r_2 + r_1)$$
$$\frac{r_1r_2}{r_2 + r_1} = R$$

15. $ax^2 + bx + c = 0$

Observe that this is standard form for a quadratic equation. The solution is given by the quadratic formula:
$$x = \frac{-b \pm \sqrt{b^2 - 4ac}}{2a}.$$

17.
$$Ax + By = C$$
$$By = C - Ax$$
$$y = \frac{C - Ax}{B}$$

19.
$$S = 2\pi r(r + h)$$
$$\frac{S}{2\pi r} = r + h \qquad \text{Dividing by } 2\pi r$$
$$\frac{S}{2\pi r} - r = h, \text{ or}$$
$$\frac{S - 2\pi r^2}{2\pi r} = h$$

21.
$$\frac{M - g}{t} = r + s$$
$$M - g = t(r + s)$$
$$\frac{M - g}{r + s} = t$$

23.
$$ab = ac + d$$
$$ab - ac = d$$
$$a(b - c) = d$$
$$a = \frac{d}{b - c}$$

25. $s = \dfrac{1}{2}gt^2$

$$2s = gt^2 \qquad \text{Multiplying by } 2$$
$$\frac{2s}{g} = t^2 \qquad \text{Dividing by } g$$
$$\pm\sqrt{\frac{2s}{g}} = t$$

27. $\dfrac{s}{h} = \dfrac{h}{t} \qquad \text{LCD} = ht$

$$ht \cdot \frac{s}{h} = ht \cdot \frac{h}{t}$$
$$st = h^2$$
$$\pm\sqrt{st} = h$$

29.
$$d = (r - w)t$$
$$\frac{d}{t} = r - w$$
$$\frac{d}{t} - r = -w$$
$$w = r - \frac{d}{t} \quad \text{or} \quad w = \frac{rt - d}{t}$$

31. $\sqrt{2x - y} + 1 = 6$

$\sqrt{2x - y} = 5$

$2x - y = 25$

$2x = 25 + y$

$x = \dfrac{25 + y}{2}$

33. $mt^2 + nt - p = 0$

$a = m,\ b = n,\ c = -p$

$t = \dfrac{-n \pm \sqrt{n^2 - 4 \cdot m \cdot (-p)}}{2 \cdot m}$

$t = \dfrac{-n \pm \sqrt{n^2 + 4mp}}{2m}$

35. $m + t = \dfrac{n}{m}$

$m(m + t) = m \cdot \dfrac{n}{m}$

$m^2 + mt = n$

$m^2 + mt - n = 0$

$a = 1,\ b = t,\ c = -n$

$m = \dfrac{-t \pm \sqrt{t^2 - 4 \cdot 1 \cdot (-n)}}{2 \cdot 1}$

$m = \dfrac{-t \pm \sqrt{t^2 + 4n}}{2}$

37. $n = p - 3\sqrt{t + c}$

$n - p = -3\sqrt{t + c}$

$(n - p)^2 = (-3\sqrt{t + c})^2$

$n^2 - 2np + p^2 = 9(t + c)$

$n^2 - 2np + p^2 = 9t + 9c$

$n^2 - 2np + p^2 - 9c = 9t$

$\dfrac{n^2 - 2np + p^2 - 9c}{9} = t$

39. $\sqrt{2t + s} = \sqrt{s - t}$

$(\sqrt{2t + s})^2 = (\sqrt{s - t})^2$

$2t + s = s - t$

$3t = 0$

$t = 0$

41. *Writing Exercise.* It is slightly easier to solve $\dfrac{1}{p} + \dfrac{1}{q} = \dfrac{1}{f}$ for f because, when clearing fractions, the multiplication on the right $\left(pqf \cdot \dfrac{1}{f}\right)$ is easier to perform than the corresponding multiplication in the first equation $\left(25 \cdot 23 \cdot x \cdot \dfrac{1}{x}\right)$.

43. $\sqrt{48} = \sqrt{16 \cdot 3} = \sqrt{16} \cdot \sqrt{3} = 4\sqrt{3}$

45. $\sqrt{6} \cdot \sqrt{15} = \sqrt{6 \cdot 15} = \sqrt{9 \cdot 10} = 3\sqrt{10}$

47. $\sqrt{20} \cdot \sqrt{30} = \sqrt{20 \cdot 30} = \sqrt{100 \cdot 6} = 10\sqrt{6}$

49. *Writing Exercise.* No; for $A \neq 0,\ B \neq 0,\ A = B$ is equivalent to $\dfrac{1}{A} = \dfrac{1}{B}$:

$A = B$

$\dfrac{1}{AB} \cdot A = \dfrac{1}{AB} \cdot B$

$\dfrac{1}{B} = \dfrac{1}{A}$

51. Substitute 8 for c and 224 for d and solve for a.

$c = \dfrac{a}{a + 12} \cdot d$

$8 = \dfrac{a}{a + 12} \cdot 24$

$\dfrac{1}{3} = \dfrac{a}{a + 12} \qquad$ Dividing by 24

$3(a + 12) \cdot \dfrac{1}{3} = 3(a + 12) \cdot \dfrac{a}{a + 12}$

$a + 12 = 3a$

$12 = 2a$

$6 = a$

The child is 6 years old.

53. $fm = \dfrac{gm - t}{m}$

$fm^2 = gm - t \qquad$ Multiplying by m

$fm^2 - gm + t = 0$

$a = f,\ b = -g,\ c = t$

$m = \dfrac{-(-g) \pm \sqrt{(-g)^2 - 4 \cdot f \cdot t}}{2 \cdot f}$

$m = \dfrac{g \pm \sqrt{g^2 - 4ft}}{2f}$

55. $V = \dfrac{k}{\sqrt{at + b}} + c$

$V - c = \dfrac{k}{\sqrt{at + b}}$

$(V - c)^2 = \left(\dfrac{k}{\sqrt{at + b}}\right)^2$

$(V - c)^2 = \dfrac{k^2}{at + b}$

$at + b = \dfrac{k^2}{(V - c)^2} \qquad$ Multiplying by $\dfrac{at + b}{(V - c)^2}$

$at = -b + \dfrac{k^2}{(V - c)^2}$

$t = -\dfrac{b}{a} + \dfrac{k^2}{a(V - c)^2} \qquad$ Dividing by a

57. When $C = F$, we have

$C = \dfrac{5}{9}(C - 32)$

$9C = 5(C - 32)$

$9C = 5C - 160$

$4C = -160$

$C = -40$

At $-40°$ the Fahrenheit and Celsius readings are the same.

Exercise Set 9.5

1. False; a complex number $a + bi$ with $b \neq 0$ is not a real number.

3. True; complex numbers $a + bi$ with $b \neq 0$ are imaginary numbers.

5. $\sqrt{-1} = i$

7. $\sqrt{-49} = \sqrt{-1 \cdot 49} = \sqrt{-1} \cdot \sqrt{49} = i \cdot 7 = 7i$

9. $\sqrt{-5} = \sqrt{-1 \cdot 5} = \sqrt{-1} \cdot \sqrt{5}$
$= i \cdot \sqrt{5} = \sqrt{5}i$

11. $\sqrt{-45} = \sqrt{-1 \cdot 9 \cdot 5} = \sqrt{-1} \cdot \sqrt{9} \cdot \sqrt{5}$
$= i \cdot 3\sqrt{5} = 3i\sqrt{5}, \text{ or } 3\sqrt{5}i$

13. $-\sqrt{-50} = -\sqrt{-1 \cdot 25 \cdot 2} = -\sqrt{-1} \cdot \sqrt{25} \cdot \sqrt{2}$
$= -i \cdot 5\sqrt{2} = -5i\sqrt{2}, \text{ or } -5\sqrt{2}i$

15. $4 + \sqrt{-49} = 4 + \sqrt{-1 \cdot 49} = 4 + \sqrt{-1} \cdot \sqrt{49}$
$= 4 + i \cdot 7 = 4 + 7i$

17. $3 - \sqrt{-9} = 3 - \sqrt{-1 \cdot 9} = 3 - \sqrt{-1} \cdot \sqrt{9}$
$= 3 - i \cdot 3 = 3 - 3i$

19. $-2 + \sqrt{-75} = -2 + \sqrt{-1 \cdot 25 \cdot 3} = -2 + \sqrt{-1} \cdot \sqrt{25} \cdot \sqrt{3}$
$= -2 + i \cdot 5\sqrt{3} = -2 + 5i\sqrt{3} \text{ or } -2 + 5\sqrt{3}i$

21. $x^2 + 25 = 0$
$x^2 = -25$
$x = \sqrt{-25} \quad or \quad x = -\sqrt{-25}$
$x = \sqrt{-1}\sqrt{25} \quad or \quad x = -\sqrt{-1}\sqrt{25}$
$x = 5i \quad or \quad x = -5i$
The solutions are $5i$ and $-5i$, or $\pm 5i$.

23. $x^2 = -28$
$x = \sqrt{-28} \quad or \quad x = -\sqrt{-28} \quad$ Principle of square roots
$x = \sqrt{-1 \cdot 4 \cdot 7} \quad or \quad x = -\sqrt{-1 \cdot 4 \cdot 7}$
$x = i \cdot 2\sqrt{7} \quad or \quad x = -i \cdot 2\sqrt{7}$
$x = 2i\sqrt{7} \quad or \quad x = -2i\sqrt{7}$
The solutions are $2i\sqrt{7}$ and $-2i\sqrt{7}$, or $\pm 2i\sqrt{7}$.

25. $t^2 + 4t + 5 = 0$
$a = 1, b = 4, c = 5$
$t = \dfrac{-b \pm \sqrt{b^2 - 4ac}}{2a}$
$t = \dfrac{-4 \pm \sqrt{4^2 - 4 \cdot 1 \cdot 5}}{2 \cdot 1}$
$t = \dfrac{-4 \pm \sqrt{-4}}{2} = \dfrac{-4 \pm \sqrt{-1}\sqrt{4}}{2}$
$t = \dfrac{-4 \pm 2i}{2}$
$t = -2 \pm i \quad$ Writing in the form $a + bi$
The solutions are $-2 \pm i$.

27. $(x - 4)^2 = -9$
$x - 4 = \pm\sqrt{-9} \quad$ Principle of square roots
$x - 4 = \pm\sqrt{-1}\sqrt{9}$
$x - 4 = \pm 3i$
$x = 4 \pm 3i$
The solutions are $4 \pm 3i$.

29. $x^2 + 5 = 2x$
$x^2 - 2x + 5 = 0$
$a = 1, b = -2, c = 5$
$x = \dfrac{-b \pm \sqrt{b^2 - 4ac}}{2a}$
$x = \dfrac{-(-2) \pm \sqrt{(-2)^2 - 4 \cdot 1 \cdot 5}}{2 \cdot 1}$
$x = \dfrac{2 \pm \sqrt{-16}}{2} = \dfrac{2 \pm i\sqrt{16}}{2}$
$x = \dfrac{2 \pm 4i}{2} = \dfrac{2}{2} \pm \dfrac{4i}{2}$
$x = 1 \pm 2i$
The solutions are $1 \pm 2i$.

31. $t^2 + 7 - 4t = 0$
$t^2 - 4t + 7 = 0 \quad$ Standard form
$a = 1, b = -4, c = 7$
$t = \dfrac{-b \pm \sqrt{b^2 - 4ac}}{2a}$
$t = \dfrac{-(-4) \pm \sqrt{(-4)^2 - 4 \cdot 1 \cdot 7}}{2 \cdot 1}$
$t = \dfrac{4 \pm \sqrt{-12}}{2} = \dfrac{4 \pm i\sqrt{12}}{2}$
$t = \dfrac{4 \pm 2i\sqrt{3}}{2} = \dfrac{4}{2} \pm \dfrac{2\sqrt{3}}{2}i = 2 \pm \sqrt{3}i$
The solutions are $2 \pm \sqrt{3}i$.

33. $5y^2 + 4y + 1 = 0$
$a = 5, b = 4, c = 1$
$y = \dfrac{-b \pm \sqrt{b^2 - 4ac}}{2a}$
$y = \dfrac{-4 \pm \sqrt{4^2 - 4 \cdot 5 \cdot 1}}{2(5)}$
$y = \dfrac{-4 \pm \sqrt{-4}}{10} = \dfrac{-4 \pm i\sqrt{4}}{10} = \dfrac{-4 \pm 2i}{10}$
$y = \dfrac{-4}{10} \pm \dfrac{2}{10}i = -\dfrac{2}{5} \pm \dfrac{1}{5}i$
The solutions are $-\dfrac{2}{5} \pm \dfrac{1}{5}i$.

35. $1 + 2m + 3m^2 = 0$
$3m^2 + 2m + 1 = 0 \quad$ Standard form
$a = 3, b = 2, c = 1$

$$m = \frac{-b \pm \sqrt{b^2 - 4ac}}{2a}$$

$$m = \frac{-2 \pm \sqrt{2^2 - 4 \cdot 3 \cdot 1}}{2 \cdot 3}$$

$$m = \frac{-2 \pm \sqrt{-8}}{6} = \frac{-2 \pm i\sqrt{8}}{6} = \frac{-2 \pm 2i\sqrt{2}}{6}$$

$$m = \frac{-2}{6} \pm \frac{2\sqrt{2}}{6}i = -\frac{1}{3} \pm \frac{\sqrt{2}}{3}i$$

The solutions are $-\frac{1}{3} \pm \frac{\sqrt{2}}{3}i$.

37. *Writing Exercise.* For $b^2 - 4ac \neq 0$, $\pm\sqrt{b^2 - 4ac}$ yields either two imaginary numbers or two real numbers. Thus, it is not possible for a quadratic equation (with real-number coefficients) to have one imaginary-number solution and one real-number solution. (If imaginary-number coefficients are allowed, then a quadratic equation can have one imaginary-number solution and one real-number solution. One example is $x^2 - (5 + i)x + 5i = 0$ with solutions 5 and i.)

39. Graph $y = \frac{3}{5}x$.

x	y	(x, y)
-5	-3	$(-5, -3)$
0	0	$(0, 0)$
5	3	$(5, 3)$

41. Graph $y = -4x$.

x	y	(x, y)
-1	4	$(-1, 4)$
0	0	$(0, 0)$
1	-4	$(1, -4)$

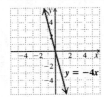

43. Graph $y = \frac{1}{2}x - 3$.

x	y	(x, y)
0	-3	$(0, -3)$
2	-2	$(2, -2)$
6	0	$(6, 0)$

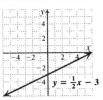

45. *Writing Exercise.* If $b^2 < 4ac$, then $b^2 - 4ac < 0$ so the solutions are imaginary.

47.
$$(x + 1)^2 + (x + 3)^2 = 0$$
$$x^2 + 2x + 1 + x^2 + 6x + 9 = 0$$
$$2x^2 + 8x + 10 = 0$$
$$x^2 + 4x + 5 = 0 \quad \text{Dividing by 2}$$

$a = 1, \, b = 4, \, c = 5$

$$x = \frac{-b \pm \sqrt{b^2 - 4ac}}{2a}$$

$$x = \frac{-4 \pm \sqrt{4^2 - 4 \cdot 1 \cdot 5}}{2 \cdot 1}$$

$$x = \frac{-4 \pm \sqrt{16 - 20}}{2} = \frac{-4 \pm \sqrt{-4}}{2}$$

$$x = \frac{-4 \pm 2i}{2} = \frac{2(-2 \pm i)}{2 \cdot 1}$$

$$x = -2 \pm i$$

The solutions are $-2 \pm i$.

49. $\dfrac{2x - 1}{5} - \dfrac{2}{x} = \dfrac{x}{2}$

We multiply by $10x$, the LCD.

$$10x\left(\frac{2x - 1}{5} - \frac{2}{x}\right) = 10x \cdot \frac{x}{2}$$

$$2x(2x - 1) - 10 \cdot 2 = 5x \cdot x$$

$$4x^2 - 2x - 20 = 5x^2$$

$$0 = x^2 + 2x + 20$$

$a = 1, \, b = 2, \, c = 20$

$$x = \frac{-b \pm \sqrt{b^2 - 4ac}}{2a}$$

$$x = \frac{-2 \pm \sqrt{2^2 - 4 \cdot 1 \cdot 20}}{2 \cdot 1}$$

$$x = \frac{-2 \pm \sqrt{-76}}{2} = \frac{-2 \pm i\sqrt{76}}{2} = \frac{-2 \pm 2i\sqrt{19}}{2}$$

$$x = \frac{-2}{2} \pm \frac{2\sqrt{19}}{2}i = -1 \pm \sqrt{19}i$$

The solutions are $-1 \pm \sqrt{19}i$.

51. Example 2(a):

Graph $y_1 = x^2$ and $y_2 = -100$.

The graphs do not intersect, so the equation $x^2 = -100$ has no real-number solutions.

Example 2(b):

Graph $y = x^2 + 3x + 4$.

There are no x-intercepts, so the equation $x^2 + 3x + 4 = 0$ has no real-number solutions.

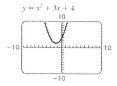

Example 2(c):

Graph $y_1 = x^2 + 2$ and $y_2 = 2x$.

The graphs do not intersect, so the equation $x^2 + 2 = 2x$ has no real-number solutions.

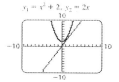

Chapter 9 Connecting the Concepts

1. $3x + 8 = 7x + 4$

$3x + 4 = 7x$

$4 = 4x$

$1 = x$

The solution is 1.

3. $3 + \sqrt{x} = 8$

$\sqrt{x} = 5$

$\left(\sqrt{x}\right)^2 = 5^2$

$x = 25$

Check:
$$\begin{array}{c|c} 3 + \sqrt{x} = 8 \\ \hline 3 + \sqrt{25} & 8 \\ 3 + 5 & \\ & \overset{?}{8 = 8} \quad \text{TRUE} \end{array}$$

The solution is 25.

5. $11 = 4\sqrt{3x+1} - 5$

$16 = 4\sqrt{3x+1}$

$4 = \sqrt{3x+1}$

$4^2 = \left(\sqrt{3x+1}\right)^2$

$16 = 3x + 1$

$15 = 3x$

$5 = x$

Check:
$$\begin{array}{c|c} 11 = 4\sqrt{3x+1} - 5 \\ \hline 11 & 4\sqrt{3 \cdot 5 + 1} - 5 \\ & 4\sqrt{16} - 5 \\ & 4 \cdot 4 - 5 \\ & 16 - 5 \\ \overset{?}{11 = 11} & \quad \text{TRUE} \end{array}$$

The solution is 5.

7. $n^2 + 2n - 2 = 0$

$n^2 + 2n = 2$

$n^2 + 2n + 1 = 2 + 1$

$(n+1)^2 = 3$

$n + 1 = \pm\sqrt{3}$

$n = -1 \pm \sqrt{3}$

The solutions are $-1 \pm \sqrt{3}$

9. $\dfrac{1}{4t} + \dfrac{t}{6} = 2$

Note $t \neq 0$

$12t\left(\dfrac{1}{4t} + \dfrac{t}{6}\right) = 12t \cdot 2$

$3 + 2t^2 = 24t$

$2t^2 - 24t + 3 = 0$

$t = \dfrac{-b \pm \sqrt{b^2 - 4ac}}{2a}$

$t = \dfrac{-(-24) \pm \sqrt{(-24)^2 - 4(2)(3)}}{2(2)}$

$t = \dfrac{24 \pm \sqrt{576 - 24}}{4}$

$t = \dfrac{24 \pm \sqrt{552}}{4}$

$t = \dfrac{24 \pm 2\sqrt{138}}{4}$

$t = \dfrac{12 \pm \sqrt{138}}{2}$

The solutions are $\dfrac{12 \pm \sqrt{138}}{2}$ or $6 \pm \dfrac{\sqrt{138}}{2}$.

11. $2\sqrt{5t+3} = 3\sqrt{2t-1}$

$\left(2\sqrt{5t+3}\right)^2 = \left(3\sqrt{2t-1}\right)^2$

$4(5t+3) = 9(2t-1)$

$20t + 12 = 18t - 9$

$20t = 18t - 21$

$2t = -21$

$t = \dfrac{-21}{2}$

Check: $$2\sqrt{5t+3} = 3\sqrt{2t-1}$$

$$2\sqrt{5\left(\frac{-21}{2}\right)+3} \;\Big|\; 3\sqrt{2\left(-\frac{21}{2}\right)-1}$$

$$2\sqrt{\frac{-105}{2}+3} \;\Big|\; 3\sqrt{-21-1}$$

$$2\sqrt{\frac{-99}{2}} \;\Big|\; 3\sqrt{-22}$$

The number $\frac{-21}{2}$ does not check since we cannot take the principle square root of a negative number. There is no solution.

13.
$$\sqrt{x+3} = x+1$$
$$\left(\sqrt{x+3}\right)^2 = (x+1)^2$$
$$x+3 = x^2+2x+1$$
$$0 = x^2+x-2$$
$$0 = (x+2)(x-1)$$
$$x+2=0 \ \text{ or } \ x-1=0$$
$$x=-2 \ \text{ or } \ x=1$$

Check: For $x=-2$
$$\frac{\sqrt{x+3} = x+1}{\begin{array}{c|c}\sqrt{-2+3} & -2+1 \\ \sqrt{1} & -1 \\ 1 \overset{?}{=} -1 & \text{FALSE}\end{array}}$$

Check: For $x=1$
$$\frac{\sqrt{x+3} = x+1}{\begin{array}{c|c}\sqrt{1+3} & 1+1 \\ \sqrt{4} & 2 \\ 2 \overset{?}{=} 2 & \text{TRUE}\end{array}}$$

The number -2 does not check. The solution is 1.

15. $(x+3)^2 = 7$
$$x+3 = \pm\sqrt{7}$$
$$x = -3 \pm \sqrt{7}$$
The solutions are $-3 \pm \sqrt{7}$

17. $t^2-100=0$
$$t^2=100$$
$$t^2=\pm\sqrt{100}$$
$$t^2=\pm10$$
or
$$t^2-100=0$$
$$(t+10)(t-10)=0$$
$$t+10=0 \ \text{ or } \ t-10=0$$
$$t=-10 \ \text{ or } \ t=10$$
The solutions are -10 and 10.

19. $5t-3 = 2-4(1-t)$
$$5t-3 = 2-4+4t$$
$$5t-3 = -2+4t$$
$$5t = 1+4t$$
$$t = 1$$
The solution is 1.

Exercise Set 9.6

1. False. For example $y=x^2$ has 1 x-intercept and $y=x^2+3$ has no x-intercepts.

3. False.

5. $y=2x^2$

We first find the vertex. The x-coordinate is
$$-\frac{b}{2a} = -\frac{0}{2\cdot2}=0.$$
We substitute 0 into the equation to find the second coordinate of the vertex.
$$x^2 = 2\cdot0^2=0$$
The vertex is $(0,0)$. The line of symmetry is $x=0$, the y-axis.

We choose some x-values on both sides of the vertex and graph the parabola.
When $x=1$, $y=2\cdot1^2=2\cdot1=2$.
When $x=-1$, $y=2\cdot(-1)^2=2\cdot1=2$.
When $x=2$, $y=2\cdot2^2+3=2\cdot4=8$.
When $x=-2$, $y=2\cdot(-2)^2=2\cdot4+3=8$.

x	y
0	0
1	2
-1	2
2	8
-2	8

7. $y=-2x^2$

Find the vertex. The x-coordinate is
$$-\frac{b}{2a} = -\frac{0}{2(-2)}=0.$$
The y-coordinate is
$$-2x^2 = -2\cdot0^2=0.$$
The vertex is $(0,0)$. The line of symmetry is $x=0$, the y-axis.

Choose some x-values on both sides of the vertex and graph the parabola.

When $x = -2$, $y = -2(-2)^2 = -2 \cdot 4 = -8$.

When $x = -1$, $y = -2(-1)^2 = -2 \cdot 1 = -2$.

When $x = 1$, $y = -2 \cdot 1^2 = -2 \cdot 1 = -2$.

When $x = 2$, $y = -2 \cdot 2^2 = -2 \cdot 4 = -8$.

x	y
0	0
−2	−8
−1	−2
1	−2
2	−8

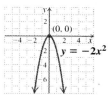

9. $y = -\dfrac{1}{3}x^2$

Find the vertex. The x-coordinate is
$$-\frac{b}{2a} = -\frac{0}{2\left(-\dfrac{1}{3}\right)} = 0.$$

The y-coordinate is
$$-\frac{1}{3} \cdot 0^2 = 0.$$

The vertex is $(0,0)$.

We choose some x-values on both sides of the vertex and graph the parabola.

x	y
−3	−3
−1	$-\dfrac{1}{3}$
2	$-\dfrac{4}{3}$
3	−3

11. $y = x^2 - 2$

Find the vertex. The x-coordinate is
$$-\frac{b}{2a} = -\frac{0}{2 \cdot 1} = 0.$$

The y-coordinate is
$$x^2 - 2 = 0^2 - 2 = -2.$$

The vertex is $(0, -2)$.

We choose some x-values on both sides of the vertex and graph the parabola.

x	y
−2	2
−1	−1
0	−2
1	−1
2	2

13. $y = x^2 - 2x + 1$

Find the vertex. The x-coordinate is
$$-\frac{b}{2a} = -\frac{-2}{2 \cdot 1} = 1.$$

The y-coordinate is
$$x^2 - 2x + 1 = 1^2 - 2 \cdot 1 + 1 = 0.$$

The vertex is $(1, 0)$.

To find the the $y-$intercept we replace x with 0:

$$y = x^2 - 2x + 1 = 0^2 - 2 \cdot 0 + 1 - 1.$$

The $y-$intercept is $(0,1)$.

We choose some x-values on both sides of the vertex and graph the parabola.

x	y
−1	4
0	1
1	0
2	1
3	4

$$y = x^2 - 3x - 10$$

$$-\frac{b}{2a} = -\frac{-3}{2(1)} =$$

$$-\frac{3}{2}$$

15. $y = x^2 + 3x - 10$

Find the vertex. The x-coordinate is
$$-\frac{b}{2a} = -\frac{3}{2 \cdot 1} = \frac{-3}{2}.$$

The y-coordinate is
$$x^2 + 3x - 10 = \left(-\frac{3}{2}\right)^2 + 3\left(-\frac{3}{2}\right) - 10$$
$$= \frac{9}{4} - \frac{9}{2} - 10 = \frac{-49}{4} = -12.25.$$

The vertex is $\left(-\frac{3}{2}, \frac{-49}{4}\right)$.

To find the the y-intercept we replace x with 0:
$$y = x^2 + 3x - 10 = 0^2 + 3 \cdot 0 - 10 = -10.$$

The y-intercept is $(0, -10)$.

We choose some x-values on both sides of the vertex and graph the parabola.

x	y
-5	0
-3	-10
$-\frac{3}{2}$	$-\frac{49}{4}$
0	-10
2	0

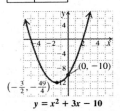

$$y = x^2 + 3x - 10$$

17. $y = -2x^2 + 12x - 13$

Find the vertex. The x-coordinate is
$$-\frac{b}{2a} = -\frac{12}{2(-2)} = 3.$$

The y-coordinate is
$$-2x^2 + 12x - 13 = -2\left(3^2\right) + 12(3) - 13$$
$$= -18 + 36 - 13 = 5.$$

The vertex is $(3, 5)$.

To find the the y-intercept we replace x with 0:
$$y = -2x^2 + 12x - 13 = -2 \cdot 0^2 + 12 \cdot 0 - 13 = -13.$$

The y-intercept is $(0, -13)$.

We choose some x-values on both sides of the vertex and graph the parabola.

x	y
0	-13
2	3
3	5
4	3
6	-13

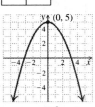

$$y = -2x^2 + 12x - 13$$

19. $y = -\frac{1}{2}x^2 + 5$

Find the vertex. The x-coordinate is
$$-\frac{b}{2a} = -\frac{0}{2\left(-\frac{1}{2}\right)} = 0.$$

The y-coordinate is
$$-\frac{1}{2} \cdot 0^2 + 5 = 0 + 5 = 5.$$

The vertex is $(0, 5)$.

We choose some x-values on both sides of the vertex and graph the parabola.

x	y
-4	-3
-2	3
2	3
4	-3

$$y = -\frac{1}{2}x^2 + 5$$

21. $y = x^2 - 3x$

Find the vertex. The x-coordinate is
$$-\frac{b}{2a} = -\frac{-3}{2 \cdot 1} = \frac{3}{2}.$$

The y-coordinate is
$$\left(\frac{3}{2}\right)^2 - 3 \cdot \frac{3}{2} = \frac{9}{4} - \frac{9}{2} = -\frac{9}{4}.$$

The vertex is $\left(\frac{3}{2}, -\frac{9}{4}\right)$.

We choose some x-values on both sides of the vertex and graph the parabola.

x	y
0	0
1	-2
2	-2
3	0

$$y = x^2 - 3x$$

23. $y = x^2 + 2x - 8$

Find the vertex. The x-coordinate is
$$-\frac{b}{2a} = -\frac{2}{2 \cdot 1} = -1.$$
The y-coordinate is
$$(-1)^2 + 2(-1) - 8 = 1 - 2 - 8 = -9.$$
The vertex is $(-1, -9)$.

To find the y-intercept we replace x with 0 and compute y:
$$y = 0^2 + 2 \cdot 0 - 8 = 0 + 0 - 8 = -8.$$
The y-intercept is $(0, -8)$.

To find the x-intercepts we replace y with 0 and solve for x.
$$0 = x^2 + 2x - 8$$
$$0 = (x + 4)(x - 2)$$
$$x + 4 = 0 \quad or \quad x - 2 = 0$$
$$x = -4 \quad or \qquad x = 2$$
The x-intercepts are $(-4, 0)$ and $(2, 0)$.

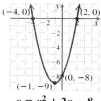

$$y = x^2 + 2x - 8$$

25. $y = 2x^2 - 6x$

Find the vertex. The x-coordinate is
$$-\frac{b}{2a} = -\frac{-6}{2 \cdot 2} = \frac{3}{2}.$$
The y-coordinate is
$$2\left(\frac{3}{2}\right)^2 - 6 \cdot \frac{3}{2} = \frac{9}{2} - 9 = -\frac{9}{2}.$$
The vertex is $\left(\frac{3}{2}, -\frac{9}{2}\right)$.

To find the y-intercept we replace x with 0 and compute y:
$$y = 2 \cdot 0^2 - 6 \cdot 0 = 0 - 0 = 0.$$
The y-intercept is $(0, 0)$.

To find the x-intercepts we replace y with 0 and solve for x.
$$0 = 2x^2 - 6x$$
$$0 = 2x(x - 3)$$
$$2x = 0 \quad or \quad x - 3 = 0$$
$$x = 0 \quad or \qquad x = 3$$
The x-intercepts are $(0, 0)$ and $(3, 0)$.

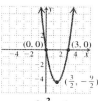

$$y = 2x^2 - 6x$$

27. $y = -x^2 - x + 12$

Find the vertex. The x-coordinate is
$$-\frac{b}{2a} = -\frac{-1}{2(-1)} = -\frac{1}{2}.$$
The y-coordinate is
$$-\left(-\frac{1}{2}\right)^2 - \left(-\frac{1}{2}\right) + 12 = -\frac{1}{4} + \frac{1}{2} + 12 = \frac{49}{4}.$$
The vertex is $\left(-\frac{1}{2}, \frac{49}{4}\right)$.

To find the y-intercept we replace x with 0 and compute y:
$$y = -0^2 - 0 + 12 = -0 - 0 + 12 = 12.$$
The y-intercept is $(0, 12)$.

To find the x-intercepts we replace y with 0 and solve for x.
$$0 = -x^2 - x + 12$$
$$0 = x^2 + x - 12 \qquad \text{Multiplying by } -1$$
$$0 = (x + 4)(x - 3)$$
$$x + 4 = 0 \quad or \quad x - 3 = 0$$
$$x = -4 \quad or \qquad x = 3$$
The x-intercepts are $(-4, 0)$ and $(3, 0)$.

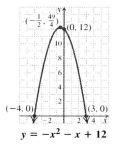

$$y = -x^2 - x + 12$$

29. $y = 3x^2 - 6x + 1$

Find the vertex. The x-coordinate is

$$-\frac{b}{2a} = -\frac{-6}{2 \cdot 3} = 1.$$

The y-coordinate is

$$3 \cdot 1^2 - 6 \cdot 1 + 1 = 3 - 6 + 1 = -2.$$

The vertex is $(1, -2)$.

To find the y-intercept we replace x with 0 and compute y:

$$y = 3 \cdot 0^2 - 6 \cdot 0 + 1 = 0 - 0 + 1 = 1.$$

The y-intercept is $(0, 1)$.

To find the x-intercepts we replace y with 0 and solve for x.

$$0 = 3x^2 - 6x + 1$$

$$x = \frac{-b \pm \sqrt{b^2 - 4ac}}{2a}$$

$$x = \frac{-(-6) \pm \sqrt{(-6)^2 - 4 \cdot 3 \cdot 1}}{2 \cdot 3}$$

$$x = \frac{6 \pm \sqrt{36 - 12}}{6} = \frac{6 \pm \sqrt{24}}{6}$$

$$x = \frac{6 \pm 2\sqrt{6}}{6} = \frac{2(3 \pm \sqrt{6})}{2 \cdot 3}$$

$$x = \frac{3 \pm \sqrt{6}}{3}$$

The x-intercepts are $\left(\dfrac{3 - \sqrt{6}}{3}, 0\right)$ and $\left(\dfrac{3 + \sqrt{6}}{3}, 0\right)$, or about $(0.184, 0)$ and $(1.816, 0)$.

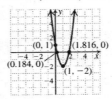

$$\boldsymbol{y = 3x^2 - 6x + 1}$$

31. $y = x^2 + 2x + 3$

Find the vertex. The x-coordinate is

$$-\frac{b}{2a} = -\frac{2}{2 \cdot 1} = -1.$$

The y-coordinate is

$$y = (-1)^2 + 2(-1) + 3 = 1 - 2 + 3 = 2.$$

The vertex is $(-1, 2)$.

To find the y-intercept we replace x with 0 and compute y:

$$y = 0^2 + 2 \cdot 0 + 3 = 0 + 0 + 3 = 3.$$

The y-intercept is $(0, 3)$.

To find the x-intercepts we replace y with 0 and solve for x.

$$0 = x^2 + 2x + 3$$

$$x = \frac{-b \pm \sqrt{b^2 - 4ac}}{2a}$$

$$x = \frac{-2 \pm \sqrt{2^2 - 4 \cdot 1 \cdot 3}}{2 \cdot 1}$$

$$x = \frac{-2 \pm \sqrt{4 - 12}}{2} = \frac{-2 \pm \sqrt{-8}}{2}$$

Because the radicand, -8, is negative the equation has no real-number solutions. Thus, there are no x-intercepts.

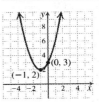

$$\boldsymbol{y = x^2 + 2x + 3}$$

33. $y = 3 - 4x - 2x^2$, or $y = -2x^2 - 4x + 3$

Find the vertex. The x-coordinate is

$$-\frac{b}{2a} = -\frac{-4}{2(-2)} = -1.$$

The y-coordinate is

$$y = 3 - 4(-1) - 2(-1)^2 = 3 + 4 - 2 = 5.$$

The vertex is $(-1, 5)$.

To find the y-intercept we replace x with 0 and compute y:

$$y = 3 - 4 \cdot 0 - 2 \cdot 0^2 = 3 - 0 - 0 = 3.$$

The y-intercept is $(0, 3)$.

To find the x-intercepts we replace y with 0 and solve for x.

$$0 = -2x^2 - 4x + 3$$

$$x = \frac{-b \pm \sqrt{b^2 - 4ac}}{2a}$$

$$x = \frac{-(-4) \pm \sqrt{(-4)^2 - 4(-2)(3)}}{2(-2)}$$

$$x = \frac{4 \pm \sqrt{16 + 24}}{-4} = \frac{4 \pm \sqrt{40}}{-4}$$

$$x = \frac{4 \pm 2\sqrt{10}}{-4} = -\frac{2 \pm \sqrt{10}}{2}$$

The x-intercepts are $\left(\dfrac{-2 - \sqrt{10}}{2}, 0\right)$ and $\left(\dfrac{-2 + \sqrt{10}}{2}, 0\right)$, or about $(0.581, 0)$ and $(-2.581, 0)$.

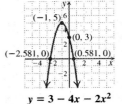

$$\boldsymbol{y = 3 - 4x - 2x^2}$$

35. *Writing Exercise.* Knowing the coordinates of the vertex of a parabola allows us to make good choices of x-values to substitute in the equation in order to find points on the graph because we know we are choosing values on both sides of the axis of symmetry. This allows us to see the shape of the parabola quickly.

37.
$$3x^2 - 4x = 3(-1)^2 - 4(-1)$$
$$= 3 \cdot 1 - 4(-1)$$
$$= 3 + 4$$
$$= 7$$

39.
$$6 - t^3 = 6 - (-2)^3$$
$$= 6 - (-8)$$
$$= 6 + 8$$
$$= 14$$

41. $(a - 9)^2 = (8 - 9)^2 = (-1)^2 = 1$

43. *Writing Exercise.* Write an equation of the form $y = k(x - p)(x - q)$.

45. a) We substitute 128 for H and solve for t:
$$128 = -16t^2 + 96t$$
$$16t^2 - 96t + 128 = 0$$
$$16(t^2 - 6t + 8) = 0$$
$$16(t - 2)(t - 4) = 0$$

$$t - 2 = 0 \quad or \quad t - 4 = 0$$
$$t = 2 \quad or \qquad t = 4$$

The projectile is 128 ft from the ground 2 sec after launch and again 4 sec after launch. The graph confirms this.

b) We find the first coordinate of the vertex of the function $H = -16t^2 + 96t$:
$$-\frac{b}{2a} = -\frac{96}{2(-16)} = -\frac{96}{-32} = -(-3) = 3$$

The projectile reaches its maximum height 3 sec after launch. The graph confirms this.

c) Since it takes 3 sec for the projectile to reach its maximum height, it will take another 3 sec for it to return to the ground. Thus, the projectile returns to the ground 6 sec after launch.

We could also do this exercise as shown below.

We substitute 0 for H and solve for t:
$$0 = -16t^2 + 96t$$
$$0 = -16t(t - 6)$$

$$-16t = 0 \quad or \quad t - 6 = 0$$
$$t = 0 \quad or \qquad t = 6$$

At $t = 0$ sec the projectile has not yet been launched. Thus, we use $t = 6$. The projectile returns to the ground 6 sec after launch. The graph confirms this.

47. a) For $r = 25$, $d = 25 + 0.05(25)^2 = 56.25$ ft

For $r = 40$, $d = 40 + 0.05(40)^2 = 120$ ft

For $r = 55$, $d = 55 + 0.05(55)^2 = 206.25$ ft

For $r = 65$, $d = 65 + 0.05(65)^2 = 276.25$ ft

For $r = 75$, $d = 75 + 0.05(75)^2 = 356.25$ ft

For $r = 100$, $d = 100 + 0.05(100)^2 = 600$ ft

b)

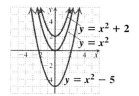

49. *Writing Exercise.*

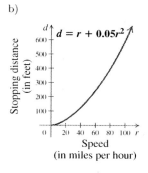

We can move the graph of $y = x^2$ up k units if $k \geq 0$ or down $|k|$ units if $k < 0$ to obtain the graph of $y = x^2 + k$.

51. $D = (p - 6)^2$

p	D
0	36
1	25
2	16
3	9
4	4
5	1
6	0

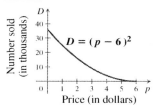

53. Graph $y = x^2 - 5$ and find the first coordinate of the right-hand x-intercept. We find that $\sqrt{5} \approx 2.2361$.

Exercise Set 9.7

---

1. True; see page 574 in the text.

3. False; see page 577 in the text.

5. Yes; each member of the domain is matched to only one member of the range.

7. Yes; each member of the domain is matched to only one member of the range.

9. No; a member of the domain is matched to more than one member of the range. In fact, each member of the domain is matched to 3 members of the range.

11. Yes; each member of the domain is matched to only one member of the range.

13. $f(x) = x + 5$

$f(4) = 4 + 5 = 9$

$f(7) = 7 + 5 = 12$

$f(-2) = -2 + 5 = 3$

15. $h(p) = -3p$

$h(-7) = -3(-7) = 21$

$h(5) = -3 \cdot 5 = -15$

$h(10) = -3 \cdot 10 = -30$

17. $g(s) = 3s - 2$

$g(1) = 3 \cdot 1 - 2 = 3 - 2 = 1$

$g(-5) = 3(-5) - 2 = -15 - 2 = -17$

$g(2.5) = 3(2.5) - 2 = 7.5 - 2 = 5.5$

19. $F(x) = 2x^2 + x$

$F(-1) = 2(-1)^2 + (-1) = 2 - 1 = 1$

$F(0) = 2(0)^2 + (0) = 0$

$F(2) = 2(2^2) + (2) = 2 \cdot 4 + 2 = 10$

21. $f(t) = (t - 3)^2$

$f(-2) = (-2 - 3)^2 = (-5)^2 = 25$

$f(8) = (8 - 3)^2 = (5)^2 = 25$

$f(\frac{1}{2}) = \left(\frac{1}{2} - 3\right)^2 = \left(-\frac{5}{2}\right)^2 = \frac{25}{4}$

23. $h(x) = |2x| - x$

$h(0) = |2 \cdot 0| - 0 = 0$

$h(-4) = |2(-4)| - (-4) = |-8| + 4 = 8 + 4 = 12$

$h(4) = |2 \cdot 4| - 4 = |8| - 4 = 8 - 4 = 4$

25. $l(x) = \dfrac{1700}{x}$

   a. $l(50) = \dfrac{1700}{50} = 34\text{yr.}$

   b. $l(85) = \dfrac{1700}{85} = 20\text{yr.}$

27. $F(x) = 2.75x + 71.48$

   a) $F(32) = 2.75(32) + 71.48$

              $= 88 + 71.48$

              $= 159.48 \text{ cm}$

   b) $F(30) = 2.75(30) + 71.48$

              $= 82.5 + 71.48$

              $= 153.98 \text{ cm}$

29. $C(62) = \dfrac{5}{9}(62 - 32) = \dfrac{5}{9} \cdot 30 = \dfrac{50}{3} = 16\dfrac{2}{3}°,$

                                or $16.\overline{6}°\text{C}$

$C(77) = \dfrac{5}{9}(77 - 32) = \dfrac{5}{9} \cdot 45 = 25°\text{C}$

$C(23) = \dfrac{5}{9}(23 - 32) = \dfrac{5}{9}(-9) = -5°\text{C}$

31. Graph $f(x) = 2x - 3$

Make a list of function values in a table.

When $x = -1$, $f(-1) = 2(-1) - 3 = -2 - 3 = -5$.

When $x = 0$, $f(0) = 2 \cdot 0 - 3 = 0 - 3 = -3$.

When $x = 2$, $f(2) = 2 \cdot 2 - 3 = 4 - 3 = 1$.

x	$f(x)$
-1	-5
0	-3
2	1

Plot these points and connect them.

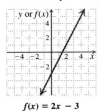

$f(x) = 2x - 3$

33. Graph $g(x) = -x + 4$

Make a list of function values in a table.

When $x = -1$, $g(-1) = -(-1) + 4 = 1 + 4 = 5$.

When $x = 0$, $g(0) = -0 + 4 = 4$.

When $x = 3$, $g(3) = -3 + 4 = 1$.

x	$g(x)$
-1	5
0	4
3	1

Plot these points and connect them.

$g(x) = -x + 4$

35. Graph $f(x) = \dfrac{1}{2}x + 1$.

Make a list of function values in a table.

When $x = -2$, $f(-2) = \dfrac{1}{2}(-2) + 1 = -1 + 1 = 0$.

When $x = 0$, $f(0) = \dfrac{1}{2} \cdot 0 + 1 = 0 + 1 = 1$.

When $x = 4$, $f(4) = \dfrac{1}{2} \cdot 4 + 1 = 2 + 1 = 3$.

x	$f(x)$
-2	0
0	1
4	3

Plot these points and connect them.

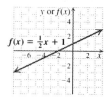

$f(x) = \frac{1}{2}x + 1$

37. Graph $g(x) = 2|x|$.

Make a list of function values in a table.

When $x = -3$, $g(-3) = 2|-3| = 2 \cdot 3 = 6$.
When $x = -1$, $g(-1) = 2|-1| = 2 \cdot 1 = 2$.
When $x = 0$, $g(0) = 2|0| = 2 \cdot 0 = 0$.
When $x = 1$, $g(1) = 2|1| = 2 \cdot 1 = 2$.
When $x = 3$, $g(3) = 2|3| = 2 \cdot 3 = 6$.

x	$g(x)$
-3	6
-1	2
0	0
1	2
3	6

Plot these points and connect them.

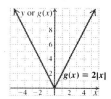

$g(x) = 2|x|$

39. Graph $g(x) = x^2$.

Recall from Section 10.5 that the graph is a parabola. Make a list of function values in a table.

When $x = -2$, $g(-2) = (-2)^2 = 4$.
When $x = -1$, $g(-1) = (-1)^2 = 1$.
When $x = 0$, $g(0) = 0^2 = 0$.
When $x = 1$, $g(1) = 1^2 = 1$.
When $x = 2$, $g(2) = 2^2 = 4$.

x	$g(x)$
-2	4
-1	1
0	0
1	1
2	4

Plot these points and connect them.

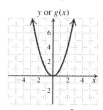

$g(x) = x^2$

41. Graph $f(x) = x^2 - x - 2$.

Recall from Section 10.5 that the graph is a parabola. Make a list of function values in a table.

When $x = -1$, $f(-1) = (-1)^2 - (-1) - 2 = 1 + 1 - 2 = 0$.
When $x = 0$, $f(0) = 0^2 - 0 - 2 = -2$.
When $x = 1$, $f(1) = 1^2 - 1 - 2 = 1 - 1 - 2 = -2$.
When $x = 2$, $f(2) = 2^2 - 2 - 2 = 4 - 2 - 2 = 0$.

x	$f(x)$
-1	0
0	-2
1	-2
2	0

Plot these points and connect them.

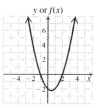

$f(x) = x^2 - x - 2$

43. The graph is that of a function because no vertical line can cross the graph at more than one point.

45. The graph is not that of a function because a vertical line can cross the graph at more than one point.

47. The graph is that of a function because no vertical line can cross the graph at more than one point.

49. *Writing Exercise.* No; since each input has exactly one output, it is not possible for a function to have more numbers in the range than in the domain.

51. $y = \frac{4}{5}x$

x	y	(x,y)
-5	-4	$(-5,4)$
0	0	$(0,0)$
5	4	$(5,4)$

53. $x - 2y < 4$

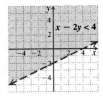

x	y	(x,y)
-2	-3	$(-2,-3)$
0	-2	$(0,-2)$
2	-1	$(2,-1)$

55. $y = x^2 - 6x + 1$

$-\frac{b}{2a} = \frac{-(-6)}{2(1)} = \frac{6}{2} = 3.$

$y = 3^2 - 6(3) + 1 = 9 - 18 + 1 = -8.$

The vertex is $(3, -8)$.

Let $x = 0$ to find the $y-$intercept

$y = 0^2 - 6(0) + 1 = 1.$

The y-intercept is $(0, 1)$.

$0 = x^2 - 6x + 1$

$x = \frac{-b \pm \sqrt{b^2 - 4ac}}{2a}$

$x = \frac{-(-6) \pm \sqrt{(-6)^2 - 4(1)(1)}}{2(1)}$

$x = \frac{6 \pm \sqrt{36 - 4}}{2} = \frac{6 \pm \sqrt{32}}{2}$

$x = \frac{6 \pm 4\sqrt{2}}{2} = 3 \pm \sqrt{2}$

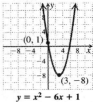

57. *Writing Exercise.* Answers will vary.

59. Graph $g(x) = x^3$.

Make a list of function values in a table. Then plot the points and connect them.

x	$g(x)$
-2	-8
-1	-1
0	0
1	1
2	8

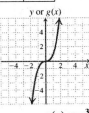

$g(x) = x^3$

61. Graph $g(x) = |x| + x$.

Make a list of function values in a table. Then plot the points and connect them.

x	$f(x)$
-3	0
-2	0
-1	0
0	0
1	2
2	4
3	6

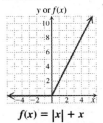

$f(x) = |x| + x$

63. Answers may vary.

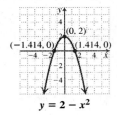

$y = 2 - x^2$

65. $g(x) = ax^2 + bx + c$

$\quad -4 = a \cdot 0^2 + b \cdot 0 + c, \text{ or } -4 = c \qquad (1)$

$\quad 0 = a(-2)^2 + b(-2) + c, \text{ or } 0 = 4a - 2b + c \quad (2)$

$\quad 0 = a \cdot 2^2 + b \cdot 2 + c, \text{ or } 0 = 4a + 2b + c \qquad (3)$

Substitute -4 for c in Equations (2) and (3).

$\quad\quad 0 = 4a - 2b - 4, \text{ or } 4 = 4a - 2b \quad (5)$

$\quad\quad 0 = 4a + 2b - 4, \text{ or } 4 = 4a + 2b \quad (6)$

Add Equations (5) and (6).

$\quad\quad 8 = 8a$

$\quad\quad 1 = a$

Substitute 1 for a in Equation (6).

$\quad\quad 4 = 4 \cdot 1 + 2b$

$\quad\quad 0 = 2b$

$\quad\quad 0 = b$

We have $a = 1$, $b = 0$, $c = -4$, so $g(x) = x^2 - 4$.

67. $g(t) = t^2 - t$

The domain is the set $\{-3, -2, -1, 0, 1\}$.

$\quad g(-3) = (-3)^2 - (-3) = 9 + 3 = 12$

$\quad g(-2) = (-2)^2 - (-2) = 4 + 2 = 6$

$\quad g(-1) = (-1)^2 - (-1) = 1 + 1 = 2$

$\quad g(0) = 0^2 - 0 = 0$

$\quad g(1) = 1^2 - 1 = 1 - 1 = 0$

The range is the set $\{0, 2, 6, 12\}$.

69. $h(x) = |x| - x$

The domain is the set $\{-1, 0, 1, 2, 3, 4, 5, 6, 7, 8, 9,$ $10, 11, 12, 13, 14, 15, 16, 17, 18, 19\}$.

$\quad h(-1) = |-1| - (-1) = 1 + 1 = 2$

$\quad h(0) = |0| - 0 = 0 - 0 = 0$

$\quad h(1) = |1| - 1 = 1 - 1 = 0$

$\quad h(2) = |2| - 2 = 2 - 2 = 0$

$\quad h(3) = |3| - 3 = 3 - 3 = 0$

$\quad\quad\quad .$

$\quad\quad\quad .$

$\quad h(19) = |19| - 19 = 19 - 19 = 0$

The range is the set $\{0, 2\}$.

Chapter 9 Review

1. Since $x^2 - 7x + 6 = 0$ is easily factored, choice (h) is correct.

3. Since $x^2 + 8x = 3$ is easily solved by completing the square, choice (g) is correct.

5. Since the term $3x^2$ in $y = 3x^2 + 4x - 7$ is positive, the parabola opens upward; choice (b) is correct.

7. In $f(2) = 5$, 2 represents the input while 5 represents the resulting output; choice (a) is correct.

9. $\quad x^2 - 10x = 1$

$\quad x^2 - 10x + 25 = 1 + 25 \qquad \left(\dfrac{-10}{2}\right)^2 = (-5)^2 = 25$

$\quad\quad (x - 5)^2 = 26$

$\quad\quad x - 5 = \pm\sqrt{26}$

$\quad\quad x = 5 \pm \sqrt{26}$

The solutions are $5 \pm \sqrt{26}$ or $5 + \sqrt{26}$ and $5 - \sqrt{26}$.

11. $\quad 5x^2 = 30$

$\quad\quad x^2 = 6$

$\quad\quad x = \pm\sqrt{6}$

The solutions are $\pm\sqrt{6}$ or $\sqrt{6}$ and $-\sqrt{6}$.

13. $\quad x^2 - 2x - 10 = 0$

$\quad\quad x^2 - 2x = 10$

$\quad x^2 - 2x + 1 = 10 + 1 \qquad \left(\dfrac{-2}{2}\right)^2 = (-1)^2 = 1$

$\quad\quad (x - 1)^2 = 11$

$\quad\quad x - 1 = \pm\sqrt{11}$

$\quad\quad x = 1 \pm \sqrt{11}$

The solutions are $1 \pm \sqrt{11}$ or $1 + \sqrt{11}$ and $1 - \sqrt{11}$.

15. $\quad 3y^2 + 5y = 2$

$\quad 3y^2 + 5y - 2 = 0$

$\quad (y + 2)(3y - 1) = 0$

$\quad y + 2 = 0 \text{ or } 3y - 1 = 0$

$\quad y = -2 \text{ or } 3y = 1$

$\quad y = -2 \text{ or } y = \dfrac{1}{3}$

The solutions are -2 and $\dfrac{1}{3}$.

17. $\quad x^2 + 1 = 0$

$\quad\quad x^2 = -1$

$\quad\quad x = \pm\sqrt{-1}$

$\quad\quad x = \pm i$

The solutions are $\pm i$ or i and $-i$.

19. $\quad (p + 10)^2 = 12$

$\quad p + 10 = \pm\sqrt{12}$

$\quad\quad p = -10 \pm 2\sqrt{3} \qquad \sqrt{12} = \sqrt{4 \cdot 3} = 2\sqrt{3}$

The solutions are $-10 \pm 2\sqrt{3}$ or $-10 + 2\sqrt{3}$ and $-10 - 2\sqrt{3}$.

21. $x^2 + x + 1 = 0$

$$x = \frac{-b \pm \sqrt{b^2 - 4ac}}{2a}$$

$$x = \frac{-1 \pm \sqrt{1^2 - 4 \cdot 1 \cdot 1}}{2 \cdot 1}$$

$$x = \frac{-1 \pm \sqrt{1 - 4}}{2}$$

$$x = \frac{-1 \pm \sqrt{-3}}{2}$$

$$x = \frac{-1 \pm i\sqrt{3}}{2} = -\frac{1}{2} \pm \frac{i\sqrt{3}}{2}$$

The solutions are $-\dfrac{1}{2} \pm \dfrac{\sqrt{3}}{2}i$.

23. $\qquad t^2 + 9 = 6t$

$t^2 - 6t + 9 = 0$

$(t - 3)(t - 3) = 0$

$t - 3 = 0$

$t = 3$

The solution is 3.

25. $3m = 4 + 5m^2$

$0 = 5m^2 - 3m + 4$

$$m = \frac{-b \pm \sqrt{b^2 - 4ac}}{2a}$$

$$m = \frac{3 \pm \sqrt{(-3)^2 - 4(5)(4)}}{2(5)}$$

$$m = \frac{3 \pm \sqrt{9 - 80}}{10}$$

$$m = \frac{3 \pm \sqrt{-71}}{10}$$

$$m = \frac{3 \pm i\sqrt{71}}{10} = \frac{3}{10} \pm \frac{\sqrt{71}}{10}i$$

The solutions are $\dfrac{3}{10} \pm \dfrac{\sqrt{71}}{10}i$.

27. $p = ct - bt$

$p = t(c - b)$

$\dfrac{p}{c - b} = t$

29. $\qquad m + n = \dfrac{p}{n}$

$$n(m + n) = n\left(\frac{p}{n}\right)$$

$mn + n^2 = p$

$n^2 + mn - p = 0$

$$n = \frac{-b \pm \sqrt{b^2 - 4ac}}{2a}$$

$$n = \frac{-m \pm \sqrt{m^2 - 4(1)(-p)}}{2(1)}$$

$$n = \frac{-m \pm \sqrt{m^2 + 4p}}{2}$$

31. $\qquad x^2 + 3 = 5x$

$x^2 - 5x + 3 = 0$

$$x = \frac{-b \pm \sqrt{b^2 - 4ac}}{2a}$$

$$x = \frac{-(-5) \pm \sqrt{(-5)^2 - 4(1)(3)}}{2(1)}$$

$$x = \frac{5 \pm \sqrt{25 - 12}}{2}$$

$$x = \frac{5 \pm \sqrt{13}}{2}$$

$$\frac{5 + \sqrt{13}}{2} \approx 4.303 \qquad \frac{5 - \sqrt{13}}{2} \approx 0.697$$

The solutions are approximately 0.697 and 4.303.

33. *Familiarize*. We make a drawing. Let $x =$ the length of the shorter leg of the right triangle. Then $x + 3 =$ the length of the longer leg, in meters.

***Translate*.** We use the Pythagorean theorem.

$$x^2 + (x + 3)^2 = 7^2$$

***Carry out*.** We solve the equation.

$x^2 + x^2 + 6x + 9 = 49$

$2x^2 + 6x - 40 = 0$

$x^2 + 3x - 20 = 0 \quad$ Multiplying by $\dfrac{1}{2}$

$$x = \frac{-b \pm \sqrt{b^2 - 4ac}}{2a}$$

$$x = \frac{-3 \pm \sqrt{3^2 - 4(1)(-20)}}{2(1)}$$

$$x = \frac{-3 \pm \sqrt{9 + 80}}{2}$$

$$x = \frac{-3 \pm \sqrt{89}}{2}$$

$$\frac{-3 + \sqrt{89}}{2} \approx 3.217 \qquad \frac{-3 - \sqrt{89}}{2} \approx -6.217$$

***Check*.** Since the length of a leg cannot be negative, -6.217 does not check. But 3.217 does check. If the smaller leg is 3.217, the other leg is $3.217 + 3$, or 6.217. Then, $(3.217)^2 + (6.217)^2$, or about $49 = 7^2$, the length of the hypotenuse.

***State*.** The legs measure 3.217 m and 6.217 m.

$a = 1, b = 4, c = -16.25$

$$w = \frac{-4 \pm \sqrt{4^2 - 4(1)(-16.25)}}{2(1)}$$

$$w = \frac{-4 \pm \sqrt{16 + 65}}{2} = \frac{-4 \pm \sqrt{81}}{2}$$

$$w = \frac{-4 \pm 9}{2}$$

$$w = -6.5 \text{ or } 2.5$$

Check. Since the length of a side cannot be negative, -6.5 does not check. But 2.5 does check. If the width is 2.5, then length is $2.52 + 4$, or 6.5. The area is 2.5×6.5, or 16.25. This checks.

State. The length is 6.5 m, and the width is 2.5 m.

19. $\sqrt{-200} = \sqrt{-1 \cdot 100 \cdot 2} = \sqrt{-1} \cdot \sqrt{100} \cdot \sqrt{2}$
$\qquad = 10i\sqrt{2} \text{ or } 10\sqrt{2}i$

21. $y = -x^2 + x - 5$

Find the vertex. The x-coordinate is
$$-\frac{b}{2a} = -\frac{1}{2(-1)} = \frac{1}{2}.$$

The y-coordinate is
$$-x^2 + x - 5 = -\left(\frac{1}{2}\right)^2 + \left(\frac{1}{2}\right) - 5 = -\frac{1}{4} + \frac{1}{2} - 5 = \frac{-19}{4}$$

The vertex is $\left(\frac{1}{2}, \frac{-19}{4}\right)$.

We choose some x-values on both sides of the vertex and graph the parabola. We make sure we find y when $x = 0$. This gives us the y-intercept.

x	y
-2	-11
-1	-7
0	-5
1	-5
2	-7

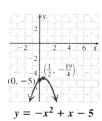

$y = -x^2 + x - 5$

23. $f(x) = \frac{1}{2}x + 1$

$f(0) = \frac{1}{2}(0) + 1 = 1$

$f(1) = \frac{1}{2}(1) + 1 = \frac{3}{2}$

$f(2) = \frac{1}{2}(2) + 1 = 2$

25. $R(t) = 30.18 - 0.06t$

$R(70) = 30.18 - 0.06(70)$

$\qquad = 30.18 - 4.2$

$\qquad = 25.98 \text{ min}$

27. Graph $g(x) = x^2 - 4$

Recall from section 9.6 that the graph is a parabola. Make a list of function values in a table.

When $x = -2$, $g(-2) = (-2)^2 - 4 = 4 - 4 = 0$.

When $x = -1$, $g(-1) = (-1)^2 - 4 = 1 - 4 = -3$.

When $x = 0$, $g(0) = (0)^2 - 4 = -4$.

When $x = 1$, $g(1) = (1)^2 - 4 = 1 - 4 = -3$.

When $x = 2$, $g(2) = (2)^2 - 4 = 4 - 4 = 0$.

x	$h(x)$
-2	0
-1	-3
0	-4
1	-3
2	0

Plot these points and connect them.

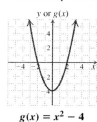

$g(x) = x^2 - 4$

29. The graph is not that of a function because a vertical line can cross the graph at more than one point.

31. $x - y = 2 \qquad (1)$

$\qquad xy = 4 \qquad (2)$

Solve equation (1) for y.

$x - y = 2 \qquad\qquad (1)$

$\qquad y = x - 2 \qquad (3)$

Substitute $x - 2$ for y in Equation (2) and solve for x.

$$x(x - 2) = 4$$

$$x^2 - 2x = 4$$

$$x^2 - 2x + 1 = 4 + 1$$

$$(x - 1)^2 = 5$$

$$x - 1 = \pm\sqrt{5}$$

$$x = 1 \pm \sqrt{5}$$

Answers for Exercises in the Appendixes

Exercise Set A

1. $t^3 + 8 = (t + 2)(t^2 - 2t + 4)$

3. $x^3 + 1 = (x + 1)(x^2 - x + 1)$

5. $z^3 - 125 = (z - 5)(z^2 + 5z + 25)$

7. $8a^3 - 1 = (2a - 1)(4a^2 + 2a + 1)$

9. $y^3 - 27 = (y - 3)(y^2 + 3y + 9)$

11. $64 + 125x^3 = (4 + 5x)(16 - 20x + 25x^2)$

13. $125p^3 + 1 = (5p + 1)(25p^2 - 5p + 1)$

15. $27m^3 - 64 = (3m - 4)(9m^2 + 12m + 16)$

17. $p^3 - q^3 = (p - q)(p^2 + pq + q^2)$

19. $x^3 + \dfrac{1}{8} = \left(x + \dfrac{1}{2} \right)\left(x^2 - \dfrac{1}{2}x + \dfrac{1}{4} \right)$

21. $2y^3 - 128 = 2(y^3 - 64) = 2(y - 4)(y^2 + 4y + 16)$

23. $24a^3 + 3 = 3(8a^3 + 1) = 3(2a + 1)(4a^2 - 2a + 1)$

25. $rs^3 - 125r = r(s^3 - 125) = r(s - 5)(s^2 + 5s + 25)$

27. $5x^3 + 40z^3 = 5(x^3 + 8z^3) = 5(x + 2z)(x^2 - 2xz + 4z^2)$

29. $x^3 + 0.008 = (x + 0.2)(x^2 - 0.2x + 0.04)$

31. *Writing Exercise.* If we simplify using FOIL, we have
$(a - b)(a^2 + b^2) = a^3 + ab^2 - a^2b - b^3$.

33. $125c^6 + 8d^6 = (5c^2)^3 + (2d^2)^3$
$= (5c^2 + 2d^2)(25c^4 - 10c^2d^2 + 4d^4)$

35. $(3x^{3a} - 24y^{3b}) = 3\left(x^{3a} - 8y^{3b} \right) = 3\left((x^a)^3 - (2y^b)^3 \right)$
$= 3(x^a - 2y^b)(x^{2a} + 2x^ay^b + 4y^{2b})$

37. $\dfrac{1}{24}x^3y^3 + \dfrac{1}{3}z^3 = \dfrac{1}{3}\left(\dfrac{1}{8}x^3y^3 + z^3 \right)$
$= \dfrac{1}{3}\left(\left(\dfrac{1}{2}xy \right)^3 + z^3 \right) = \dfrac{1}{3}\left(\dfrac{1}{2}xy + z \right)\left(\dfrac{1}{4}x^2y^2 - \dfrac{1}{2}xyz + z^2 \right)$

Exercise Set B

1. For 13, 21, 18, 13, 20, the mean is the average of these 5 numbers
$$\frac{13 + 21 + 18 + 13 + 20}{5} = \frac{85}{5} = 17.$$
The mean is 17.

Put the numbers in order:

13   13   18   20   21
$\uparrow$

The median is 18.

Since 13 appears twice, the mode is 13.

3. For 3, 8, 20, 3, 20, 10, the mean is the average of these 6 numbers
$$\frac{3 + 8 + 20 + 3 + 20 + 10}{6} = \frac{64}{6} = 10.\overline{6}$$
The mean is $10.\overline{6}$.

Put the numbers in order:

3   3   8   10   20   20
   $\uparrow$ $\uparrow$

To find the median, we average 8 and 10:
$$\frac{8 + 10}{2} = \frac{18}{2} = 9.$$
The median is 9.

Since 3 appears twice, and 20 appears twice, the modes are 3 and 20.

5. For 4.7, 2.3, 4.6, 4.9, 3.8, the mean is the average of these 5 numbers
$$\frac{4.7 + 2.3 + 4.6 + 4.9 + 3.8}{5} = \frac{20.3}{5} = 4.06.$$
The mean is 4.06.

Put the numbers in order:

2.3   3.8   4.6   4.7   4.9
$\uparrow$

The median is 4.6.

Since no number appears more than once, there is no mode.

7. For 234, 228, 234, 228, 234, 278, the mean is the average

of these 6 numbers

$$\frac{234 + 228 + 234 + 228 + 234 + 278}{6} = \frac{1436}{6} = 239.\overline{3}.$$

The mean is $239.\overline{3}$.

Put the numbers in order:

228   228   234   234   234   278
                ↑   ↑

The average of 234 and 234 is 234 is 234, which is the

median.

Since 234 appears three times, it is the mode.

9. The average is

$$\frac{0 + 19 + 25 + 75 + 104 + 51 + 5 + 0}{8} = \frac{279}{8} = 34.875.$$

The mean is 34.875.

Put the numbers in order:

0   0   5   19   25   51   75   104
             ↑   ↑

The average of 19 and 25 is

$$\frac{19 + 25}{2} = 22.$$

The median is 22.

Since 0 appears twice, it is the mode.

11. The average is

$$\frac{87 + 90 + 86 + 86}{4} = \frac{349}{4} = 87.25.$$

The mean is 87.25 in.

Put the numbers in order:

86   86   87   90
     ↑   ↑

The average of 86 and 87 is

$$\frac{86 + 87}{2} = 86.5.$$

The median is 86.5 in.

Since 86 appears twice, 86 in. is the mode.

13. The average is

$$\frac{254 + 202 + 184 + 269 + 151 + 223 + 258 + 222 + 202}{9}$$
$$= \frac{1965}{9} = 218.\overline{3}.$$

The average score is $218.\overline{3}$.

Put the numbers in order:

151   184   202   202   222   223   254   258   269
                             ↑

The median is 222.

Since 202 appears twice, it is the mode.

15. The average $= \dfrac{\text{home runs}}{\text{years}}$, so to find $h =$ the number of

home runs in 22 years,

$$34\frac{7}{22} = \frac{h}{22}$$
$$755 = h$$

Aaron had 755 home runs in 22 years.

To find the number of home runs in 21 years,

$$35\frac{10}{21} = \frac{h}{21}$$
$$745 = h$$

Aaron had 745 home runs in 21 years.

To find the number of home runs in the final year we take

the difference, $755 - 745 = 10$. Hank Aaron hit 10 home

runs in his final year.

17. For 18, 21, 24, a, 36, 37, b to have a median of 30, then a

must be 30.

To find b, we find the average

$$\frac{18 + 21 + 24 + 30 + 36 + 37 + b}{7} = 32$$
$$\frac{166 + b}{7} = 32$$
$$166 + b = 224$$
$$b = 58$$

Then a is 30 and b is 58.

Exercise Set C

1. The set of whole numbers 8 through 11 is $\{8, 9, 10, 11\}$.

3. The set of odd numbers between 40 and 50 $\{41, 43, 45, 47, 49\}$.

5. $\{x|$ the square of x is $9\} = \{-3, 3\}$

7. Since 5 is an odd number, $5 \in \{x|x$ is an odd number$\}$ is true.

9. Since skiing is a sport, skiing $\in$ the set of all sports is true.

11. Since 3 is not a member of $\{-4, -3, 0, 1\}$, the statement $3 \in \{-4, -3, 0, 1\}$ is false.

13. Since $\frac{2}{3}$ is a rational number, the statement $\frac{2}{3} \in \{x|x$ is a rational number$\}$ is true.

15. Since 0 is not a member of $\{-3, -2, -1, 1, 2, 3\}$, the statement $\{-1, 0, 1\} \subseteq \{-3, -2, -1, 1, 2, 3\}$ is false.

17. Since the set of integers are a subset of the set of rational numbers, the statement

The set of integers $\subseteq$ The set of rational numbers

is true.

19. $\{a, b, c, d, e\} \cap \{c, d, e, f, g\} = \{c, d, e\}$

21. $\{1, 2, 3, 4, 6, 12\} \cap \{1, 2, 3, 6, 9, 18\} = \{1, 2, 3, 6\}$

23. $\{2, 4, 6, 8\} \cap \{1, 3, 5, 7\} = \emptyset$

25. $\{a, e, i, o, u\} \cup \{q, u, i, c, k\} = \{a, e, i, o, u, q, c, k\}$

27. $\{1, 2, 3, 4, 6, 12\} \cup \{1, 2, 3, 6, 9, 18\} = \{1, 2, 3, 4, 6, 9, 12, 18\}$

29. $\{2, 4, 6, 8\} \cup \{1, 3, 5, 7\} = \{1, 2, 3, 4, 5, 6, 7, 8\}$

31. *Writing Exercise.* Set-builder notation can be used to write infinite sets, whereas roster notation cannot.

33. Since the set of whole numbers is a subset of the set of intergers, the union of the set of integers and the set of whole numbers is the set of integers.

35. The union of the set of rational numbers and the set of irrational numbers is the set of real numbers.

37. Since none of the set of rational numbers is in the set of irrational numbers and vice versa, the intersection of the set of rational numbers and the set of irrational numbers is the empty set, or $\emptyset$.

39. a) $A \cup \emptyset = A$

b) $A \cup A = A$

c) $A \cap A = A$

d) $A \cap \emptyset = \emptyset$

41. $\emptyset \subseteq A$ is true.

43. $A \cap B \subseteq A$ Reading from left to right, the intersection of A and B is a subset of A is true.

45. $A \cap (B \cup C) = (A \cap B) \cup (A \cap C)$ is true.

For example, if $A = \{1, 3, 5\}, B = \{4, 5, 6\}$, and $C = \{1, 4, 9\}$.

If $B \cup C = \{4, 5, 6\} \cup \{1, 4, 9\} = \{1, 4, 5, 6, 9\}$ then $A \cap (B \cup C) = \{1, 3, 5\} \cap \{1, 4, 5, 6, 9\} = \{1, 5\}$. If $A \cap B = \{1, 3, 5\} \cap \{4, 5, 6\} = \{5\}$ and $A \cap C = \{1, 3, 5\} \cap \{1, 4, 9\} = \{1\}$ then $(A \cap B) \cup (A \cap C) = \{5\} \cup \{1\} = \{1, 5\}$. Examples will vary.